Biophysical Techniques

Biophysical Techniques

Iain D. Campbell

St John's College and Department of Biochemistry
University of Oxford

OXFORD
UNIVERSITY PRESS

OXFORD
UNIVERSITY PRESS

Great Clarendon Street, Oxford OX2 6DP

Oxford University Press is a department of the University of Oxford.
It furthers the University's objective of excellence in research, scholarship,
and education by publishing worldwide in

Oxford New York

Auckland Cape Town Dar es Salaam Hong Kong Karachi
Kuala Lumpur Madrid Melbourne Mexico City Nairobi
New Delhi Shanghai Taipei Toronto

With offices in

Argentina Austria Brazil Chile Czech Republic France Greece
Guatemala Hungary Italy Japan Poland Portugal Singapore
South Korea Switzerland Thailand Turkey Ukraine Vietnam

Oxford is a registered trade mark of Oxford University Press
in the UK and in certain other countries

Published in the United States
by Oxford University Press Inc., New York

British Library Cataloguing in Publication Data

Data available

Library of Congress Cataloguing in Publication Data
Library of Congress Control Number: 2011944050

Typeset by Techset Composition Ltd, Salisbury, UK
Printed in Italy
on acid-free paper by
L.E.G.O. S.p.A.—Lavis TN

ISBN 978-0-19-964214-4

10 9 8 7 6 5 4 3 2 1

I dedicate this book to Raymond who started me on this particular journey and Karin who, for over 40 years, has supported and tolerated various selfish journeys like this one.

PREFACE

Raymond Dwek and I wrote *Biological Spectroscopy* nearly 30 years ago, now long out of print. Completion of that project was mainly due to Raymond's energy and drive. Competing demands for our time meant that we did not find the energy to revise and update it in subsequent years. The available biophysical tools and techniques have, however, evolved and expanded in an astonishing way in recent decades. We can now see single molecules in living cells and make high resolution movies of the ribosome in action. We have ready access to databases of genes and sequences from many hundreds of organisms and over 70 000 sets of macromolecular coordinates. Microscopy has been revolutionized and computers are now ubiquitous and extraordinarily powerful. I thus decided to write a new book with somewhat different emphasis and scope to *Biological Spectroscopy*. Some elements of the old are still discernible and I will always be thankful to Raymond for his contributions to those.

My aim here is to emphasize the general principles and practicalities of a wide range of biophysical techniques. I focus mostly on the analytical toolbox; this means that some production and preparative techniques get less attention than they deserve. Space limitations and the ready availability of alternative sources also led me to give relatively few illustrative examples; readers can now find numerous up-to-date specific examples related to their own interests by using PubMed and other powerful search engines.

A selection of additional recent examples is, however, given in the online "Journal Club" supplement (see panel below). My main goal is to impart a basic understanding of how the tools work, so that their limitations and advantages can be appreciated. I hope practicing molecular biologists and students will be able to gain useful insight into the wide array of powerful tools now available and how they can be applied to their problems. This book aims to complement the excellent available molecular cell biology and biochemistry texts that describe the numerous exciting insights into biological molecules that have been obtained by applying biophysical tools.

I have taught this subject for over 30 years and many students and colleagues have contributed to my enjoyable, educational journeys through the subject. My thanks go to them. Thanks too to Elspeth Garman, Justin Benesch, Sandra Campbell, Michèle Erat, Philip Fowler, Kate Heesom, Mark Howarth, Nick Price, Jason Schnell, and David Staunton for their input. Jonathan Crowe of Oxford University Press has also been particularly helpful and constructive; those errors that remain are, of course, entirely my responsibility.

www.oxfordtextbooks.co.uk/orc/campbell/
The Online Resource Centre to accompany *Biophysical Techniques* includes:

- figures from the book in electronic format, ready to download;
- a series of Journal Clubs—discussion questions, perfect for seminar or tutorial use, built around a selection of journal articles that build on and augment topics covered in the book.

Register for lecturer access by visiting the URL given above.

CONTENTS

Some frequently used abbreviations   x

1   Introduction   1

What are "biophysical techniques"?   2
What questions can biophysical
techniques answer?   2
Which technique to use?   3
Organization of this book   3

2   Molecular principles   5

2.1   Molecules and interactions   6
Introduction   6
Box 2.1   Atoms and elements   6
Box 2.2   Kinetic model of gases   7
Non-covalent interactions   7
Box 2.3   Electrostatics, dielectrics, and polarity   8
Box 2.4   Molecular biology tools and
base pairing   11
Binding   11

3   Transport and heat   17

3.1   Diffusion, osmosis, viscosity, and friction   18
Introduction   18
Diffusion   18
Osmosis   19
Application of a force to a molecule in solution   21
Viscosity   21
Box 3.1   The frictional coefficient, f, depends
on molecular shape   23
3.2   Analytical centrifugation   26
Introduction   26
Some basic principles of sedimentation   26
Sedimentation velocity   28
Sedimentation equilibrium   30
Density gradient sedimentation   31
3.3   Chromatography   36
Introduction   36
Theory   37
Chromatography techniques   38
Quantitative chromatography   40
3.4   Electrophoresis   44
Introduction   44
Theory   44

Experimental   45
Some electrophoresis systems   46
3.5   Mass spectrometry   55
Introduction   55
Ionization   55
Ion sorting/analysis   57
Applications of mass spectrometry   62
3.6   Electrophysiology   72
Introduction   72
Membrane potential   72
Action potentials   75
Experimental   76
Propagation of an action potential
in a neuron   79
3.7   Calorimetry   84
Introduction   84
Isothermal titration calorimetry   84
Differential scanning calorimetry   87
Box 3.2   Heat capacity   89

4   Scattering, refraction, and diffraction   93

4.1   Scattering of radiation   95
Introduction   95
Scattering theory   96
Box 4.1   Weight average molecular weight   97
Turbidity   97
Solution scattering and molecular shape   97
Box 4.2   Radii of gyration, R_G, and hydration, R_H 98
Dynamic light scattering   103
Box 4.3   Correlation functions   103
**4.2   Refraction, evanescent waves,
and plasmons**   107
Introduction   107
Box 4.4   Classical optics   108
Evanescent waves   110
Surface plasmon resonance   111
Box 4.5   The streptavidin/biotin complex   114
4.3   Diffraction   119
Introduction   119
Principles of diffraction   119
Box 4.6   Single particle imaging with X-ray laser   119
Diffraction experiments   122
Interpretation of diffraction data   125
Other crystallography techniques   132
Achievements of crystallography   133

5 Electronic and vibrational spectroscopy   139

5.1 Introduction to absorption and emission spectra   140
Introduction   140
Energy states   140
Absorption   141
Box 5.1 Transition dipole moments and transition probability   142
Emission   144
Box 5.2 The laser   145

5.2 Infrared and Raman spectroscopy   148
Introduction   148
IR spectra and applications   150
Raman scattering   155
Applications of Raman spectroscopy   157

5.3 Ultraviolet/visible spectroscopy   162
Introduction   162
Measurement of electronic spectra   162
Electronic energy levels and transitions   163
Absorption properties of some key chromophores   164
Applications of UV/visible spectra   167
Box 5.3 Isosbestic points   169
Properties associated with the direction of the transition dipole moment   169
Monitoring rapid reactions   171

5.4 Optical activity   176
Introduction   176
The phenomenon   176
Measurement   177
Applications of CD spectra   178

5.5 Fluorescence   185
Introduction   185
Physical basis of fluorescence   185
Measurement   186
Fluorophores   187
Environmental effects on fluorescence   188
Fluorescence anisotropy   191
Förster resonance energy transfer (FRET)   193
Exploitation of fluorescence sensitivity   196
Box 5.4 Immunofluorescence   196
Box 5.5 Fluorescence in situ hybridization (FISH)   196
Phosphorescence   197

5.6 X-ray spectroscopy   203
Introduction   203
Theory   203
Measurement   204
Edge spectra   205
EXAFS   205
Analytical uses of X-ray emission   206

6 Magnetic resonance   209

6.1 Nuclear magnetic resonance   210
Introduction   210
Box 6.1 Magnetism   210
The NMR phenomenon   210
Measurement   212
The spectral parameters in NMR   214
Applications of NMR   223
Other NMR applications   236

6.2 Electron paramagnetic resonance   243
Introduction   243
Measurement   243
Spectral parameters   244
Box 6.2 Spin labels   246
Spectral anisotropy   247
Applications of EPR   248

7 Microscopy and single molecule studies   257

7.1 Microscopy   258
Introduction   258
Factors that influence resolution   259
Box 7.1 Diffraction at a slit   259
The optical microscope   260
The fluorescence microscope   262
Electron microscopy   266
Box 7.2 X-ray tomography   270
Box 7.3 The contrast transfer function (CTF)   272
The scanning electron microscope   273
Scanning probe microscopy   275

7.2 Manipulation and observation of single molecules   279
Introduction   279
Manipulation by force   279
Fluorescence methods   282

8 Computational biology   289

8.1 Computational biology   290
Introduction   290
Mathematical modeling of systems   290
Bioinformatics   292
Molecular modeling   296
Box 8.1 Force and potential energy   298

9 Tutorials   309

1 Biological molecules   310
2 Thermodynamics   312
3 Motion and energy of particles in different force fields   314
4 Electrical circuits   317
5 Mathematical representation of waves   318
6 Fourier series, Fourier transforms, and convolution   319
7 Oscillators and simple harmonic motion   321
8 Electromagnetic radiation   323
9 Quantum theory and the Schrödinger wave equation   325
10 Atomic and molecular orbitals, their energy states, and transitions   327
11 Dipoles, dipole–dipole interactions, and spectral effects   330

Appendices 332

A1 Prefixes, units, and constants 332
  A1.1 Prefixes 332
  A1.2 Units 332
  A1.3 Constants 332
A2 Some mathematical functions 332
  A2.1 Trigonometry 332
  A2.2 Vectors 332
  A2.3 Complex variables 333

  A2.4 Logarithms, exponentials,
    and trigonometric functions 333
  A2.5 Calculus 333
A3 Basic statistics 334

Solutions to Problems 335

Index 346

SOME FREQUENTLY USED ABBREVIATIONS

AFM atomic force microscopy

ANS 8-anilinonaphthalene-1-sulfonate

ATP adenosine triphosphate

ATR attenuated total reflectance

AUC analytical ultracentrifuge

BRET bioluminescence resonance energy transfer

CASP critical assessment of structure prediction

CCD charge coupled device

CD circular dichroism

CE capillary electrophoresis

CG coarse grain

CID collision induced dissociation

COSY correlated spectroscopy

CSA chemical shift anisotropy

CTF contrast transfer function

DAPI 4′,6-diamidino-2-phenylindole

DEER double electron–electron resonance

DIC differential interference contrast

DLS dynamic light scattering

DNP dinitrophenyl

DSC differential scanning calorimetry

EM electron microscopy

EMSA electrophoretic mobility shift assay

ENDOR electron nuclear double resonance

EPR electron paramagnetic resonance

ESI electrospray ionization

ESR electron spin resonance

EXAFS extended X-ray absorption fine structure

FACS fluorescence-activated cell sorting

FCS fluorescence correlation spectroscopy

FEG field emission gun

FEP free energy perturbation

FISH fluorescence *in situ* hybridization

FLIM fluorescence lifetime imaging microscopy

FLIP fluorescence loss in photobleaching

fMRI functional magnetic resonance imaging

FPLC fast protein liquid chromatography

FRAP fluorescence recovery after photobleaching

FRET Förster resonance energy transfer

FT Fourier transform

FT-ICR Fourier transform ion cyclotron resonance

FTIR Fourier transform infra-red

GFP green fluorescent protein

GST glutathione S-transferase

HOMO highest occupied molecular orbital

HPLC high performance liquid chromatography

HSQC heteronuclear single quantum coherence

ICAT isotope-coded affinity tagging

IEF isoelectric focusing

IM ion mobility

IMAC immobilized metal ion affinity chromatography

INEPT insensitive nuclei enhanced by polarization transfer

IR infra-red

ITC isothermal titration calorimetry

iTRAQ isobaric tagging for relative and absolute quantification

LC liquid chromatography

LUMO lowest unoccupied molecular orbital

MAD multiple wavelength anomalous dispersion

MALDI matrix assisted laser desorption ionization

MAS magic angle spinning

MC Monte Carlo

MD molecular dynamics

MIR multiple isomorphous replacement

MM molecular mechanics

MRI magnetic resonance imaging

MRS magnetic resonance spectroscopy

MS mass spectrometry

NA numerical aperture

NADH nicotinamide adenine dinucleotide

NMR nuclear magnetic resonance

nOe nuclear Overhauser effect

NOESY nuclear Overhauser effect spectroscopy

ORD optical rotatory dispersion

PAGE polyacrylamide gel electrophoresis

PALM photoactivated localization microscopy

PCR polymerase chain reaction

PDB protein data bank

PELDOR pulsed electron-double resonance

PGE pulsed field gel electrophoresis

PIXIE particle-induced X-ray emission

PRC photosynthetic reaction center

PSF point spread function

rf radiofrequency

rmsd root mean square deviation

SANS small angle neutron scattering

SAXS small angle X-ray scattering

SDS sodium dodecyl sulfate

SEC size exclusion chromatography

SEM scanning electron microscope

SERS surface enhanced Raman scattering

SILAC stable isotopic labeling by amino acids in cell culture

SPR surface plasmon resonance

STED stimulated emission depletion

TAP tandem affinity chromatography

TEM transmission electron microscopy

TIRF total internal reflectance fluorescence

TOF time of flight

TROSY transverse relaxation optimized spectroscopy

UV ultraviolet

XAS X-ray absorption spectroscopy

XRF X-ray fluorescence

ZFS zero field splitting

1 Introduction

Although the life we see around us appears very diverse (ants to elephants—Figure 1.1), it is remarkably similar at the level of molecules. It is thus especially rewarding to explore life in the atom-to-cell range (0.1 nm to 50 μm; Figure 1.1). To understand life, we need to identify the molecules involved, measure their concentrations and interaction partners, define their structure, and their location in the cell. As such small things cannot be observed with the naked eye, we need a range of techniques to visualize and monitor them. Human skill and ingenuity have given us an extraordinarily powerful set of physical and mathematical tools to study molecules; these are the subject of this book.

To function properly, molecular assemblies need appropriate *dynamic* and *energetic* properties, as well as the correct location and structure. The molecules have to be in the right place at the right time and their chemical reactions and rates have to be carefully controlled. Some biological time scales of interest are shown in Figure 1.2. Exploration of the $1\,s{-}10^{-14}\,s$ time scale range is especially useful for understanding molecular mechanisms.

Energy and its transformations, sunlight to ATP for example, are essential for life. Humans need around 6×10^6 J per day of energy intake, while breaking a single molecular bond needs about 7×10^{-19} J ($\sim$400 kJ mol^{-1}). Chemical reactions will not proceed unless the reactants have enough energy to surmount a threshold energy barrier. We need to explore the energy associated with essential life processes and how environmental factors, such as pH and temperature, influence them.

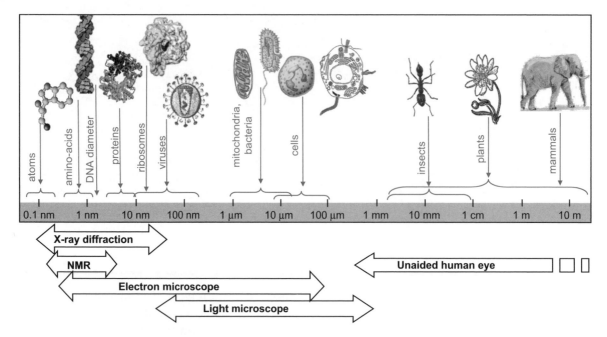

Figure 1.1 An illustration of the dimensions of life-related objects, presented on a logarithmic scale. The dimensions accessible to some different techniques are indicated by horizontal arrows.

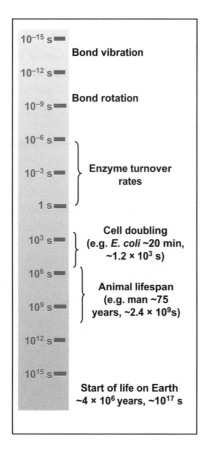

Figure 1.2 Some relevant biological time scales.

What are "biophysical techniques"?

According to the Biophysical Society website **(http://www.biophysics.org/)**, **biophysics** is "that branch of knowledge that applies the principles of physics and chemistry, the methods of mathematical analysis, and computer modeling to understand how biological systems work… (it) seeks to explain biological function in terms of the structures and properties of specific molecules." This definition includes a huge range of topics and techniques: transport, hydrodynamics, spectroscopy, diffraction, microscopy, computational modeling, and electrophysiology all fall within this definition. Biophysical tools are diverse, powerful, and *complementary*; they exploit developments in technology and help us apply insights from the laws of physics and chemistry to all aspects of molecular and cellular biology.

What questions can biophysical techniques answer?

As mentioned above, a complete description of molecules in a living cell requires information about their concentration at particular locations, their structure, and their dynamic and energetic properties. Studies of cell structure, structural cell biology, began when early microscopes, developed around 350 years ago, gave the first glimpse of single cells. Since then, increasingly sophisticated instruments have revealed ever more information about cell structure. Visualization of the large molecules found in cells, structural molecular biology, began only around 60 years ago when atomic resolution models of DNA and proteins were first constructed. Like the studies of cells, these studies of macromolecules have advanced remarkably in recent years and high resolution structures of many thousands of them are now known (see for example the Protein Databank (PDB) website **www.rcsb.org**). Although structural cell biology and structural molecular biology largely evolved as separate disciplines, there is an increasing awareness that they are more and more convergent. Ever better resolution can be achieved in studies of intact cells and ever larger structures and complexes can be seen at the molecular level. A new structural biology is emerging that encompasses both molecular and cellular approaches.

In addition to structural studies, biophysical techniques play a key role in the exploration of other properties of cellular components. There are a multitude of carefully regulated enzyme-catalyzed reactions to be monitored and understood. The dynamic processes of diffusion, ion transport, protein folding, and conformational changes occur on a wide range of time scales. Complex interaction networks between molecules need to be dissected but also reconstructed to give us a holistic view of cell properties. Biophysical methods also play a key role in studies of disease and they contribute significantly to drug discovery programs and the detailed analysis of pathologic states.

system will often only be reached by trial and error and in consultation with experts. The information obtained from the various techniques is also *complementary* and a detailed study of any biological system will usually involve application of several different tools and cross-checking of the various results obtained.

To illustrate the range and scope of the methods let us look at a few possible situations. We may be interested in a large membrane protein with a single amino-acid substitution that causes an inherited disease. The activity of the protein could be studied in cells by electrophysiology techniques (Chapter 3.6); crystallography (Chapter 4.3) could determine its structure; and molecular dynamics simulations (Chapter 8) could be combined with the electrophysiology to explore the influence of mutations on the passage of ions across the membrane. The shape in solution of a large complex containing RNA and many proteins could be studied by analytical centrifugation (Chapter 3.2) and small-angle X-ray scattering (Chapter 4.1). The reaction mechanism and ligand geometry around a metal at the active site of an enzyme could be studied by UV/visible spectroscopy (Chapter 5.3), X-ray absorption spectroscopy (Chapter 5.6), and electron paramagnetic resonance (Chapter 6.2).

A slightly more detailed example is given in Table 1.1. Imagine that careful comparison, using 2D gels (Chapter 3.4), of the proteins in normal and cancer cell lines identified a protein X that appears at much higher levels in the cancer cells than in normal cells. How might we explore the properties and characteristics of protein X? Possible approaches are listed in Table 1.1, together with an indication of the relevant chapters where the various techniques are described.

Which technique to use?

Faced with a bewildering array of tools and techniques, how do we choose the right one to give us the information we want? The right choice will depend on many variables, including the composition of the system being studied, its environment (*in vivo* or *in vitro*), the amount of sample available, and the question being asked. The right answer for a particular

Organization of this book

The task of covering all of the biophysical tools in a reasonable space is clearly a daunting one; I have, however, tried to distil out the main features and principles of the various methods, and to explain how they work and what their limitations are. I assume throughout that you, the reader, are relatively familiar with the goals and themes of molecular biology.

Table 1.1 Possible approaches to the study of protein X (~25 kDa) detected in a 2D gel (Chapter 3.4)

Information sought about X	Technique	Chapter
Identification and amino-acid sequence	Analyze spot in 2D gel using mass spectrometry	3.5
Homologues and properties (pI, mol. wt., etc.)	Bioinformatics tools	8
Purify and characterize, (after cloning and expression, e.g. in *E. coli*)	Chromatography and electrophoresis	3.3, 3.5
Physical properties/preliminary structure analysis	Mass spectrometry, CD, FTIR, centrifugation, DLS	3.5, 5.4, 5.2, 3.2, 4.2
Function (e.g. catalytic activity/kinetics)	UV/visible coupled assay	5.3
Identify binding partners (e.g. Y)	Affinity chromatography, mass spectrometry	3.3, 3.5
Characterize binding of Y	SPR, ITC, fluorescence, ^{1}H-^{15}N NMR	4.3, 3.7, 5.5, 6.1
Structure determination of X and XY complex	Crystal structure of X and XY complex; NMR	4.3, 6.1
Location in cell	Express with GFP tag; fluorescence microscopy	7.1

I make fewer assumptions about your knowledge of mathematics, physics, and chemistry, and try to cover basic concepts in those subjects in "tutorials", "boxes", and appendices. Some specific papers and reviews are also discussed in more detail in the online "Journal Club", which you can find at **www.oxfordtextbooks.co.uk/orc/campbell/**.

I have grouped techniques together where there are similarities. Some of the groupings are rather arbitrary but in other cases there is clear synergy and shared physical concepts. Examples include: absorption/emission spectra (infrared, UV/visible absorption, circular dichroism, and fluorescence); magnetic resonance techniques (electron and nuclear); scattering and diffraction. Many of these involve the interaction of electromagnetic radiation with matter, which is one broad definition of spectroscopy. Other groupings are: microscopy, with manipulation of single molecules by optical tweezers and atomic force microscopy; molecular transport (electrophoresis, chromatography, analytical ultracentrifugation, mass

spectrometry), and computational methods (bioinformatics, molecular dynamics simulations, and molecular visualization).

The historical development of the various biophysical tools is of considerable interest; primitive early anatomical dissection tools evolved into sophisticated techniques that can observe even single molecules. Some tools have become established as standard laboratory techniques in a remarkably short time, e.g. surface plasmon resonance (Chapter 4.2) while others have a long and fascinating history, e.g. microscopy. I have tried to indicate some historical aspects of the techniques in the introductions to the various chapter; these brief references are supplemented by a sprinkling of quotations.

The chapters are designed to stand alone although cross referencing is recommended, especially to the tutorial material at the back of the book. The introductory Chapter 2.1 provides some basic concepts, nomenclature, and background material for use in later chapters.

2 Molecular principles

> *"Living organisms are composed of lifeless molecules that conform to all the laws of chemistry but interact with each other in accordance with another set of principles—the molecular logic of the living state."*
>
> Albert Lehninger, 1975

> *"Life is order, death is disorder. A fundamental law of Nature states that spontaneous chemical changes in the universe tend toward chaos. But life has, during milliards of years of evolution, seemingly contradicted this law. With the aid of energy derived from the sun it has built up the most complicated systems to be found in the universe. The beauty we experience when we enjoy the exquisite form of a flower or a bird is a reflection of a microscopic beauty in the architecture of molecules."*
>
> Bo Malmström, 1992

> *"Nothing in life is certain except death, taxes, and the second law of thermodynamics."*
>
> Seth Lloyd, 2004

Unlike other parts of this book, this chapter is not about tools and techniques; it is an introduction to some key concepts that are important for understanding other parts of the book. The modern molecular biologist often lacks basic knowledge of physics and chemistry, while those with a more physical training often do not understand the language of molecular biology. The aim here is to bridge some of these gaps. Supplementary information is also provided in "boxes", "tutorials", "further reading", appendices, and an online "Journal Club". Many of you may be able to skip this material; others may use it as a reminder of things forgotten.

2.1 Molecules and interactions

Introduction

We have learned a great deal in the last 100 years about the molecules that assemble to make a living cell. Their nature and composition are largely known (Tutorial 1). A major remaining challenge facing the molecular biologist in the 21st century is to unravel the myriad specific and dynamic interactions and reactions, to understand how these are regulated and how the property of *life* emerges from this mix. Although the cell is amazingly complex, the various interactions, reactions, and structures all obey the laws of chemistry and physics. This chapter sets out some of these laws.

Physicists use the word matter to describe anything that takes up space and has mass. The matter we are concerned with is the living cell, which consists of many molecules in many different environments. The molecular properties outlined in Tutorial 1 are sufficient for understanding many effects but some observed properties need to be explained at the level of atoms (Box 2.1) and quantum mechanics (Tutorial 9).

Matter can be described in terms of concentration, structure, energy, and dynamics and these are interrelated (some relevant size and dynamic scales were shown in Figures 1.1 and 1.2).

One example of the interrelationship between structure and energy is illustrated by the solution of the Schrödinger wave equation for a particle of mass m, constrained to be in a one-dimensional box of width L; the allowed energies that the particle can have are quantized and related to L (Tutorial 9). Another example of the interrelationship between structure and energy is the folding of a chain of amino acids in aqueous solution to form a protein with a well-defined three-dimensional shape that corresponds to an energy minimum (see Figure 2.1.7). Other examples include the aggregation of lipids to form a membrane bilayer or the assembly of

---

Box 2.1 Atoms and elements

While particle physicists explore a world occupied by quarks and leptons, it is sufficient for most aspects of biophysics to consider an atom (~10^{-10} m diameter) as a dense nucleus (~10^{-15} m diameter) made of protons (charge $= +1$) and neutrons (charge $= 0$), surrounded by electrons (charge $= -1$). The neutron and the proton have approximately the same mass, while the mass of the electron is much less (see Appendix A1.3). The distribution of electron clouds around the nucleus can be described by atomic orbitals (Tutorial 10); different electronic energy states have different shaped orbitals. The proton number, Z, is the number of protons in the nucleus. The nucleon number, A, is the total number of protons plus neutrons. A chemical element consists of one *type* of atom, distinguished by its Z value. Isotopes have the same Z number but different values of A. For example, $^{12}_{6}C$, the carbon-12 isotope, has 6 neutrons and 6 protons in the nucleus while the isotope $^{13}_{6}C$ has 6 protons and 7 neutrons.

The lightest elements (hydrogen, helium, lithium, and beryllium) were made during the "big bang" that formed the Universe. The other (~88) elements were created in nuclear reactions in dying stars.

The four most abundant elements in living cells are oxygen (65%), carbon (18.5%), hydrogen (9.5%), and nitrogen (3.3%); the numbers refer to the approximate % by weight. Important other elements in life forms include calcium, phosphorus, potassium, sulfur, sodium, chlorine, and trace metals such as iron, copper, zinc, and molybdenum.

Box 2.2 Kinetic model of gases

Some of the relationships between motion, diffusion, and bulk properties can be illustrated by considering simple gas kinetics. A gas consists of a large number of small particles, moving randomly with a distribution of velocities, colliding with each other and with the walls of their container. The pressure, P, produced by collisions of the particles with the walls, can be predicted from consideration of Newton's laws of motion (Tutorial 3): $P = nM\hat{u}^{2}/V$, where $\hat{u}$ is the average velocity of n particles, each with mass M in a volume V.

The *spread* of particle velocities in a gas is described by the Maxwell–Boltzmann distribution, as illustrated in Figure A. The fraction, f, of molecules that have velocities in the range v to $v + \Delta v$ is given by $f = F(v)\Delta v$ where $F(v) = 4\pi v^{2}(M/2\pi RT)^{3/2} \exp(-Mv^{2}/2RT)$. This equation looks rather complicated but the observation (Figure A) that the bell-shaped distribution of velocities broadens and shifts to higher velocity on increasing the temperature (T), or reducing the mass (M) of the particles involved, is intuitively expected. This kind of velocity distribution explains many phenomena relevant to molecules in cells, including their diffusion and the dependence of reaction rates on temperature. (Diffusion and Boltzmann distributions are further discussed in Chapters 3.1 and 5.1).

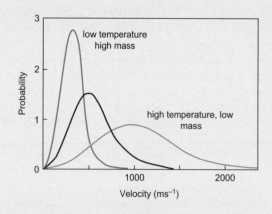

Figure A An illustration of the velocity distributions observed in a gas, as predicted by the Maxwell–Boltzmann law at different temperatures and for different particle masses.

---

numerous proteins and RNA to form the ribosome, the protein-manufacturing machine in the cell.

Dynamics and energetics are also interrelated; molecular vibrations and the random movements of molecules in solution (rotational and translational diffusion) increase with temperature (Box 2.2), a property that means that reactions are more likely to occur at higher temperature.

Non-covalent interactions

The forces involved in the formation of many complexes in the living cell are "non-covalent"—they do not result in the formation of a chemical bond. As will be described in more detail below, the key non-covalent interactions are: i) charge–charge (ionic) interactions; ii) hydrogen bonds; iii) Van der Waals forces; and iv) the hydrophobic effect. These define the structures formed by the key molecules of life, such as proteins and DNA. Their weak, transient nature also facilitates rapid reversible recognition between interacting molecules in complexes; for example the P and L complex that forms in water, as depicted in Figure 2.1.1.

Ionic interactions

Positive and negative charges attract, while like charges repel each other. As described in Box 2.3, the

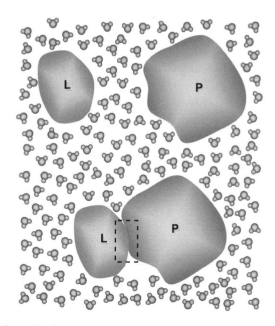

Figure 2.1.1 Representation of two molecules, L and P, in water; they form a complex with an interface at the region marked with a dotted box.

Box 2.3 Electrostatics, dielectrics, and polarity

Charges (q) attract or repel each other with an electric force that, according to Coulomb's law, is $F \propto q_1q_2/r^2$ where r is the charge separation. In a vacuum, $F = q_1q_2/4\pi\varepsilon_o r^2$ where ε_o is the permittivity of free space (values of physical constants are given in Appendix A1.3). In some situations, we wish to know the potential energy, U, between two charges rather than the force. This can be calculated by bringing one of the charges up from infinity to a charge separation distance, r_o; the result, which can be obtained by simple integration—see Appendix A2.5—is $U = q_1q_2/4\pi\varepsilon_o r_o$.

If the charges are in a medium rather than a vacuum, the interaction between them can be reduced by ε, a factor known as the dielectric constant or relative permittivity of the medium (the dielectric) so that $U = q_1q_2/4\pi\varepsilon\varepsilon_o r$. This dielectric effect arises if electric charges in the medium shift slightly from their average equilibrium positions to create an internal electric field that partly compensates the applied external field (Figure A). The value of ε depends on whether, or not, the molecules in the dielectric are polar. Polar molecules have a dipole moment—see Tutorial 11; water, for example, is polar because it contains an uneven charge

distribution and it has a relatively high value of ε (~80). Methane is an example of a non-polar molecule with no significant charge separation and a low value of ε (<2).

Note: the frequency dependence of ε depends on the ability of molecules in the medium to reorient in response to an oscillating applied electric field (Chapter 4.2).

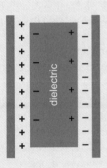

Figure A A dielectric material between two metal plates becomes partly polarized so as to oppose an electric field applied between the plates.

potential energy between two charges q_1 and q_2 can be described by the equation $q_1q_2/4\pi\varepsilon\varepsilon_o r$, where ε_o is the permittivity of free space and ε is the dielectric constant of the medium. In aqueous solution, most ions are surrounded by several partially charged water molecules (a hydration shell); these water molecules shield the ionic charges from each other. This shielding by water is reflected in a high value of the dielectric constant, ε (Box 2.3). While ε is high in aqueous solution (~80), it is much lower in the interior of a macromolecule (~4) because the interior ions are not hydrated; buried ionic interactions are thus much stronger than those on the surface (see Problem 2.1.2).

Hydrogen bonds

A hydrogen bond (H-bond) results from an attractive interaction between a hydrogen atom and an electronegative atom. Nitrogen and oxygen are termed electronegative because they tend to attract electrons from other atoms; for example, oxygen attracts electrons from carbon in a C=O bond, making O slightly negative and C slightly positive, thus introducing a

dipole (Tutorial 11). Other examples of induced dipoles are $N^{\delta-}H^{\delta+}$ and $O^{\delta-}H^{\delta+}$. These dipoles can interact with each other to make H-bonds, for example $NH^{\delta+}{-}^{\delta-}O{=}C$. The strength of this weak electrostatic interaction between partial charges on the atoms H and O is around 20 kJ mol^{-1} which is about 1/20th of the strength of a covalent bond. The H-bond strength depends on the separation between donor and acceptor (typically ~0.3 nm) and on their relative geometry, with the strongest bond arising when the donor and acceptor dipoles are in line. H-bonds can occur between different molecules (intermolecular), or within a single molecule (intramolecular). H-bonding is very important in defining the properties of water, as well as the structure of large molecules like DNA and proteins. It is, however, not usually a significant driving force for complex formation. This can be understood by considering the situation in Figure 2.1.1: those atoms on the surface of free L that have H-bonding potential will be H-bonded to water. When they form the PL complex, these atoms with H-bonding potential will, instead, H-bond to a group on P, giving no net change in H-bonding upon complex formation.

Van der Waals forces

Interactions, named after Johannes van der Waals (1837–1923), can arise from induced dipole moments in non-polar molecules. Such forces have two components, an attractive term caused by "dispersion" forces, sometimes known as "London" forces, and a repulsive term arising from electron overlap when the atoms are close. The attraction components occur when temporary fluctuations in neighboring electron clouds produce favorable compensating charge distributions in an atom pair as they approach each other (Figure 2.1.2).

These attraction and repulsion terms can be calculated using quantum mechanics (Tutorial 9); the repulsion term is found to vary as r^{-12} (that is, when r is small, the repulsion is very large) while the attractive term varies as r^{-6} (the closer the electron clouds are the more favorable the interaction). Van der Waals interactions can thus be described as the sum of these attractive and repulsive terms with a function of the form $Ar^{-12} - Br^{-6}$, where A and B are constants; the energy curve generated, illustrated in Figure 2.1.2, is called the "Lennard-Jones" potential.

The very steep rise of the repulsion term makes it possible to think of atoms as having definite dimen-sions, described by a van der Waals radius, corresponding to the minimum of the curve in Figure 2.1.2. Some important van der Waals radii are: H = 0.12 nm; C = 0.17 nm; N = 0.15 nm; O = 0.15 nm.

Repulsive forces are very important in the determination of the allowed conformations of macromolecules and their complexes (see Figure 2.1.5 below). The attractive van der Waals forces can be significant when they are numerous, but they are weaker than H-bonds.

The hydrophobic effect

The hydrophobic effect is exemplified by the fact that solutes without charges or H-bonding potential (e.g. oils) gather together, rather than mix with water. The hydrophobic effect depends on the structure of liquid water. Fluctuating structures with extensive hydrogen bonding networks can be formed in water; these structures, similar to that found in ice, are enhanced locally by the introduction of a hydrophobic group. This enhanced order is thermodynamically unfavorable (the entropy decreases—see Tutorial 2) so the whole system tries to rearrange itself to reduce this increase in order. One way to do this is to minimize the surface area of the hydrophobic groups by clustering them together (this minimization of surface area is why oil and water separate). Note that, in Figure 2.1.1, water is expelled from the interface region between P and L when the PL complex is formed. The driving force for complex formation comes from an *overall* tendency of the system, including water, to become more disordered. This overall increase in disorder (entropy) occurs in spite of some increase in local order resulting from macromolecular folding or complex formation. The hydrophobic effect can be explained by the second law of thermodynamics which states that the overall entropy of a system increases (Tutorial 2).

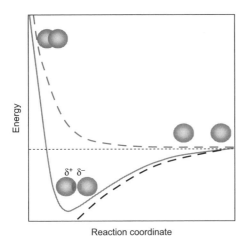

Reaction coordinate

Figure 2.1.2 Representation of the energy profile resulting from van der Waals forces. As two non-polar atoms approach each other (the "reaction coordinate" gets smaller) an attractive force can be generated by fluctuating dipoles set up in the electron clouds. As the atoms get closer still, there is strong repulsion between them when the electron clouds overlap. The r^{-6} attractive and r^{-12} repulsive terms are shown as black and cyan dotted lines, respectively.

Water as a solvent

It should not have escaped your notice that the properties of water (a partially charged, polar molecule, $^{\delta+}H–O^{\delta-}–H^{\delta+}$, with a dipole moment) play a key role in the thermodynamics of non-covalent interactions. Water strongly influences the strength of ionic interactions by shielding charges and its ability to H-bond reduces the net contributions that H-bonds make to a macromolecular structure made between folded and unfolded states; above all, however, it leads to the hydrophobic effect, which is the most important energetic driving force in the formation of complexes and macromolecular structure.

Non-covalent interactions in proteins

Although the main driving force for protein folding in aqueous solution is the hydrophobic effect, the local geometry in the folded state is defined by other non-covalent forces. Favorable intramolecular H-bonding patterns lead to the formation of α-helices and β-sheets; the H-bonding pattern found in α-helices, for example, is indicated in Figure 2.1.3. In β-sheets, the observed H-bonding is between NH and CO groups on adjacent strands.

Van der Waals forces, especially the repulsion terms, are also important in defining allowed bond rotation angles. Consider the situation in Figure 2.1.4. As ϕ is changed, steric clashes can occur between groups (e.g. between f and a); the energy of the molecule thus changes as ϕ changes. In general, the staggered position in Figure 2.1.4 will have a lower energy than the eclipsed conformation, but the exact torsion angle dependence of the energy depends on the nature of the groups at positions a–f.

In a protein there are three important backbone torsion angles; these are defined as ϕ, ψ, and ω, as shown in Figure 2.1.5. The peptide bond is usually planar, so ω is fixed at 180° (or 0°) but the ϕ and ψ angles are variable, and some of these angles are more favored than others. A Ramachandran plot is shown in Figure 2.1.6. This plot, of ψ against ϕ, is a way of visualizing conformations that are energetically favorable. Ramachandran first calculated "allowed" ψ/ϕ angles (those with low energy) by considering a tripeptide made up of hard spheres. Figure 2.1.6

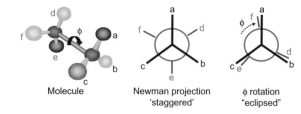

Figure 2.1.4 Two groups, abc and efg, are shown at each end of a rotatable bond. The middle view is the same molecule, drawn as a Newman projection (end-on view). A torsion angle rotation of ϕ = 60° takes the molecule from a "staggered" to an "eclipsed" conformation, as shown on the right.

shows the most favored regions of ψ/ϕ space, in cyan, calculated using a Lennard-Jones potential. Some relaxation of the conditions expands the allowed regions to include the gray areas in the plot. The small size of the glycine sidechain (R = H; Tutorial 1) allows ψ/ϕ angles outside the normal regions (e.g. $α_L$). For a right-handed alpha helix ϕ = –57°, ψ = –47°, and ω = 180°. For parallel strands in a β-sheet ϕ = –118°, ψ = +113°, and ω = 180°. For an anti-parallel β-sheet ϕ = –139° ψ = +135°, and ω = 180°.

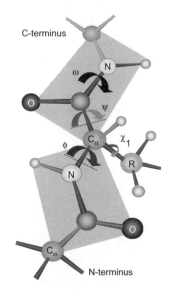

Figure 2.1.5 Definition of some angles in a protein chain. The convention is that the angles rotate clockwise as we proceed in the direction N terminus → C terminus.

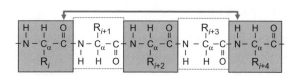

Figure 2.1.3 The observed H-bonding pattern in an α-helix is from the CO group of amino acid *i* to the NH of amino acid *i* + 4 (see also Tutorial 1).

Non-covalent interactions in nucleic acids

As with proteins, the structure of double-stranded DNA is governed by a balance of non-covalent forces.

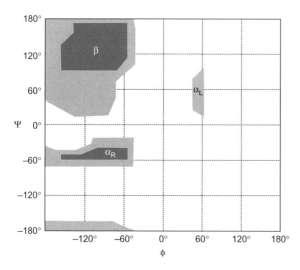

Figure 2.1.6 A Ramachandran plot, showing energetically allowed regions of $\phi\psi$ space in cyan. Less favorable regions are shown in gray. The positions corresponding to right- and left-handed helices and β-sheets are indicated. α_L corresponds to a left handed helix.

In aqueous solution, this balance favors double, rather than single, strands. Stabilizing forces in the DNA double-helix are Watson–Crick hydrogen bonds (Figure 2.1.7) and "stacking" of adjacent nucleotides. Hydrophobic effects dominate these stacking interactions but van der Waals forces and electrostatic interactions between partial charges also contribute. Electrostatic repulsion between phosphate groups is likely to be a destabilizing force and there will be an entropic cost in going from a flexible backbone to a relatively rigid double helix.

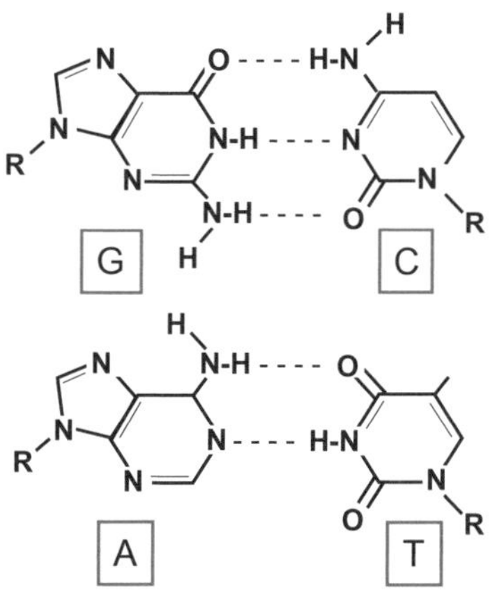

Figure 2.1.7 Classic Watson–Crick H-bonding patterns between G–C and A–T base pairs (see also Tutorial 1).

> **Box 2.4** Molecular biology tools and base pairing
>
> A wide range of **molecular biology** methods exploit specific base-pairing recognition in DNA and RNA. Examples include **DNA microarrays**, DNA sequencing, and **fluorescence** *in situ* hybridization (see Box 5.5; Chapter 5.5). Microarrays consist of thousands of microscopic spots of oligonucleotides, each containing tiny amounts (10^{-12} moles) of a specific DNA sequence, known as **probes**. These interact with single stranded nucleic acids in a sample to form complexes, or **hybrids**, when the sequences match (are **complementary**). The formation of hybrids can be detected with fluorescent probes (Chapter 5.5) allowing sequence identity to be assessed and specific sequences detected. Hybridization methods can be carried out in arrays, in solution, or with an immobilized component on a gel or nitrocellulose paper.

The fact that the balance between base-paired and non-base-paired states lies in favor of pairing at normal temperatures is, of course, essential for life as we know it. The specific pairing that arises in DNA and RNA has also led to many powerful molecular biology tools (see Boxes 2.4 and 5.5).

Binding

We now consider how the interactions between molecules (e.g. between L and P in Figure 2.1.1) can be quantified; we wish to be able to compare affinities of different molecules for each other, to find out if interactions are cooperative, and to measure the relative amount of L that binds to P (the stoichiometry).

Binding to a single site

For the simple equilibrium reaction $P + L \rightleftharpoons PL$, we can define a **binding**, or **association**, constant for the forward reaction ($P + L \rightarrow PL$) as $K_A = ([PL]/[P][L])$ where [ ] represents concentrations.

We can also define a **dissociation constant** $K_D = 1/K_A = ([P][L]/[PL])$ for the reverse reaction ($PL \rightarrow P + L$).

The *fraction of protein occupied by ligand* is $\tilde{n} = ([PL]/[P] + [PL])$.

Substituting for $[PL] = ([P][L]/K_D)$ gives $\tilde{n} = ([P][L]/K_D ([P] + [P][L]/K_D)) = ([L]/K_D + [L])$.

This equation, $\tilde{n} = [L]/(K_D + [L])$, often known as the Langmuir equation, describes the protein occupancy with increasing [L]. It is shown as a plot in Figure 2.1.8. It follows from this equation that $K_D = [L]$ when $\tilde{n} = \frac{1}{2}$.

Binding to multiple independent sites

For a protein with multiple binding sites:

$$P + n\text{L} \rightleftharpoons \text{PL} + (n-1)\text{L} \rightleftharpoons \text{PL}_2 +$$
$$(n-2)\text{L} \rightleftharpoons .... \rightleftharpoons \text{PL}_n$$

We then have $\tilde{n} = ([\text{PL}] + 2[\text{PL}_2] + ... + n[\text{PLn}])/$ [total protein]. If the n sites are *independent and identical*, then $\tilde{n} = n[L]/(K_D + [L])$ which is the same form as the Langmuir equation but now with maximum occupancy n rather than 1.

Non-independent (cooperative) binding

Many biological interactions do not follow the Langmuir model illustrated in Figure 2.1.8. Cooperative behavior is often observed where binding at one site influences binding at other sites. If the ligand binding affinity increases as more sites become occupied we have *positive* cooperativity. A famous example of positive cooperativity is oxygen binding to the four sites on hemoglobin where the oxygen binds progressively more tightly as more binding sites become occupied. *Negative* cooperativity, where the interactions weaken with increasing occupancy, can also occur. Cooperative behavior can be detected by direct plots as illustrated in Figure 2.1.9 (see also discussion of Scatchard and Hill plots below).

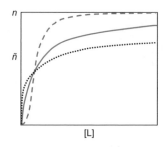

Figure 2.1.9 Direct plot for n non-interacting sites (cyan). For negative cooperativity, the black dotted lines are observed, while positive cooperativity gives the gray dashed lines.

Proton binding, pH, pK$_a$, and buffers

The ionization of a weak acid, HA, can be described by the equilibrium $\text{HA} \rightleftharpoons \text{H}^+ + \text{A}^-$. The acid dissociation constant is defined as $K_a = [\text{H}^+][\text{A}^-]/[\text{HA}]$. Rearranging this equation gives $[\text{H}^+] = [\text{HA}]K_a/[\text{A}^-]$. Taking logs to the base 10 gives $\log_{10}[\text{H}^+] = \log_{10}K_a + \log_{10}[\text{HA}]/[\text{A}^-]$ (see Appendix A2.4 for definition of logs and how to manipulate them). Finally, by changing the signs and defining $\text{pH} = -\log_{10}[\text{H}^+]$ and $\text{pK}_a = -\log_{10}K_a$ we obtain $\text{pH} = \text{pK}_a + \log_{10}([\text{A}^-]/[\text{HA}])$, which is known as the Henderson–Hasselbalch equation, an equation that relates the pH of a solution to the pKa and the $[\text{A}^-]/[\text{HA}]$ ratio. Note that the pK$_a$ value is the pH value where $[\text{HA}] = [\text{A}^-]$ (since $\log(1) = 0$). A plot of the fraction in the [HA] form, as a function of pH is shown in Figure 2.1.10 (see Problem 2.1.5).

A buffer is a solution containing either a weak acid and its salt, or a weak base and its salt. Such a solution has a capacity to resist changes in pH. The buffer capacity is a maximum when $\text{pH} = \text{pK}_a$.

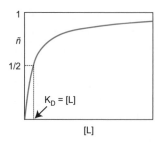

Figure 2.1.8 A "direct" or Langmuir plot of the fraction of protein occupied by ligand ($\tilde{n}$) against ligand concentration [L].

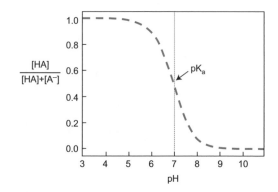

Figure 2.1.10 A plot of the fraction of [HA] as a function of pH for a group with a pK$_a$ value of 7.0.

The isoelectric point of a system is the pH at which there is no net charge on a molecule. For example, the isoelectric point, pI, for a molecule with two ionizing groups, pK_{a1} and pK_{a2}, is pI = ½ $(pK_{a1} + pK_{a2})$.

Analysis of binding data

This book will describe numerous ways of collecting binding data, including equilibrium dialysis (Chapter 3.1), analytical ultracentrifugation (Chapter 3.2), calorimetry (Chapter 3.7), surface plasmon resonance (Chapter 4.2), and spectroscopy (Sections 5 and 6). Once collected, the binding data have to be analyzed in a way that yields insight into a binding model and gives best estimates for the parameters describing that model. Analysis can be done in various ways. For example, the Langmuir equation, derived above, can be rearranged to give a Scatchard plot, $(\tilde{n}/[L]) = (n/K_D - \tilde{n}/K_D)$, where a plot of $\tilde{n}/[L]$ (y) against $\tilde{n}$ (x) is linear, with the form $y = mx + c$; the intercept on the x-axis is n and the slope is $-1/K_D$ (Figure 2.1.11). When analyzing data it is also important to obtain an estimate for how good a fit the model parameters (n and K_D) are to the data. Appropriate statistical analysis of the binding data is thus also required, as briefly described below.

Linear regression is a statistical method to find values for the slope and intercept of a straight line that come closest to the experimental data (see Appendix A3). The method does, however, have limitations: it assumes that the standard deviation of the data is the same at every value of x, an assumption that is often invalid. The experimental values of x and y in a Scatchard plot are, for example, not independent ($x = \tilde{n}$ and $y = \tilde{n}/[L]$) and this violates another assumption of linear regression. If any cooperativity is present in the data, regression to a straight line is also not appropriate (see below).

Non-linear regression is a general technique to fit any model to experimental data. This is the method of choice for most forms of data analysis. A mathematical description of the model is assumed, for example the Langmuir equation or some more complex representation of binding. Non-linear regression finds the values of parameters that generate a model curve that comes closest to the experimental data (it minimizes the sum of the squares of the vertical distances between the data points and the curve). The Langmuir curve, for example, can be fitted to determine the variable parameters, n and K_D. Cooperative behavior can also be treated, although more parameters are then required. Non-linear regression uses fairly complicated iterative procedures but this complexity is hidden from most users of fitting programs. Typically, we start with an initial estimated value for each variable in an equation, generate the curve defined by those initial values, calculate the sum-of-the squares of the vertical distances of the experimental points from the curve, adjust the variables to make the curve come closer to the data and repeat this until no more improvement is observed. Non-linear regression analysis also gives appropriate estimates for the errors in the various parameters used to describe a particular model.

Other analysis plots

While most data are now analyzed using non-linear regression fitting programs, it is often convenient to apply other data representations. The Scatchard plot discussed above is, for example, sensitive to cooperative effects, showing concave-down for negative and concave-up curvature for positive cooperativity, as illustrated in Figure 2.1.11.

An alternative plot, the Hill plot (Figure 2.1.12) is $\ln(\tilde{n}/(n - \tilde{n}))$ against $\ln[L]$. The n sites are half saturated when $\ln(\tilde{n}/(n - \tilde{n})) = 0$; the slope of the Hill plot (h) at this point is a measure of the degree of positive cooperativity; h cannot be greater than n, the number of binding sites on P (see also Problem 2.1.4).

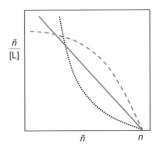

Figure 2.1.11 Scatchard plots for n non-interacting sites (cyan). For negative cooperativity, the black dotted line is observed, while positive cooperativity gives the gray dashed line.

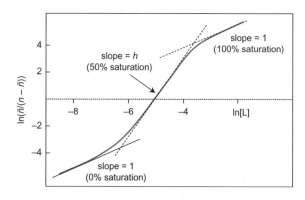

Figure 2.1.12 A Hill plot ($\ln(\bar{n}/(n - \bar{n}))$ against $\ln[L]$). At very low concentrations of L, all the P sites are empty and Langmuir-type binding is observed with a slope of 1. At high [L], all the sites but one are occupied and again non-cooperative binding is observed. At intermediate concentrations, where there is partial occupancy of sites, binding at one site can increase the affinity at neighboring sites. This leads to a slope >1 in the Hill plot with a maximum observed when half the sites are occupied. The Hill coefficient, h, is defined as the slope at 50% saturation.

 ## Summary

1 Cellular components can be described in terms of their concentration, structure, energy, and dynamics. These four quantities are interrelated. Quantum theory is required to describe many events involving atoms and molecules.

2 Non-covalent interactions depend on: ionic interactions, hydrogen bonds, Van der Waals forces, and the hydrophobic effect. The hydrophobic effect, which depends on water structure, is the main driving force for macromolecular assembly and folding.

3 Protein conformation and DNA base pairing arise from favorable non-covalent interactions including H-bonding. A Ramachandran plot describes energetically allowed regions in a protein, in terms of two backbone dihedral angles, ψ and ϕ.

4 The binding of a ligand, concentration [L], to a protein, can be quantified by the relationship $K_D = 1/K_A = [P][L]/[PL]$, where K_D and K_A are dissociation and association constants.

5 The Henderson–Hasselbalch equation $pH = pK_a - \log_{10}([A^-]/[HA])$ describes the relationship between pH, pK_a and fraction of the acid species, [HA].

6 Analysis of binding data is best done using non-linear regression methods but information can also be obtained using Scatchard and Hill plots.

 ## Further reading

Book covering basic biophysical principles

Atkins, P., and de Paula, J. *Physical Chemistry for the Life Sciences* (2nd edn) Oxford University Press, 2010.

Major biochemistry and molecular cell biology texts

Berg, J. M., Tymoczo, J.L., and Stryer, L. *Biochemistry* (6th edn) WH Freeman, 2006.

Alberts, B., Johnson, A., Lewis, J., Raff, M., Roberts, K., and Walter, P. *Molecular Biology of the Cell* (5th edn) Garland, 2008.

The hydrophobic effect

Chandler, D. Interfaces and the driving force of hydrophobic assembly. *Nature* 437, 640–7 (2005).

? Problems

2.1.1 One of the results of gas kinetic theory is that the pressure, p, of a gas of molar mass M in a volume V is $pV = (nM\hat{u}^2)/3$, where $\hat{u}$ is the root-mean-square (rms) gas velocity, defined as the square root of the mean of the square of the velocities. This equation can be compared with Boyle's law for an ideal gas: $pV = nRT = N_A kT$ (the various constants and their values are given in Appendix A1.3). What is the temperature dependence of $\hat{u}$ and the rms gas kinetic energy? What is $\hat{u}$ for nitrogen molecules in air at 25°C (see Box 2.1 and refer to Tutorial 3 for a discussion of kinetic energy)?

2.1.2 Calculate the potential energy between two opposite charges that are 0.28 nm apart: i) on the surface and ii) in the interior of a protein. Assume the dielectric constant on the surface is 80 and in the interior 4. Compare your answers with the strength of a typical protein H-bond ($\sim$20 kJ mol^{-1}) (see Box 2.3).

2.1.3 Why is the pattern of forbidden regions in a Ramachandran plot (Figure 2.1.5) unsymmetrical for amino acid sidechains other than glycine?

2.1.4 Use direct, Scatchard, and Hill plots to analyze the following data obtained for a ligand, L, binding to a monomeric (m) and a tetrameric (t) form of P. Assume [L] = free concentration. Comment on the results.

[L] (μM)	0.25	1.0	2.5	4	7	10	13	16	19	25
ñ (m)	0.18	0.46	0.70	0.79	0.86	0.90	0.92	0.935	0.95	0.96
ñ (t)	0.0	0.05	0.18	0.40	1.08	2.0	2.72	3.12	3.42	3.69

2.1.5 Derive the relationship between pH and pK_a and the concentrations of acidic and basic forms of a compound. ATP hydrolysis proceeds as follows at pH 8.0:

$$\text{ATP}^{4-} + 2\text{H}_2\text{O} \rightarrow \text{ADP}^{3-} + \text{HPO}_4^{2-} + \text{H}_3\text{O}^+$$

A 1 mM solution of ATP was hydrolyzed by an enzyme in a 0.1 M tris buffer at pH 8.0, 37°C. Calculate the pH value at the end of the reaction. What would the final pH have been if 0.01 M tris buffer had been used? (Assume the pK_a is 8.1 for tris and 7.2 for ADP and HPO$_4{}^{2-}$.)

3 Transport and heat

"We, on the other hand, must take for granted that the things that exist by nature are, either all or some of them, in motion."

Aristotle, ~300 BC

"Heat is a motion; expansive, restrained, and acting in its strife upon the smaller particles of bodies."

Francis Bacon, 1620

"Impressed force is the action exerted on a body to change its state either resting or moving uniformly straight forward."

Isaac Newton, 1687

"Life is like riding a bicycle. To keep your balance you must keep moving."

Albert Einstein, ~1950

This section explores a number of important, but quite diverse, techniques. It contains seven chapters that are relatively self-contained; the first six involve molecular transport in various forms, where the driving force for net molecular movement can arise from a concentration gradient (diffusion, osmosis, membrane transport), flow (viscosity, chromatography), a sedimentation force (centrifugation), or an electric field (electrophoresis, mass spectrometry). The molecule studied can be in the solution state (centrifugation), or in a vacuum (mass spectrometry). Some general aspects of transport are covered in the introductory Chapter 3.1. The final chapter in this section describes the analysis of reactions by the measurement of small heat changes (calorimetry); this is rather different from the other chapters but, as pointed out by Francis Bacon nearly 400 years ago (see above quotation), heat is also related to molecular motion.

3.1 Diffusion, osmosis, viscosity, and friction

Introduction

The cell contains large numbers of ions, metabolites, and macromolecules. The various macromolecules have unique properties, including mass, size, shape, and charge, that influence the way they move in response to flow and an applied driving force. Variation in movement is the basis of diverse separation techniques that have become essential tools in everyday biochemistry. Analysis of the movement also provides valuable information about many macromolecular properties including ion channels in membranes. In this introductory chapter, we consider diffusion, osmosis, dialysis, and viscosity.

Diffusion

Diffusion arises from the random thermal motion of a liquid, or a gas (see Box 2.2). Diffusion is an essential mechanism for molecules to get in and out of cells and to reach their correct location. It causes net movement of molecules from a region of high concentration to one of lower concentration. Diffusion requires kinetic energy from the environment but cellular energy (e.g. ATP) is not required.

Robert Brown noted the random motion of pollen grains in a solution in 1827; this **Brownian motion** arises from solvent collisions with small visible particles. In 1905, Einstein predicted that the diffusional motion of a particle in a fluid, at temperature T, would be characterized by a diffusion coefficient $D = kT/f$, where k is Boltzmann's constant, T is the temperature, and f is the **frictional coefficient** (see Box 3.1 and Tutorial 3) of the particle. A diffusing particle undergoes a **random walk**. In time t, it travels a distance $\bar{a} = \sqrt{2Dt}$, where D is the diffusion coefficient (Figure 3.1.1); $\bar{a}$ is often called the **root mean square (rms) displacement**. For the small amino acid glycine, molecular weight 75 Da, D is ~1×10^{-9} m^2 s^{-1}, while for a

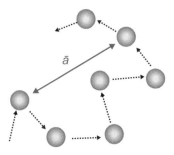

Figure 3.1.1 Random walk and root mean square displacement ($\bar{a}$) of a diffusing particle as a function of time.

protein, molecular weight 66 kDa, D~ 0.06×10^{-9} m^2 s^{-1}. This means that in 1 second, a glycine molecule diffuses ~4.5 μm, while for the protein $\bar{a}$ ~1.1 μm. Although the diffusion rate diminishes with size, single cells are still small enough to experience significant diffusional effects on their motion. Note that D has dimensions of area divided by time (m^2 s^{-1}; often quoted in units of cm^2 s^{-1}).

Concentration gradients

Diffusion tends to break down any concentration gradient in a solution. As illustrated in Figure 3.1.2, the concentration becomes equal throughout the solution after some time because of random diffusional

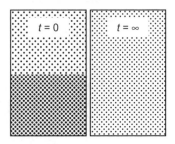

Figure 3.1.2 Breakdown of a concentration gradient as a function of time.

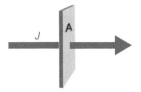

Figure 3.1.3 The flux J is the net diffusional transport of material per unit time across a unit cross-sectional area (mol m^2 s^{-1}).

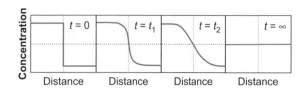

Figure 3.1.5 Fick's second law predicts the time and distance dependence of a concentration gradient. The concentration distance distribution is shown at increasing time intervals (compare Figure 3.1.2).

walks (more molecules start off in the high concentration zone so more wander in the direction of low concentration than to higher concentration). The way the concentration changes with time can be described by Fick's laws of diffusion. J, the flux, is the number of particles passing through a window of area A in a given time interval (Figure 3.1.3).

Fick's first law describes the dependence of the diffusional flux on the concentration gradient; mathematically this is written as $J = -D\mathrm{d}c/\mathrm{d}x$, where $\mathrm{d}c/\mathrm{d}x$ is the concentration gradient, as illustrated in Figure 3.1.4. (Note the flux decreases with increasing distance from the boundary, hence the negative sign.)

Fick's second law of diffusion, $\mathrm{d}c/\mathrm{d}t = D\mathrm{d}^2c/\mathrm{d}x^2$, gives a mathematical description of the breakdown of concentration gradients as a function of time, as illustrated in Figure 3.1.5 (relevant mathematical concepts are given in Appendix A2.5). The diffusion constant D can be obtained by monitoring the rate at which the concentration gradient disappears. Diffusion effects

are important for understanding many experiments described in this section, for example sedimentation velocity (Chapter 3.2).

Osmosis

A special case of diffusion is when water moves across a semi-permeable membrane, from an area of low solute concentration to an area of high solute concentration. This flow of water is called osmosis. It can cause an increase in pressure across the membranes that can, for example, cause cells to lyse (burst).

Semi-permeable membranes play a number of roles. In cells, they are vital for separating cellular contents from the general environment and for enabling cells to communicate with that environment. Unlike biological cell membranes, which contain pumps (Chapter 3.6), artificial membranes only act passively but they are still useful analytical tools as they pass molecules selectively on the basis of size. They can be used in dialysis (see below) and filtration to purify macromolecules and to measure binding between small membrane permeable molecules and impermeable ones.

Osmotic pressure is defined as the pressure that must be applied to a solution to keep its solvent in equilibrium with pure solvent (Figure 3.1.6). Osmotic pressure can be explained by the fact that the chemical potentials (Tutorial 2) on both sides of the membrane have to be equal at equilibrium; i.e. $\mu_A^o(P) = \mu_A(P+\pi) = \mu_A(P) + RT\ln[A]$ where $[A]$ is the concentration (mol L^{-1}) of the solute component A in Figure 3.1.6. It follows for dilute solutions and no Donnan effect (see below) that $\pi = RT[A]$ (see, for example, Price et al. in further reading).

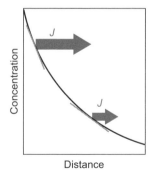

Figure 3.1.4 Fick's first law predicts that the flux, J, is proportional to the concentration gradient. The diagram shows concentration decreasing with distance from the origin and the length of the cyan arrow indicates the flux magnitude.

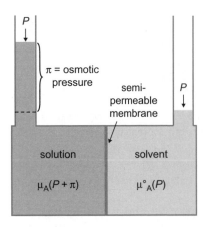

Figure 3.1.6 Illustration of an osmometer, a device to measure the osmotic pressure, π. At equilibrium across the semi-permeable membrane, the chemical potential on the solvent side of the semi-permeable membrane (μ_A^o), at pressure P, equals the chemical potential on the solution side (μ_A) at pressure $P + \pi$. The solution contains both solvent and solute particles that cannot cross the membrane.

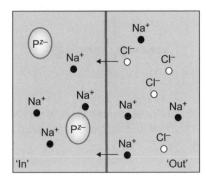

Figure 3.1.7 Illustration of the Donnan effect before equilibrium. A protein, P, with z negative charges is contained in the left compartment by a membrane that is permeable to Na^+ and Cl^- ions but not to protein. Initially P^{z-}, balanced by Na^+ ions, is on the left and NaCl is on the right. Cl^- ions will move down a concentration gradient from right to left, followed by Na^+ ions to maintain neutrality. In the absence of P^{z-}, $[Cl^-(in)] = [Cl^-(out)]$ but in presence of P^{z-}, $[Cl^-(in)] \neq [Cl^-(out)]$.

The Donnan effect

The macromolecules on one side of the membrane can be charged. It is important to take account of the effect of those charges when considering osmotic pressure. A macromolecule with a net charge $-z$ will cause ions with net charge $+z$ to associate with it to maintain electrical neutrality. The resulting uneven distribution of charge across the membrane is called the Donnan effect. As discussed above, osmotic pressure depends on concentration ($RT\ln[A]$) rather than on molecular weight, so small changes in ion concentration can have a large effect on the observed osmotic pressure. There are two ways to reduce an uneven balance of small ions across the membrane: one is to operate at a pH value where the macromolecule is electrically neutral, i.e. at the isoelectric point when the pH = pI; another is to operate at a high concentration of ions compared to the macromolecule concentration (see worked example).

Worked example

In two equal compartments, separated by a semi-permeable membrane, a 2 mM solution of protein, with a net charge of −2, is added to one side while 50 mM NaCl is added to the other side of the membrane (see

Figure 3.1.7). What are the equilibrium concentrations of ions on the two sides of the membrane?

Answer: There will be a flow of ions from right to left to neutralize the charges on the protein, causing a reduction in concentration in the right-hand compartment (assume this is y). In the general situation with protein concentration p, charge z, and NaCl concentration x, we have, at equilibrium, the following concentrations: $[Na^+(in)] = zp + y$; $[Na^+(out)] = x - y$; $[Cl^-(in)] = y$; $[Cl^-(out)] = x - y$. We also know that the chemical potentials of NaCl on the two sides of the membrane have to be equal at equilibrium; i.e. $\mu_{Na^+(in)} + \mu_{Cl^-(in)} = \mu_{Na^+(out)} + \mu_{Cl^-(out)}$. As each chemical potential is related to $RT\ln[X]$ we have $[Na^+(in)][Cl^-(in)] = [Na^+(out)][Cl^-(out)]$ (remember $\ln AB = \ln A + \ln B$; Appendix A2.4). Substituting then gives $(zp + y)(y) = (x - y)(x - y)$ which gives $y = x^2/(zp + 2x)$. In the particular example here, with $z = 2$, $p = 2 \times 10^{-3}$ M, and $x = 50 \times 10^{-3}$ M we get: $y = 50 \times 50 \times 10^{-6}/(2 \times 2 \times 10^{-3} + 2 \times 50 \times 10^{-3}) = 2.5 \times 10^{-4}/(104 \times 10^{-3}) = 24$ mM. Hence, at equilibrium, $[Na^+(in)] = 28$ mM; $[Na^+(out)] = 26$ mM; $[Cl^-(in)] = 24$ mM; $[Cl^-(out)] = 26$ mM.

Note: if the initial concentration of NaCl was 200×10^{-3} M, $y = 200 \times 200 \times 10^{-6}/(2 \times 2 \times 10^{-3} + 2 \times 200 \times 10^{-3}) = 4 \times 10^{-2}/(404 \times 10^{-3}) = 99$ mM. Hence, at equilibrium $[Na^+(in)] = 103$ mM; $[Na^+(out)] = 101$ mM; $[Cl^-(in)] = 99$ mM; $[Cl^-(out)] = 101$ mM so Donnan "error" effects are less at higher salt concentration.

Dialysis

Dialysis is a technique based on the diffusion of solutes across a semi-permeable membrane. These membranes can have specified pore sizes, corresponding to different molecular weights, e.g. 20 kDa, 50 kDa, etc. Consider a solution that contains a macromolecule P, separated from solvent by a bag made of a semi-permeable material, as shown in Figure 3.1.8. We assume the bag membrane passes ions and L, but not P. The system can be considered as having two phases, outside and inside. P and PL only exist inside the bag, while ions and L will be everywhere; i.e. we have $[L] = [L]_{free}$ outside while $[L] = [L]_{free} + [L]_{bound}$ inside the bag. $[L]_{bound}$ can therefore be found by taking the difference. With a suitably labeled ligand we can thus deduce the fraction bound at different values of total concentration and hence the affinity and the stoichiometry of the interaction between the small molecule and macromolecule (see Chapter 2.1 and Problem 3.1.7). Dialysis is also widely used in purification procedures, for example to change the composition of the solution in which a protein is bathed.

Application of a force to a molecule in solution

In several kinds of solution experiment, there is net movement of molecules. Examples above were diffusion in a concentration gradient and osmosis. As we will see, these effects are important in electrophysiology (Chapter 3.6). Examples where movement occurs because a force is applied to a molecule include: a *gravitational force in sedimentation experiments* (Chapter 3.2) and an *electrical field* in both electrophoresis (Chapter 3.4) and mass spectrometry (Chapter 3.5). On application of the force, the molecule will accelerate for a short time (for most cases of interest here, this short acceleration time can usually be neglected) and then it will achieve its "terminal" or equilibrium velocity, where the applied force in one direction is balanced by a frictional force, $-fv$, in the opposite direction (Box 3.1 and Tutorial 3); f is the frictional coefficient and v is the velocity (the frictional drag is directly related to the velocity).

In solution, the effective mass depends on the mass of solvent displaced; $m_{eff} = m(1 - \bar{v}\rho)$ where m is the mass of the particle, $\bar{v}$ is the partial-specific volume, and ρ is the solvent density (Figure 3.1.9). In the simple case of a free falling sphere, the gravitational force is mass × acceleration $= m_{eff}\,g = m(1-\bar{v}\rho)g = fv$. (Note that the molecule will also undergo random diffusional motion with or without the applied force.)

Viscosity

Viscosity describes a fluid's internal resistance to flow; it is a measure of fluid friction or the ease with which a particle can move in a fluid. In 1851, George Stokes derived the relationship $F = 6\pi r\eta v$, where F is the frictional force exerted on small spherical particle, radius r, moving at velocity v in a fluid with viscosity η. As $v = F/f$ at the terminal velocity (Figure 3.1.9), it follows that the frictional coefficient for a sphere is $f_o = 6\pi r\eta$ (Box 3.1). We also know that $v = m(1 - \bar{v}\rho)g/f$ (see previous section) so $v = m(1 - \bar{v}\rho)g/6\pi r\eta$. This relationship allows the viscosity of a solution to be calculated from first principles by measuring the velocity at which a sphere passes through liquid of density ρ. The viscosity coefficient η (eta) is measured in units of $kg\,m^{-1}s^{-1}$ (Pa s), although units of poise ($g\,cm^{-1}\,s^{-1}$) are often used.

A viscometer is an instrument used to measure the viscosity of a fluid. Two types of viscometer are shown in Figure 3.1.10—a glass device named after

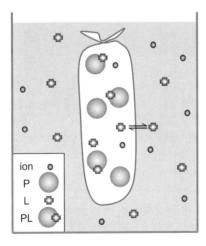

Figure.3.1.8 A dialysis membrane bag containing macromolecules suspended in a solvent. The semi-permeable (cyan) membrane allows small molecules (e.g. L) and ions to pass, but not the macromolecules.

Wilhelm Ostwald which is used to time a fluid falling through a capillary, and a rotational viscometer which measures the torque (Tutorial 3) on the inner tube caused by rotation of the outer tube. Viscosity values are usually derived by comparison with the viscosity of known standards using the formula $\eta_1/\eta_2 = \big((1 - \bar{v}\rho_2)/(1 - \bar{v}\rho_1)\big)(v_1/v_2)$ where 1 and 2 refer to the observed parameters in the two different solutions.

The effect of macromolecules on viscosity

Adding macromolecules to a solution changes the viscosity in a way that depends on the concentration, size, and shape of the added macromolecule. The viscosity of a solution is increased less by compact globular macromolecules than extended or unfolded ones. Measurement of viscosity can thus be used to measure protein denaturation (it was used, for example, by Anfinsen in his classic experiments on the folding/unfolding transitions of ribonuclease—see *Science* 181, 223–30, 1973). It can also be used to measure the transition between single stranded and double stranded DNA, to observe interactions between small molecules and DNA, and to distinguish between circular and linear DNA forms. The intrinsic viscosity [η] is defined as the limit of $(\eta - \eta_o)/c\eta_o$ as c, the solute concentration, approaches zero (η_o is the viscosity in the absence of solute; [η] has units of inverse concentration). For molecules like DNA there is a direct relationship between molecular weight, M, and intrinsic viscosity; it is found that [η] scales approximately as $M^{1.05}$ in the range $10^4 \leq M \leq 2 \times 10^6$.

Viscoelasticity

Viscous materials, like treacle, resist shear flow when a stress is applied. Elastic materials undergo strain when stretched and a rapid return to their original state once the stress is removed. Viscoelastic materials have both elastic and viscous properties. Long molecules like DNA exhibit viscoelastic behavior—they stretch when subjected to sheer and then relax back to a more compact structure. The rotating viscometer in Figure 3.1.10 can be used to measure viscoelasticity as the rotation first creates a shear force that causes the molecules to elongate

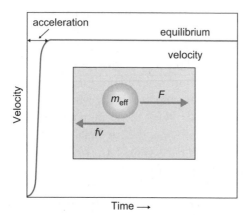

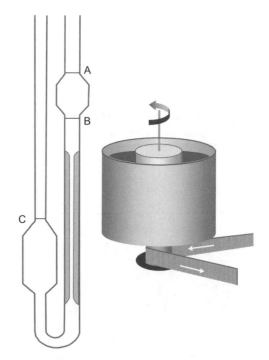

Figure 3.1.9 The behavior of a particle in solution, subjected to a force F. The cyan line shows the velocity as a function of time. In the gray box, a particle mass m_{eff} travelling at its "terminal" velocity is subject to balanced forces of F and fv after a brief acceleration period.

Figure 3.1.10 Two types of viscometer. The glass capillary Ostwald device is used to time the rate at which a fluid passes between the marks at A and B. In the rotating viscometer, shown on the right, rotation of the outer cylinder produces a measurable torque on the inner cylinder.

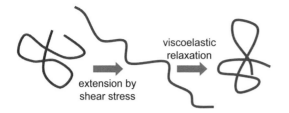

Figure 3.1.11 Representation of viscoelastic behavior, which can be detected with the rotating viscometer in Figure 3.1.10.

(Figure 3.1.11); if the drive is then switched off, the molecules recoil causing a reverse movement of the inner cylinder which is a measure of the viscoelasticity. The viscoelastic properties of DNA correlate well with molecular weight over a wide range and this method has been used to measure the molecular weight of very large DNA fragments. The viscoelastic properties of weight-bearing biological materials such as cartilage are also of considerable interest. (Note that atomic force microscopy (AFM)—Chapter 7.2—can be used to measure the elastic properties of single molecules.)

The Reynolds number and cell movement

Viscosity has interesting effects in the micron-scale world occupied by bacteria and single cells. In the world with which we are familiar, a walrus can flip its flippers and progress for some considerable time in water before any more thrust is required; in other words it has a large moment of inertia and the viscosity effects are relatively low. In contrast, swimming bacteria have a low inertia and relatively high viscosity so they will stop almost immediately if any thrust is removed. A measure of this effect is given by Reynolds number, Re, which depends on the *ratio* of *inertial* to *viscosity* effects. It equates to a dimensionless quantity $Re = vL\rho/\eta$, where v is the velocity, L is a measure of particle size, ρ is the solution density, and η the viscosity. The swimming walrus will have a large Re value ($> 10^5$), while the Re value will be low for a bacterium ($< 10^{-4}$); see Problem 3.1.6. Small bacteria are also buffeted by solvent and diffusion effects; their (flagellar) motors must overcome these effects, as well as viscosity, if they are to make significant progress towards finding a meal.

Box 3.1 The frictional coefficient, f, depends on molecular shape

The frictional coefficient, f, mentioned in the viscosity and diffusion discussions above, can also be determined experimentally by centrifugation (Chapter 3.2) or dynamic light scattering (Chapter 4.1) experiments. f is related to the diffusion coefficient, D, by $f = kT/D$ or $RT/N_A D$, where T is the absolute temperature, k is Boltzmann's constant, R is the gas constant, and N_A is Avogadro's number. Stokes showed that $f_o = 6\pi r\eta$ for a sphere. Jean-Baptiste Perrin (Nobel Prize, 1926) calculated adjustments to f for shapes that deviate from a sphere. This information can tell us something about particle shape in solution. Defining ellipsoid shapes by the ratio of the axes ($p = a/b$) as shown in Figure A, we have $p > 1$ for prolate ellipsoids (cigar shaped), $p < 1$ for oblate ellipsoids (disc shaped), and $p = 1$ for a sphere. Some examples of calculated f/f_o ratios are: prolate, when $a/b = 2$, $f/f_o = 1.044$ and when $a/b = 10$, $f/f_o = 1.543$; oblate, when $a/b = 1/2$, $f/f_o = 1.042$ and when $a/b = 1/10$; 1.458. In general, $f/f_o > 1$. The relationships between shape and friction will be discussed further in Chapter 3.2.

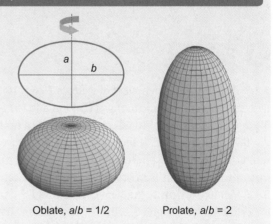

Oblate, $a/b = 1/2$ Prolate, $a/b = 2$

Figure A Illustration of different ellipsoid shapes defined by axes lengths a and b. The 3D shapes are obtained by rotation around the a axis.

 Summary

1 Molecules in solution undergo random diffusional motion and have a root mean square path length of $(2Dt)^{1/2}$ where D is the diffusion coefficient.
2 Diffusion results in a net flux of molecules that reduces a concentration gradient to zero after some time. This concentration-dependent diffusion behavior is described by Fick's first and second laws.
3 Osmosis is diffusional flow in a direction that reduces a concentration gradient across a semi-permeable barrier. This flow can lead to the generation of osmotic pressure. Osmotic pressure can be influenced by charges on the impermeable molecules (the Donnan effect).
4 Dialysis with a semi-permeable membrane can be used to purify macromolecules and to measure their binding to small molecules.
5 Viscosity describes a fluid's internal resistance to flow. It is described by the viscosity coefficient, η, which is sensitive to macromolecular size and shape.

 Further reading

Books covering basic biophysical principles

Atkins, P., and de Paula, J. *Physical Chemistry for the Life Sciences* (2nd edn) Oxford University Press, 2010.

Price, N.C., Dwek, R.A., Ratcliffe, R.G., and Wormald, M.R. *Principles and problems in Physical Chemistry for Biochemists* (3rd edn) Oxford University Press, 2001.

Tinoco, I., Sauer, K., Wang, J.C., and Puglisi, J.D. *Physical Chemistry: Principles and Applications in Biological Sciences* (4th edn) Prentice Hall, 2001.

Van Holde, K.E., Johnson, W.C., and Ho, P.S. *Principles of Physical Biochemistry* (2nd edn) Prentice Hall, 2005.

? Problems

3.1.1 The diffusion coefficient of water is 2.26×10^{-9} m^2 s^{-1} at 25°C. How long does it take for a water molecule to travel (i) 1 cm and (ii) 1 μm in an unstirred solution?

3.1.2 Calculate the frictional coefficient, f, for a protein with $D = 12 \times 10^{-11}$ m^2 s^{-1} and $\bar{v} = 0.71$ kg L^{-1} at 25°C. If the protein's molecular weight is 15 kDa, calculate its volume, radius (assume it is a sphere, volume $4\pi r^3/3$), and expected frictional coefficient f_o (assume the viscosity is mPa$= 10^{-3}$ kg m^{-1} s^{-1}). Compare f with f_o (see Box 3.1).

3.1.3 A sucrose density gradient of -0.2 mol L^{-1} cm^{-1} is set up for a centrifugation experiment (see Chapter 3.2; the "$-$" sign means that the gradient increases from top to bottom). Using Fick's first law, calculate the amount of sucrose passing through a 1 cm^2 area in 1 min (assume this is unstirred and before centrifugation starts; $D = 5.2 \times 10^{-10}$ m^2 s^{-1}).

3.1.4 What is the osmotic pressure of a solution containing 18 g glucose in 1 L solution at 20°C? What would the new osmotic pressure be if 0.1M MgCl$_2$ was added to the solution?

3.1.5 In an equilibrium dialysis experiment to determine the binding of a drug to an enzyme inside the dialysis bag, radioactively labeled drug was used to measure the free and total concentrations. Using the information about Scatchard plots in Chapter 2.1, calculate the stoichiometry and K_D from the following data obtained at an enzyme concentration of 50 μM.

Total (mM)	0.125	0.26	0.38	0.61	1.14	2.15
Free (mM)	0.09	0.197	0.297	0.50	0.98	1.99

3.1.6 Calculate Reynolds number, for (i) a motile *E. coli* cell with dimension $L = 1$ μm, velocity $v = 10$ μm s^{-1} and (ii) a human with $L = 1$ m, $v = 1$ m s^{-1}. Assume for water that $\rho = 10^3$ kg m^{-3} and $\eta = 10^{-3}$ kg m^{-1} s^{-1}. What is the significance of your results?

3.2 Analytical centrifugation

Introduction

Hand-driven centrifuges were used in the middle ages to separate milk products. Today, a variety of preparative centrifuges, ranging from small desktop to large scale devices, are in routine use in almost all molecular biology laboratories. We focus here on the analytical ultracentrifuge (AUC), where the behavior of the sedimenting molecules is visualized. The first successful AUC was built by Theodor Svedberg in 1924. This was a formidable technical achievement; a rotor was spun in a low-pressure hydrogen atmosphere to minimize heating; interactions with the sample cell walls were reduced by making the cells "sector" shaped (like a radially cut segment from a round cake) and the resulting sedimentation behavior was monitored photographically. Svedberg and his collaborators went on to develop most of the theory required to analyze ultracentrifugation in the late 1920s. Interest in analytical ultracentrifugation waned in the period 1970–1990 but the introduction of new instruments in the 1990s led to a great resurgence and numerous applications. The method is particularly good at monitoring macromolecular interactions, both self-association and association with other molecules.

There are two basic types of sedimentation experiment: equilibrium and non-equilibrium. In the non-equilibrium experiment, a boundary is created which migrates under the sedimentation force. An analogy is a swimming pool full of a similar swimmers who, at $t = 0$, are randomly distributed in the pool. If they all then start at the same time to swim towards one end of the pool at a constant velocity, the pool will empty at one end with the formation of a boundary between solvent and "swimmers". The concentration of the moving band of swimmers will remain constant but the boundary will move steadily towards one end of the pool.

An additional factor—diffusion—has to be considered to explain sedimentation experiments. Random diffusional movements of the molecules (swimmers) result in the dissipation, with time, of the concentration gradient generated at the boundary by sedimentation (Chapter 3.1). In sedimentation equilibrium experiments, a balance is set up between diffusion and sedimentation so that no net change is observed with sedimentation time. Versions of sedimentation and sedimentation equilibrium experiments exist for both "normal" solutions and those where a density gradient is created.

Some basic principles of sedimentation

Instrumentation

A schematic diagram of an analytical ultracentrifuge is shown in Figure 3.2.1; typical sample cells are shown in Figure 3.2.2. A sedimentation force is generated by placing a sample in a spinning rotor. The behavior of the sample is then monitored by an optical system.

Optical systems

Various optical methods can be used to monitor the sedimentation behavior during the experiment, as now described.

Absorbance

The most versatile and sensitive method to measure sedimentation behavior is to monitor the absorbance across the cell at different values of r. (Note that absorbance properties and instrumentation for UV visible spectra are described in more detail in Chapter 5.3.) A xenon lamp is used as the light source (Figure 3.2.1) and a grating splits this into

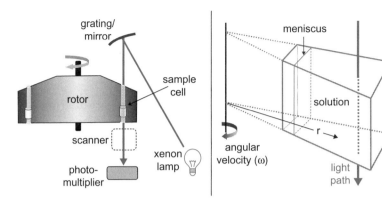

Figure 3.2.1 An AUC instrument showing a simplified optics system (left) and an expanded view of a sector-shaped sample cell.

light with a range of wavelengths that can be scanned. Protein and nucleic acid concentrations can be measured with good signal-to-noise ratio at concentrations as low as 10 µg mL^{-1}. For samples containing two or more components with different absorption spectra, data can be collected at multiple wavelengths in the same experiment. Figure 3.2.3 shows illustrative absorbance plots of a sedimenta-

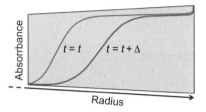

Figure 3.2.3 Plots of absorbance (concentration) against distance in a sector-shaped cell measured at two different times (t and $t + \Delta$) during a sedimentation experiment.

tion boundary at two time intervals (see also below for further discussion of the shape of these curves).

Rayleigh interference optics

Here, double sector cells (Figure 3.2.2) are used. One of the sectors contains the sample, the other contains buffer only. Refractive index is proportional to concentration (Chapter 4.2) and thus the light path is different through the sample and the buffer. If the light passing through the two cells is then combined, the path differences generate interference fringes, as shown in Figure 3.2.4 (light interference effects are discussed in Chapter 4.1). Interference optics can be used to detect macromolecules at high concentra-

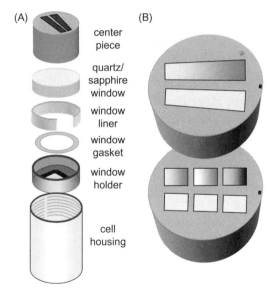

Figure 3.2.2 Some components of a typical AUC sample holder are shown in (A). The window, the liner, gasket, and holder are repeated above the center piece. Once the sample is inserted, the whole assembly is carefully tightened to a measured value of torque. (B) Shows an expanded view of two types of center piece. The upper one is for Rayleigh interference optics (see below) while the lower one, with 3 × 2 shorter cells, is used for sedimentation equilibrium experiments. As viewed, the sample is loaded in the top cells and a reference buffer is loaded in the lower ones.

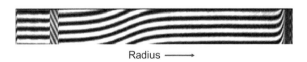

Figure 3.2.4 An example of interference fringes generated by a double sector cell system. Fringe shifts are seen at the sample boundary and the bottom of the cell (compare the shape of the central fringe with one of the curves shown in Figure 3.2.3).

tions and to monitor molecules such as polysaccharides that do not have a suitable light-absorbing center (a chromophore—see Chapter 5.3).

Fluorescence

A significant recent advance in AUC analysis is the introduction of fluorescence detection. Fluorescence optics can detect concentrations in the low-nanomolar range (Chapter 5.5) which is considerably lower than is possible with absorbance optics.

Sedimentation velocity

The Svedberg equation

Measurement of the sedimentation properties of molecules in solution tells us something about their shape, mass, and interactions with themselves and with other components. The spinning rotor in Figure 3.2.1 generates a sedimentation force on a particle of $m\omega^2 r$ (where m is the particle mass, ω the rotor angular velocity, and r the distance from the center of rotation—see Tutorial 3). In solution, the particle displaces some solvent, so the sedimentation force acts on an effective mass, $m_{eff} = m(1 - \bar{v}\rho)$ (see Chapter 3.1) that is less than m; ρ is the solvent density and $\bar{v}$ is the partial-specific volume of the macromolecule (this means that $m\bar{v}$ is the volume of liquid displaced). $\bar{v}$ can be measured, but it is often calculated from knowledge of the macromolecular composition. $\bar{v}$ for proteins is in the range 0.70–0.75; for polysaccharides it is 0.59–0.65; for RNA it is 0.47–0.55; and for DNA it is 0.55–0.59 L kg^{-1}. As ρ is ~ 1.0 kg L^{-1} for water, $(1 - \bar{v}\rho) \sim 0.27$ for proteins. At the "terminal" velocity, the sedimentation force is balanced by the frictional force and the mean velocity, v, of a particle, mass m, will be constant. As illustrated in Figure 3.1.9 and discussed in Chapter 3.1, this behavior can be described by the equation $v = m(1 - \bar{v}\rho)\omega^2 r/f$. The frictional coefficient, f, is related to the diffusion coefficient, D, by $f = kT/D$ or $RT/N_A D$ (see Box 3.1). As the molecular weight $M = N_A m$, where N_A is Avogadro's number, this equation can be written as $v = DM(1 - \bar{v}\rho)\omega^2 r/RT$. If we define the sedimentation coefficient as $s = v/\omega^2 r$, then we obtain the Svedberg equation: $s/D = M(1 - \bar{v}\rho)/RT$, which describes the sedimentation behavior in terms of D and M.

The sedimentation coefficient (s)

The sedimentation coefficient, s, is measured in units of Svedberg (S) where 1 S = 10^{-13} seconds. s can be measured by viewing the sedimenting molecules at increasing time intervals. As shown in Figures 3.2.3 and 3.2.6, a boundary is created that proceeds towards higher r values with increasing time. This boundary also becomes less steep with time because diffusion breaks down the concentration gradient that is formed by sedimentation.

The sedimentation coefficient, s, of a molecule tells us something about its size and shape. Ribosomal particles, for example, are described in terms of s values. The intact *E. coli* ribosome is classified as 70 S while the large and small ribosomal subunits are denoted 50 S and 30 S. (Note the Svedberg units (S) and the fact that the component parts do not add up to the value for the intact unit because s depends both on size *and* shape—see below.)

The concentration dependence of s

As the sample concentration increases, macromolecules will generally give smaller observed s values because of interference from other molecules in the sedimentation path and non-ideal solution behavior. For compact globular proteins this is a relatively small effect. For extended molecules, e.g. RNA or an unfolded protein, there can, however, be a large decrease in s with increasing concentration (Figure 3.2.5). If association occurs $(A + A \rightleftharpoons A_2)$, then higher s values are observed at higher concentration, as the amount of the higher mass complex (A_2) increases with increasing concentration (Figure 3.2.5).

Note It is often useful to transform the experimentally measured value of sedimentation coefficient, s_{exp}, to a value that would have been observed under "standard" conditions of water at 20°C, using the relationship $s_{20,w} = s_{exp}(\eta_{exp}(1 - \bar{v}\rho_{20,w})/(\eta_{20,w}(1 - \bar{v}\rho_{exp})$, where η and ρ refer to the viscosity and density of the buffer. This transformation allows ready comparisons between results obtained under different conditions. When the observed value is also extrapolated to zero concentration the s value, corrected for temperature and viscosity effects, is denoted $s^o_{20,w}$.

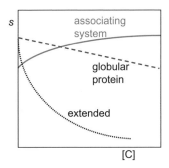

Figure 3.2.5 Illustration of the concentration dependence of *s* for different kinds of macromolecule.

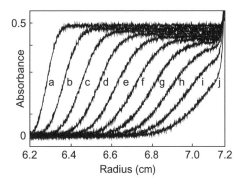

Figure 3.2.6 A sedimentation velocity experiment monitored using absorbance optics. The scans (a–j) were taken at 40 min intervals for a solution containing a protein of 50 kDa (sedimentation coefficient $s = 4.0$ S), rotor speed 5000 rpm (revolutions per minute). *r* is the distance across the sample cell (see Figures 3.2.1 and 3.2.3). (These data were simulated using SEDFIT.)

The Lamm equation

An equation describing the time dependence of the concentration, *c*, in an AUC experiment can be derived by combining the Svedberg equation and Fick's diffusion laws (Chapter 3.1). This is the Lamm equation: $dc/dt = (1/r)(d/dr)[rD\, dc/dr - s\omega^2 r^2 c]$. It is a differential equation (Appendix A2.5) that can be solved numerically. A full description of the concentration behavior also requires calculation of the radial dilution effect caused by the sector-shaped cell, as well as "end effects" due to accumulation of the material at the bottom of the cell, and depletion at the meniscus. Although these features make the Lamm equation difficult to solve, it can be done and there are excellent simulation programs available to fit experimental data (e.g. SEDFIT **http://www.analyticalultracentrifugation.com**); examples of Lamm equation fits to data are shown in Figures 3.2.6, 3.2.7, and 3.2.10. These fits allow the variables *s* and *D* to be determined for various sedimenting species.

Shape and *f/f*₀

We saw in Chapter 3.1 (Box 3.1) that the frictional coefficient, *f*, of an object moving through a solution is related to its shape. We also saw above that the sedimentation velocity is related to $m(1 - \bar{v}\rho)/f$. The coefficient, *f*, can thus be evaluated from a sedimentation experiment if *m* and $\bar{v}$ are known (these values can often be calculated, e.g. from known amino-acid composition). If the object volume has a shape other than a sphere, *f/f*₀ is greater than 1 (Box 3.1). *f/f*₀ ranges between 1.3 for globular hydrated proteins to about 2.0 for elongated or glycosylated proteins.

Comparison of the experimental values of *f/f*₀ with those derived from theoretical models can thus give useful information about possible shapes of the sedimenting molecule. A number of computer programs (e.g. HYDROPRO **http://leonardo.inf.um.es/macromol/programs/programs.htm**) are available that will calculate *f/f*₀ for any shape using hydrodynamic bead modeling, which represents a macromolecule by an assembly of beads. The apparent hydration of a particle (δ_{app}) can also be accounted for using the equation $P = f/f_0[1 + (\delta_{app}/\bar{v}\rho)]^{-1/3}$, where *P* is called the Perrin coefficient.

Polydispersity

Sedimentation velocity measurements can be used to analyze mixtures of species and to detect interactions between those species. Experimental curves of the sort shown in Figure 3.2.5 are functions of concentration, time, and radial distance; they can be fitted by solutions to the Lamm equation to give estimates for *s* and *D*. We can write these solutions to the Lamm equation in shorthand as $\chi(s, D, r, t)$ at a particular concentration.

For a mixture of species, the observed sedimentation curves can again be simulated with the Lamm equation. Figure 3.2.7 shows the same kind of sedimentation velocity experiment as in Figure 3.2.6 but this time there is a mixture of two proteins rather than one. In the case of a system with only two components,

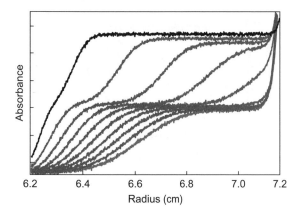

Figure 3.2.7 A sedimentation velocity experiment on a 50:50 mixture of two proteins, one with $M = 25$ kDa the other with $M = 90$ kDa. A series of concentration profiles are shown at different times. Note the observed separation of the two species which allows the sedimentation and diffusion behavior of each species to be measured separately. If the two proteins interact, a third boundary associated with the complex may also be observed.

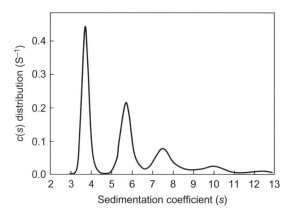

Figure 3.2.8 Sedimentation analysis of a recombinant form of vitamin-K dependent anticoagulant protein that can exist as multimers and monomers in solution. A distribution of recombinant protein forms is seen, with the most abundant having an s value ~3.8 (adapted with permission from *J. Thromb. Haemost.* 4, 385–91, (2006)).

A and B, the experimental data can be represented by the sum of two separate solutions to the Lamm equation multiplied by the component concentrations, c_A and c_B, i.e. $c_A \chi(s_A, D_A, r, t) + c_B \chi(s_B, D_B, r, t)$.

The approach that was applied to two species in Figure 3.2.7 can be expanded to treat n components and any experimentally observed sedimentation pattern can be fitted to a sum of Lamm plots calculated for each of the different species in solution. Experiments on mixed species, of the kind illustrated in Figure 3.2.7, can also be analyzed to give distribution plots showing the number of particles within a particular range of molecular weights or s values. Plots of $c(s)$, the amount of material in a sample, with a sedimentation coefficient between s and $s + \delta s$ give information about the distribution of observed sedimentation coefficients, as shown in Figure 3.2.8.

Ligand binding and association

Plots of $c(s)$ against s of the kind shown in Figure 3.2.8 are useful for quantifying association events as well as polydispersity, as the amounts of free and bound species present in a sedimenting solution can be measured. In these experiments, faster-migrating complexes remain in a pool of slower sedimenting species so that association and dissociation events can take place throughout the time course of the experiment. The data set from a single sedimentation run contains information about a wide range of concentrations that exists between the meniscus and the bottom of a centrifugation cell. It is therefore possible to deduce a binding curve and calculate a K_D value (Chapter 2.1) from one run although it is advisable to run mixtures at several different loading concentrations to give as wide a spread of free and bound concentrations as possible. Note that the observed sedimentation behavior depends on the lifetime of the complexes relative to the timescale of the experiment and different kinds of averaging can be observed for different on/off rates for complex formation.

Fluorescence detection systems (Chapter 5.5) are useful in this kind of experiment as low concentrations and thus tight binding can be monitored. Association can also be measured in sedimentation equilibrium experiments (see below).

Sedimentation equilibrium

We have been discussing experiments where sedimentation causes net movement of a boundary at a rate that can be measured. At lower rotor speeds, we can have a situation where the sedimentation of molecules down the centrifuge cell is balanced by their diffusion back up the cell (Figure 3.2.9).

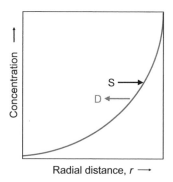

Figure 3.2.9 Illustration of the balance between sedimentation and diffusion forces that occurs in a sedimentation equilibrium experiment.

A sedimentation equilibrium experiment is illustrated in Figure 3.2.10. It can be seen that it can take a long time to achieve equilibrium (48 hours in this case). The experiment in Figure 3.2.10 used a standard length cell; shorter columns can also be used and these have the advantage that equilibrium is achieved more rapidly—more samples can also be analyzed with multiple short sample holders of the kind shown in Figure 3.2.2.

Sedimentation and diffusion both depend on molecular diffusion, defined by D. When sedimentation and diffusion are in equilibrium, the D term cancels out. The concentration distribution in the cell can thus be derived by assuming $dc/dt = 0$

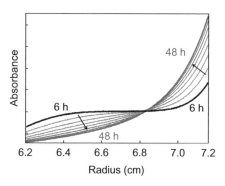

Figure 3.2.10 Approach to equilibrium for a molecule of molecular weight 50 kDa. The rotor speed is 10 000 rpm. Initially, the concentration is constant across the cell. At relatively short sedimentation times (e.g. ~6 h) the profile changes with time but after ~48 hours, diffusion and sedimentation are in balance and there is no further change in concentration profile with time. The arrows show the direction the curves move with time.

(equilibrium) and by eliminating D from the Lamm equation. This gives, for a single, ideal, non-associating solute, a curve described by the equation: $M(1 - \bar{v}\rho)\omega^2 = 2RTd(\ln c)/dr^2$, where M is the solute molar weight, ω is the angular velocity of the rotor, and c is the concentration of the solute at a radial distance r from the axis of rotation. A plot of $\ln c$ against r^2 thus gives a straight line with a slope proportional to M.

Historically, the sedimentation equilibrium experiment was used to measure molecular weights but mass spectrometry has largely replaced this role. The main applications of sedimentation equilibrium experiments now are the detection of complexes and self-association and the quantification of binding between species. As with sedimentation velocity experiments, a wide concentration range is typically established between the meniscus and the bottom of the cell in a single equilibrium experiment. Generally, K_D values in the range 10^{-4} to 10^{-8} M for self-association can be studied. The method is illustrated in Figure 3.2.11.

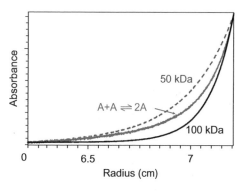

Figure 3.2.11 A sedimentation equilibrium experiment carried out on a monomer/dimer self-associating system with monomer molecular weight 50 kDa. The dotted gray line is the curve obtained for pure monomer; the solid black line is for pure dimer. The cyan line is what is observed experimentally. At low r values (where the concentration is low) the cyan line lies near the monomer line while at high r and high concentration it lies near the dimer line. Numerical fits to the cyan line using combinations of the dotted and solid line give the association constant for the monomer/dimer equilibrium.

Density gradient sedimentation

Versions of non-equilibrium (velocity) and equilibrium experiments also exist for sedimentation in density gradients. In our previous discussion we generally assumed that the term $M(1 - \bar{v}\rho)$ was positive and that centrifugation takes place in the direction of increasing radial distance, r. Whether this occurs in practice depends on the value of $\bar{v}\rho$. If $\bar{v}\rho > 1$ then the centrifugation force will be in the opposite (decreasing r) direction because the displaced solvent is heavier than the particle. As we will see, density gradients can be set up in different ways so that ρ is in fact an experimental variable.

Zonal centrifugation

Non-equilibrium density gradients can be generated by placing layer upon layer of gradient media, such as sucrose or inert commercial materials such as Ficoll, in a tube where the heaviest layer is at the bottom and the lightest at the top (Figure 3.2.12). Different types of gradient can be formed by adjusting the levels in the sucrose 1 and sucrose 2 chambers in Figure 3.2.12. The sample to be separated is then placed on top of the pre-formed gradient and centrifuged, usually with a swing-out rotor. This rate-zonal method (the equivalent of sedimentation velocity) separates macromolecules with different values of $M(1 - \bar{v}\rho)$.

The sedimenting particles form "zones" of material with the same mass and density, as shown in Figure 3.2.12. The run is terminated before any of the separated particles reach the bottom of the tube. The sedimentation coefficient, s, of a zone can be estimated by comparison with standards. Common preparative applications of sucrose density gradients include separation of cellular organelles or proteins. After centrifugation, the zones are usually collected as a series of fractions; this can be done by puncturing the bottom of each centrifuge tube and allowing the contents to drip out. The contents of the fractions can then be analyzed by a wide range of assays and analytical techniques.

Equilibrium or isopycnic gradients

In the isopycnic ("equal density") technique, the sample is prepared by uniformly mixing the macromolecules of interest with a gradient-forming material, usually a concentrated solution of cesium chloride (for DNA experiments, a CsCl concentration of ~1.7 g mL^{-1} is used). The sample is then subjected to a long sedimentation run until diffusion and sedimentation are in equilibrium. The Cs$^+$ ions generate a $\ln c/r^2$ profile (compare Figure 3.2.9) from the meniscus to the base, thus forming a density gradient throughout the centrifugation cell. Compared with the small Cs$^+$ ions, the macromolecules diffuse relatively slowly and they form bands rather than a wide distribution in the cell (see below). Macromolecules sediment to the position in the centrifuge cell where $(1 - \bar{v}\rho) = 0$, i.e. where their buoyant density, ρ_o, matches that of the solvent and the net sedimentation force is zero (Figure 3.2.13).

Purification and analysis of macromolecules by this isopycnic method is facilitated by the fact that the buoyant density ρ_o is different for different materials: for proteins, ρ_o ~1.3; for RNA, ρ_o ~ 2; for mitochondria, ρ_o ~1.1; and for ribosomes, ρ_o ~1.6 (ρ is in units of kg L^{-1}). For DNA, ρ_o depends on the base content; it is around 1.69 when the GC base content is 30% and

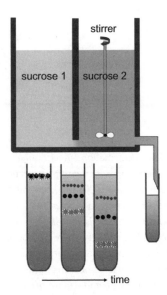

Figure 3.2.12 Illustration of a rate-zonal centrifugation experiment. The top picture shows how a sucrose density gradient can be formed by mixing dense (sucrose 2) and less dense (sucrose 1) solutions. The bottom picture shows how a mixture of three different components, layered onto the sucrose gradient at $t = 0$, migrates as zones through the pre-formed gradient as a function of time.

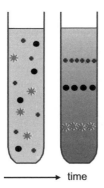

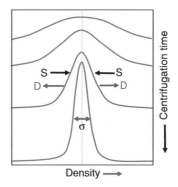

Figure 3.2.13 Representation of equilibrium/isopycnic density gradient sedimentation. At $t = 0$ there is a homogeneous mixture of three different kinds of macromolecule and CsCl (gray). After centrifugation for some time the Cs^+ is distributed throughout the cell, with a $\ln c/r^2$ profile. The three macromolecular types will sediment to a position where $(1 - \bar{v}\rho) = 0$ and remain there, in equilibrium.

Figure 3.2.14 In an equilibrium density gradient, a macromolecule experiences balanced forces of sedimentation and diffusion which produce a band. The band width, σ, is proportional to $1/M$.

1.73 when it is 70%. The band width of the macro-molecules in the isopycnic experiment depends on their molecular weight. The concentrating tendency of centrifugation (towards the position $1 = \bar{v}\rho_o$ is opposed by diffusion (Figure 3.2.14)); the net result is that, at equilibrium, the width of the band is inversely related to the molecular weight, M (Figure 3.2.14).

were grown for several generations in a medium containing the normally rare nitrogen isotope, ^{15}N. The *E. coli* cells with only ^{15}N in their DNA were then transferred to a medium containing the abundant, lighter isotope, ^{14}N, and were allowed to divide. DNA was extracted from these cells and subjected to density gradient centrifugation in CsCl. DNA containing the ^{15}N isotope could thus be separated from DNA containing the ^{14}N isotope. This allowed the progress of cell division to be monitored, as shown in Figure 3.2.15, giving a clear and elegant demonstration of semiconservative DNA replication.

Example The Meselson–Stahl experiment

Ultracentrifugation in a CsCl gradient had a spectacular introduction. A 1958 publication by two young post-doctoral workers, Matthew Meselson and Frank Stahl, elegantly proved that DNA replicated semi-conservatively (one-half of each new DNA molecule is old; one-half is new).

To quote from the original paper: '*We anticipated that a label which imparts to the DNA molecule an increased density might permit an analysis of this distribution by sedimentation techniques. To this end, a method was developed for the detection of small density differences among macromolecules. By use of this method, we have observed the distribution of the heavy nitrogen isotope ^{15}N among molecules of DNA following the transfer of a uniformly ^{15}N-labeled, exponentially growing bacterial population to a growth medium containing the ordinary nitrogen isotope ^{14}N.*' This experiment did not win a Nobel Prize but it has been called "the most beautiful experiment in biology". *E. coli*

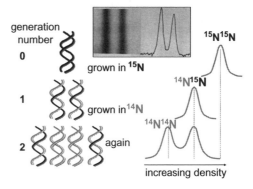

Figure 3.2.15 The classic Meselson and Stahl experiment designed to show that DNA replicates semi-conservatively. The detection method used was equilibrium density gradient sedimentation in CsCl. The inset is a photograph of the UV absorbance and its densitometer trace taken from the original paper (reproduced with permission from *PNAS* 44, 671–82 (1958)).

Summary

1 Analytical ultracentrifugation measures the sedimentation behavior of macromolecules. It can be a powerful tool for characterizing macromolecules and their binding properties to themselves and other ligands.

2 Measurement of the sedimentation velocity, described by the Svedberg equation, $s/D = M(1 - \bar{v}\rho)/RT\}$, gives information about the size, shape, and interactions of macromolecules. s is measured in Svedberg units (10^{-13} seconds).

3 In a sedimentation equilibrium experiment, diffusion and sedimentation forces are balanced; the result is that the concentration distribution in the centrifuge cell only depends on the effective molecular weight. Analysis of the sedimentation profile can also be used to quantify macromolecular association, e.g. dimer formation.

4 Density gradient centrifugation can be carried out using a *pre-formed* gradient. The sedimenting sample passes through the gradient in zones; this provides a powerful way of purifying and characterizing the sample.

5 Equilibrium density gradients can be formed by suspending the sample in a concentrated salt such as CsCl and centrifuging until the system is in sedimentation/diffusion equilibrium. The molecules then form bands, centered at the position where $(1 - \bar{v}\rho)$ is zero.

 Further reading

Manufacturer website with useful application material and descriptions of equipment
https://www.beckmancoulter.com/wsrportal/wsr/research-and-discovery/products-and-services/centrifugation/index.htm

Sedimentation protocols
Brown, P.H., Balbo, A., and Schuck, P. Characterizing protein–protein interactions by sedimentation velocity analytical ultracentrifugation. *Curr. Protoc. Immunol.* Chapter 18: Unit 18.15 (2008).
Balbo, A., Brown, P.H., Braswell, E.H., and Schuck, P. Measuring protein–protein interactions by equilibrium sedimentation. *Curr. Protoc. Immunol.* Chapter 18: Unit 18.8 (2007).

 Problems

3.2.1 Given the following concentration dependence of an enzyme's sedimentation coefficient, estimate the corrected sedimentation velocity $s^{\circ}_{20,w}$: $s^{0.2\%}_{20,w} = 7.72$ S; $s^{0.3\%}_{20,w} = 7.65$ S; $s^{0.4\%}_{20,w} = 7.48$ S; $s^{0.5\%}_{20,w} = 7.38$ S (where superscripts = % protein). Calculate the molecular weight of the enzyme at 20°C using the following parameters: $D = 4.75 \times 10^{-11}$ m^2 s^{-1}; $\bar{v}_; = 0.73$ L kg^{-1}; $\rho = 1$ kg L^{-1}.

3.2.2 (i) A solution of an enzyme in pH 7.0 buffer was sedimented to equilibrium at 20°C at a rotational speed of 4908 rpm (revolutions per minute). The enzyme concentrations at various distances (r cm) from the axis of rotation were determined using interference optics which gives fringe deflection (J) proportional to concentration. The results were:

J	2.514	2.651	2.812	2.974	3.108	3.307	3.497	3.699	3.912	4.150	4.437	4.688
r (cm)	6.925	6.943	6.962	6.981	6.999	7.018	7.036	7.054	7.072	7.091	7.110	7.129

Determine the molecular weight (M) of the enzyme given that $\bar{v}$ = 0.740 L kg^{-1} and ρ = 1.014 kg L^{-1}. Look out for difficult units here—for example convert to radians per second rather than rpm.

(ii) The sedimentation coefficient, $s^{o}_{20,w}$, measured in the same buffer at pH 7.0, was 8.0 S. At pH 3.0, the sedimentation coefficient decreased to 3.5 S and the molecular weight, as measured by sedimentation equilibrium, was 81 kDa. The molecular weight was also determined by polyacrylamide gel electrophoresis in sodium dodecyl sulfate under reducing conditions (Chapter 3.4). A single component was observed with a molecular weight of 40 kDa. Comment on the results.

3.2.3 Under sedimentation equilibrium conditions with a protein of 50 kDa, calculate the rotation speed in rpm required to achieve a concentration ratio of 100 between the meniscus (r_m = 6.7 cm) and bottom of the cell (r_b = 7.0 cm). Assume $\bar{v}$ = 0.73 L kg^{-1}, ρ = 1 kg L^{-1}, and T = 298 K.

3.2.4 A protein is centrifuged in a pre-formed sucrose density gradient with a swing-out rotor. Estimate the ratio of observed sedimentation coefficients and velocities at the top and bottom of the tube (see figure), given the following information: $\bar{v}$ = 0.73 L kg^{-1}; $\rho_{sucrose}^{10\%}$ = 1.0381; $\rho_{sucrose}^{25\%}$ = 1.1036; $\eta_{sucrose}^{10\%}$ = 1.167 cP; $\eta_{sucrose}^{25\%}$ = 2.201 cP (density units kg L^{-1}; viscosity centipoise).

3.2.5 Explain the plot showing sedimentation equilibrium CsCl density gradient results obtained for DNA isolated from *E. coli* and from crab.

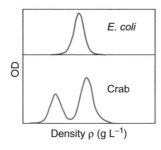

3.2.6 The plot shows the concentration dependence of the apparent molecular weight (M_{app}) of a DNA-binding protein in sedimentation equilibrium experiments. A similar shaped curve was observed at three different loading concentrations (0.25, 0.5, and 1.0 g/L). Comment on the data.

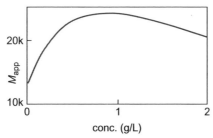

3.3 Chromatography

Introduction

Chromatography refers to a broad range of methods that separate and analyze mixtures of molecules. The components to be separated are distributed between two phases: a stationary phase and a mobile phase which percolates through the stationary phase. Chromatography literally means "color writing". The German chemist Runge used paper to analyze dyes in the mid-1880s but the first chromatography experiments are usually attributed to the Russian botanist Mikhail Tsvet who used filter paper to analyze plant pigments at the beginning of the 20th century. Martin and Synge then demonstrated the potential of chromatography in the 1940s by separating amino acids in a column—methods that facilitated Fred Sanger's efforts to sequence insulin. Chromatography is now an essential tool in the purification of a wide range of macromolecules but it can also be used to characterize molecular weights and to identify interactions.

The key experiments

Figure 3.3.1 illustrates the three main types of chromatography used in biological applications: size exclusion, which separates on the basis of size; ion exchange, which separates on the basis of charge; and affinity chromatography, which separates on the basis of binding preferences. They all involve a stationary phase associated with the column and the mobile phase associated with the solvent flowing through the column.

Terminology

It is useful to begin with a few definitions of words used in this field. An analyte (or solute) is the chemical entity being analyzed. A chromatograph usually refers to the apparatus used. A detector is a device that monitors the concentration of the analyte (see also below). A chromatogram is a graphical presentation of the response. The mobile phase is the fluid which percolates through the stationary bed; it may be a liquid or a gas. The eluent is another word for the mobile phase. The effluent is the mobile phase that exits the column. The stationary phase may be a solid, gel, or a liquid. The support is the solid that holds the stationary phase; the stationary medium may be *bonded* to the support or *immobilized* on it.

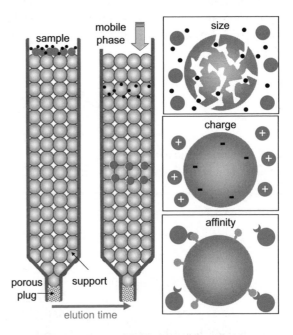

Figure 3.3.1 Illustration of the basic methodology in liquid chromatography. The mobile phase is an analyte-containing solvent that passes through the support matrix. The support contains beads with different properties. Three examples are illustrated: a bead that separates on the basis of size (the black molecules can enter the pores while the cyan ones cannot); beads that separate on the basis of charge, where positive analyte is attracted to the negative surface of the bead; and affinity-based separation, where a specific binding site for the protein is provided by the support. The example on the column is size exclusion.

The void volume is the volume in the column that is filled with the mobile phase. A zone, or band, is a region of the chromatographic bed where sample components are located.

Theory

We can describe the way the components in a chromatography system are distributed by assuming that the analyte A is in equilibrium between the stationary and mobile phases: $A_{stationary} \rightleftharpoons A_{mobile}$. The equilibrium is described by the equilibrium constant, or partition coefficient, $K = [A]_{stationary}/[A]_{mobile}$. The time taken for the mobile phase to pass through the column is τ_M (Figure 3.3.2). The time taken for an analyte peak, X, to reach the detector after sample injection is the called the retention time (τ_R^X). (Retention times are directly related to retention volumes (V_R).) The elution rate of A through a column is defined by the retention factor $k'_A = (\tau_R^A - \tau_M)/\tau_M$ where τ_R^A and τ_M can be obtained directly from an experimental chromatogram (Figure 3.3.2). When the elution rate is fast, the retention factor is small ($\tau_R^A \approx \tau_M$), and the A peak will be poorly resolved from the mobile phase peak. The resolution (R_s) between two peaks in a chromatogram is defined as $R_S = 2(\tau_R^A - \tau_R^B)/(W_A + W_B)$.

The theoretical plate model

The chromatographic column can be considered to contain a large number of separate layers, called theoretical plates. Equilibration of the sample between the stationary and mobile phases can be considered to occur in these plates. This model was used by Martin and Synge in their pioneering chromatography studies in the 1940s. The plate theory concept gives a measure of the efficiency of a column by defining a plate number, N. The N value for a column can be found from the empirical equation $N = 16\tau_R^2/W^2$; where W is the width of a chromatogram peak at the baseline (Figure 3.3.2); an equivalent formula is $N = 5.55\tau_R^2/\omega_{1/2}^2$ where $\omega_{1/2}$ is the peak width at half height. The higher the N value, the better the resolution, with narrower peaks (smaller W) and longer retention times (τ_R). Empirically it can also be shown that $N \approx 0.4 L/p$ where L is the column length and p is the particle size, i.e. longer columns and smaller particles have higher N values. Another parameter used to define the column is $H = L/N$, 'the height equivalent to a theoretical plate'; note that the lower the value of H, the better the resolving power of the column. Modern support media have very high values of N/unit length, e.g. for Superose 6 this is ~30 000 m^{-1}.

The van Deemter equation

A plot of plate height against the velocity of the mobile phase, u_m, is called a van Deemter plot (Figure 3.3.3). The equation $H = A + B/u_m + Cu_m$ gives a practical description of chromatography behavior and resolution. The equation takes account of the three main ways in which chromatography peaks are broadened. The relative contributions of these terms are described by the constants A, B, and C. The A term is related to eddy diffusion, which is caused by non-uniform packing of the columns and the ability

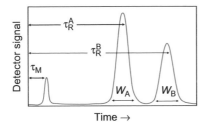

Figure 3.3.2 A schematic chromatogram showing the retention times (τ_R) for peaks A and B and the time for the mobile phase to pass through the column (τ_M). The base widths, W, for the A and B profiles are also shown.

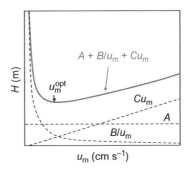

Figure 3.3.3 A van Deemter plot showing the dependence of the A, B, and C terms on the velocity of the mobile phase, u_m. (The A, B, and C components are shown as dotted lines and the sum of A, B, and C is shown in cyan). There is a value of u_m that gives a minimum value for H (maximum efficiency).

of different molecules to take different flow paths; this term is independent of u_m. The B term is related to longitudinal diffusion, which is caused by the tendency for any concentration gradient to be broken down by diffusion (Fick's law; Chapter 3.1); when u_m is high, there is less time for this effect to occur so it is inversely proportional to u_m. The C term is related to resistance to mass transfer, which is caused by the finite time required for partitioning between mobile and stationary phases; this term is proportional to u_m as the higher the velocity the poorer the mass transfer. The van Deemter plot thus predicts a curved plot for H against u_m that has a minimum value of H that corresponds to the optimum flow rate u_m^{opt} for the mobile phase (Figure 3.3.3). In practice, the analysis time can be speeded up by increasing the flow rate beyond the optimum (e.g. to $\sim 3 \times u_m^{opt}$) as this often gives acceptable separation but in a shorter time.

Chromatography techniques

Many different chromatography techniques are available; some of these are briefly summarized here.

The support and columns

The stationary phase is immobilized on support matrix particles packed in a column container. There have been considerable advances in support media over the years and a wide range of support materials are available commercially. Most are carbohydrate-based, consisting of chemically crosslinked polymeric beads that have good flow properties and can be cleaned for repeated use. Examples include dextrans (Sephadex—GE Healthcare), polyacrylamide (Bio-Gel P—Bio-Rad), and agarose (Sepharose—GE Healthcare/Bio-Gel A—Bio-Rad). Because of their large pores, agarose beads are good for separating very large molecules. Superdex (GE Healthcare) is a composite matrix of dextran and agarose that combines the good gel filtration properties of crosslinked dextran (Sephadex) with the physical and chemical stability of highly crosslinked agarose, to produce separation media with high selectivity and high resolution. The columns that hold the matrix in place are designed for high flow rates at greater than atmospheric pressure. High pressure (performance) liquid chromatography (HPLC) requires steel columns and chromatographic media which are able to handle very high pressures. A guard column is often placed before the separating column to filter or remove unwanted species.

Detectors

A wide range of detectors can be used to detect the analyte in the effluent; this can be collected in a series of test tubes and/or monitored continuously as it emerges from the column. The most common method is to monitor UV/visible absorbance (Chapter 5.3) but refractive index (Chapter 4.2), light scattering (Chapter 4.1), or ionic conductivity can also be used. Some of these methods, e.g. refractive index and conductivity, are especially useful for compounds, such as carbohydrates, that do not contain a suitable chromophore. A mass spectrometer can also be used to detect and identify the eluted analytes (Chapter 3.5).

Ion exchange

Ion-exchange chromatography (Figure 3.3.1) separates molecules based upon their charge. The stationary phase consists of a matrix support that has charged groups. The choice of ion exchanger depends on the charge on the molecule to be separated. Cation exchangers have negative charge while anion exchangers have positive charge. Strong ion exchangers retain their charge over a wide range of pH (e.g. $R–SO_3^-$) while weak ion exchangers are only charged over a limited pH range; for example, diethylaminoethyl, DEAE, is a commonly used weak anion exchanger with a pKa ~ 9.5. Once the separated species has been retained on the column, an eluting aqueous buffer is used to "wash" it off the column for collection. The eluting mobile phase weakens the ionic interaction by changing the pH or increasing the salt concentration (Problem 3.3.2).

Chromatofocusing

Chromatofocusing is a technique where a stable pH gradient is set up in an ion exchange column using buffered amphoteric media (this kind of methodology and the concepts involved are dealt with in more detail in the isoelectric focusing section in Chapter 3.4). Proteins have zero net charge when the buffer pH matches their isoelectric point (pH = pI; see Chapter 2.1). Proteins have a net positive charge when the buffer pH < pI and thus they will not bind to an anionic column. If pH > pI, however, they will be negatively charged and will bind. The pH gradient thus leads to a distance-dependent behavior that results in a focusing effect on the column; this causes the proteins to elute in order of their pI value.

Size exclusion chromatograph (SEC)

This technique, also called *gel-filtration or molecular sieve* chromatography, separates macromolecules based on size and shape (Figure 3.3.1). The support matrix for SEC consists of beads with pores of defined size. Larger molecules, which cannot penetrate the pores, migrate through the spaces that separate the beads faster than the smaller molecules, which penetrate the pores. Consequently, larger molecules exit the column first. This is the only chromatographic technique which does not involve binding of the analyte to a support. It is not only a powerful separation tool but it can also give information about molecular weight and interactions as discussed further below in the section on *quantitative chromatography*.

Reverse-phase chromatography

Originally "normal" phase chromatography referred to a polar stationary phase and a non-polar organic mobile phase. These days, the mobile phase is more likely to be a mixture of polar solvents (reverse-phase) while the stationary phase in HIC and HPLC (see below) is hydrophobic. The length of time spent in the hydrophobic stationary phase (the retention time) can then be increased by adding more water to the mobile phase to strengthen the hydrophobic interactions or can be decreased by adding more organic solvent, such as acetonitrile, to the eluent.

Hydrophobic interaction chromatography (HIC)

HIC is complementary to ion exchange chromatography and gel filtration. HIC and reverse-phase chromatography media are designed to absorb hydrophobic patches on macromolecules. The solvent-accessible non-polar groups on the surface of the biomolecules interact with hydrophobic groups (e.g. alkyl) covalently attached to the gel matrix. Adsorbents for HPLC (see below) are more hydrophobic than HIC adsorbents.

High performance (pressure) liquid chromatography (HPLC)

Here, a pump propels the mobile phase and analyte at high pressure (up to several hundred atmospheres) through columns which are densely packed with small particles (3–20 μm diameter). HPLC can use a variety of stationary phase supports. A typical hydrophobic stationary phase would be silica bonded to RMe_2SiCl, where R is a hydrocarbon chain of defined length (e.g. $C_{18}H_{37}$ or C_8H_{17}). Reverse phase HPLC is widely used in analytical and preparative separations of peptides and low molecular weight proteins that are stable in aqueous-organic solvents.

Fast protein liquid chromatography (FPLC)

FPLC is a form of liquid chromatography similar to HPLC but operating at lower pressure. The mobile phase velocity is carefully controlled by pumps. A wide range of relatively large volume, pre-packed FPLC columns are available, including those for ion exchange, size exclusion, hydrophobic interaction, and affinity chromatography. FPLC is particularly useful for purifying large amounts of material.

Affinity chromatography

Affinity chromatography is a powerful purification technique which potentially allows one-step purification of the target molecule (Figure 3.3.1). A specific ligand (a molecule that binds to the target protein) is immobilized on a support. The method is in widespread use with recombinant protein expression techniques, where a purification tag is used to select the protein of interest from an impure mixture. Elution from the column can be achieved, for example, by adding a competing ligand or by changing the binding properties. Tandem affinity protocols are discussed in more detail below. A common example of single step purification is to use the peptide glutathione attached to a support matrix; this ligand selectively binds to proteins that have been expressed, fused to glutathione S-transferase. The protein can then be eluted from the column by adding free glutathione.

Immobilized metal ion affinity chromatography (IMAC)

IMAC is a variant of affinity chromatography that retains proteins with an affinity for metal ions on the column. Immobilized metal ions on the column, such as nickel or cobalt, bind selectively to certain groups on a macromolecule. IMAC is particularly useful for the purification of recombinant proteins tagged with histidines (a sequence of six histidines attached to the N- or C-terminus of a protein is common), or for proteins containing phosphorylated groups. The protein of interest can be eluted by changing the pH or, for histidine-tagged proteins, adding imidazole, which competes for the metal ion and displaces the protein from the column.

Tandem affinity purification (TAP)

TAP is an efficient method for isolating proteins and various complexes from crude cell extracts. The gene coding for a protein of interest is fused to a recombinant TAP tag (attached either to the N- or C-terminus of the protein) and is expressed using standard DNA cloning procedures. The TAP tag is a composite of two different binding motifs plus a protease cleavage site (Figure 3.3.4). Purification of the tagged protein is carried out in two consecutive, high-affinity chromatography steps. These combined steps usually give a virtually pure protein complex.

A specific example is shown in Figure 3.3.4. In this case, the TAP tag was engineered to exploit two known biological interactions: (i) the binding between a peptide (CP) and a protein called calmodulin; and (ii) the binding between a protein (protein A) and immunoglobulin, IgG. The calmodulin peptide and protein A were fused together with an amino-acid sequence containing a cleavage site for a specific protease (tobacco etch virus, TEV). This TAP tag was attached to a "bait" protein for which unknown binding partners were sought. The engineered bait with its TAP tag was mixed with an

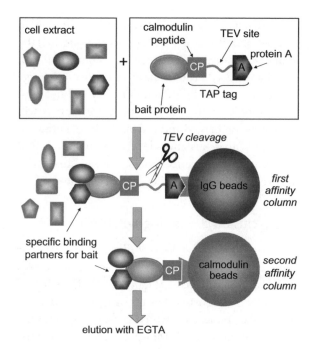

Figure 3.3.4 Illustration of tandem affinity purification (TAP). There are two key affinity purification steps, one involving a protein A/IgG interaction, the other a peptide CP/calmodulin interaction. The figure is based on a protocol published in *Nat. Biotech.* 17, 1030–32 (1999).

entire cell extract to "fish" for proteins that bind to the bait. This mixture was then passed through a column containing IgG beads that bind specifically to protein A in the TAP tag; this process caused the bait and any component(s) of the cell extract that bind to that bait to be sequestered on the column (by the binding of protein A to IgG). TEV protease was then added to cleave the bound protein complexes. Another round of binding was carried out on a second column, containing beads which bind specifically to the CP part of the TAP tag. The complex was then eluted by removing the calcium, and binding capacity, from the calmodulin with the Ca^{2+} chelating agent, EGTA.

Quantitative chromatography

The main chromatography applications discussed so far have been preparative and empirical. The method can, however, also be used to quantitatively analyze mass, shape, and interactions.

Direct estimation of molecular weight

Size exclusion chromatography (SEC) can give estimates of analyte molecular weight (*M*). The larger the molecule, the faster it elutes, and so the smaller the elution volume. SEC requires calibration of the elution profile for a particular support matrix with analytes of known mass, M. An example of a calibration curve is shown in Figure 3.3.5; the elution profile of the unknown species is then compared to the calibration curve to deduce its molecular weight (see Problem 3.3.3). The partition coefficient between mobile and stationary phases is better represented by the Stokes radius—the effective radius of the solvated analyte (Chapter 3.1)—than by *M*. Thus, proteins with unusual shapes, or those that have undergone post-translational modification such as glycosylation, often run anomalously.

Different support media also have different characteristics. In the example shown in Figure 3.3.5, a dextran matrix with a molecular weight range of 5–600 kDa was used (Sephadex G-200). Sephadex G-50, in contrast, is good for analyzing peptides in the range 1.5–30 kDa. SEC of very high molecular weight particles (e.g. DNA and viruses) can be done with agarose gels with a large pore size (e.g. Sepharose). For flexible molecules, the size parameter governing partition is the radius of gyration, R_G, (see Chapter 4.1) rather than the effective Stokes radius.

Characterization of self-association

As a sample migrates down a column, there is progressive dilution of the analyte because of diffusion

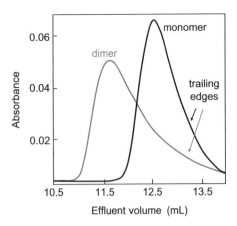

Figure 3.3.6 A size exclusion chromatography analysis of a DNA binding ATPase that dimerizes in the presence of ATP. The elution profile was calibrated using cytochrome c (12.4 kDa: 14.3 mL); carbonic anhydrase (29 kDa: 12.4 mL); and BSA (66 kDa: 10.3 mL). Note the relatively sharp leading edge and long trailing edges, especially in the dimer profile. The absorbance was monitored at 280 nm (data derived with permission from *PNAS* 104, 20326–31 (2007)).

down a concentration gradient from the sample zone (the band layered on the column) to the surrounding solvent. For a monomer/dimer equilibrium ($A + A \rightleftharpoons A_2$) this dilution will increase the concentration of A relative to A_2 as the sample zone proceeds down the column. The size exclusion properties of the column also mean that the larger molecular weight species (A_2) migrates faster than A. The net result is that the leading edge of the sample zone, dominated by A_2, is sharpened. In contrast, the trailing edge is broadened by the presence of both A and A_2. SEC can thus be a useful tool for detecting the presence of self-association and for quantifying the association, as shown by the examples in Figures 3.3.6 and 3.3.7.

An extension of the method illustrated in Figure 3.3.6 is "frontal chromatography". If sufficient solute solution is applied to the column, an elution profile is generated where the concentration of the trailing edge is approximately constant (a "plateau region" is created). The leading edge in Figure 3.3.7 shifts to the left as more of the dimer is formed at higher protein concentrations (Figure 3.3.7). It is thus possible to estimate the K_D for a monomer/dimer equilibrium by measuring the leading edge position as a function of concentration (see caption to Figure 3.3.7).

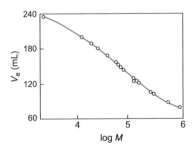

Figure 3.3.5 Logarithmic calibration plot for the direct determination of molecular mass from the retention time (elution volume, V_e) using size exclusion chromatography of proteins on Sephadex G-200. (Data taken from *Biochem. J.* 96, 595–606 (1965).)

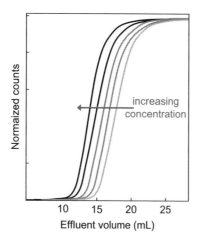

Figure 3.3.7 Radioactively labeled protein (a λ phage repressor) was monitored by SEC using a zone of analyte large enough to generate a plateau region at larger elution volumes. The leading boundaries of five separate zones, formed at different plateau concentrations, are shown; from light to dark gray/right to left the concentrations were: 5.01×10^{-11} M; 1.07×10^{-10} M; 4.98×10^{-9} M; 1.02×10^{-8} M; and 1.02×10^{-6} M. The boundary positions shift toward lower elution volumes with increasing concentration, reflecting the increasing average molecular weight as the repressor forms more dimers. The dissociation constant for the monomer/dimer equilibrium was measured to be 5.6nM from these curves (data derived with permission from *Anal. Biochem.* 196, 69–75 (1991)).

Interactions between dissimilar species

SEC methods have also been used to characterize chemical equilibria involving different reactants as well as self-association. Such interactions can range from a small drug binding to an enzyme to the

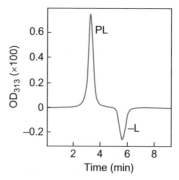

Figure 3.3.8 A SEC study of the anticoagulant, warfarin, binding to human serum albumin (HSA). The concentration of PL could be measured from the absorbance at 313 nm where the warfarin has a strong absorbance peak compared to the protein (data from *J. Chromatogr.* 167, 159–70 (1978)).

formation of a large protein–nucleic acid complex. One way of quantifying binding (e.g. $L + P \rightleftharpoons PL$) is the Hummel and Dreyer method. A known quantity of $P + L$ is layered as a zone on a size exclusion column. This is then eluted with a buffer containing a constant concentration of ligand [L]. From Chapter 2.1 we know that the fraction bound $\check{n} = [PL]/([P] + [PL]) = [L]/(K_D + [L])$ where K_D is the dissociation constant. If [L] in the effluent is measured, then a positive elution peak will be observed for the PL complex which migrates faster than L, along with a slower migrating negative peak, corresponding to the quantity of L that is withdrawn from the solvent (Figure 3.3.8). P and L will remain in equilibrium throughout the separation and both [PL] and [L] can be estimated from the elution profile, thus allowing K_D to be determined (Chapter 2.1). An example of a Hummel and Dryer experiment is shown in Figure 3.3.8.

 ## Summary

1 Chromatography is a powerful tool for separating mixtures, purifying DNA, and characterizing proteins. It involves partition between a mobile and a stationary phase.

2 The three most common forms of chromatography used in molecular biology are size exclusion, ion exchange, and affinity chromatography. A wide range of suitable support matrices is available for these applications. Many variations of these three approaches are possible.

3 An understanding of practical chromatography is aided by the concept of theoretical plates and the van Deemter plot.

4 Affinity chromatography and its variants such as IMAC (immobilized metal ion affinity chromatography) and TAP (tandem affinity purification) are extremely powerful for protein purification.

5. Size exclusion chromatography can be used to measure molecular weights and to study self-association and association with other molecules.

 ## Further reading

Useful websites

http://www.gelifesciences.com/aptrix/upp01077.nsf/content/protein_purification
http://www.chromatography-online.org/
http://www.waters.com/webassets/other/corp/lp/chroma.html?xcid = 9440_20100409

General description of biophysical methods
Sheehan, D. *Physical Biochemistry* (2nd edn) Wiley, 2009.

More advanced review
Winzor, D.J. Analytical exclusion chromatography. *J. Biochem. Biophys. Meth.* 56, 15–52 (2003).

 ## Problems

3.3.1 What different properties of a protein can be used in a purification scheme? For each property, describe a chromatography technique that could be used for separation.

3.3.2 In eluting a protein from an ion-exchange column, what type of pH gradient (high to low or low to high) should be used to elute (i) from an anion exchange column (ii) from a cation exchange column?

3.3.3 You wish to separate a DNA binding protein, D, from three other proteins A, B, and C. The estimated pI and M values of the four proteins, from bioinformatics analyses, were: A = 7.4, 80 kDa; B = 3.5, 21.5 kDa; C = 7.9, 23 kDa; D = 7.8, 20 kDa. Suggest suitable chromatography experiments to separate (i) D from A, (ii) D from B. Can you suggest a protocol for separating D from an ABCD mixture?

3.3.4 DNA can be eluted from DEAE by increasing the pH to 10.5. Why?

3.3.5 Using Figure 3.3.5, estimate the molecular weight of a protein with an elution volume of 180 mL.

3.3.6 Estimate the theoretical plate parameters N and H for the monomer peak in Figure 3.3.6 assuming the column was 10 cm long and that 10 mL took 5 min to elute. How might the resolution of the column be improved?

3.3.7 How do you expect van Deemter curves to differ when using support matrix beads with diameters of 10 μm and 3 μm?

3.4 Electrophoresis

Introduction

Migration rates of molecules through a matrix depend on their charge, size, and shape. Electrophoresis exploits the differential mobility of molecular ions in an applied electric field. The method is relatively simple and inexpensive, with good resolving power, so it has become a major tool for the separation and characterization of proteins and nucleic acids. It is very versatile, with many possible arrangements of the apparatus and a wide range of migration media. Electrophoresis in two dimensions can resolve thousands of proteins present in a cell lysate.

Early electrophoresis experiments by Arne Tiselius (Nobel Prize, 1948) in the 1930s were performed with protein solutions in a quartz U-tube. Protein boundaries, created by the application of a voltage, were monitored by photography. This moving boundary method can, however, only isolate the fastest migrating component in a mixture and the method was largely replaced in the 1950s by zone electrophoresis where layers of macromolecules pass through gel matrices. Discontinuous gels were introduced by Ornstein and Davis in 1959. Further developments included the introduction of denaturing agents by Weber and Osborn in 1969, isoelectric focusing, 2D gels, and agarose in the 1970s followed by capillary electrophoresis (CE) and pulse field gel methods. These various techniques and some applications are introduced below.

Theory

Charges on macromolecules

Electrophoresis depends on charge. In proteins the charges on the N- and C- termini and on the sidechains depend on the pH of the solution and the pK_a

values of the various R groups (Tutorial 1). Proteins have an isoelectric point (pI) (the pH value where the number of positively-charged groups equals the number of negatively-charged groups—see Chapter 2.1). Depending on the amino-acid composition, protein pI values can range from 1 to 11. At pH values lower than the pI, proteins have a net positive charge while they have a net negative charge when pH > pI. Some covalent modifications, e.g. phosphorylation, also change the net protein charge. Nucleic acids have a uniform charge distribution along their length as the phosphate backbone gives one negative charge per nucleotide at neutral pH (see also below).

Migration

As was shown in Figure 3.1.7, a particle in solution, subjected to a force, F, has a terminal velocity $v = F/f$, where f is the frictional coefficient. In electrophoresis, the force on a particle in an electric field E is qE, where the charge, $q = ze$ (Tutorial 3); z is the number of charges and e is the elementary charge in coulombs. We can thus define an electrophoretic mobility, the rate a species moves, as $u = v/E = ze/f$. From this equation, it might seem that we could use u to obtain information about molecular weight and shape just as in ultracentrifugation. There are, however, difficulties in using electrophoresis quantitatively. One of these arises because the molecule of interest is in a buffer with numerous other charged particles; these form an ionic atmosphere that distorts the applied electric field. Electrophoresis in free solution has also been replaced by more practical zonal methods where a band of molecules migrates through a matrix in a way that depends strongly on the matrix properties. Electrophoresis is therefore mainly used empirically, comparisons being made with molecular weight standards or markers.

In some cases, the charge distribution along a macromolecular chain is approximately *constant*. This applies to DNA and denatured proteins coated with an anionic detergent like sodium dodecyl sulfate (SDS; see also below). Progress through the gel is then dominated by the molecular sieving properties of the entangled gel network; longer fragments are slowed down by frictional forces in the network more than shorter ones. If the charge distribution along the length is constant, the result is that mobility then depends mainly on molecular weight.

Electrical properties

Some points arising from electrical circuit theory (Tutorial 4) are worth noting for electrophoresis experiments. The electric field E experienced by a molecule is V/d, where V is the applied voltage and d is the separation between the points where the voltage is applied. Short gels therefore have higher E values than long ones for the same applied voltage and thus species will generally migrate faster through shorter gels. In gels with regions of different resistance (e.g. in discontinuous gels—see below), E will be different in different regions because the voltage drop over a region depends on the resistance in that region ($V = RI$; where R and I are the resistance and current). Finally, the power dissipated as heat in a system is V^2/R (or I^2R). Heating effects in electrophoresis can be considerable and care has to be taken to maintain the system at an even temperature to avoid distortion of the bands in the gel.

Experimental

Apparatus

Figure 3.4.1 illustrates a typical vertical slab electrophoresis unit. Molecules migrate through a gel, separated by two baths containing buffer. Electrodes in the baths pass electric current from a high-voltage power supply through the gel. If the electrodes are arranged in such a way that the upper bath is negative (cathode) and the lower bath is positive (anode), the system is known as an anionic system. Voltage applied in the opposite direction gives a cationic system. The gels can be placed vertically, as shown in Figure 3.4.1, or horizontally and they can either be tubular (formed in glass tubes) or flat (formed between two glass plates). Flat (slab) gels allow multiple samples to be run at the same time and a variety of methods to visualize the bands are readily applied (see below). Slab gels have thus become standard and a range of precast gels can be purchased for an assortment of apparatus, with different sizes and applied voltages.

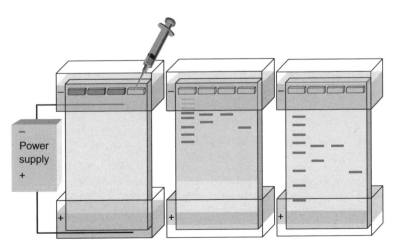

Figure 3.4.1 Schematic view of a vertical anionic slab gel electrophoresis experiment. Samples (volume ~20 μL, concentration 1 μg) are injected into the four loading wells. Once the voltage is switched on negatively charged proteins or DNA move progressively toward the positive anode (middle and right hand views).

Gel media

Electrophoresis separations are generally performed in a non-conducting matrix formed from acrylamide or agarose.

Acrylamide

Polyacrylamide gels are formed from the polymerization of acrylamide with N,N'-methylene-bis-acrylamide—a crosslinking agent. The polymerization is initiated by addition of a catalyst (e.g. ammonium persulfate and tetramethylethylenediamine). The gels created are neutral, hydrophilic, three-dimensional networks of long crosslinked hydrocarbons. The separation of molecules within a gel depends on the relative pore size, which is determined by the total amount of acrylamide present (%T) and the amount of crosslinker (%C). Higher gel concentrations (%T) give smaller pore size and a better ability to separate smaller molecules. The optimum %T depends on the molecular weight of the protein to be separated: 5%T gels (with %C = 2.6) are useful for proteins ranging from 50 to 300 kDa, while 10%T gels work well in the 15–70 kDa range (see also discussion of native gels below).

Agarose

A natural polymer called agarose, made from a repeating disaccharide, makes suitable gel matrix material, especially for DNA fragments. A gel is made by dissolving agarose powder in boiling buffer solution. The solution is then cooled to approximately 55°C and poured into a mold where it solidifies. A 0.7% agarose gel is good for resolving large DNA fragments (5–10 kb) while a 2% gel is suitable for smaller fragments (0.2–1 kb). An example is given in Figure 3.4.2.

Visualization

After electrophoresis, the molecules in the gel have to be made visible. The most common dye used to detect DNA or RNA bands is *ethidium bromide*. It fluoresces under UV light when intercalated into DNA (or RNA). Newer, commercially available dyes with similar optical properties but with lower toxicity are also available. *Coomassie brilliant blue*, the most popular protein stain, is an anionic dye that can

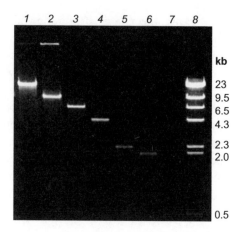

Figure 3.4.2 DNA fragments run on a 1% agarose gel. Lane 8 shows a set of DNA markers with their sizes in kilobases shown on the right. Lanes 1–7 are runs of isolated DNA fragments. Visualization was with ethidium bromide (see below).

usually detect a ~50 ng of protein in a band. *Silver* staining which is about 50-fold more sensitive than Coomassie stain can also be used. Here, silver ions are chemically reduced to insoluble silver metal granules in the vicinity of the protein molecules.

Some visualization techniques which are collectively termed "blotting" are selective for a particular protein or nucleic acid sequence. For example, western blotting uses antibodies to detect specific proteins, Southern blotting uses a DNA probe to detect complementary DNA (see also Boxes 2.4 and 5.5), and northern blotting detects RNA.

Some electrophoresis systems

Discontinuous gels

It is possible to increase the resolving power of gel electrophoresis by using more than one kind of matrix in a gel. Proteins migrating from one matrix to another, in discontinuous (disc) gels, can be concentrated in sharp bands at the boundary between the two gels. A common arrangement is to have a stacking gel and a resolving, or running, gel (Figure 3.4.3). Typically, the stacking gel contains a buffer, pH ~ 7, with relatively low salt concentration (~150mM), while the resolving gel has higher pH (~9) and higher salt concentration (~350 mM). The

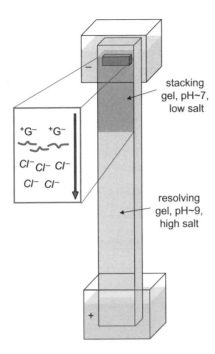

stacking gel, pH~7, low salt

$^+G^-$ $^+G^-$

Cl^- Cl^- Cl^-
Cl^- Cl^-

resolving gel, pH~9, high salt

Figure 3.4.3 Schematic view of one section of a discontinuous electrophoresis gel system, showing the stacking and resolving gels. In an expanded region of the stacking gel the idea of isotachophoresis (see text) is illustrated where the cyan protein is sandwiched between slowly running glycine zwitterions and rapidly migrating chloride ions.

stacking gel is made with lower acrylamide concentration so that it has a larger pore size than the resolving gel. There is also higher resistance and greater force, E, on the molecules in the stacking gel (see "theory" section above) because of the lower salt concentration. This larger value of E, together with the larger pore size, means that samples are more mobile in the stacking gel than in the resolving gel.

The following steps occur in discontinuous gels: (i) sample in a **running buffer** is applied to a well in the stacking gel (a typical running buffer has pH ~9 and a high concentration of glycine, mainly in the negatively charged form, $NH_2CH_2COO^-$); (ii) glycine enters the **stacking buffer** (pH ~7) where it mainly becomes the neutral zwitterion ($NH_3^+CH_2COO^-$) with no net migration force; (iii) the negative proteins (or DNA) in the added sample migrate in the stacking gel but not as fast as the Cl^- ions in the buffer; (iv) a "sandwich" is thus formed with Cl^- ions leading, protein/DNA in the middle followed by a few $NH_2CH_2COO^-$ ions (see inset in Figure 3.4.3);

(v) this sandwich carries the protein in a narrow band until the resolving gel is reached, where the $NH_2CH_2COO^-$ concentration greatly increases due to proton loss from the NH_3^+ group at pH ~9. The different buffering and higher acrylamide properties of the resolving gel then separate the protein/DNA on the basis of charge and size.

Denaturing gels/SDS-PAGE

Use of sodium dodecyl sulfate (SDS)-polyacrylamide gel electrophoresis (PAGE) with discontinuous gels was first described by Ulrich Laemmli in a 1970 paper that is one of the most highly cited in biology. SDS is an anionic detergent; it dissociates oligomeric proteins into constituent subunits and unfolds them (Figures 3.4.4 and 3.4.5). The SDS binds along the polypeptide chain, forming negatively charged complexes with fairly constant charge to mass ratios (~1.4 mg SDS binds per mg of protein). The sample is usually heated to encourage denaturation and exposed to a reducing agent to break disulfide bonds, e.g. 2-mercaptoethanol (Figure 3.4.5) or dithiothreitol. Because the charge/weight ratio is constant, the migration rate through a gel is determined predominantly by the protein molecular weight (M). A plot of $\ln(M)$ against relative mobility closely follows a straight line, but marker proteins of known molecular weight are usually used for calibration (see Problem 3.4.1).

Native gels

Native (non-denaturing) electrophoresis can provide useful information about macromolecular structure, e.g. subunit composition. As proteins usually retain their native conformation in these gels, functional properties, such as enzyme activity, can be detected with sensitive assays and antibodies. The interpretation of native gels is, however, more complex than for SDS-PAGE; the migration rate in the

Note The sandwiching of ions between ions of different electrophoretic ability is called isotachophoresis, which literally means "migration with the same speed". Other buffer combinations can be used in discontinuous gels including those that operate at pH ≤ 9.00.

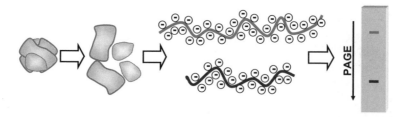

Figure 3.4.4 Illustration of SDS-PAGE electrophoresis. SDS breaks up subunit interactions and unfolds the subunits. Disulfide bonds are broken by the addition of 2-mercaptoethanol. The SDS coats the unfolded proteins in negative charge so that the migration rate is inversely proportional to molecular weight.

gel depends on charge, which is protein specific because proteins have different pI values (see also isoelectric focusing below). As illustrated in Figure 3.4.6, the migration rate is also strongly dependent on the nature of the gel.

At low gel concentrations (large pore size), proteins mainly migrate on the basis of their charge. In contrast, at higher gel concentrations (e.g. the 8%T, 2.67%C gel in Figure 3.4.6) there is a strong sieving effect and the migration rate depends both on charge and size. The dependence on pore size can be exploited to give more quantitative information with native gels. Plots of ln R (R is the relative electrophoretic mobility) against acrylamide concentration (%T) are approximately linear. These "Ferguson" plots allow the charge and mass of a protein to be estimated (Figure 3.4.7). The intercept, when %T = 0, depends only on charge (no sieving effects) while the slope depends on the mass of the migrating species (depends on sieving effects).

Electrophoretic mobility shift assays (EMSA)

The interaction of proteins with DNA is essential for many key cellular processes including replication and transcription. A useful technique for studying protein/DNA interactions is the electrophoretic

mobility shift assay (EMSA), commonly called a band-shift assay (Figure 3.4.8). EMSA is based on the observation that protein/DNA complexes migrate through a non-denaturing polyacrylamide gel more slowly than free DNA fragments or double-stranded oligonucleotides.

Band-shift assays are carried out by first incubating a protein (or a relatively crude nuclear or cell extract) with a labeled DNA fragment (e.g. with radioactive ^{32}P or a fluorescent label) containing the putative protein binding site. The reaction products

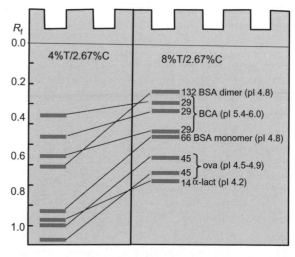

Figure 3.4.6 Effect of pore size on relative migration rate (R) in native gels. The diagrams show the migration patterns of a set of proteins. On the left, a 4%T, 2.67%C gel was used, while the pattern on the right was obtained from a 8%T, 2.67%C gel. The lines connect bands representing the same proteins in the two gels. Note the large relative mobility shifts between the two gels types. (BSA is bovine serum albumin, BCA is bovine carbonic anhydrase, ova is ovalbumin, and α-lact is α-lactalbumin (adapted with permission from *Essential Cell Biol.* 1, 197–268 (2003)).

Figure 3.4.5 The chemical formulae of SDS and 2-mercaptoethanol.

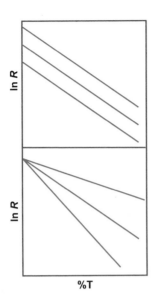

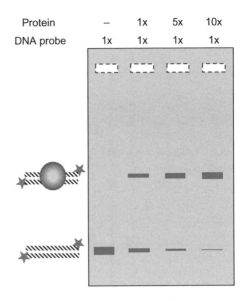

Figure 3.4.7 Schematic representation of Ferguson plots for (top) three proteins with same mass but different charge and (bottom) three proteins with the same charge but different mass. The slopes of the plots mainly depend on the sieving properties of the gel which depends on protein mass. The intercepts on the y-axis depend on the relative mobility (R) of the protein in the absence of the matrix, which depends on the protein charge.

Figure 3.4.8 Schematic representation of an EMSA experiment. Lane 1 has labeled DNA probe alone while lanes 2, 3, and 4 have increasing concentrations of protein so that the amount of the protein/DNA complex increases from left to right. Measurements of the amounts in the complex as a function of concentration indicate the affinity of the protein for the DNA.

are then analyzed on a non-denaturing polyacrylamide gel. During electrophoresis, the protein/DNA complexes are resolved from free DNA, and the amounts in the bands can be measured to give information about the equilibrium between bound and free DNA in the original sample. Protein/RNA and protein/peptide interactions can also be studied using similar methodology.

Isoelectric focusing

In isoelectric focusing (IEF), a pH gradient is established along the length of the gel. Such a stable pH gradient can be generated by ampholytes (molecules that can act as both an acid and a base). Polycarboxylic acid polyamines (molecular weight range 300–500) are the most commonly used ampholytes as they have good buffering capacity across a broad pH range. A mixture of ampholytes, with different pI values, is subjected to an electric field; if the ampholyte is in a region where pH > pI, it will have a negative charge while it will carry a positive charge if it is in a region with pH < pI. They thus migrate through the gel until

pI = pH (Figure 3.4.9). Provided the ampholyte mixture is well designed, they produce a stable, smooth pH gradient over a defined range. Commercial systems are available that cover either a broad (e.g. 2–12) or a narrow (e.g. 4–6) pH range across the gel.

Now consider a protein in a gel with an established pH gradient and with ampholyte concentration much greater than protein concentration. As with the ampholytes, there is no net charge and thus no net electrophoretic mobility of the protein when pI = pH. Proteins will thus migrate in different directions through the pH gradient in the gel until they reach the point where pI = pH. This gives rise to focusing, as proteins are continually directed to bands centered at their pI value (Figure 3.4.9). IEF is thus an equilibrium system where a run is continued until protein movement ceases.

2D gels

Cell or tissue extracts contain thousands of proteins, too many to resolve by one-dimensional techniques. Separation on the basis of two parameters (e.g. size and isoelectric point) lowers the probability of overlap,

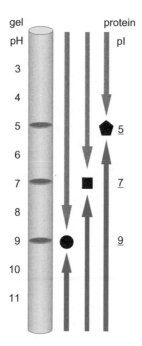

Figure 3.4.9 Illustration of an isoelectric focusing experiment on three proteins with pI values of 5, 7, and 9. In this case, the ampholytes in the tube gel have produced a stable pH gradient from pH 3 to 11. A protein stops migrating when it enters a region where pH = pI. If pH < pI the protein will be positively charged; if pH > pI it will be negative. The protein thus migrates either up or down until it reaches a position where pH = pI (gray and cyan arrows).

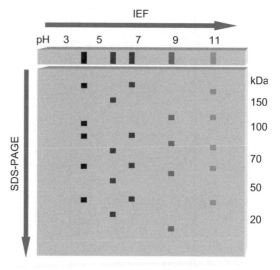

Figure 3.4.10 Schematic view of a 2D gel electrophoresis experiment where an isoelectric focusing strip is run first then laid on top of an SDS-PAGE gel and run again in the directions shown.

and allows the resolution of thousands of proteins on one gel. A one-dimensional gel is first used to separate by one parameter (e.g. pI); commercially available strips with a stabilized pH gradient are usually used in this step. This first IEF gel then serves as the sample for a second dimension which separates on the basis of another parameter, e.g. molecular weight in SDS-PAGE. Figure 3.4.10 illustrates an IEF/SDS-PAGE two-dimensional (2D) gel system.

Experimental problems with high-resolution 2D electrophoresis include missing proteins and lack of reproducibility. One way of improving reproducibility is to use difference gel electrophoresis (DIGE). DIGE involves pre-labeling proteins from different samples with different fluorescent tags (e.g. Cy2, Cy5—see Chapter 5.5). The samples are then combined and analyzed in the same 2D gel. A given protein from either sample will co-migrate to an identical spot position because the added tags are carefully matched for charge and mass. Scanning the gel at the excitation wavelengths of the different fluorescent tags provides images that can be overlaid because the proteins have been separated under identical conditions in the same gel; this allows accurate evaluation of the spot patterns from the different samples so that subtle variation in protein expression levels can be detected, e.g. between a normal and cancer cell line.

The excellent resolution obtained with 2D gels is not their only advantage; single proteins can be efficiently extracted for mass spectrometry analysis (see Figure 3.5.17 in Chapter 3.5). A number of annotated protein 2D gel images are available at **http://expasy. org/ch2d/**.

Pulsed field gel electrophoresis

Using standard electrophoresis methods, DNA molecules larger than ~20 kb show essentially the same mobility in a gel, making separation impossible. This problem arises because large DNA fragments elongate and align themselves with the applied field as they migrate toward the anode. In 1983, Schwartz and Cantor developed pulsed field gel electrophoresis (PFGE) to overcome this limitation. Periodic switching of the orientation of the electric field direc-

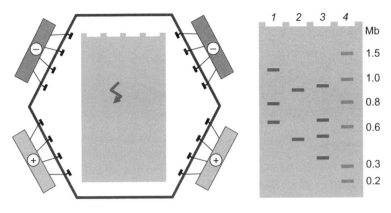

Figure 3.4.11 PGFE system where the applied field is changed by 120° every 90 seconds for ~24 hours. The stained gel on the right shows the kind of DNA megabase range separable by this technique (lane 4 is a marker lane).

tion forces the DNA to reorient, relax, and realign in the new field direction. Large DNA fragments require more time to reverse direction in an electric field than small DNA fragments so that alternating the direction of the current during gel electrophoresis allows DNA fragments of up to ~2 Mb to be resolved. A variety of PFGE techniques have been developed but all the systems apply fields in varying directions. Figure 3.4.11 shows one popular apparatus—a hexagonal gel box allows the angle of the applied fields to be switched by 120° relative to the agarose gel.

Capillary electrophoresis (CE)

If electrophoresis is carried out in small-diameter (~50 μm) capillaries, high currents can be used because heat is efficiently dissipated. Capillary electrophoresis (CE) allows efficient and rapid sepa-

ration of ions in ways that are complementary to other forms of electrophoresis.

Zonal CE is the most widely used technique. A small volume of sample is injected into the buffer-filled capillary at the positive end. The compounds are separated on the basis of their relative charge and size, and migration is driven by charge alone. The separated components are detected near the negative end of the capillary (Figure 3.4.12). Fused silica capillaries are usually used; these produce an electrical double layer at the capillary surface due to the attraction of positively charged cations in the buffer to negative silicate groups on the glass capillary wall. In the presence of an electric field, the cations in this layer move toward the cathode and drag the solvent with them, producing an **electro-osmotic flow** (Figure 3.4.13). The resulting flow profile is flatter than the usual parabolic shape in liquid flow because the driving force is stronger near the wall; this flow

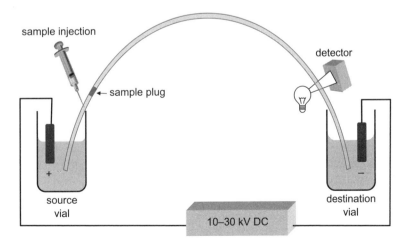

Figure 3.4.12 Capillary electrophoresis apparatus.

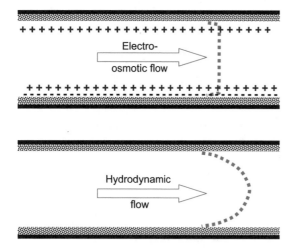

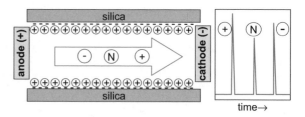

Figure 3.4.14 Effect of electro-osmotic flow in CE. Positive ions in the buffer sweep all analyte components towards the cathode, whether neutral or charged. Positive ions elute first followed eventually by negative ions (picture on right).

Figure 3.4.13 Illustration of the profile of electro-osmotic flow compared with hydrodynamic flow, as generated by a pump.

profile leads to good separation properties. All components, whether charged or neutral, are swept towards the negative electrode by the electro-osmotic flow. Positive compounds, with additional electrophoretic force in the cathode direction, elute first, negatively charged compounds elute last, and neutral molecules elute as a single band with the same velocity as the electro-osmotic flow (Figure 3.4.14).

CE is generally used to separate small molecules, including amino acids, peptides, and drugs (Figure 3.4.15). CE uses detectors similar to those employed in HPLC (Chapter 3.3); these include absorbance, fluorescence, and mass spectrometry. CE methods can analyze very small volumes (1–50 nL).

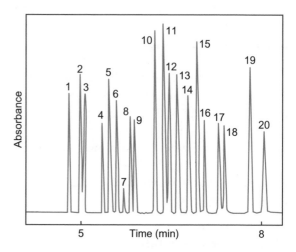

Figure 3.4.15 A CE electropherogram of a mixture of 20 drugs in aqueous solution, each at a concentration of 1 µg/mL. All are resolved. The drugs shown include: amphetamine (5), methamphetamine (6), ephedrine (9), codeine (13), salbutamol (15), and verapamil (19) (adapted with permission from Beckman Coulter application note A-1859A).

➕ Summary

1 Gel electrophoresis is in widespread use for separating electrically charged molecules. It is a key technique in molecular biology laboratories, giving separation and purification of DNA and RNA; it is also very powerful for characterizing protein preparations.

2 Polyacrylamide and agarose are useful gel support media. Discontinuous gels give improved resolution. Gel electrophoresis is usually followed by staining or blotting to identify the separated molecules.

3   DNA and RNA have an even distribution of charge. Fragments up to about 20 kb thus electrophorese is a manner that is related directly to their molecular weight.

4   Sodium dodecyl sulfate-coated proteins migrate in a manner related to their molecular weight on polyacrylamide gels (SDS-PAGE); the proteins are denatured by the SDS and reducing agents.

5   Electrophoresis behavior of proteins in non-denaturing conditions can give useful complementary information to SDS-PAGE, e.g. about subunit composition. Electrophoretic mobility shift assays (EMSA) can be used to detect protein nucleic acid interactions.

6   Isoelectric focusing (IEF) separates proteins on the basis of their pI value (the pH where the protein is electrically neutral).

7   IEF run in one dimension, plus SDS-PAGE in another, gives 2D gels which are a potent high resolution diagnostic tool for analyzing cell and tissue content.

8   Very high molecular weight DNA fragments can be separated by varying the migration direction in pulsed field gel electrophoresis (PGE).

9   Capillary electrophoresis is very useful for analyzing small molecules.

 ## Further reading

Tutorial on DNA electrophoresis
http://www.teachersdomain.org/asset/hew06_int_electroph/

Comprehensive book
Westermeier, R. *Electrophoresis in Practice* (3rd edn) Wiley, 2001.

Reviews
Herschleb, J., Ananiev, G., and Schwartz, D.C. Pulsed-field gel electrophoresis. *Nat. Protocols* 2, 677–84 (2007).
Westermeier, R., and Schickle, H. The current state of the art in high-resolution two-dimensional electrophoresis. *Arch. Physiol. Biochem.* 115, 279–85 (2009).

 ## Problems

3.4.1 The figure overleaf shows three lanes of a representative SDS-PAGE gel, with bands stained by Coomassie blue. In lane 1 there are markers with molecular weights 66, 45, 36, 29, 24, 20, and 14.2 kDa. In lane 2 is a total *E. coli* cell extract in which a plasmid had been inserted to express a protein fused to glutathione S-transferase; GST. Lane 3 shows the over-expressed protein after purification with glutathione affinity chromatography. From the marker lane information, show that a plot of distance migrated (cm) against $\ln(M)$ gives an approximate straight line. If the GST molecular weight is 26 kDa, what is the molecular weight of the expression product?

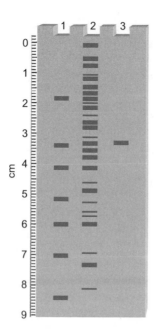

3.4.2 The electric field migration velocity of a protein in solution was measured at several values of pH as shown in the table. Explain the results and estimate the protein pI. Positive values of velocity correspond to migration towards the cathode.

pH	4	4.5	5	5.5	6.0
velocity (μM s^{-1})	0.55	0.19	−0.19	−0.5	−0.8

3.4.3 How would you expect DNA fragments of 0.5, 1, 30, and 50kb to run on a 1% agarose gel?

3.4.4 How might the equilibrium constant of a protein/DNA interaction be estimated from electrophoresis experiments?

3.4.5 Suggest explanations for the following observations in slab gel experiments: (i) "smile" profile of bands across the gel (i.e. edges run slower than center); (ii) in an SDS-PAGE gel a protein of interest runs slower than expected but it appears to have the correct mass in a mass spectrometer; (iii) the bands for protein A in a gel are smeared out.

3.4.6 What is the electro-osmotic effect in capillary electrophoresis? Why do all species migrate to the cathode? What is the general order of migration time for cations, anions, and neutrals? Can you think of ways that it might be possible to improve the resolution of the neutral species, for example?

3.5 Mass spectrometry

Introduction

Mass spectrometry (MS) is a technique that measures the mass/charge ratio of particles (this simply means that a protein with mass 25 kDa (m) and charge +5 (z) will be defined by $m/z = 5000$). Mass spectrometers can analyze a wide range of materials with great precision and sensitivity. The method is employed in chemical analysis, pharmacology, the environmental sciences, and, increasingly, in the biological and biomedical sciences, where it can: (i) identify proteins; (ii) sequence proteins; (iii) detect post-translational modifications; (iv) give information about macromolecular conformation; and (v) characterize large, non-covalent complexes.

The foundations of mass spectrometry began with the work of JJ Thomson in Cambridge in the early 20th century. Thomson discovered that electrical discharges in gases produced ions that, when passed through electromagnetic fields, had trajectories that were mass and charge dependent. Thomson's student, Aston, then went on to design mass sensitive instruments that led to the discovery of isotopes. Nobel Prizes were awarded to Thomson in 1906 and Aston in 1922.

The first commercial mass spectrometers appeared in the 1940s with instruments that mainly used magnetic and electric fields to deflect the ions. Early attempts to separate ions by differences in their velocities (time of flight) were also going on at this time. The next major technical development was ion trapping and quadrupole mass filters that uses both radiofrequency and direct-current components to separate ions (Wolfgang Paul, Nobel Prize 1989). A major advance for biological applications was when John Fenn (Nobel Prize 2002) developed the electrospray ionization (ESI) technique. Matrix-assisted laser desorption ionization (MALDI), where molecules are "desorbed"/ionized by a laser, was also developed in

the late 1980s by Hillenkamp and Karas; Koichi Tanaka (Nobel Prize, 2002) also developed laser desorption.

The main stages, involved in mass spectrometry are shown in Figure 3.5.1; these involve the production, analysis, and detection of molecular ions, followed by data analysis. These stages will now be discussed in turn.

Ionization

A number of ionization methods have been developed over the years. These include electron ionization, where a beam of electrons is directed against a gaseous sample, and fast atom bombardment (FAB), where a beam of gaseous atoms (e.g. Ar, Xe) is directed against a sample in a liquid matrix. For biological samples, however, the most useful ionization techniques are currently:

- matrix-assisted laser desorption ionization (MALDI) and
- electrospray ionization (ESI).

These two methods are complementary—both can ionize molecules with mass >1 MDa; MALDI ionizes samples from the solid state while ESI ionizes liquids and is thus better suited for direct input from a chromatography system (Chapter 3.3) and for measuring a series of samples rapidly. Both methods are very sensitive but MALDI is more so, being able to detect 10^{-18} mol; MALDI is more salt tolerant than ESI but, as we will see below, ESI is very good for detecting native states and different conformations.

MALDI

In MALDI, the ionization is initiated by irradiation of the sample with a laser beam. A matrix of

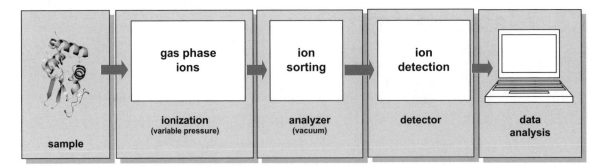

Figure 3.5.1 The main stages involved in mass spectrometry are ionization, analysis, and detection.

crystallized organic molecules is used to help protect the molecule (analyte) from the irradiation and to facilitate vaporization and ionization by absorbing the laser irradiation. Examples of these molecules are shown in Figure 3.5.2.

A typical matrix might be 20 mg mL^{-1} sinapinic acid in a mix of acetonitrile and water. A drop of this solution, with the analyte dissolved in it, is placed on a metal plate. The solvents evaporate, leaving matrix with analyte spread throughout the crystals in a MALDI "spot". As shown in Figure 3.5.3, an applied laser pulse vaporizes the matrix containing the analyte thus creating ions which can be extracted by a high voltage and passed to a mass analyzer. The laser irradiation can be applied to the spot, either in a vacuum or at atmospheric pressure (AP-MALDI).

ESI

The ESI method was developed by John Fenn and colleagues in the 1980s; a liquid containing the macromolecule of interest is sprayed through a capillary to create a fine aerosol (see Figure 3.5.4). Typical solvents

for electrospray ionization are prepared by mixing water with volatile organic solvents such as acetonitrile. Additives (<2%) that can be used to improve the sample properties include formic acid and triethylamine. An inert gas such as nitrogen can be used to facilitate droplet formation and desolvation (Figure 3.5.4). The aerosol then enters the first vacuum stage of a mass spectrometer through a capillary.

Most of the droplets formed by ESI are empty but some contain protein; solvent evaporation from these charged droplets makes them unstable because of the increasing build-up of charge as the droplet size diminishes. This instability results in droplet fission and the production of offspring droplets that

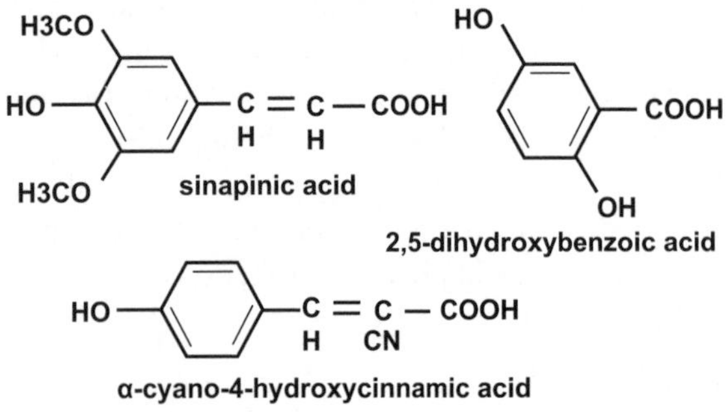

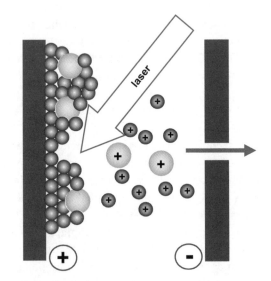

Figure 3.5.3 Illustration of MALDI: the molecules of interest (cyan) are mixed with the matrix and subjected to a laser pulse, often from a nitrogen laser with an irradiation wavelength of 337 nm. The applied voltage then accelerates these molecules and passes them to the mass analyzer (cyan arrow).

Figure 3.5.2 Three examples of molecules used to form a matrix and aid the MALDI ionization process.

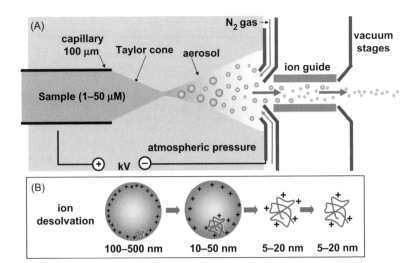

Figure 3.5.4 Illustration of electrospray ionization. (A) Charged droplets are formed in a fine spray at atmospheric pressure. These are accelerated with a high voltage and injected through a capillary into a vacuum. Solvent evaporation is encouraged with a gas flow and sometimes with heating in the ion guide. (B) The droplets explode as the solvent evaporates, leading to the formation of macromolecular ions with variable charge.

contain, on average, one analyte ion with multiple charges. These charged species can be positive or negative although, for proteins, positive ions are usually selected for acceleration by the applied voltage. Current views suggest that small ions are liberated via an "ion evaporation mechanism", where ions are released directly from the droplet; larger ions probably form by a "charged residue mechanism" where the droplet breaks up into successively smaller droplets with solvent evaporation proceeding until only a "dry" ion is left.

For larger molecules, such as proteins, ESI mainly forms multiply-charged species. As mass spectrometers measure the m/z ratio rather than mass directly, the production of species with a high z value is advantageous as molecules with large m can then be measured in instruments with a relatively small mass range (see also below).

The terms "micro-electrospray" and "nanospray" are used to describe different ESI instruments, depending on the diameter of the capillary tip. Typical nanospray devices use glass capillaries with internal diameters of ~10 μm from which solution normally flows at rates <100 nL min^{-1}. Micro-electrospray uses larger capillaries (10–100 μm) with slightly higher flow rates.

Ion sorting/analysis

Once the ions are formed by MALDI or ESI they pass to a mass analyzer where they are separated on the basis of their m/z ratio. A variety of mass analyzers are in use, each with different properties, such as mass range, resolution, and sensitivity. There are two main types of analyzer: beam analyzers, where the ions leave the ion source in a beam and pass through an analyzing field to the detector, and trapping analyzers, where the ions are trapped in the analyzing field. The main types of analyzer are briefly described below, but first we discuss some theoretical aspects of ion deflection and separation.

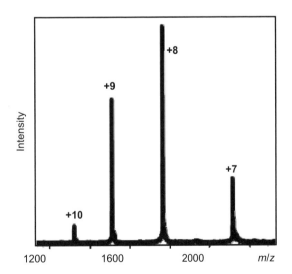

Figure 3.5.5 Illustration of a mass spectrum of a small protein with different charges, produced by ESI. Most of the ions have 8 positive charges but those with 7, 9, and 10 charges are also evident.

Deflection by electric and magnetic fields (sector analysis)

Deflection by electric and magnetic fields (Figure 3.5.6) was initially the predominant way to analyze ions in mass spectrometry. As described in Tutorial 3, the deflection of an ion, charge q, and mass m, by both electric (E) and magnetic (B) fields depends on m/q. The dimensionless mass-to-charge ratio, m/z, is usually used to describe mass spectra rather than m/q; z is the number of elementary charges (e) on the ion ($z = q/e$). The deflection by B depends on the ion velocity v and produces circular motion, while the E deflection is velocity independent and the path is parabolic. Electric and magnetic sector deflections are thus often used in tandem because they are sensitive to somewhat different properties of the ion beam.

Time of flight

Ions with different mass can have different transit times when passing through a "drift region" of a mass spectrometer. This is exploited in **time of flight** (TOF) instruments. A simple version of a TOF instrument, coupled to a MALDI source, is shown in Figure 3.5.7. After a laser pulse and extraction of the ions by an electric field, $\mathbf{E}_1$, a field, $\mathbf{E}_2$, accelerates the ions into the drift region with an energy of qV, where q is the ion charge and V is the applied voltage ($E = V/d$ where d is the distance travelled—see Tutorial 3). The kinetic energy of the ion is $1/2mv^2$, where m is the mass and v the velocity. Thus, $1/2mv^2 = qV$ and $v = (2qV/m)^{1/2}$. If L is the length of the drift tube, the transit time will be $t = L/v = L/(2qV/m)^{1/2} = Lm^{1/2}/$

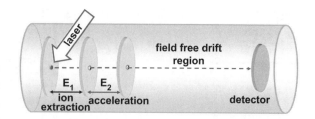

Figure 3.5.7 Illustration of TOF ion analysis. After ion extraction (see also Figure 3.5.6), ions are first accelerated and then allowed to drift in a region with no applied fields.

$(2qV)^{1/2}$. This means that the transit time is proportional to $(m/z)^{1/2}$.

TOF instruments often include a **reflectron** (Figure 3.5.8), where a series of charged rings or grids act as an ion mirror. This increases the path length, L, and the electrode arrangement helps compensate for any spread in kinetic energies of the ions, thus improving the instrument resolution.

TOF instruments can measure m/z ratios of virtually any size—this is significantly above the range of most other mass analyzers. This combination of high m/z range and compatibility with MALDI makes TOF analysis very powerful for protein studies.

Quadrupole analyzers

A quadrupole mass analyzer consists of four parallel cylindrical rods (Figure 3.5.9). It filters ions on the basis of their m/z ratio using oscillating electric fields applied to the rods. Each opposing rod pair is connected together, and radiofrequency (rf) and direct (DC/non-oscillating) voltages are applied together to these pairs. Ions with a certain m/z value

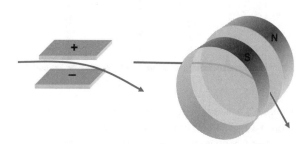

Figure 3.5.6 Deflection of ions by electric and magnetic fields.

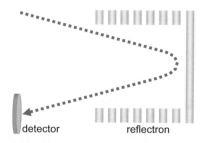

Figure 3.5.8 A reflectron in a TOF instrument.

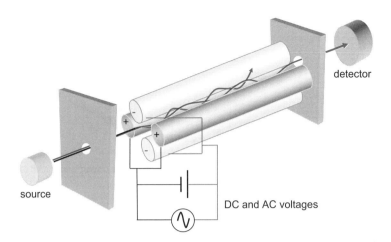

only reach the detector at particular values of applied voltage. Ions with different *m/z* values have unstable trajectories and collide with the rods. A range of *m/z*-values can be scanned by varying the applied voltages.

Quadrupole traps

So far, we have discussed analyzers that act on a moving ion beam; a powerful alternative method is to trap the ions selectively. The quadrupole mass analyzer (above) applies electric fields in two dimensions (Figure 3.5.9); in quadrupole traps, electric fields are applied in three dimensions. This is usually done using two endcap electrodes with convex shaped surfaces that face each other and a ring-shaped electrode between these endcaps (Figure 3.5.10). The shape of these three electrodes is designed to trap the ions in the space between them when certain electric fields are applied. An inert gas is often used to help stabilize the trapped ions.

The trapped ions oscillate with characteristic frequencies that depend on the geometry of the trap and the applied voltages (see also FT-ICR below). Applied voltages tuned to these frequencies cause certain ions to pick up increasing amounts of kinetic energy; these ions are activated and ejected from the trap in the *z*-direction. Sweeping the frequency of the applied AC voltage causes ion ejection in a mass-selective manner. Ion traps are sensitive at high *m/z* values and their mass accuracy is similar to quadrupole analyzers. Ion traps are popular because of their sensitivity, ease of use, and relatively low cost.

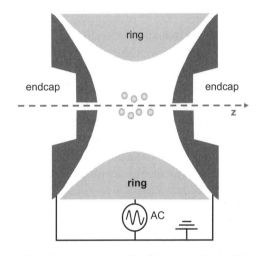

Figure 3.5.10 A quadrupole (Paul) ion trap. AC and DC voltages are applied to the central ring electrode that lies between two electrically earthed endcap electrodes. Ions remain in the trap until they are excited by a particular applied AC frequency.

Fourier-transform ion cyclotron resonance (FT-ICR)

These devices use a strong magnetic field to determine the *m/z* of an ion. In the various ion beam analyzers discussed above, the ions have energies in the keV range but the energies in FT-ICR are much lower. At such low kinetic energies the ions become trapped, oscillating around an applied magnetic field, B_o, with a frequency ($f = B_o q / 2\pi m$—see Tutorial 3) that is inversely related to *m/z*. Initially, little signal is observed because the radius of the motion is very small. Excitation of a range of *m/z* values is achieved

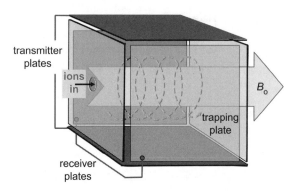

Figure 3.5.11 Illustration of an ion cyclotron resonance device. The trapped ions (cyan) are exposed to a strong magnetic field (B_o). The oscillation of ions is recorded and converted to a frequency spectrum with a Fourier transform.

Figure.3.5.12 An illustration of an orbitrap with ions cycling around a central electrode.

by a swept radio-frequency (rf) pulse, applied across transmitter plates (see Figure 3.5.11). When the applied frequency matches the ion oscillation frequency (resonance), ions with a particular m/z value are excited to a higher energy orbit where their oscillations induce an alternating current in a pair of receiver plates. The frequency of this induced current is the "cyclotron frequency" of the ions and the intensity of the oscillation is proportional to the number of resonating ions. On removal of the excitation, the ions drop back down to their natural orbit. The oscillations of all the ions are detected at once, thus producing a complex frequency vs. time spectrum that contains all the signals. This is called a free induction decay which can be Fourier transformed to give an m/z spectrum (this approach is discussed in more detail in Chapter 6.1 and Tutorial 6). Very high mass resolution and mass accuracy can be achieved with FT-ICR. The high magnetic fields required are generated using a superconducting magnet (see Chapter 6.1).

Orbitrap

In an orbitrap, ions are injected into a shaped electrode system with an electric potential applied between inner and outer electrode surfaces. The ions become trapped because their electrostatic attraction to the inner electrode is balanced by centrifugal forces generated by the cycling ions (Figure 3.5.12). The ions also move back and forward along the central axis of the trap. By sensing the ion oscillation frequency, as in FTICR-MS, the trap can be used as a mass analyzer. Orbitraps have a high mass accuracy, good resolving power, a high dynamic range, and no large magnet is required.

Tandem mass spectrometry

Tandem (MS/MS) mass spectrometers have more than one analyzer. These include hybrid systems, where different types of analyzer are combined to enhance and extend the analysis, e.g. for structural or sequencing studies (see below). Compounds can be introduced one by one into a mass spectrometer using chromatography (Chapter 3.3). Alternatively, a tandem mass spectrometer (MS/MS) can be used, where the first stage of separation is achieved with one mass analyzer (MS1), followed by further analysis of the separated ions in a second mass analyzer (MS2). This is a common arrangement in protein sequencing which may also involve the production of peptide fragments between MS1 and MS2 (see Figure 3.5.13).

Fragmentation

A collision chamber containing a neutral gas, such as argon (Figure 3.5.13), is often introduced to produce

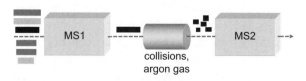

Figure 3.5.13 A MS/MS system where a compound is selected in MS1 and then exposed to collisions in a gas filled chamber. Resulting fragments are then analyzed in MS2.

peptide fragments. When an ion collides with a neutral gas, enough of the kinetic energy can be transferred to cause chemical bonds to break. This is called collision-induced dissociation (CID). Other fragmentation procedures that can be employed include exposure to light (photo-dissociation) or a high-current electron beam (electron capture dissociation, ECD). Here, we will focus on CID which produces molecular fragments in a relatively predictable way. For peptide fragmentation, CID mainly gives rise to CO-NH bond breakage, producing b and y ions of the sort shown in Figure 3.5.14. The mass difference between two adjacent b or y ions gives information about particular amino acid residue types (Tutorial 1). Note that the protein backbone bonds (NH–CH and CH–CO) and amino-acid sidechains can also be broken to give different ion fragments.

If a doubly charged peptide ion is fragmented, b and y ions are both formed. When singly charged ions fragment, either a b ion or a y ion is formed. It is then said that a parent (or precursor) ion (m_1^+) is split to give a daughter ion (m_2^+) and a neutral species (n). A variety of analysis procedures can be used, but the most common is the product ion scan which detects product ion fragments (m_2^+) arising from m_1^+. Neutral ions cannot be detected directly and they have to be found by difference analysis (a procedure sometimes called neutral loss). An example would be a mass loss of 98; this corresponds to a phosphate group (H_3PO_4) so this neutral loss value could be used to deduce the presence of a phosphopeptide. Note that only one or two proteolytic fragments of a target are usually required to get enough

sequence data to identify a protein. With post-translational modifications and multiple fragments, a more careful analysis will be required.

Separation by ion mobility

The separation methods discussed so far have relied on m/z variation alone. In ion mobility (IM) systems, the ions are directed through a drift tube while contained within a stack of ring electrodes designed to generate a relatively uniform electrostatic field along the axis of ion propagation (Figure 3.5.15). As the ions pass through the tube, they collide with a neutral gas. The aim of the gas here is not to fragment the ions as discussed above, rather it is to slow them down. Ions with a larger cross-section experience more collisions and therefore emerge more slowly from the drift tube than ions with a smaller surface area.

If an ion with a particular value of m/z has several shapes or conformations, a distribution of drift times is observed. This can be detected in an instrument that combines an IM device (sensitivity to shape) with MS (measurement of m/z), as shown in Figure 3.5.16.

Other experimental factors

Vacuum systems

To minimize unwanted effects of gas collisions, mass spectrometers operate under high vacuum conditions, typically 10^{-2}–10^{-5} Pa. There are usually two pumping stages: the first stage is a mechanical pump that provides rough vacuum down to 0.1 Pa and a high vacuum second stage that uses pumps that

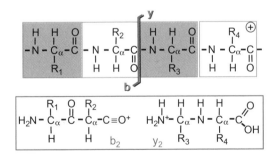

Figure 3.5.14 Fragmentation patterns obtained using collision-induced dissociation (CID). Shown below is the chemistry of the end groups obtained for b and y ions from two amino acids.

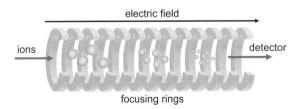

Figure 3.5.15 An ion mobility instrument separates ions by charge and shape. The ions pass through a region of relatively high gas pressure; larger ions have more collisions than small ones therefore reach the detector later.

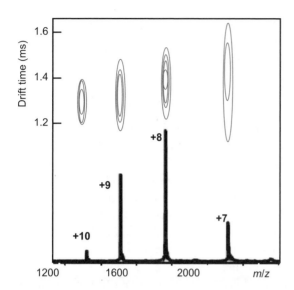

Figure 3.5.16 A plot of drift times seen for different charge states in a small protein (the same as that shown in Figure 3.5.5), obtained in an ESI-IM-TOF mass spectrometer. The black lines indicate a series of ions with different m/z values. The cyan contour plots of the observed ion distributions above each peak indicate the "conformational space" sampled by each of the four different charge states. It can be seen that charge state +8 appears more compact than charge state +7.

essentially blow out gas molecules (diffusion or turbo-molecular pumps), or use a very cold surface where gases solidify (cryo-pumps). ICR instruments have particularly high vacuum requirements.

Detectors

The number of ions leaving the mass analyzers is typically small but modern detectors are very sensitive. Ion detectors, usually some kind of photon or electron multiplier, record either the charge induced or the current produced when an ion passes by or hits a surface. Particularly sensitive are micro-channel plate (MCP) detectors, consisting of arrays of tiny hollow glass tubes, each operating as an individual electron multiplier.

Data analysis

The final component of a mass spectrometer is the data system/computer. This is used to control the instrument, analyze the data, and perhaps search many thousands of reference spectra to identify the spectrum of an unknown compound (see Problem 3.5.5).

Applications of mass spectrometry

Here, we will illustrate some of the many applications of the mass spectrophotometer, from relatively straightforward mass analysis to proteomics and conformational states of native proteins.

Mass analysis

The ability of mass spectrometers to measure the mass of a macromolecular with great precision is extremely powerful. The mass accuracy obtained is typically better than 0.01%! It is possible to compare calculated and observed molecular weight with an accuracy of ~1 Da, thus aiding identification and characterization. Any wanted or unwanted covalent modification can also be detected (see Table 3.5.1). It is even possible to say whether an SH group is reduced or oxidized in an S–S bond.

Different ionization methods lead to a different distribution of ions in mass spectra. In MALDI,

Table 3.5.1 Some mass changes observed after protein post-translational modification

Modification	Mass change	Modification	Mass change	Modification	Mass change
S–S bond	−2.016	Acetylation	43.037	Farnesylation	204.356
Methylation	14.027	Carboxylation	44.010	Myristoylation	210.36
Met oxidation	15.999	Phosphorylation	79.98	pyridoxyl-phosphate	231.145
Hydroxylation	15.999	Sulfation	80.06	palmitoylation	238.414
formylation	28.010	hexoses (Fru, Gal, Glc, Man)	162.14	ADP-ribosylation	541.305

ionization generally occurs by the addition of a single, or possibly two, H^+ ions to the molecule, so the spectrum consists of a dominant peak at the full mass of the protein with a secondary peak at $m/2$. ESI gives more complex spectra because multiple ions with different charge states are produced. A typical mass spectrum, obtained using ESI, is shown in Figure 3.5.17. The addition of a positive charge corresponds to the addition of one H^+; the formula $m/z = (M + nH^+)/n$ thus applies, where M is the molecular weight and nH^+ is the number of positive charges on each peak (the mass of a proton, H^+, is 1.008 Da). For example, the m/z value for the peak with $z = 18$ in Figure 3.5.17 is 942.75. Thus, $M = 18 \times 942.75 - 18.144 = 16\,951.4$ Da. In fact, the computer in the spectrometer is programmed to fit the data over all peaks to obtain the best fit for the mass. The number of charges on the ions is not known initially, but it can be calculated if the assumption is made that any two adjacent peaks differ by one charge unit (Problem 3.5.3 illustrates how the above formula for m/z can be used).

If we look at Figure 3.5.17, we see a number of small peaks on the high mass side of the main peaks. These can arise from: (i) the presence of isotopes, for example a typical carbon-containing sample has around 1% of ^{13}C and 99% ^{12}C; and (ii) the presence

of other ions that fly with the protein ion in the mass spectrometer, for example Na^+ (22), K^+ (38), and $H_3PO_4^+$ (97) are often seen as protein adducts, where the numbers in brackets correspond to the extra ion masses (minus the mass of the proton the ion replaces). The biggest limitation in resolving two proteins is the width of the distribution that arises from the presence of different isotopes. For a 100-kDa protein, this isotope distribution is approximately 50 mass units wide.

The high accuracy of MS is valuable in a wide range of applications. One is to monitor hydrogen exchange between protein groups with solvent water. Labile hydrogens in a protein (typically those attached to nitrogen in amide groups) exchange with deuterium atoms when the protein is placed in a solution of heavy water (^2H_2O). The increase in protein mass as the exchange occurs can be measured with high-resolution MS. The specific location of the deuterium incorporation can also be determined by monitoring the ^2H incorporation in peptide fragments that are produced after the labeling reaction (see above). The rate of hydrogen exchange gives information about the solvent accessibility of various parts of molecule, and thus the tertiary structure, folding mechanisms, and ligand protection (see also the H-exchange example given in Figure 5.2.6).

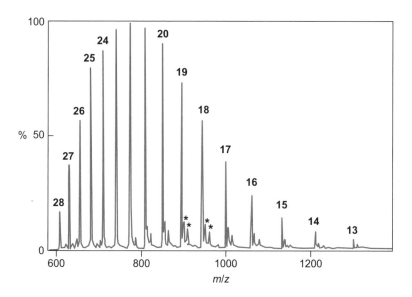

Figure 3.5.17 A mass spectrum of myoglobin, obtained from a solvent with aqueous acetonitrile (1:1) and 0.1% formic acid. The numbers indicate the values of z for the different peaks. Some of the adduct peaks are marked with an *.

Proteomics and protein sequencing

The term proteomics was coined in 1994 by Marc Wilkins who defined it as '*the study of proteins, how they are modified, when and where they are expressed, how they are involved in metabolic pathways and how they interact with one another*'. In analyzing the protein content of a cell we want to know the identity of each protein, determine its concentration, and see if it is modified in any way. The mass spectrometer has become the main tool for proteomic analysis.

Protein identification

Let us consider how a protein spot on a 2D gel (Chapter 3.4) might be identified. One way is to excise the spot and subject it to enzyme digestion to give a series of peptide fragments—trypsin is a useful enzyme as it cleaves proteins after the positively charged residues Lys or Arg. These fragments can be separated by liquid chromatography (LC) and analyzed by MS. It may be possible to identify the protein already at this stage from the mass spectra of the LC separated peptides. If identification is not yet possible the peptides can be subjected to tandem mass spectrometry and fragmentation of the peptides (see Figures 3.5.13 and 3.5.14) to give sequence information and probable protein identification (Figure 3.5.18) after comparison of the peptide and/or fragment ions with mass databases (see **http://www. genebio.com/products/phenyx/aldente/principle. html** and Problem 3.5.5).

Quantification

In proteomics it is valuable to quantify absolute and relative amounts of protein in a sample. Some idea of amounts can be obtained from 2D gels but overlap or low abundance often precludes reliable analysis. The most commonly sought quantitative information is changes in protein levels between two samples, e.g. between cell populations exposed or not to a growth factor. Two main approaches are used to quantify results from mass spectrometry. One is to use a label-free approach and the other uses isotope labels.

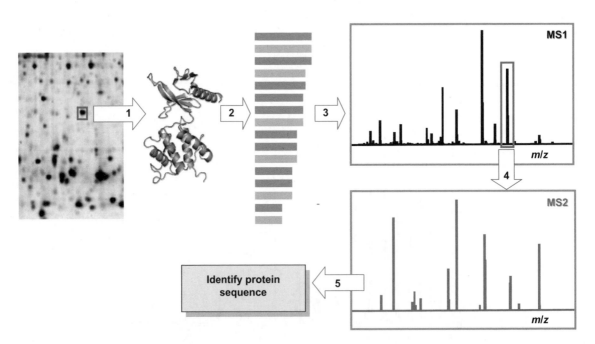

Figure 3.5.18 Proteomics analysis with two mass spectrometers, MS1 and MS2. (1) a spot is excised from a 2D gel; (2) the protein is subjected to proteolytic enzyme cleavage, e.g. by trypsin; (3) the multiple peptide fragments produced are run in MS1; (4) one selected fragment is subjected to collisional fragmentation (see Figure 3.5.13) and run in MS2 to give a fragmentation fingerprint (Figure 3.5.14) that allows the protein to be identified.

Numerous variations of these procedures are available and kits with suitable reagents can be purchased. In the label-free approach, the two proteomes to be compared, let us call them "control" and "sample", are analyzed by separate LC–MS/MS experiments. Peptides are identified in different LC runs from their retention times and their m/z values. In principle, this allows the quantification of all of the peptides detected from a biological sample. Changes in relative protein abundance are computed from the signal intensities of the extracted ions.

An alternative is to use isotope labeling approaches where "heavy" stable isotopes (e.g. ^2H, ^{13}C, ^{15}N) are incorporated into the "sample" and normal "light" isotopes (^1H, ^{12}C, ^{14}N) into the "control". The isotopes can be introduced at different stages of the preparation procedure—at the cellular, protein, or peptide level—by metabolism, chemistry (e.g. cysteine labeling), or enzymes (e.g. ^{18}O labeling), respectively (see, for example, Problem 3.5.6). One of these quantification methods (SILAC) is discussed in more detail below (see also Problem 3.5.6).

In stable isotopic labeling by amino acids in cell culture (SILAC) experiments, two cell populations, the "control" and the "sample" are grown under identical conditions apart from a specific stimulant added to the "sample" along with a different isotope label in the growth medium. Numerous labeling strategies are possible but particularly convenient are the amino acids $^{13}C_6$-Arg and $^{12}C_6$-Arg. When added to a growth medium deficient in these amino acids, the metabolic machinery of the cells labels all the cellular proteins with these amino acids (Figure 3.5.19). (Differences between the incorporation rates of ^{13}C or ^{12}C isotopes into cellular proteins are usually insignificant.)

Let us assume that the control is grown in a medium with a "light" label while the sample is grown in medium with a "heavy" label. Control and sample cell populations are then mixed and processed together, using the same protein separation and purification methods, thus eliminating any quantification errors due to unequal sampling. Peptide fragments are produced using the enzyme trypsin, which cleaves on the C-terminal side of Arg or Lys. The peptides containing the light and heavy amino acids are chemically identical but they can then be easily distinguished by MS analysis; for

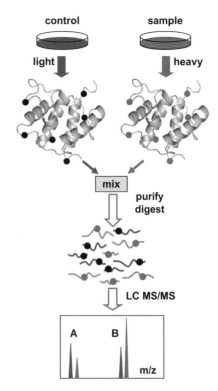

Figure 3.5.19 Illustration of the SILAC method where control cells are grown in normal (light) media and the sample cells in isotope labeled (heavy) media. Differences in protein levels between control and sample can be quantified using MS of mixed peptide fragments from the two experiments. In the illustration, Arg-containing peptides appear in pairs; the protein labeled A is expressed at a lower level in the sample than in the control, while B is expressed at a higher level in the sample. The mass of the cyan peaks is greater than the gray peaks because each Arg is 6 Da heavier.

example, Arg has 6 carbons, so peptides from the sample contain Arg that is 6 Da heavier than Arg in peptides from control proteins. Based on the relative peak intensity of the isotopic peptide pairs, differential protein expression can thus be quantified in different samples (see also below).

Example A proteomics study of protein–protein interactions involved in EGF signaling

When epidermal growth factor (EGF) binds to the EGF receptor (EGFR) a number of intracellular changes occur, triggered by auto-phosphorylation of the intracellular region of the EGFR. Several proteins bind phosphorylated EGFR, leading to the assembly of a signaling complex via multiple specific protein-protein interactions. As described in Chapter 3.3, protein interaction partners can be identified by affinity chromatography. In the illustrative example, shown in Figure 3.5.20, a bead coated with glutathione S-transferase (GST) was attached to a Src homology 2 (SH2) domain that binds specifically to phosphorylated proteins, such as the EGFR. As shown in Figures 3.5.19 and 3.5.20, the control was grown in ^{12}C-arginine-containing medium (gray dish), whereas the sample was grown with ^{13}C-Arg medium (cyan). The sample was also stimulated with EGF.

The contents of lysed cells were affinity purified with the GST beads attached to an SH2 domain. Some proteins bound non-specifically to the GST beads but some bound directly to the EGFR which is phosphorylated by EGF stimulation.

As the EGFR contains more than one phosphorylation site, complexes of the sort GST-SH2-EGFR-X were also detected where X binds to phosphorylated EGFR. This affinity purification procedure identified 228 proteins, 28 of these were found to be selectively enriched by EGF stimulation. EGFR and a protein called Shc had especially large differential ratios between sample and control. This was interpreted as direct SH2 binding to phosphorylated EGFR, as well as EGFR binding to Shc (Figure 3.5.20).

This example shows the power of affinity purification methods, coupled with MS. Such methods are being increasingly used to detect a wide range of protein-protein interactions in many organisms. Some of the data accumulated from these interaction networks (the interactome) can be found at **http://www.pathguide.org/**.

Native mass spectrometry

Soon after the introduction of electrospray it was realized that intact proteins and even protein complexes could be monitored by MS. The term "native" MS is often applied to MS studies of intact native proteins and their complexes (compare native gels in Chapter 3.4). While it does not yield high resolution structural information, the MS approach is relatively rapid and very sensitive (picomole amounts of protein complexes can be analyzed). It can also measure several species present in a mixture at the same time and provide information about the stoichiometry of protein complexes that is complementary to that obtained from other biophysical methods.

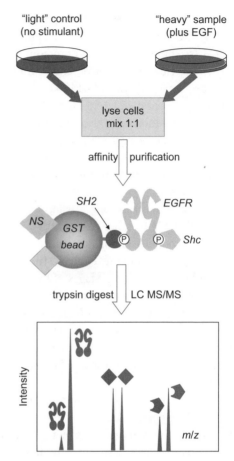

Figure 3.5.20 Illustration of the SILAC-based strategy applied to EGF stimulation. The cells in the "sample" were exposed to EGF, lysed, and mixed with the lysate of unlabeled, untreated "control" cells. The combined cell lysates were affinity-purified with GST beads fused to an SH2 domain which binds the activated, phosphorylated form of the EGFR directly, and associated EGFR binders indirectly. Proteins eluted from the beads were digested with trypsin and the resulting peptides then analyzed by MS. Non-specific (NS, ◆) binding proteins show a 1:1 ratio between sample and control states while EGFR and Shc are elevated in the sample, as shown in the schematic MS spectrum (adapted and redrawn from *Nat. Biotechnol.* 21, 315–8 (2003)).

ESI-MS is sensitive to conformation

ESI generates multiple charged species where the number of charges on the protein depends on the exposure of the protein groups to solvent. For example, more charges can be added to an unfolded chain than a compact folded one (Figure 3.5.21).

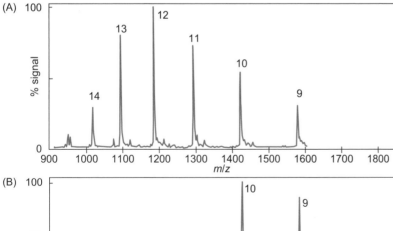

Figure 3.5.21 Schematic charge distribution on a folded and an unfolded protein. Note the larger number of charges on the unfolded form.

Example ESI of α-lactalbumin

Figure 3.5.22 shows the ESI-MS spectra of the calcium binding protein, α-lactalbumin, obtained (a) under conditions that denature the protein and (b) conditions where it flies in the mass spectrometer in its native conformation. In the native state, the mass is 38.8 Da greater than in the denatured state. The mass of Ca^{2+} is 40; taking into account the displacement of an H^+ ion, the extra mass can be assigned to a Ca^{2+} flying with the protein ion in the native state. Note also that the distribution of peaks is moved to lower m/z values in the denatured state because that state has more charges (higher z), as is predicted in Figure 3.5.21.

Example ESI of a large multimeric complex

A more complex example is shown in Figure 3.5.23. This is a study of a protein from *Helicobacter pylori*, a bacterium that infects approximately half of the world's population, causing a variety of gastroduodenal diseases. The bacterium produces a large amount of the enzyme **urease** to hydrolyze urea, thus helping to neutralize its acid environment in the gut. *Helicobacter pylori* urease consists of 26.5 kDa (α) and 61.7 kDa (β) subunits, but these assemble to form a complex of over 1 MDa, an architecture that is believed to help it survive in its harsh environment. ESI TOF was used to measure the molecular mass of the urease complex, showing that it is a dodecamer consisting of 12 α and 12 β subunits. The dodecamer was found to disassemble readily into $(\alpha\beta)_3$ subunits, revealing the building blocks of the assembly.

Ion mobility

IM-MS was introduced above (Figure 3.5.14). The method is based on the rate at which a large protein ion migrates through a region of neutral buffer gas under the influence of a weak electric field. IM-MS can identify and characterize protein complexes not only from their mass but also from their conformation. A spread in drift times is observed for each charge state due to the presence of protein ions in more than one conformation. The center position of the drift time distribution gives some structural information about the size

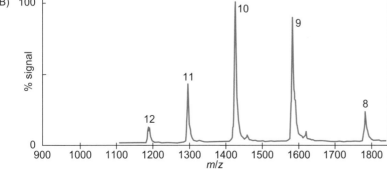

Figure 3.5.22 ESMS of α-lactalbumin under different initial conditions: (A) 50°C acetonitrile/formic acid: $M = 14\,178.04 \pm 0.83$ (denatured state); (B) water pH 5.0: $M = 14\,216.8 \pm 1.8$ (native state) (adapted with permission from *Nature* 372, 646–51 (1994)).

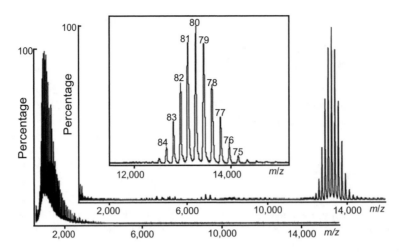

Figure 3.5.23 Mass spectra of *H. pylori* urease. The lower spectrum (*m/z* < 3000) is of denatured urease. This was electrosprayed from aqueous 50% (vol/vol) acetonitrile containing 0.1% (vol/vol) formic acid; this spectrum arises from the sum of the α and β monomer distributions. The upper spectrum of native urease (*m/z* ~13 500) was obtained using an aqueous ammonium acetate solution. This shows multiple positively charged ions of the intact $\alpha_{12}\beta_{12}$ urease with a measured mass of 1063.4 ± 1.0 kDa. The upper inset is an expanded region from the native spectrum that shows the calculated charges (adapted and redrawn with permission from *Nat. Methods* 5, 927–33 (2008)).

of the complex (it can be related to a "collisional cross-section" in units of nm²).

Example Ornithine carbamoyl transferase (OCT) structure

The enzyme, OCT, was structurally analyzed using mass spectrometry, as shown in Figure 3.5.24. The OCT protein complex was first examined by ESI-MS under conditions that preserve the subunit interactions. This gave a mass consistent with a dodecamer. Organic solvents were then added to destabilize the non-covalently bound sub-complexes but without denaturing the subunits. Four major subcomplexes were observed and assigned (on the basis of mass) to the intact dodecamer, as well as nonamer, hexamer, and trimer units of the complex (Figure 3.5.24). IM-MS was used to record the drift time distributions. Combining this information with molecular modeling of the expected shape of different assemblies allowed the overall arrangement of the subunits within the various assemblies to be determined, as depicted in Figure 3.5.25. The conclusion is that the enzyme is assembled to a dodecamer from trimeric subcomplexes.

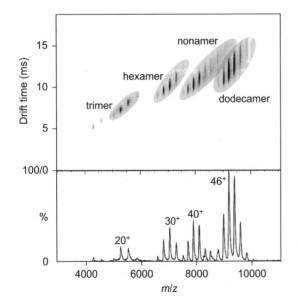

Figure 3.5.24 Mass spectra and ion mobility contour plots of ornithine carbamoyl transferase (OCT). Data were acquired in the presence of 33% acetonitrile to destabilize the complex (adapted with permission from *Structure* 17, 1235–43 (2009)).

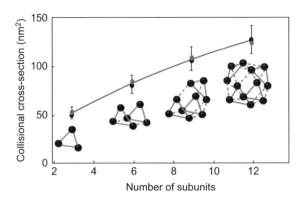

Figure 3.5.25 Collision cross-sections (CCS) determined experimentally from the IM-MS drift time data in Figure 3.5.24, plotted against the number of subunits in the subcomplexes. The CCS values calculated using models of the illustrated topological arrangements (cyan circles) are also shown. These values are in good agreement with the experimental data (black circles) (adapted with permission from *Structure* 17, 1235–43 (2009)).

 ## Summary

1 Mass spectrometers produce, analyze, and detect molecular ions. Mass spectrometry has many applications, including macromolecular characterization, proteomics, and the analysis of large non-covalent complexes.

2 The two most useful ion production methods for biological applications are matrix-assisted laser desorption ionization (MALDI) and electrospray ionization (ESI).

3 A wide range of mass analyzers are in use to separate ions on the basis of their mass to charge ratio (m/z). Beams of ions can be separated by electric and magnetic fields or by the time taken to reach the detector (time of flight, TOF). They can also be selected in a quadrupole filter. Alternatively, ions can be trapped and released in a way that depends on m/z. The characteristic frequencies of trapped oscillating ions can also be detected.

4 A major application of MS is the characterization of macromolecules, for example the measurement of mass and the detection of post-translational modifications.

5 MS is a very powerful tool in analyzing the amount and nature of proteins in a cell (the proteome). It can be used to identify and sequence proteins using peptide fragmentation and tandem MS/MS. The amount of protein can also be quantified; for example, using isotope labeling of a sample.

6 MS, coupled with tandem chromatographic affinity purification methods, is very useful for mapping and characterizing protein–protein interaction networks.

7 Structural information about protein conformation and non-covalent macromolecular complexes can be obtained from ESI done under different ionizing conditions (e.g. native and denatured) and from ion-mobility MS.

 ## Further reading

Useful websites
http://masspec.scripps.edu/book_toc.php
http://www.astbury.leeds.ac.uk/facil/MStut/mstutorial.htm

Reviews

Glish, G.L., and Vachet, R.W. The basics of mass spectrometry in the twenty-first century. *Nat. Rev. Drug Discovery* 2, 140–50 (2003).

Yates, J.R., Ruse, C.I., and Nakorchevsky, A. Proteomics by mass spectrometry: approaches, advances, and applications. *Annu. Rev. Biomed. Eng.* 11, 49–79 (2009).

Ong, S.E., and Mann, M. A practical recipe for stable isotope labeling by amino acids in cell culture (SILAC). *Nat. Protoc.* 1, 2650–60 (2006).

Heck, A.J.R. Native mass spectrometry: a bridge between interactomics and structural biology. *Nat. Methods* 5, 927–33 (2008).

Ruotolo, B.T., Benesch, J.L., Sandercock, A.M., Hyung, S.J., and Robinson, C.V. Ion mobility-mass spectrometry analysis of large protein complexes. *Nat. Protoc.* 3, 1139–52 (2008).

 ## Problems

3.5.1 Using the table in Tutorial 1 and Table 3.5.1, calculate the mass change expected for (i) a protein with two mutations Gly/Ser and His/Gln, and (ii) a peptide that is acetylated and phosphorylated.

3.5.2 A MALDI TOF mass spectrum gave five main peaks with m/z values and relative intensities as follows:

m/z	10 176.5	17 659.7	20 350.8	35 317.3	40 702.5
Intensity (%)	40	20	100	80	10

Can you say anything about the molecular weights and properties of the protein(s) involved?

3.5.3 A schematic mass spectrum, obtained using ESI, is shown in the figure. Find the number of charges associated with each m/z peak and the molecular weight of the protein. Estimate the error involved in the mass determination.

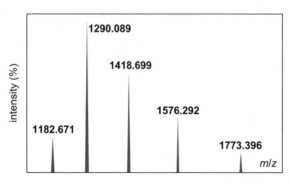

3.5.4 The masses of two dipeptide b_2 ions in a fragmentation experiment were found to be 129 Da and 230 Da. What are the two amino-acid pairs?

3.5.5 There are powerful online databases available for identifying peptide fragments. For example, access the Aldente site (**http://expasy.org/tools/aldente/**) and try to identify the protein from which the following peptides masses were produced by trypsin treatment: 748.4, 920.5, 1271.6, 1378.8, 1484.9, 1502.6, 1562.8, 1606.8, 1815.8, 1853.9.

3.5.6 The SILAC technique was introduced in the discussion of proteomic quantification. Other related quantification methods include isotope-coded affinity tagging (ICAT) and isobaric tagging for relative and absolute quantification (iTRAQ). (Isobaric means same weight.) An example of an iTRAQ tag is illustrated in the figure. Note that the reporter group and the balance group are designed to produce the same overall mass for different tags. As shown in the lower figure, a series of peptide samples can be tagged at their N-terminus with different "145" tags. During MS/MS, the tags are cleaved to produce fragments that report the relative abundances of peptides from each sample. What advantages do you think this kind of labeling might have?

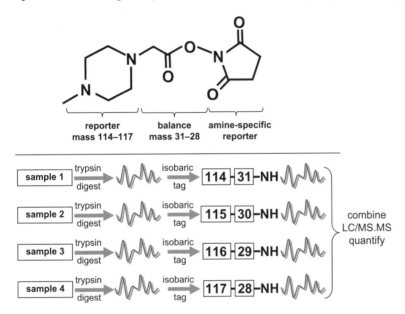

3.6 Electrophysiology

Introduction

Electrophysiology is the study of the electrical properties of biological cells and tissues. Valuable information can be obtained by measuring voltages and currents in a wide range of systems, ranging from single ion channels to beating hearts. Measurement of the electrical activity of neurons is very important in neuroscience and recordings of electric signals are routinely used in hospitals to study the heart (electro-cardiograms) and the brain (electro-encephalograms). The detected signals arise from the movement of ions across biological membranes, e.g. action potentials in nerve cells. At least 400 different genes for ion channels have been identified, and an increasing number of diseases are being correlated with ion channel dysfunction.

In the late 18th century, Luigi Galvani noticed that the application of electrical stimuli to exposed frog muscle resulted in twitching of the legs. A good explanation for such effects had to wait until the early 1950s when Alan Hodgkin and Andrew Huxley (Nobel Prize, 1963) conducted a series of key studies on large squid axons, using voltage clamp techniques. Their work gave rise to a model for nerve electrical activity that is still valid. Another advance was the introduction of the patch-clamping method in the 1970s by Erwin Neher and Bert Sakman (Nobel Prize, 1991). This technique allowed currents to be monitored at the level of individual membrane channels and led to important discoveries about neuronal function. These, and other electrophysiology methods, are explored in this chapter.

Membrane potential

The membrane potential is a voltage difference that is found between the inside of a cell and its surroundings; it is negative inside the cell and exists in all mammalian cells, even when they are in their "resting" state. It can be measured using a small glass microelectrode inserted into a cell (Figure 3.6.1). (Note that Tutorial 4 explains some of the ideas and nomenclature that are used to describe electrical circuits.) The membrane potential provides battery power to operate a variety of "molecular devices" embedded in the membrane; it can also be used to transmit signals in electrically excitable cells, such as neurons. The value of the resting potential is dependent on the cell type. For example, for a typical neuron, it is around –70 mV, while it is more negative in skeletal muscle cells. The equilibrium value reached depends on the ion transport properties of the membrane, as we will now briefly consider.

Ion channels and pumps

Biological membranes, mainly formed from a lipid bilayer, provide an ion-impermeable barrier around cells. Embedded in the bilayer are protein channels that regulate the flow of ions. Protein channels have a water-filled pore that passes ions with considerable selectively. They can, for example, distinguish Na^+ and K^+ ions which have the same charge and only differ in radius by ~0.04 nm. Ions can move rapidly, single file, through the channels and their passage can be controlled by a gate, which may be opened or closed by electrical, chemical, or mechanical signals.

Some of the ways that ions cross a membrane, along with some of the terminology used, are illustrated in Figure 3.6.2. One important active transport system, the Na^+K^+ pump, is also illustrated. This uses energy from ATP to pump three Na^+ ions out of the cell for every two K^+ ions pumped in. This Na^+K^+ pump is said to be electrogenic because it generates a net charge flow, a small continuous ionic current across the membrane, as a result of its activity. As described below, however, the membrane potential is mainly generated by the passive movement of ions

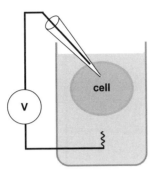

Figure 3.6.1 Measurement of membrane potential with a microelectrode inserted into a cell.

Table 3.6.1 Approximate concentrations of intra- and extracellular ions found in mammalian cells (note these are "free" concentrations; ions can also bind to macromolecules)

Ion	Intracellular concentration	Extracellular concentration
Na^+	15 mM	140 mM
K^+	140 mM	4.5 mM
Ca^{2+}	100 nM	1.25 mM
Mg^{2+}	2 mM	2 mM
H^+	63 nM (pH 7.2)	40 nM (pH 7.4)
Cl^-	10 mM	120 mM
HCO_3^-	15 mM	30 mM

and not by the pump directly. The pump mainly helps to regulate the cell's osmotic balance.

Ion asymmetry

Passive and active ion transport result in an asymmetric distribution of ions across the cell membrane. Approximate values for the main ionic concentrations observed in mammalian cells are given in Table 3.6.1. The ion gradients are carefully controlled and used in various ways by living cells to help them survive in, and communicate with, their environment.

The resting potential and the Nernst equation

The equilibrium distribution of ions across the membrane sets the resting potential. Ion pumps create concentration gradients but there is a tendency for these gradients to dissipate because of diffusion (Chapter 3.1). The extent to which dissipation occurs depends on the membrane permeability of the different ions. It

turns out that the membrane is more permeable to K^+ than other ions, so K^+ ion efflux is the dominant process in setting the resting potential. Equilibrium is reached when the electrically-driven K^+ flux into the cell (by the Na^+K^+ pump) equals the diffusion-driven K^+ flux out. The voltage at which this equilibrium point occurs is called the K^+ ion equilibrium potential (E_K). The equilibrium potential for any ion can be calculated from the values of the inside and outside concentrations using the Nernst equation $E_K = RT\ln([K^+]_{out}/[K^+]_{in})/zF$ (see Tutorial 2). Using the values of R and F given in Appendix A1.3, and assuming physiological temperature (310 K), the Nernst equation for an ion with a single charge ($z = 1$) can be written as $E_K \sim 26.7\ln([K^+]_{out}/[K^+]_{in})$ mV. Using the $[K^+]_{out}$ and $[K^+]_{in}$ values shown in Figure 3.6.3, this formula then

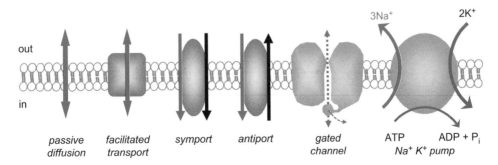

Figure 3.6.2 Illustration of some of the ways molecules and ions can cross lipid membranes via passive, active, or coupled (symport/antiport) transport. The Na^+K^+ pump is also shown.

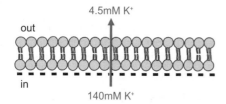

4.5mM K$^+$

out

in

140mM K$^+$

Figure 3.6.3 The passive efflux of K$^+$ ions is very important in setting the value of the resting membrane potential. A negatively charged layer of ions is formed at the inner membrane surface.

predicts a resting membrane potential, $E_K = -91.8$ mV. The K$^+$ imbalance between "in" and "out" faces of the membrane (Figure 3.6.3) thus creates a negative membrane potential and a micro-environment close to the membrane that is occupied by negatively charged anions such as Cl$^-$ which have a lower membrane permeability than K$^+$.

More detailed analysis of membrane potentials

The Goldman equation

Unsurprisingly, the assumption about the totally dominant effect of K$^+$ permeability on the membrane potential reached is an over-simplification. In fact, there is also some Na$^+$ and Cl$^-$ permeability. A more realistic estimate of the resting membrane potential E_M is given by the Goldman (or constant field) equation. If P_{Na}, P_K and P_{Cl} are the relative permeabilities of the cations and anions, and [ ] represents their concentrations, this equation can be written as:

$$E_M = \frac{RT}{F} \ln\left(\frac{P_{Na}[Na^+]_{out} + P_K[K^+]_{out} + P_{Cl}[Cl^-]_{in}}{P_{Na}[Na^+]_{in} + P_K[K^+]_{in} + P_{Cl}[Cl^-]_{out}} \right).$$

Taking the relative ion permeabilities as $P_K = 1$, $P_{Na} = 0.04$, and $P_{Cl} = 0.45$, for K$^+$, Na$^+$, and Cl$^-$ and the temperature as $T = 310$ K, substitution of the values in Table 3.6.1 gives:

$$E_M = 26.7 \ln\left(\frac{1 \times [4.5]_{out} + 0.04 \times [140]_{out} + 0.45 \times [10]_{in}}{1 \times [140]_{in} + 0.04 \times [15]_{in} + 0.45 \times [120]_{out}} \right)$$
$$= -70.1 \text{ mV}.$$

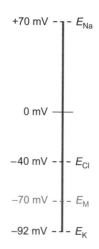

+70 mV — E_{Na}

0 mV —

−40 mV — E_{Cl}

−70 mV — E_M

−92 mV — E_K

Figure 3.6.4 Summary of approximate equilibrium ion potentials for K$^+$, Na$^+$, and Cl$^-$, as well as a net resting membrane potential (E_M). What this means in terms of Na$^+$ and K$^+$, is that if only Na$^+$ could move across the membrane, the membrane potential would move to E_{Na} (+70 mV). If only K$^+$ could transverse the membrane, the membrane potential would move to E_K (-92 mV). When the membrane is conductive to both Na$^+$ and K$^+$, the membrane potential comes to a point between E_K and E_{Na}, but closer to E_K (because $P_K > P_{Na}$).

Note that this value is different from E_K, the equilibrium potential arising from K$^+$ movement alone (see also problems at end of this chapter). Figure 3.6.4 summarizes typical equilibrium ion potentials for K$^+$, Na$^+$, and Cl$^-$, as well as a typical observed value for E_M.

Other factors can also influence the observed resting potential. For example, the Donnan effect (Chapter 3.1) occurs when large impermeable ions like proteins have a net charge that perturbs the equilibrium ion distribution.

An electrical model of membrane potentials

Electric circuit models, such as illustrated in Figure 3.6.5, can help us predict and understand the properties of cell membranes (see Tutorial 4). At the simplest level, it is convenient to describe the membrane by a resistance to ion flow of R_M (or conductance $1/R_M = G_M$). Ions also generate charge on the membrane, which acts as a capacitor, with capacitance C_M, that stores the membrane potential E_M. Because the

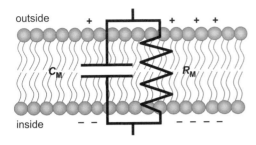

Figure 3.6.5 Representation of a membrane's electrical properties by a capacitance C_M and a resistance R_M. The potential across C_M is E_M.

membrane is very thin (~4 nm), the charge separation is stabilized by mutual electrostatic attraction.

Ion transport can thus be represented as a circuit with a current passing through resistors with conductance G (Figure 3.6.6). The equilibrium membrane potential of each ion (e.g. E_K) is considered as equivalent to a battery voltage. Each permeable ion is subject to the electric field of the membrane (E_M) and its own equilibrium potential. (We showed above, using the Nernst equation, that E_K was ~ –92 mV.) The driving force for the current is the *difference* between these two electromotive forces ($E_M - E_K$). The flow of ions (current, I) can then be described by Ohm's law ($I = VG$) for each ion type. For K^+ we have $I_K = (E_M - E_K)G_K$, where G_K is the membrane conductance to K^+ ions. Similar equations can be written for other ions, e.g. $I_{Na} = (E_M - E_{Na})G_{Na}$. At rest $I_{Na} + I_K = 0$, thus $(E_M - E_{Na})G_{Na} = (E_K - E_M)G_K$ or $E_M = (E_K G_K + E_{Na} G_{Na})/(G_K + G_{Na})$; this equation is sometimes known as the chord conductance equation.

We can thus calculate E_M with the chord conductance equation, using conductances and voltages, or with the Goldman equation, using relative permeabilities and ion concentrations.

Action potentials

The resting membrane potential tells us about the situation when a cell is quiescent. An action potential occurs when the membrane potential rapidly rises and falls in response to a stimulus (touch, taste, etc.). Action potentials are generated by voltage-gated ion channels which open if a stimulus increases the membrane potential above a defined threshold value. On opening, these channels allow an inward flow of positive ions which produces a further rise in the membrane potential thus causing more voltage-gated channels to open and more current to flow. This cascade results in a large increase in membrane potential, as illustrated in Figure 3.6.7. The voltage-gated ion channels then close, allowing the membrane potential to return towards its resting level; this falling phase of an action potential can be enhanced by other types of voltage-gated ion channel that open in this period. The amplitude of an action potential may reach a peak potential of up to +50 mV, called an overshoot (Figure 3.6.7). The duration of an action potential is highly variable between different cell types, but it typically lasts a few milliseconds.

A model for the kind of action potential illustrated in Figure 3.6.7 was first put forward by Alan Hodgkin and Andrew Huxley in 1952. Their model entails initial opening of Na^+ channels followed by an opening of K^+ channels. This pattern of channel opening is illustrated in Figure 3.6.8, where the Na^+ and K^+ conductances of a membrane are shown as a function of time. (As we will see below this conductance information can be extracted from voltage clamp experiments.)

This model can be understood in terms of our discussion of membrane potentials above (see Figure 3.6.4). The membrane potential for Na^+ is about +70mV, so an increase in G_{Na} during the depolarization phase leads to a large increase in E_M followed by

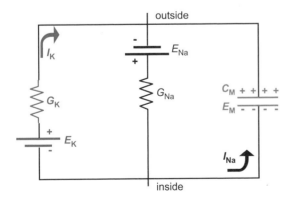

Figure 3.6.6 An electrical circuit model of Na^+ and K^+ ion potentials (E_{Na}, E_K) conductances (G_{Na}, G_K), currents (I_{Na}, I_K), and a membrane capacitance (C_M).

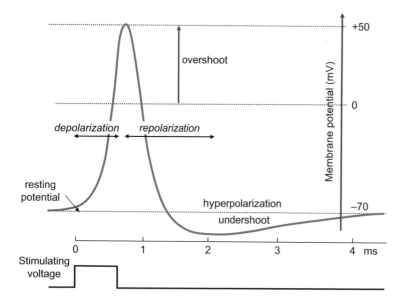

Figure 3.6.7 Illustration of an action potential with some nomenclature. Depolarization of the membrane potential is caused by an applied stimulating voltage above a certain stimulus threshold level. This is followed by repolarization and, sometimes, a hyperpolarization or "undershoot".

a decrease when an increase in G_K leads to repolarization and, eventually, a return to the resting state.

Action potentials in animal cells can involve calcium channels, as well as sodium and potassium ones. An example of an action potential in a cell in which both calcium and sodium channels are active is shown in Figure 3.6.9. As in Figure 3.6.7, the initial upstroke is dominated by Na^+ ion flow into the cell, followed by a longer lasting inflow of Ca^{2+} ions. Repolarization is dominated by the flow of K^+ out of the cell, as shown in Figure 3.6.7.

Experimental

Electrophysiology involves the measurement of currents and voltages across the cell membrane. Two types of "clamp" experiment, where the experimenter can set the membrane potential at a chosen value have been particularly important, as described below.

Voltage clamp

The voltage clamp method allows ion currents to be measured while the membrane voltage is held at a set level. As many of the channels of interest are voltage gated (their activity depends on the potential they experience), it is important to be able to manipulate the voltage independently of the ionic currents. The methodology was introduced in the 1940s by Cole and Marmount, but it was Hodgkin and Huxley who, after performing a series of voltage clamp experiments on the giant squid axon, presented a

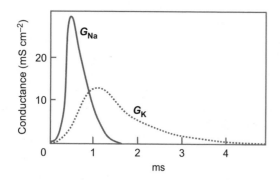

Figure 3.6.8 The time dependence of the channel conductance of Na^+ and K^+ during an action potential.

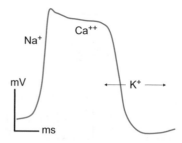

Figure 3.6.9 An example of a more complex action potential than shown in Figure 3.6.7. In this case, it is generated by calcium, sodium, and potassium channels operating in concert.

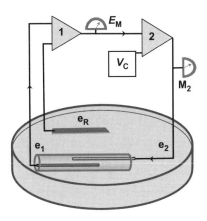

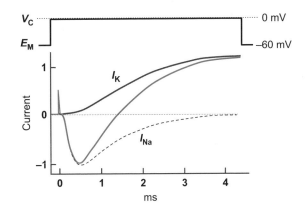

Figure 3.6.10 A voltage clamp experiment. Electrodes e_1 and e_2 are inserted in a giant squid axon (cyan). The resting membrane potential, E_M, is measured between e_1 and e_R, a reference electrode in the bathing solution. The triangles 1 and 2 represent amplifiers. The experimenter sets the control voltage V_C. The difference $E_M - V_C$ is detected in amplifier 2 and a current is fed to e_2 to eliminate the difference; this feedback loop is rapid. The meter M_2 measures the current fed through e_2.

Figure 3.6.11 A voltage clamp experiment. V_C is changed from –60 to 0 mV at $t = 0$. The cyan curve represents the normalized applied error current. A membrane capacitance spike occurs at $t = 0$, caused by the application of V_C. The cyan curve otherwise arises from the sum of two components, one from I_{Na} (dotted gray curve) and one from I_K (continuous gray line). The rise and fall of the I_{Na} current illustrates that both activation and deactivation of the Na$^+$ channels occurs.

model to explain the ion flow observed during an action potential.

The voltage clamp experimenter sets a command potential or control voltage, V_C, (Figure 3.6.10) and the voltage clamp uses negative feedback to maintain the cell at this voltage. Whenever the cell deviates from V_C, an amplifier generates an error signal and sends a current into the cell to reduce the error signal to zero. The clamp circuit thus produces a current, equal and opposite to the ionic current. This error current can be measured to give an indication of the currents flowing across the membrane.

Figure 3.6.11 shows the result of a voltage clamp experiment. At $t = 0$, a depolarizing voltage clamp is applied, stepping from a potential of –60 mV to 0 mV. This is observed to produce an inward (negative-going) current followed by an outward (positive-going) current.

It is possible to resolve the main current response in Figure 3.6.11 into two components: one arising from the Na$^+$ current (I_{Na}), the other from the K$^+$ current (I_K). This resolution into components can be achieved with further experiments, for example by reducing the concentration of extracellular sodium, which changes I_{Na}. Specific inhibitors can

also be used to block the voltage-gated channels. For example, tetrodotoxin from the puffer fish blocks the voltage-gated sodium channel and tetraethylammonium blocks the voltage-gated potassium channel.

The voltage clamp method is still widely used to study ionic currents in neurons and other cells. Variations of the method include the **current clamp**, where the membrane potential is free to vary while the current is kept constant with a feedback loop. Another powerful method for measuring ion movement is the patch clamp, as described below.

Patch clamp

The patch clamp technique was invented by Erwin Neher and Bert Sakmann in the late 1970s. A pipette tip diameter, as small as 1 μm, can be formed by heating and pulling a glass or quartz capillary tube. The tip size can be treated with different coatings for

Note When a rapid change in voltage (E_M to V_C) is applied at $t = 0$ across the membrane, this results in a current spike because of a charge rearrangement across the membrane capacitance; see Figure 3.6.11 and Problem 3.6.1.

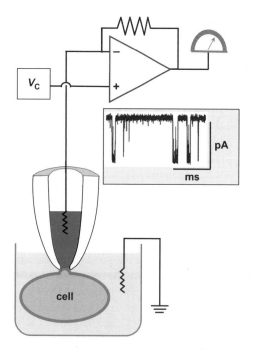

Figure 3.6.12 Illustration of a patch clamp experiment. The triangle represents an amplifier. The potential inside the pipette, related to V_C, may either be held at a steady level or changed in a step-wise fashion. A typical output (ms pulses of picoamps) resulting from ion channels switching between open and closed states, is shown in the inset.

currents that result from a single ion channel opening can then be measured with a very sensitive electronic amplifier connected to the pipette. As with the voltage clamp method, the patch clamp method allows experimental control of the membrane potential by setting the value of V_C (Figure 3.6.12).

A wide range of possible experimental arrangements make the patch clamp method very versatile. Figure 3.6.12 is an example of the **cell-attached recording method**; a more detailed illustration of this membrane and pipette tip arrangement is shown in Figure 3.6.13A.

The situation in Figure 3.6.13B arises when a stronger suction pulse is applied to the pipette, puncturing a hole in the membrane so that cytoplasm and pipette interior are continuous. This **whole-cell recording method** allows measurements of electrical signals from the entire cell and provides a convenient way of injecting drugs and other compounds into the system. This mode gives stable intracellular recording but some important components of the intracellular fluid can be diluted more than is desirable.

Once a tight seal has formed between the membrane and the glass pipette, small pieces of membrane can be detached from the cell without disrupting the seal. If a pipette that is in the cell-attached configuration is raised, a small vesicle of membrane remains attached to the tip. Exposure to air causes the vesicle to rupture, exposing a small patch of membrane with its intracellular surface exposed (Figure 3.6.13C). This is called the **inside-out patch** configuration. It

different types of experiment. The pipette is placed on the surface of an intact cell with a small amount of suction to make a seal between the pipette and the cell membrane. This seal is tight enough that all the ions must flow into the pipette. The small electrical

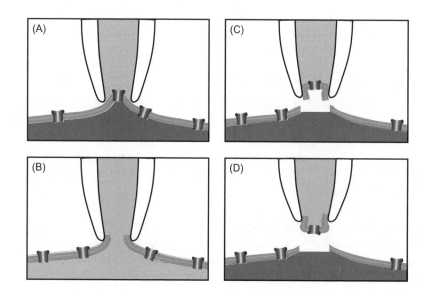

Figure 3.6.13 Illustration of possible patch clamp arrangements: (A) the cell-attached method; (B) the whole-cell method; (C) the inside-out and (D) the outside-out patch methods.

allows single-channel currents to be measured in a variety of media; the effect of various intracellular molecules on ion channel function can thus be investigated. It is also possible, with more manipulation, to produce a membrane patch with the extracellular surface exposed (Figure 3.6.13D); this arrangement is well suited to studies of the effects of extracellular chemical signals, such as neurotransmitters. This is called the outside-out patch configuration.

Arrangements other than those shown in Figure 3.6.13, are also possible; these include the perforated patch, where small holes are made in the patch with pore-forming agents so that large molecules, such as proteins, stay inside the cell while ions can pass through the holes freely.

Patch clamp experiments have shown that the macroscopic currents measured by Hodgkin and Huxley arise from the combined effect of thousands of microscopic currents, through individual channels. These experiments have confirmed that voltage-gated Na^+ and K^+ channels are responsible for the macroscopic conductances and currents that lead to an action potential. Measurements of the behavior of single ion channels have also greatly increased our general understanding of the voltage-gating mechanisms and kinetics of ion channels.

Propagation of an action potential in a neuron

Now that we have discussed resting and action membrane potentials, we can consider how impulses are propagated in the nervous system. Special cells called neurons form the main components of the nervous system which includes the brain and the spinal cord. Neurons, which transmit information by electrical and chemical signaling, have three main parts. The cell body, or soma, contains a nucleus and has multiple dendrites that receive messages from other nerve cells. One axon emerges from the cell body: it is a long, single fiber that transmits messages from the cell body to the dendrites of other neurons or to other body tissues, such as muscles. The axon termini have synaptic knobs (boutons) that release neurotransmitters (small molecules such as glutamate) and stimulate the next neuron. A protective covering called the myelin sheath covers most axons. Myelin is composed of Schwann cells in the peripheral nervous system (oligodendrocytes in the central nervous system) that wrap around the axon to form a thick fatty layer that prevents ions from entering or escaping the axon. This insulation facilitates the propagation of nerve signals. Some diseases, e.g. multiple sclerosis, cause loss of the myelin sheath, resulting in diverse symptoms that depend on the functions of the affected neurons.

The motor neuron shown in Figure 3.6.14 is designed to carry electrical signals from left to right. Signal is collected in the dendrite region, propagated along the axon, and output at the boutons in the axon termini. The amplitude of the action potential is constant—it can travel at diverse speeds (0.5–120 m s^{-1}) along axons that can be many centimeters in length.

Refractory periods and firing rate

A neuron that emits an action potential is said to "fire", giving a train of "spikes". After each action potential there is an absolute refractory period when no further action potentials can be generated, no matter how strong the stimulus; this is followed by a relative refractory period where an abnormally high stimulus is required, followed, eventually, by full recovery to the normal stimulus level (Figure 3.6.15).

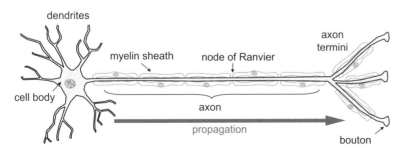

Figure 3.6.14 A schematic view of a neuron.

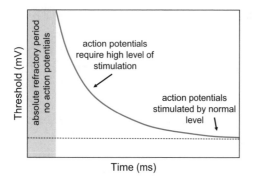

Figure 3.6.15 Time-dependence of the voltage stimulation threshold after firing of an action potential (cyan line). For a short time, called the absolute refractory period, no stimulus is large enough (the gray area). The required threshold level then reduces exponentially to the normal level shown by the dotted line.

A consequence of the time-dependent threshold level after an action potential (Figure 3.6.15) is that the firing rate is specified by the level of stimulation. A new action potential will only fire when the threshold level drops back to the stimulus level. If the applied stimulus is higher, the off-period will be shorter. The firing rate is thus dependent on the level of stimulation, with a higher stimulation voltage leading to more rapid firing (see Figure 3.6.16).

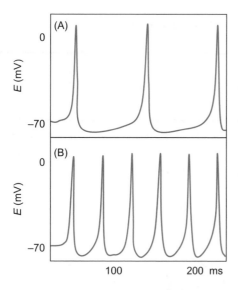

Figure 3.6.16 Action potential firing observed for two different levels of stimulation with (B) > (A). The higher stimulation level in (B) gives a faster firing rate. The amplitude of the potentials is not much affected as this is a property of the membrane potentials rather than the stimulation.

As well as influencing the firing frequency, the refractory property ensures that the action potential cannot move backwards and is propagated in only one direction along an axon (this unidirectional property is called orthodromic). In myelinated axons, conduction is saltatory—the action potential jumps nearly instantaneously from one node of Ranvier (unmyelinated spaces ~ 2 μm long) to the next—a property that greatly increases the propagation efficiency. Unmyelinated axons have voltage-gated sodium channels along the entire length of the membrane, while myelinated axons only have these channels in the nodes of Ranvier. Note also that without working voltage-gated channels, the axon produces distorted, slowly propagating signals that decay after only a few millimeters.

Cable theory

Lord Kelvin developed a model for transatlantic telegraph cable transmission that was shown to be applicable to neuron conductance by Hodgkin and colleagues in the 1940s. The neuron can be treated as an electrically passive, cylindrical transmission cable. There is a flow of charge along the neuron, with some losses due to leakage. A simple cable structure, like an axon, can be represented by a circuit of the sort shown in Figure 3.6.17. The voltage across the membrane can then be described as a function of time and distance along the neuron using a differential equation (Appendix A2.5) that depends on R_M, R_L, and C_M. The solution to this equation has two key parameters; one is the time constant of the response, $\tau = R_M C_M$ (see also Problem 3.6.3); another is a distance constant, $\lambda = (R_M/R_L)^{1/2}$, the larger λ is, the further the current will flow. In an axon, typical τ and λ values are in the range 1–5 ms and 0.5 mm–0.5 cm, respectively. The parameters λ and τ give insight into the conduction properties of a neuron. For example, if the internal resistance per unit length, R_L, is lower in one axon than in another (e.g. because it has a larger radius), the distance constant λ becomes longer and the action potential will propagate faster. If the transmembrane resistance R_M increases (e.g. because of fewer channels) the average leakage current across the membrane will diminish, causing $\lambda = (R_M/R_L)^{1/2}$ to become longer, again causing the action potential to move faster.

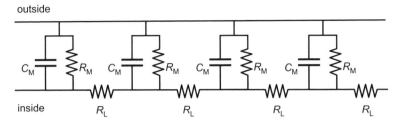

Figure 3.6.17 Equivalent circuit for an axon cable. C_M is the membrane capacitance, R_M is the transmembrane resistance, and R_L the longitudinal resistance.

Measurement of propagation

Propagating signals can be measured by inserting a series of microelectrodes along an axon. An example of propagation in a passive axon is shown in Figure 3.6.18. Injection of a pulse of current at the first electrode produces a response that has rise and fall time constants related to $\tau = R_M C_M$. The peak voltage decays with distance with a time constant related to λ. (Note that an active axon action potential propagates in one direction and does not decay as long as the initial stimulus is greater than the threshold value.)

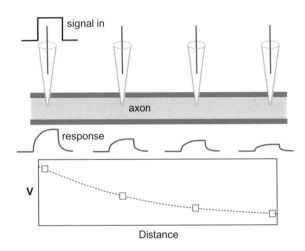

Figure 3.6.18 Measurement of propagation along a passive axon. The response signals and the decay with distance are shown.

Advanced methods

A number of new methods are becoming available to study neural circuits, including super-resolution microscopy (Section 7) and imaging of neural voltages. A method that is causing considerable excitement is "optogenetics" where light is used to drive and/or halt neuronal activity with millisecond precision. The field started in 2005 when teams at Stanford and Frankfurt reported that light-sensitive microbial ion pumps (e.g. members of the opsin family—Chapter 5.3) can cause voltage changes when they are genetically engineered into neurons and irradiated with light. Photostimulation of specific classes of neurons can then be carried out non-invasively with good spatial and temporal resolution. The earliest optogenetic experiments involved using these light-irradiated microbial proteins to control the movements of small organisms such as nematode worms and fruit flies but the technique is increasingly being used to control behavior in mammals (an example of this methodology is given in Journal Club).

➕ Summary

1 Electrical activity in cells and tissues results from the movement of ions through ion channels in membranes.
2 There is a voltage difference across the membrane called the membrane potential E_M in resting cells. E_M mainly arises from passive leakage of K^+ ions. Membrane potentials can be calculated using the Nernst, Goldman, and chord conductance equations.
3 The electrical properties of membranes can be modeled using simple electrical circuit theory with capacitors and resistors.

4 An action potential is generated by the opening and closing of gated ion channels, as proposed by Hodgkin and Huxley. Different cells have different patterns of action potential firing.

5 Extremely sensitive electrical methods are available to detect ion currents, including voltage and patch clamping. These methods have allowed detailed studies of single channel ion flow in membranes.

6 The firing rate of a neuron depends on the stimulus level.

7 The propagation of signal along a neuron can be modeled by cable theory.

 ## Further reading

Useful websites

http://stg.rutgers.edu/stg_lab/protocols/The%20axon%20Guide.pdf
http://www.unm.edu/~toolson/435ap.html

Books

Hille, B. *Ion Channels of Excitable Membranes* (3rd edn) Sinauer, 2001.

Levitan, I.B., and Kaczmark, L.K. *The Neuron* (3rd edn) Oxford University Press, 2002.

Kandel, E.R., Schwartz, J.H., and Jessel, T.M. *Principles of Neural Science* (4th edn) McGraw-Hill, 2000.

Reviews

Peterka, D.S., Takahashi, H., and Yuste, R. Imaging voltage in neurons. *Neuron* 69, 9–21 (2011).

Szobota, S., and Isacoff, E.Y. Optical control of neuronal activity. *Annu. Rev. Biophys.* 39, 329–48 (2010).

 ## Problems

3.6.1 What are the main factors that influence the movement of ions across a membrane? Why are the equilibrium potentials for K^+ and Na^+ negative and positive when both are positive ions?

3.6.2 Use the Nernst equation and the values in Table 3.6.1 to calculate E_{Ca} and E_{HCO3} at 37°C. What is the electrochemical gradient for Ca^{2+} when the membrane potential is –70 mV?

3.6.3 The membrane acts as a capacitor. The charge stored on a capacitor is $Q = V \times C$ where Q = charge (coulombs), V = voltage and C is capacitance (farads). If the capacitance of the membrane is 10^{-6} farad cm^{-2}, how many moles of charge are stored/cm^2 on a membrane where E_M = –100 mV? (Note that units of capacitance (farad) are CV^{-1} (coulombs volt $^{-1}$) while units of the Faraday constant are C mol^{-1}.)

3.6.4 In an action potential, what is meant by the terms activation and deactivation? Do the Na^+ (in) and K^+ (out) currents have the same activation rates?

3.6.5 Using information in Tutorial 4 and Appendix A2.5, show that the time constant of the response to a step in voltage, V_C, is $R_M C_M$ (see diagram).

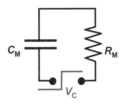

3.6.6 Use the Goldman equation to estimate the changes in relative permeability that take place in an action potential starting at $E_M = -70$ mV and overshooting to $+50$ mV at 37°C (assume that only Na^+ and K^+ are involved and that the resting concentrations given in Table 3.6.1 apply).

3.7 Calorimetry

Introduction

Calorimetry (Latin *calor* = heat) measures the heat produced by chemical reactions or physical changes. Joseph Black (1728–1799) is usually said to be the founder of the method but Antoine Lavoisier (1743–1794) designed a calorimeter to measure the metabolic heat produced by a guinea pig around the same time. As heat is either generated or absorbed in nearly every chemical process, calorimetry can be used as a universal detection system; unmodified and unlabeled material can be monitored in almost any solution conditions. There have been great technical advances in the last 20 years and microcalorimeters are now available with enough sensitivity to measure nanowatt changes. The information these instruments provide about the thermodynamics of systems can be used to tell us about the structure and interactions of proteins, nucleic acids, lipids, and drugs.

Two main types of calorimetry are used to study biological systems: isothermal titration calorimetry (ITC) and differential scanning calorimetry (DSC). ITC measures heat changes when a complex is formed at constant temperature, while DSC measures the heat flow associated with molecular changes brought about by changing the temperature. These two methods are considered in this chapter.

Isothermal titration calorimetry

Molecular recognition and the formation of specific complexes are key processes in the living cell. The formation of complexes involves energy changes (Chapter 2.1). ITC is a good way of detecting such energy changes and the method has found widespread application in the study of protein–ligand, protein–nucleic acid, and protein–protein interactions. ITC instruments measure small heat changes at constant temperature. Key thermodynamic parameters (see Tutorial 2) can be obtained for a binding process in a single experiment.

Instrumentation

A typical ITC instrument is shown in Figure 3.7.1. The most common operational mode is to add increasing amounts of a ligand into a sample at fixed

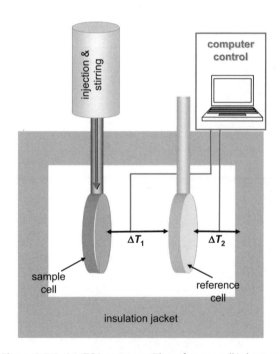

Figure 3.7.1 An ITC instrument. The reference cell is kept at a constant temperature by careful control of ΔT_2 (typically ΔT_2 is ~8°C above the jacket temperature). When an injection is made into the sample cell, the change in heat (endothermic or exothermic) associated with binding results in a change in temperature in the sample cell. Power (heat/s) is applied to return the cells to the same temperatures ($\Delta T_1 = 0$) and this power is recorded.

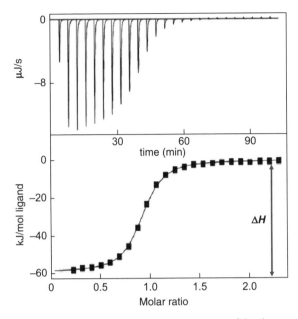

Figure 3.7.2 Raw data (top) and an integration of the data (below). The integrated curve is obtained by applying corrections derived from information obtained in control experiments (e.g. the effect of dilution is obtained by titrating a ligand into the buffer alone).

time intervals and to compensate for any heat lost or gained in the reaction by adjusting the applied power. There are two adjacent cells, a reference and a sample cell, surrounded by a thermal insulation jacket. The reference cell is kept at a constant temperature by feedback control to the jacket that keeps the temperature difference between reference cell and the jacket (ΔT_2) constant.

As a chemical reaction takes place, any temperature difference between the sample cell and the reference cell (ΔT_1) is sensed and power is fed in to keep $\Delta T_1 = 0$. The power supplied is measured and sent as data to a computer. Typical raw data are shown in Figure 3.7.2 for a ligand (L) added to a receptor (P); each injection is accompanied by energy given out (exothermic) and the spikes in Figure 3.7.2 (top) correspond to the enthalpy changes, ΔH_{obs}, resulting from the interaction between L and P. As the course of injections continues, the sample binding sites

gradually saturate and the exothermic effect disappears because no new interactions take place.

Binding measurements

As discussed in Chapter 2.1, the reaction $L + P \rightleftharpoons PL$ can be described by the equilibrium constant $K_A = [PL]/[P][L]$ where $\Delta G° = -RT\ln K_A$. Provided the concentrations of L and P are known, the total enthalpy change (ΔH), the binding constant (K_A), and the stoichiometry (n, the number of moles of L that bind to P) can all be determined by curve-fitting to the ITC data using a suitable model. The entropy change (ΔS) can also be calculated from $\Delta G = \Delta H - T\Delta S$. Table 3.7.1 illustrates the kind of data that can be obtained.

The full thermodynamic description given by ITC can give new insight into a reaction and the binding properties. For example, the Ile and Val ligands of the SH2 domain shown in Table 3.7.1 have fairly similar affinities (K_A). However, the substitution of Val for Ile increases the $T\Delta S$ value (more favorable), while the ΔH term is *reduced* (less favorable). From structural knowledge, it is known that there is a pocket in the SH2 domain that is designed to fit an Ile sidechain. One can thus speculate that the smaller Val sidechain fits less well than Ile and has more flexibility than Ile when bound; in other words, the binding of Ile is more entropically unfavorable than Val due to being more ordered in the complex, but Ile fits better to the site than Val, resulting in a more favorable enthalpy change (Figure 3.7.3).

Notes (i) Results in this field are often quoted in units of calories rather than joules (1 calorie = 4.184 joules; see Appendix A1.3). (ii) The first injection in these experiments (Figure 3.7.2) often gives an anomalously low value for ΔH.

Table 3.7.1 ITC measurement of the interaction of an SH2 domain from Src kinase, with two peptides. The peptides had the sequence Ac–GluProGln(pTyr)GluGluXxxProIleTyrLeu–NH$_2$, where Xxx was Ile or Val; pTyr is phosphotyrosine. Data from *Prot. Sci.* 9, 1975–85 (2000)

Peptide sequence	$K_A \times 10^6$ M^{-1}	$\Delta G°$ (kJ mol^{-1})	$\Delta H°$ (kJ mol^{-1})	$T\Delta S°$ (kJ mol^{-1})
Ile	10.8	−40.1	−38.7	1.4
Val	6.24	−38.8	−28.6	10.2

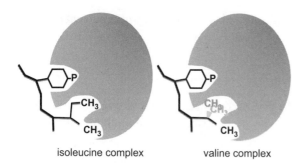

Figure 3.7.3 Isoleucine- and valine-containing peptides binding to the Src SH2 domain. Ile fits better than Val in the site so Val is more able to move and the entropic cost of binding is less.

⋮ **Example** RNA duplex formation

⋮ Although ITC has been used widely to study protein–
⋮ ligand interactions, the versatility of the method is
⋮ illustrated by the example shown in Figure 3.7.4. This
⋮ is an investigation of an RNA duplex (two comple-
⋮ mentary RNA strands that form a double-stranded
⋮ complex). Note that thermodynamic parameters, as
⋮ well as the affinity and stoichiometry, can again be
⋮ obtained from this one experiment.

Experimental design

In designing an ITC experiment, it is important to choose appropriate experimental conditions. There are two main requirements for success: (i) that a significant, measurable value of ΔH_{obs} is obtained for the initial injection; (ii) that a suitable variation in ΔH_{obs} is obtained with subsequent injections. The first requirement depends on the concentration and properties of the reaction being monitored, as well as the instrument sensitivity. For example, if the enthalpy of the reaction is low, then a higher concentration will be required to obtain a measurable value of ΔH_{obs}. The second requirement relates to the curvature of the ΔH_{obs} vs. molar ratio plot (Figure 3.7.5) which depends on the properties of the reaction. Figure 3.7.5 shows simulations of ITC plots that would be obtained under different conditions; the variable is a dimensionless constant $c = nK_A P_t$ where P_t is the total macromolecular concentration in the ITC cell, and K_A is the association constant for the binding reaction: $nL + P \rightleftharpoons nPL$. Plots of normalized heat produced per injection against mole

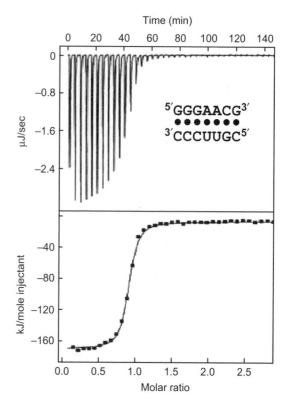

Figure 3.7.4 Data from an ITC experiment on RNA. A 75 μM solution of one RNA strand was titrated into 1.4 mL of 5 μM of the complementary strand. The fit to the curve gave parameters: $\Delta H° = 192.7$ kJ mol^{-1}; $\Delta S° = 623$ J K^{-1}mol^{-1}; $K_A = 4.3 \times 10^7$ M^{-1}, and $n = 0.91$ (adapted with permission from *J. Am. Chem. Soc.* 126, 6530–1 (2004)).

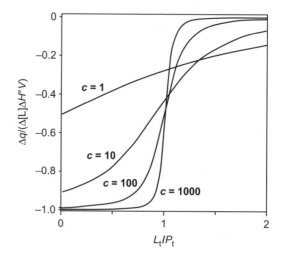

Figure 3.7.5 The shape of ITC titrations as a function of c (the product of the receptor concentration and the association constant). $\Delta q/\Delta[L]\Delta H°V$ is the heat change per mole where V is the cell volume (adapted from *J. Am. Chem. Soc.* 125, 14859–66 (2003)).

fraction of ligand for different values of c can be used to predict a practical "window" where viable experiments can be performed. Figure 3.7.5 shows that c values in the range 10–500 will lead to curve fitting that is sensitive to parameters but when $c > 1000$ (the curve becomes like a step function) and when $c < 1$ (the curve is almost linear) parameter sensitivity is then low. For typical practical values of P_t (1 μM–10mM), the available K_A range using standard ITC is 10^4–10^8 M^{-1}.

For low values of K_A, quite large amounts of sample may be required to fulfill these conditions (Problem 3.7.2). When the fitting is insensitive, some parameters, e.g. the stoichiometry, are often set to some reasonable value rather than being determined. The practical range of experimentally accessible K_A values can sometimes be expanded by using a higher affinity ligand that competes with the ligand to be investigated.

ITC kinetics

The ability of ITC to be a universal measuring method—a method in which no chromophore or labeling is required and opaque solutions are not a barrier—is also of value in studying enzyme kinetics. As shown in Figure 3.7.6, enzyme catalysis involves a binding step followed by product formation. The dependence of rate on substrate concentration, [S], for a simple enzyme reaction is described by the Michaelis–Menten equation (reaction rate $R = k_{cat}$[S]/$(K_M + $[S]$)$ where k_{cat} is the turnover rate and K_M is the Michaelis constant).

ITC measures the thermal power (dq/dt) generated by the enzymatic conversion of substrate to product. The heats involved in product formation are usually much larger than for the binding interactions so the binding contribution is usually neglected. The amount of heat involved in converting n moles of substrate to product is $q = n\Delta H_t = [P]_t V_0 \Delta H_t$ where ΔH_t is the total molar enthalpy for the reaction (J mol^{-1}); $[P]_t$ is the total concentration of product generated and V_0 is the volume of the ITC cell. Rearrangement of this equation gives the rate of the reaction $R = $ dq/d$t(V_0 \Delta H_t)$. ΔH_t and dq/dt can be determined experimentally by ITC. A complete Michaelis–Menten curve can be generated from one ITC experiment as a wide range of [S] values can be sampled during the experiment ([S] goes from a high value to zero as it is consumed by the reaction). Rate measurements following a single injection of substrate (typically at concentrations much greater than the K_M) can be made by monitoring the thermal power (dq/dt) as the substrate is consumed. [S] and [E] can be calculated from knowledge of the values of the initial substrate and enzyme concentrations and the ITC injection volumes.

Figure 3.7.7 illustrates the calorimetric response in a trypsin hydrolysis experiment. A rapid decrease in instrumental thermal power occurs initially because the hydrolysis is exothermic. The thermal power then remains nearly constant for several minutes (400–1000 s) because the reaction occurs at the maximal rate (k_{cat}[E$_0$]), where [E$_0$] is the total enzyme concentration. Substrate depletion then causes a decrease in rate and the power returns to the initial baseline after ~1400 s. The whole process is slower when inhibitor is present. K_M and k_{cat} can be obtained by fitting the Michaelis–Menten equation to the observed rate vs [S] curves.

Differential scanning calorimetry

Differential scanning calorimetry (DSC) measures heat changes that occur during a controlled increase or decrease in temperature of an experimental system. Watson and O'Neill developed modern DSC

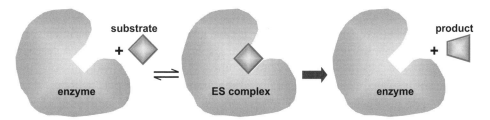

Figure 3.7.6 Illustration of a simple enzyme–substrate interaction, followed by product formation.

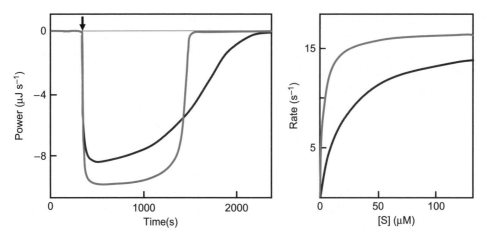

Figure 3.7.7 Trypsin-catalyzed hydrolysis of a substrate with enzyme alone (cyan lines) and with enzyme in presence of an inhibitor (gray lines). Substrate, S, was injected at the time indicated by the arrow. Conversion of thermal power to rate gives plots of rate against [S] as shown on the right. As described in the text, this analysis leads to the determination of both K_M and k_{cat}. The uninhibited reaction gave values of $K_M = 4$ μM and $k_{cat} = 16$ s^{-1} (adapted with permission from *Anal. Biochem. 296, 179–87 (2001)*).

methodology in the 1960s; it has since become a powerful analytical tool that can be used to measure directly the stability and unfolding of proteins, lipid membranes, or nucleic acids.

Instrumentation

DSC determines the heat capacity, C_p, of a molecule in aqueous solution, as a function of temperature (C_p is simply the amount of energy required to raise the sample temperature by 1 degree K—see Box 3.2 and Problem 3.7.1). A typical concentration of a macromolecule in a sample might be ~100 μM while the solvent water concentration is much higher (~55M); this "dynamic range" problem is compounded by the fact that water has a high heat capacity due to its extensive H-bonding interactions. To distinguish the very large solvent signal from the very small signal of interest, two cells are carefully compared, one containing the molecule to be studied (e.g. a protein) in a buffer solution, the other containing only the buffer (Figure 3.7.8). Differences in the heat capacity between the two samples give the heat capacity contribution of the protein alone. Heaters are used to transfer heat energy to the cells to change the temperature in a thermally insulated chamber. Generally, a DSC analysis involves increasing the overall temperature linearly with time while maintaining the reference and sample cells at the same temperature

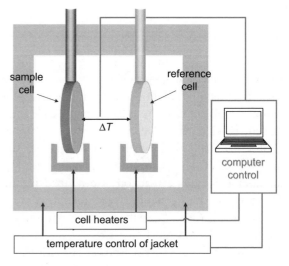

Figure. 3.7.8 Illustration of a differential scanning calorimeter. The overall temperature is raised or lowered but ΔT is kept at a minimum by applying different amounts of heating to the two sample cells.

throughout the experiment. Different power application will be required to the two cells and the difference gives the required experimental data. Because there will also always be slight differences in the volume and shape of the two cells, careful controls are required. The DSC experiment thus starts by loading the cell and collecting "buffer/buffer" runs. Protein is then loaded in the sample cell and "protein/ buffer" runs are collected. Buffer/buffer runs are subtracted from the protein/buffer runs to account for any small differences between cells.

Folding/unfolding transitions

A major application of DSC is the measurement of macromolecular unfolding due to heat denaturation. Proteins often undergo a simple transition, from a native, biologically active conformation (N), to a denatured conformation (D) with increasing temperature—a property they share with many other biological macromolecules (e.g. DNA). If we follow a property of the protein that is sensitive to a conformational transition (see elsewhere in this book, e.g. Figure 5.3.16), then we obtain a sigmoidal curve as N converts to D. The observed changes, which are usually reversible on cooling, arise from changes in the non-covalent interactions in the structure (Chapter 2.1). The N and D states are in reversible equilibrium and the concentrations of [N] and [D] change with temperature. At any one temperature we can define the equilibrium constant, $K_{eq} = [D]/[N]$ or $\Delta G° = -RT\ln K_{eq}$. The temperature at which the concentrations of D and N are equal is the "melting" temperature, T_m, which is a measure of a protein's thermal stability; at this temperature $K_{eq} = 1$ and $\Delta G = 0$.

DSC measures the heat capacity (C_p) as a function of temperature (Figure 3.7.9). Unfolding arises after the disruption of the numerous interactions that maintain the native protein structure. The unfolding process is usually endothermic, so energy has to be provided to keep the temperature of the sample cell at the same value as that of the reference cell. As well as the heat capacity changes arising from the unfolding process, there are contributions to C_p from temperature-induced changes in the pure native (N) and pure denatured (D) states of the protein. These contributions can be estimated from the values observed on either side of the transition peak (Figure 3.7.9).

After correction for native and denatured values, the area under the C_p vs. T curve is a measure of the excess energy required to denature the protein; remember C_p is related to enthalpy by $\Delta H = \int C_p dT$ (Box 3.1). If we know the concentration of the protein solution and the volume of the calorimeter cell, we can calculate ΔH (J mol^{-1}). In a single DSC experiment, we can thus determine the transition midpoint, T_m, the enthalpy (ΔH), and the heat capacity change (ΔC_p) associated with unfolding. An example showing DSC studies of lysozyme unfolding is given in Figure 3.7.9.

Complex formation

The value of T_m that is observed when a macromolecule unfolds is sensitive to environment (e.g. pH) and to the kind of complex formed. Usually, complex formation stabilizes the protein (Chapter 5.3) and raises the T_m. In the example shown in Figure 3.7.10, the complex has higher T_m and ΔC_p values than those seen for the isolated ligand or protein.

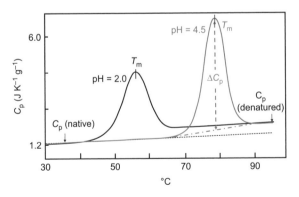

Figure 3.7.9 A DSC calorimeter output for protein denaturation. Plots of C_p against temperature are shown for the enzyme lysozyme at pH values of 2.0 and 4.5. The T_m values are shown for each curve, as well as the approximately linear regions for pure native and denatured states. ΔC_p for unfolding at pH 4.5 is also shown (adapted with permission from *J. Mol. Biol.* 86, 665–84 (1974)).

Box 3.2 Heat capacity

When a substance is heated, the temperature usually rises (an exception is when a phase change occurs, e.g. water at its boiling point). The size of the temperature increase depends on the heat capacity of the substance. Heat capacity, at constant pressure, C_p, is defined as $C_p = q/\Delta T$ where q is the energy supplied as heat and ΔT is the rise in temperature. The units of heat capacity are J K^{-1} mol^{-1}. Heat capacity is related to enthalpy by the relation $C_p = \Delta H/\Delta T$ (from $\Delta H \approx \Delta q$ at constant pressure). An alternative way of expressing this is $\Delta H = \int C_p dT$, the integral of C_p over the relevant range of temperature.

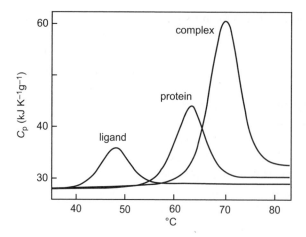

Figure 3.7.10 DSC data for the molar heat capacity of free ligand, free protein, and a high affinity protein–ligand complex.

 ## Summary

1 Isothermal calorimetry (ITC) measures heat changes at constant temperature. It is especially useful for measuring complex formation. It can yield ΔH, ΔS, K_D, and stoichiometry (n) of binding in a single experiment. It can also be used to measure enzyme kinetics.

2 Differential scanning calorimetry (DSC) measures heat changes that occur during a controlled increase or decrease in temperature. It is especially useful in monitoring heat denaturation of macromolecules, yielding the "melting" temperature T_m, the enthalpy change (ΔH), and heat capacities of denaturation in a single experiment. Information about domain structure and the influence of ligands on thermal stability can be obtained.

 ## Further reading

Useful websites
http://www.microcal.com/
http://www.tainstruments.com/

Book
Cooper, A. *Biophysical Chemistry* Royal Society of Chemistry, 2011.

Reviews
Ladbury, J.E. Application of isothermal titration calorimetry in the biological sciences: things are heating up! *Biotechniques* 37, 885–7 (2004).

Privalov, P.L., and Dragan, A.I. Microcalorimetry of biological macromolecules. *Biophys. Chem.* 126, 16–24 (2007).

Krell, T. Microcalorimetry: a response to challenges in modern biotechnology. *Microbial Biotechnol.* 1, 126–136 (2008).

❓ Problems

3.7.1 Sophisticated calorimeters have been described in this chapter, ones that can measure microscopic temperature changes (see next problem) and minute (nW) heat differences, but consider first a crude experiment in a "coffee-cup" calorimeter: a polystyrene container with a lid and a thermometer. Consider a chemical reaction that occurs in 100 grams of water at an initial temperature of 25.0°C. As a result of the reaction, the temperature of the water increases to 31.0°C. Determine the heat change. Is the reaction endothermic? (The amount of heat required to raise the temperature of 1 g of water by 1°C is 4.18 J g^{-1} °C^{-1}.)

3.7.2 A 25-kDa protein has a thermal unfolding transition with $\Delta H = 50$ kJ mol^{-1}. What temperature difference would you expect to be generated between reference and sample cell in a DSC experiment with 0.5 mg protein in a 1 mL sample container?

3.7.3 In designing an ITC experiment, a protein–ligand interaction is believed to form a 1:1 complex and have $K_D \sim 100$ µM. Estimate how much protein might be required if the molecular weight is 30 kDa and the cell volume is 1.4 mL (see Figure 3.7.5).

3.7.4 Figure 3.7.10 is an example of a "thermal shift" assay where a ligand stabilizes the folded state, causing an increase in the folding temperature, T_m (see also Figure 5.5.8). If there is a two-state native/disordered equilibrium with $K_f = [D]/[N]$, derive an approximate relationship describing the change in stability. Assume the ligand (L) binds only to the native (N) form of the protein and that its dissociation constant $K_D = [N][L]/[NL]$ (see Chapter 2.1). You can also assume the free ligand concentration is high compared to the protein concentration.

3.7.5 Interpret the DSC data below, obtained from a 25-kDa protein that, from sequence alignment, showed homology to two immunoglobulin-like domains.

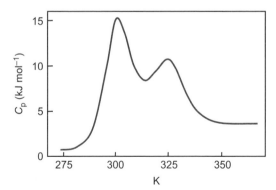

4 Scattering, refraction, and diffraction

"When light is scattered by particles which are very small compared with any of the wavelengths, the ratio of the (intensity) of the vibrations of the scattered and incident light varies .... as the inverse fourth power."

J.W. Strutt, later Lord Rayleigh. On the light from the sky, its polarization, and colour. *Phil. Mag.* 41, 107–20 (1871)

"We can scarcely avoid the inference that light consists in the transverse undulations of the same medium which is the cause of electric and magnetic phenomena."

James Clerk Maxwell, 1890

"Light brings us the news of the Universe."

William Bragg, 1933

"A great advantage of X-ray analysis as a method of chemical structure analysis is its power to show some totally unexpected and surprising structure with, at the same time, complete certainty."

Dorothy Crowfoot Hodgkin, Nobel lecture, 1964

As explained by Maxwell, electromagnetic radiation can be thought of as waves of electric and magnetic fields that propagate in a direction transverse to the oscillations (see Tutorial 8). When electromagnetic radiation interacts with matter, several processes can occur; one of these is scattering, which is dealt with in this section. Scattering results from secondary radiation, produced when oscillations are induced by electromagnetic radiation striking an object. Our ability to observe most objects arises because light is scattered (reflected) from these objects. As we will see here, scattering can also give us useful information about molecules in suspensions and crystals.

Reflection and refraction are fundamental properties of light that can be explained by scattering, as well as simple geometrical laws. Lord Rayleigh's theory of scattering, published in 1871, was the first correct explanation of why the sky is blue. Localized "evanescent" waves, generated when light undergoes total internal reflection, have several important effects that can be exploited in biophysical methods. One of them is surface plasmon resonance which is discussed in Chapter 4.2.

Scattering from ordered arrays gives interpretable diffraction patterns. The wavelength of light is usually too long to study atomic details of molecules by this method but the serendipitous discovery of X-rays by Wilhelm Röngten in 1895, followed by the structure determination of crystalline common salt in 1913 by Lawrence Bragg led to the astonishingly successful modern discipline of X-ray crystallography. The related discipline of neutron diffraction arose from the discovery of neutrons by James Chadwick in 1932.

4.1 Scattering of radiation

Introduction

Electromagnetic radiation can be considered as a stream of photons moving at the speed of light. These photons have no mass and can be treated as waves. There is a spectrum of electromagnetic radiation, ranging from low-energy radiowaves to high energy γ-rays; this spectrum can be expressed in terms of energy, wavelength, or frequency (Tutorial 8).

With various practical modifications, Figure 4.1.1 represents the basis of many of the experiments that are discussed in this book (see Sections 4, 5, 6, and 7). Scattering, which is dealt with in this section, is one result of the interaction of electromagnetic radiation with matter. Another is transitions induced between molecular energy levels, which leads to absorption (e.g. visible spectra) and emission (e.g. fluorescence); these topics are dealt with in Section 5. Light-induced chemical events (photochemistry) can also arise (e.g. photosynthesis) but here we mainly wish to explore how electromagnetic radiation can be used to tell us about the structure, energy, and dynamics of molecules.

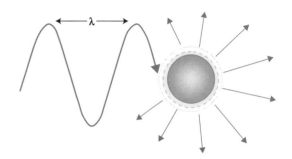

Figure 4.1.2 Electromagnetic radiation, wavelength λ, produces oscillations in a particle; secondary radiation is then produced in all directions.

Scattering

As shown schematically in Figure 4.1.2, incident electromagnetic radiation of intensity I and wavelength λ causes a particle to oscillate. The oscillations give rise to secondary radiation, which can give information about the concentration, shape, and motion of the scattering particle. This section will explain how scattering gives rise to various effects, including **refraction** and **diffraction**, and how such effects can give useful information about molecules.

There are three main classes of scattering experiment that will be explored in this chapter. As illustrated in Figure 4.1.3, these monitor: (A) a reduction in transmitted intensity (turbidity); (B) the angular

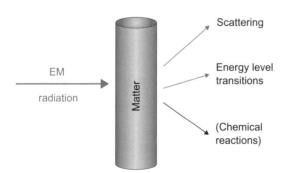

Figure 4.1.1 Electromagnetic radiation incident on a sample can give rise to scattering, transitions between energy levels and photochemistry.

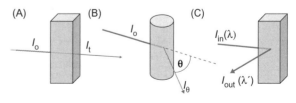

Figure 4.1.3 Some classes of scattering experiment: (A) where the intensity of light is reduced "in line" (turbidity); (B) where the light intensity is measured as a function of the angle θ; (C) where the wavelength dependence of the output light is measured.

dependence of the scattering radiation; and (C) changes in wavelength (or frequency), measured at a fixed angle to the incident beam (usually 90°).

Scattering theory

Much of scattering theory can be modeled by the response of an oscillator to a driving force (Tutorial 7). First consider an isolated particle where the applied wavelength, λ, is greater than the size of the scatterer, e.g. electrons in an atom, irradiated by visible light. The driving force is the electric component of the electromagnetic radiation which we can describe as an oscillation with amplitude E_o and frequency $\nu = c/\lambda$, $E_o\cos(2\pi\nu t)$ (see Tutorials 5 and 8). Applying this wave to an atom will induce an oscillating displacement of the electrons with respect to the nucleus. (The nucleus is essentially unaffected by most electromagnetic radiation because it is much heavier than the electron.) The magnitude of the induced electron displacement will be proportional to the molecular polarizability (α), which depends on the size and shape of the electron cloud and how easily it can be distorted by the applied field. The induced oscillation in the electron cloud is then proportional to $\alpha E_o\cos(2\pi\nu t)$. This creates an oscillating dipole (Tutorial 11) with amplitude $\alpha^2 E_o^2$ that transmits electromagnetic radiation in all directions like an aerial. It was shown by Lord Rayleigh that this kind of scattering can be described by the equation: $R_\theta = (I_\theta(1 + \cos^2\theta)/I_o r^2) = 8\pi^4 a^2/\lambda^4$; R_θ is known as the Rayleigh ratio; the incident light intensity is I_o; and I_θ is the observed scattering at a distance r and angle θ (Figure 4.1.3B).

Three results of this equation should be noted: (i) the dependence on molecular polarizability (α^2) which increases strongly with particle size; (ii) the angular dependent term $(1 + \cos^2\theta)$ which means that scattering is present at all θ values; and (iii) the λ dependence (λ^{-4}) which means that short wavelength blue light is scattered much more than red—this results in the sky being blue when viewed at large angles to the sun while, in the evening, it looks red in the direction of the sun because red is scattered less and is thus better transmitted than blue.

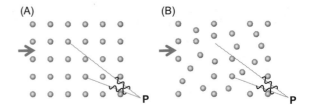

Figure 4.1.4 For an infinite array of static scatterers (A) it is always possible to choose two scatterers that give rise to destructive interference at P (see also Figure 4.1.5). In the solution state (B), where the position of the scatterers fluctuates with time, canceling does not always occur.

Scattering from particle arrays

Consider first an infinite array of particles that are small compared to λ (Figure 4.1.4); scattering from this array explains many of the observations in classical optics, such as reflection and refraction (this idea will be expanded further in Chapter 4.2; see also Figure 4.2.2 and Box 4.4). The optical observations can be explained in terms of interference between electromagnetic radiation scattered from different particles in the array. Figure 4.1.5 reminds us that when coherent waves interact, the sum can be additive or destructive. In general, it is possible, in a static ordered array, to choose pairs of particles where the phase difference between the light scattered from the two particles is 180°. The net result is that there is no observed scattering from the array at angles other than $\theta = 0$ and certain angles of reflection and refraction (Box 4.4).

In solution, rather than a static array, motion of the particles (Brownian motion) causes the local concentration to vary, or fluctuate, as a function of time. Canceling of light from pairs of scatterers is then no

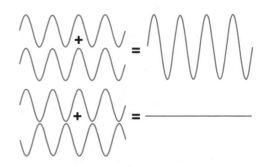

Figure 4.1.5 Two coherent waves of the same phase add constructively while two waves of opposite phase subtract.

longer always found and finite scattering will be observed (Figure 4.1.4) at all angles. It can be shown that this concentration-dependent scattering is related to the product of the concentration, c, and the molecular weight, M, or $R_\theta = KcM$, where K is a constant. If there is a mixture of species in solution the value of M that should be used is the weight average M_w (Box 4.1). As we will see in the turbidity section, this ability to obtain information about c and M_w is very useful.

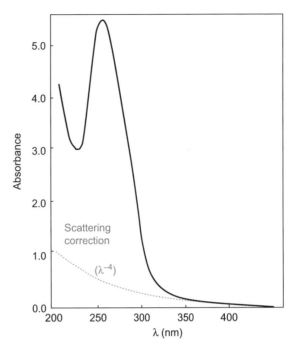

Figure 4.1.6 Absorbance from a solution of T7 phage. The apparent absorbance at 269 nm is 5.4 but the "tail" at high values of λ suggests that both scattering and absorbance are occurring. Extrapolating the λ^{-4} dependence, observed when $\lambda > 350$ nm, to values when $\lambda < 350$ nm gives a scattering correction of ~0.4 absorbance units at 260 nm.

Box 4.1 Weight average molecular weight

High molecular weight components in a mixture give relatively strong scattering (remember scattering is proportional to α^2 and α increases with size). If there is a distribution of molecular weights rather than single species in a solution, an appropriate way to describe molecular weight in a scattering experiment is to use a weight average, defined as $M_w = \sum_i N_i M_i^2 / \sum_i N_i M_i$. Note that any large component (dust, aggregates, and bubbles) will contribute disproportionately to the scattering, so great care has to be taken in preparing samples—see Problem 4.1.2.

Turbidity

Scattering produces a loss of transmitted intensity due to deflection of the radiation to angles other than $\theta = 0$ (see Figure 4.1.3A). This loss of transmitted intensity is called turbidity. Some optically transparent materials, such as water, transmit most of the light that falls on them, while in others the solution can appear opaque because of turbidity, e.g. milk.

The transmitted intensity can be described by $I_t = I_o \exp(-\tau l)$, where l is the path length of the light in the sample holder. The turbidity coefficient, τ, which is analogous to the absorption coefficient in Beer's law (this is discussed in detail in Chapter 5.3) is proportional the Rayleigh ratio $R_\theta = KcM$, which depends on the number of scatterers, their mass, and λ^{-4}. This information can be used in many ways, for example to monitor the density of multiplying bacteria in a cell growth medium. The λ^{-4} dependence can also be used to correct absorbance measurements when scattering is present, as illustrated in Figure 4.1.6.

Another simple, but powerful, example of turbidity measurements is to follow membrane transport. The volume of cells and organelles depends on the osmotic forces in the solution. When small molecules, such as sugars or amino acids are added to a suspension of cells, there is an osmotic imbalance across the cell membrane (see Chapter 3.1). This results in a rapid flow of water out of the cell, causing cell shrinkage. This is followed by an expansion in volume as water flows back into the cell along with the small molecule being transported. These cell-volume changes (shrinkage and then expansion) are reflected in turbidity changes. Figure 4.1.7 shows the cell expansion part of transport-induced volume/turbidity changes.

Solution scattering and molecular shape

Scattering from larger particles

So far, we have considered the scattering from particles that are small compared with the wavelength, λ.

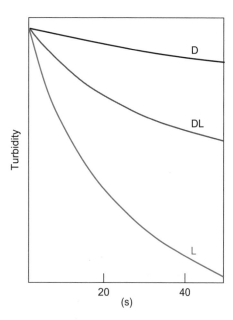

Figure 4.1.7 The transport of threonine into mitochondria, as measured by changes in absorbance as a function of time. Solutions of aminoacids (D, L, and DL mixtures) were added to a suspension of mitochondria at $t = 0$. The time course clearly shows that the most abundant amino-acid type found in nature (L) is transported more rapidly into the cell than D (adapted from *Biochemistry* 16, 5116–20 (1977)).

In solution, such particles will be distributed randomly so the scattering from them is uncorrelated. If, however, λ is of similar magnitude to the particle size, then the scattered radiation produced from different points on one particle can give rise to interference effects because the scattering points have correlated motion. Consider the situation in Figure 4.1.8 where scattering at P arises from points A and B on one scattering object. There is a path length difference (PD) in

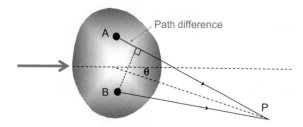

Figure 4.1.8 Scattering from a large particle, whose size is $\geq \lambda$. There is a path difference between scattering from A and B that depends on the observation angle θ from the incident beam (incident irradiation direction is shown as a cyan arrow).

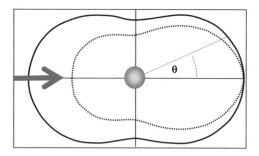

Figure 4.1.9 The angular dependence of scattering caused by electromagnetic radiation (cyan arrow). If $\lambda >$ particle size, then the scattering intensity has a $1 + \cos^2\theta$ dependence (solid line). If $\lambda \sim$ particle size, then interference effects cause the scattering to drop off faster with increasing θ (dotted line).

the light scattered from A and B; this is zero for forward scattering ($\theta = 0$) but the PD increases with increasing θ. Such interference effects lead to an **angular dependence of scattering** that can give information about the **particle shape**. (Interference effects are discussed in more detail later, e.g. Figures 4.3.8 and 7.1.2). The $(1 + \cos^2\theta)$ angular dependence of scattering predicted by the Rayleigh ratio is thus incorrect when λ is similar to the particle size. A more realistic representation of observed scattering is given by the dotted curve shown in Figure 4.1.9.

Note that as $\theta \to 0$, we get the scattering predicted by Rayleigh. It can also be shown that, for small θ, an approximate description of the angular dependent scattering is given by $I_\theta = \exp\left(-Q^2 R_G^2/3\right)$ where $Q = 4\pi\sin(\theta/2)/\lambda$, and R_G is the **radius of gyration** of the particle (see Box 4.2). This relationship suggests

---

Box 4.2 Radii of gyration, R_G, and hydration, R_H

R_G is used to indicate the dimensions of a particle in scattering and hydrodynamic measurements (Chapter 3.2). For a polymer with i segments of mass m, a distance r from the molecular center of mass, R_G is defined as $R_G^2 = \sum_i m_i r^2 \sum_i m$. This equation can be evaluated for various shapes. For example, for a sphere of radius r, $R_G = r(3/5)^{1/2}$; for a cylindrical rod of length L and radius r, $R_G^2 = r^2/2 + L^2/12$. In practice, a parameter called the **hydrodynamic radius**, R_H, is measured in most experiments rather than R_G, because macromolecules are surrounded by a water layer. R_H depends on the system studied, but typically $R_H \sim 1.3 R_G$.

that a plot of ln I_θ against Q^2 will give a straight line with slope $R_G^2/3$. This is known as a Guinier plot which is useful for obtaining an estimate of R_G.

Scattering using light, X-rays, and neutrons

As we saw in Figure 4.1.9, interference effects arise if λ ~ particle size. Information about particle shape and size can then be obtained from the angular dependence of scattering. In the solution state, these large scatterers will be randomly oriented so there is extensive averaging of the resultant scattering. Solution scattering experiments thus only give relatively low resolution information.

Light scattering (λ ~ 500 nm) can be used to give information about the shape and mass of scatterers (see also dynamic light scattering dealt with later in this chapter). Commercial instruments are available that detect the light intensity scattered at multiple angles (multi-angle light scattering; MALS). These instruments are surprisingly sensitive and they can detect particles with radii as small as 1 nm, much less than the wavelength. As explained above, solution scattering for small particles is described by the equation $R_\theta = KcM$. For larger particles, this equation only applies strictly when $\theta = 0$ and because of non-ideal solution behavior it only applies strictly at low concentration. A double extrapolation ($\theta \rightarrow 0$ and $c \rightarrow 0$) of light scattering data is therefore carried out to obtain a good estimate of M. This double extrapolation is called a Zimm plot.

X-ray and neutron scattering

The wavelength of light is generally too long to obtain detailed information about the shape of most biological particles of interest. More detailed information can be obtained using X-rays and neutrons because they have a wavelength, λ ~ 0.1 nm, which satisfies the condition $\lambda < R_G$ for most macromolecules and complexes. Small-angle X-ray and small-angle neutron scattering, abbreviated to SAXS and SANS, respectively, are thus widely used to determine the shape of particles in solution. (Neutrons are scattered by the dense nuclei at the center of atoms while X-rays are scattered by the surrounding electrons.)

Various parameters are used to describe the efficiency of neutron and X-ray scattering from particles; these include scattering length and scattering density (the total of scattering lengths in the macromolecule divided by its volume). For X-rays, the scattering length of the elements increases linearly with the number of electrons they contain (the atomic number), but there is a large variation of neutron scattering length with atomic number, and some values are *negative* (see Table 4.1.1). This variability in neutron scattering lengths results partly from resonance effects (Tutorial 7). An explanation of the negative value for ^{1}H is that it corresponds to a situation where λ, the wavelength of the incoming radiation λ, is $<\lambda_o$ the resonant wavelength of the scatterer (i.e. above resonance). In such cases the scattered wave is 180° out of phase with respect to the incoming radiation; if $\lambda > \lambda_o$, then the scattering will have the same phase and the scattering length will be positive.

SAXS experiments

A schematic view of a SAXS experiment is shown in Figure 4.1.10. The scattered X-rays are detected by an area detector (Chapter 4.3) and the observed angular dependence of the scattering is translated into an I vs. Q plot which tells us about the particle size and shape ($Q = 4\pi\sin(\theta/2)/\lambda$; see above). Samples are typically ~60 μL, with a concentration of 0.1–10 mg mL^{-1}. Scattering from the solvent is significant, and careful controls to correct for this and other

Table 4.1.1 Scattering lengths of some atoms for X-rays and neutrons for typical applied wavelengths. Note the systematic increase for X-rays and the negative value for neutron scattering from ^{1}H

Atom	Isotope	Atomic number	X-ray scattering length (fm)	Neutron scattering length (fm)
Hydrogen	^{1}H	1	2.8	-3.74
	^{2}H	1	2.8	6.67
Carbon	^{12}C	6	16.9	6.65
Nitrogen	^{14}N	7	19.7	9.4
Oxygen	^{16}O	8	22.5	5.8
Phosphorus	^{31}P	15	42.3	5.1
Lead	^{208}Pb	82	230.4	9.5

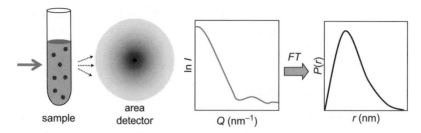

Figure 4.1.10 A schematic view of a SAXS experiment. The scattering intensity, *I*, is detected by an area detector (discussed in Chapter 4.3) as a function of *Q* which is a measure of the angular deviation from the center of the detector. (The intense spot in the center of the detector is from the direct beam and this is usually blocked by a protective "beam stop".) A Fourier transform of *I* vs. *Q* gives *P*(*r*) vs. *r*, a representation of the scatterer dimensions in real space (Chapter 4.3).

artifacts are required. Well-behaved (mono-disperse/non-aggregated) samples are also required.

The scattering experiment shown in Figure 4.1.10 has many similar features to diffraction experiments. *Q* can be considered as reciprocal space and the *I* vs. *Q* function can be converted to a real space representation, *P*(*r*) vs. *r*, by a Fourier transformation. (The concepts of real space, reciprocal space, and Fourier transforms are discussed extensively in Chapter 4.3 and Tutorial 6.) The position where the *P*(*r*) function crosses the *r* axis is related to the maximum dimensions of the scatterer.

We have indicated how R_G can be determined from the angular dependence of scattering at small angles using a Guinier plot. The angular dependence at larger angles gives more information about shape (Figure 4.1.11). Expected scattering profiles for any assumed model shape can be calculated using various available computer programs that generate calculated curves. A simple example of how observed scattering can be fitted to computed models is shown in Figure 4.1.11. This illustrates a variety of possible conformations for a Y-shaped antibody molecule.

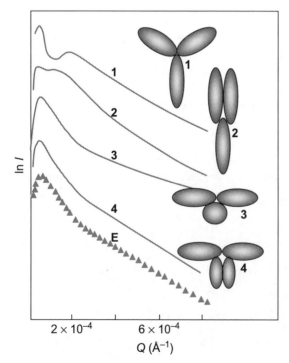

Figure 4.1.11 The observed experimental scattering of an antibody molecule as a function of *Q* is shown by cyan triangles. The computed scattering curves, 1–4, are shown for a series of model shapes (offset from each other for illustrative purposes). It can be seen that shape 4 gives the best fit to the experimental data (adapted with permission from *Biochemistry* 9, 211–9 (1970)).

Example Mysosin head

The results of a SAXS study of the energy-transducing component of the **actomyosin motor** are shown in Figure 4.1.12. These SAXS studies on a sub-fragment of the myosin "head" indicated that conformational changes occur on the addition of ATP. Models of shapes of the myosin head that best fit the scattering data are shown as $P(r)$ vs. r plots.

Note that even when $\lambda < R_G$ it is only possible to obtain information about the overall shape of a system in solution (e.g. the relative orientation of protein domains in a multi-domain protein) because of averaging of the scattering from all the orientations of a particle in solution. If there is also extensive motion within the scatterer itself (e.g. a protein with disordered regions) then the scattering patterns are harder to interpret. Some plots are, however, useful for emphasizing sample flexibility. For example, a Kratky plot (IQ^2 against Q) can help identify an unfolded protein, as shown in Figure 4.1.13.

SANS and contrast matching

Neutron scattering provides an alternative to SAXS for obtaining information about the shape of molecules in solution. There are better sources and

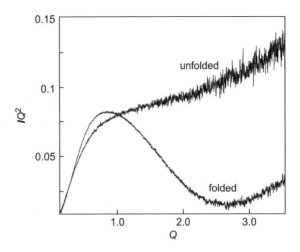

Figure 4.1.13 Schematic Kratky plot for a flexible "unfolded" protein compared with one from a folded protein.

detectors for SAXS than SANS, so SAXS is usually the preferred method, but the large and different neutron scattering from hydrogen and deuterium (Table 4.1.1) can be exploited in SANS to give **contrast matching** (or **contrast variation**). Figure 4.1.14 shows the scattering length density for water and various biological macromolecules as a function of the H_2O/D_2O ratio. As can be seen, there is a very

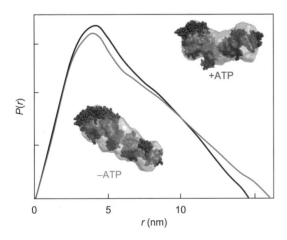

Figure 4.1.12 $P(r)$ plots for X-ray scattering from the myosin head. The cyan curve corresponds to the situation without ATP and the gray curve is with ATP. $P(r)$ gives direct low-resolution information about the shape of the scatterer. The intercept on the r axis suggests a maximum size for the molecule of about 16 nm (adapted with permission from *J. Mol. Biol.* 392, 420–35 (2009).

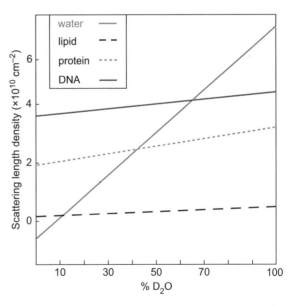

Figure 4.1.14 The effect of varying the H_2O/D_2O ratio on the scattering-length density. The cyan line shows the very large change observed for water. The match points occur where the lines cross, e.g. for DNA it is around 65%, while for proteins it is around 43%.

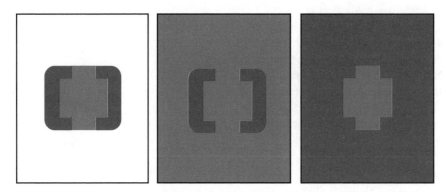

Figure 4.1.15 Illustration of contrast matching. In a system made from gray and cyan components, changing the solvent to match either cyan or gray will give scattering that is predominantly from one type of component.

large change in scattering density between H_2O, and D_2O, so changing the H_2O/D_2O ratio in a sample can be used to vary the scattering density of the solvent over a wide range. At certain ratios of H_2O to D_2O, called match points, the scattering from the molecule will equal that of the solvent, and thus be eliminated when the scattering from the buffer is subtracted. For example, the match point for proteins is ~43% D_2O; scattering from the protein will be thus be indistinguishable from that of the buffer in a 43% mixture of D_2O/H_2O.

When a particle is made from different materials—for example, ribosomes made from RNA and protein, chromatin made from DNA and protein, or membranes made from lipid, protein, and carbohydrate—the various parts can be studied selectively by adjusting the H_2O/D_2O ratio to the various match points (Figure 4.1.15). Note that the scattering density of macromolecules also changes somewhat with percentage D_2O because some of the hydrogens in –OH and –NH groups exchange with solvent, but this effect is small compared with the changes in scattering at different H_2O/D_2O ratios.

Example The ribosome

An example of a contrast matching SANS experiment is given in Figure 4.1.16. The aim here was to measure the separation between specific proteins in reconstituted ribosomes. Selected pairs of **protonated** ribosomal proteins were reconstituted into **deuterated** 50S subunits from *E. coli* ribosomes. The rRNA was derived from cells grown in 76% D_2O and deuterated protein was derived from cells grown in 84% D_2O. These values were experimentally chosen so that the neutron scattering from the deuterated proteins and the RNA was nearly the same. The neutron scattering was recorded in a reconstitution buffer containing approximately 90% D_2O, which matched the partially deuterated protein and rRNA in the reconstituted ribosome sample. The net result was that everything except the pair of *protonated* proteins was nearly "invisible". The observed buffer-corrected scattering was then relatively simple to interpret and the distance between the centers of the two protonated proteins could be deduced. In this way, 50 separation distances between proteins within the large ribosomal subunit were determined; some of the protein positions obtained are shown in Figure 4.1.16.

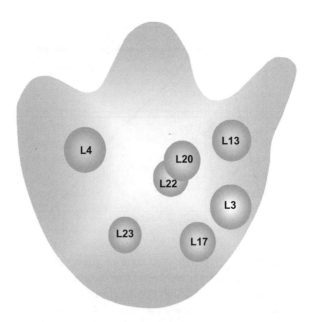

Figure 4.1.16 Model of the 50S subunit. The positions of the seven proteins (centers of the corresponding spheres) are based on distance data derived from SANS data (adapted with permission from *EMBO J*. 11, 373–8 (1992)).

Dynamic light scattering

As discussed above (see Figure 4.1.4), particles or macromolecules suspended in a liquid medium undergo Brownian motion which causes fluctuations in local concentration. This results in local variations in refractive index (see Chapter 4.2) and fluctuations in the intensity of the scattered light. If a coherent laser source (see Box 5.2) is used as the irradiation light source then these time-dependent fluctuations can be analyzed. This is the basis of a technique known as dynamic light-scattering (DLS). (DLS is also sometimes known as quasi-elastic and photo-correlation spectroscopy.) DLS is very useful for giving information about particle size and distribution in solutions.

The observed fluctuations are related to the diffusion rates of the scattering particles and thus to their diffusion coefficients, D. Large particles diffuse more slowly than small ones so the scattering fluctuations from large particles will contain fewer high-frequency components than from small particles. One way to analyze random diffusional processes is to use correlation functions (Box 4.3).

Box 4.3 Correlation functions

We can define the time scale of random molecular motion with a correlation function $g(\tau)$. This is related to the probability of finding a correlation between the orientation of a molecule and some defined external direction as a function of time. After a short time interval (δt), the molecule will have a similar orientation or position to the $t = 0$ position but this correlation will diminish with increasing time because of random motion (Figure A).

It is often found that $g(\tau)$ decreases exponentially with time and thus $g(\tau)$ has the form $\exp(-t/\tau_c)$ where the time constant τ_c is defined as the correlation time. $g(\tau)$ vs.τ plots are often presented on a logarithmic scale so appear as shown in Figure B.

τ_c is inversely proportional to temperature and is proportional to the molecular weight. For a macromolecule of hydrodynamic radius R_h in a solution with viscosity η_o, the rotational correlation time $\tau_c \approx 4\pi R_h^3 \eta_o / 3kT$ where k is the Boltzmann constant and T the temperature. τ_c is also related the diffusion coefficient, D (Box 3.1). Correlation functions are important features of a number of biophysical methods and will be discussed again in Sections 6 and 7.

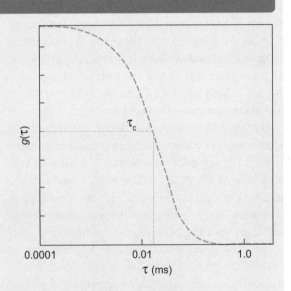

Figure B A typical logarithmic plot of a correlation function as might be obtained in a DLS experiment.

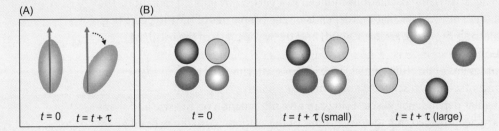

Figure A Representation of how molecular motion can be characterized by a correlation time: (A) the position of a molecule undergoing random *rotational* motion at times t and $t = \tau$; (B) four molecules undergoing *translational* diffusion at short and long values of τ. It can be seen that the correlations between positions is high at short τ but low at large values of τ.

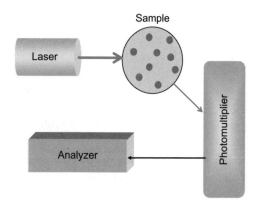

Figure 4.1.17 A typical experimental set up in a dynamic light scattering experiment.

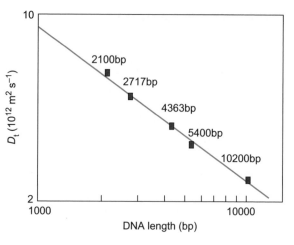

Figure 4.1.18 Diffusion coefficients, measured by DLS, for superhelical plasmid DNA as a function of their length (adapted with permission from *Methods Enzymol.* 211, 430–48 (1992)).

A typical DLS experimental set up is shown in Figure 4.1.17. An intense monochromatic light beam from a laser is scattered by the sample. The fluctuations in the scattered light are analyzed to obtain the correlation function which is a measure of the diffusion coefficient, D. D and the hydrodynamic radius R_H are related by $D = kT/6\pi\eta_o R_H$, where η_o is the sample viscosity; see Boxes 3.1 and 4.2.

Dynamic light scattering (DLS) instruments are simple to use: macromolecules (typical concentrations 0.5–2 mg mL^{-1}) are placed in a microcuvette (cell volume 10–20 μL^{-1}). The measurement time is usually less than one minute. DLS can measure particle sizes in the nanometer to micrometer range, although measurement is more difficult for small particles because the scattered light intensity is weak.

DLS has many applications. It can measure polydispersity and the presence of aggregates in protein samples and nanoparticle suspensions. Crystallographers also find DLS a valuable tool for screening protein samples for sample aggregation. DLS has also been used to detect early cataract formation in the eye. An example of DLS experiments on DNA is given in Figure 4.1.18.

Example Plasmid DNA shape

The translational diffusion coefficients for plasmid DNA of different lengths were measured by DLS. The observed correlation between DNA length and diffusion coefficient was interpreted in terms of relatively rigid elongated superhelical DNA structures in solution.

Summary

1 Scattering arises from oscillations induced in a particle by applied radiation. The size of these induced oscillations depends on the properties of the molecule (e.g. the polarizability of the electron cloud) and the wavelength of the incident radiation.

2 Measurements of turbidity, which arises from scattering, can be used to measure the concentration of a scatterer and indicate its relative size.

3 The angular dependence of scattering can give information on the size and shape of a particle, provided that the dimensions of the scatterer (usually defined by the radius of gyration, R_G) are comparable to the wavelength of the incident radiation.

4 X-rays and neutrons are very useful for obtaining molecular shape information because of their short wavelength ($\lambda \sim 0.1$ nm).

5 X-rays are scattered by electrons and the scattering increases linearly with atomic number. Small-angle X-ray scattering in solution is known as SAXS.

6 Neutrons are scattered by nuclei and the scattering varies widely between isotopes. As the scattering from hydrogen and deuterium are very different, the scattering from solvent water can be matched to different components of a heterogeneous biological sample, for example making a protein "invisible" in a protein–DNA complex. Small-angle neutron scattering is known as SANS.

7 Motion of the scatterer can be detected by analysis of the scattered light (dynamic light scattering—DLS). The diffusion coefficient of the scatterer can be obtained by analyzing the fluctuations in the scattered intensity caused by Brownian motion.

 ## Further reading

Useful websites
http://www.neutron.anl.gov/
http://www.wyatt.com/solutions/hardware/hardware.html

Reviews

Koch, M.H., Vachette, P., and Svergun, D.I. Small-angle scattering: a view on the properties, structures, and structural changes of biological macromolecules in solution. *Q. Rev. Biophys.* 36, 147–227 (2003).

Svergun, D.I. Small-angle X-ray and neutron scattering as a tool for structural systems biology. *Biol. Chem.* 391, 737–43 (2010).

Lipfert, J., and Doniach, S. Small-angle X-ray scattering from RNA, proteins, and protein complexes *Annu. Rev. Biophys. Biomol. Struct.* 36, 307–27 (2007).

Santos, N.C., and Castanho, M.A. Teaching light scattering spectroscopy: the dimension and shape of tobacco mosaic virus. *Biophys. J.* 71, 1641–50 (1996).

 ## Problems

4.1.1 What is the apparent molecular weight of a molecule in a light-scattering experiment when its true molecular weight is 40 kDa but there is a 0.1% contamination with particles of molecular mass 10^3 kDa? (see Box 4.1).

4.1.2 A certain DNA molecule can exist either in a single stranded circular form or in a rigid double-stranded form, 250 nm long. In an experiment with light irradiation with $\lambda = 400$ nm, calculate the ratio of the scattering produced by each of these forms when (i) $\theta = 10°$; and (ii) when $\theta = 30°$. (Assume R_G for a rod is $L/\sqrt{12}$, while for a circle it is equal to the radius.)

4.1.3 Tomato bushy stunt virus is an RNA–protein virus found to have a neutron scattering match point of 44.1% D_2O. Given that the match points of the RNA and the protein are 70% D_2O and 41% D_2O, respectively, and assuming that the H–D exchange properties of RNA and protein are the same, estimate the volume ratio of RNA to protein from the measured match point.

4.1.4 The figure shows a Guinier plot obtained in a SAXS experiment on freshly prepared bovine serum albumin (lower black line) and a sample after 8 hours incubation (dashed upper line). Explain the data.

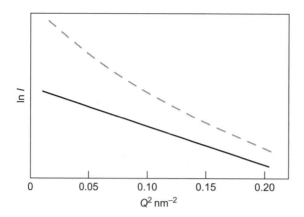

4.1.5 The figure shows the R_G values measured in SAXS experiments on a small protein as a function of (a) the concentration of the denaturant guanidinium hydrochloride and (b) the variation with time after the removal of denaturant. Discuss.

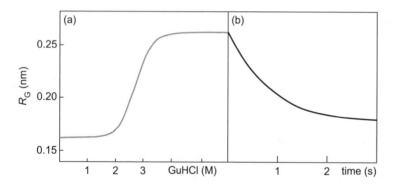

4.1.6 In an experiment on calf thymus DNA ($M \sim 15 \times 10^6$ Da) using DLS, a time constant of 10 μs was observed together with a spectrum of slower time constants in the range 0.5–18 ms. Can you interpret these data? (At room temperature, the rotational diffusion coefficient of a spherical molecule is given approximately by $10^{13}/(5M)$ s^{-1}.)

4.2 Refraction, evanescent waves, and plasmons

Introduction

When light goes from one transparent medium to another, for example from air to water, the light beam appears to deviate from a straight line (Figure 4.2.1), a phenomenon known as refraction. Ptolemy made a list of angles observed in going from air to water as early as 140 AD and Arabian scholars developed this approach but the first clear mathematical description of refraction angles was given by Willebrord Snell in 1621. This was followed, in the 1660s, by Isaac Newton's investigations of the refraction of light by a glass prism. He showed that white light could be broken up into colored rays (the rainbow), each ray being defined by a different deviation angle. These early observations and later developments have led to a very good understanding of the geometry of light rays optics. When light strikes a mirror, for example, it is *reflected* and the angle of incidence equals the angle of reflection ($\theta_1 = \theta_1'$ in Figure 4.2.1). While geometric optics (Box 4.4) is very powerful, non-linear effects have to be introduced to explain some observations; the generation of a localized evanescent wave when light undergoes total internal reflection is one of these. As will be described, evanescent waves turn out to be very useful in several biophysical techniques.

The refractive index (n) and Snell's law

The refractive index, n, defines the apparent change in velocity when light passes from one medium to another; $n = c_0/c_m$, where c_0 is the velocity in a vacuum and c_m is the apparent velocity in a medium. A consequence of the different velocities of light in media with different refractive indices (n_1 and n_2) is that light changes its angle of direction in the new medium (Figure 4.2.1 and Problem 4.2.1). Snell's law relates the angles of incidence (θ_1) and refraction (θ_2) via the formula $n_1\sin\theta_1 = n_2\sin\theta_2$.

When $\theta_2 \geq 90°$, all the light is reflected rather than refracted—this is called total internal reflection. The "critical angle", θ_c, occurs when $\theta_2 = 90°$, i.e. when $n_1\sin\theta_c = n_2\sin\pi/2 = n_2$ (see Box 4.4).

Scattering explains reflection and refraction

Classical optics can explain most of the observed behavior of light rays in terms of geometric rules, as illustrated in Box 4.4. These observed effects are, however, due to scattering events occurring at the level of atoms and molecules. When atoms in a medium are irradiated they scatter light in all directions, as discussed in Chapter 4.1. The net effect of scattering from a large number of atoms and interference between the different scattered waves is what gives rise to the phenomena observed in Figures A–D (Box 4.4).

When an electromagnetic wave interacts with a static array of molecular scatterers, e.g. in a slab of glass, the scattering from each molecule is related to its polarizability (α) and, as discussed in Chapter 4.1, the scattering will mostly cancel, except at certain

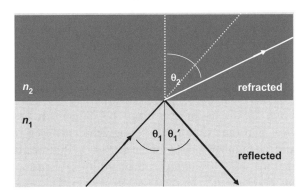

Figure 4.2.1 A representation of reflection and refraction as light passes from a medium with refractive index n_1 to one with a higher refractive index n_2. The angle of reflection is $\theta_1 = \theta_1'$ and the refraction angle is θ_2.

Box 4.4 Classical optics

A few reminders about classical optics are given in Figures A–D.

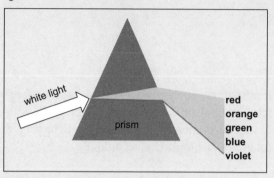

Figure A A prism disperses white light into its constituent colors.

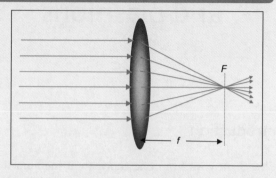

Figure B A convex lens, of focal length f, focuses parallel light at the principal focus F.

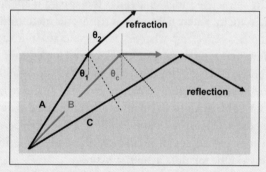

Figure C Depending on the position of the object with respect to the principal focus, a convex lens can produce a virtual, upright, magnified image. The dimensions are given by the "lens-maker's" formula $1/u + 1/v = 1/f$.

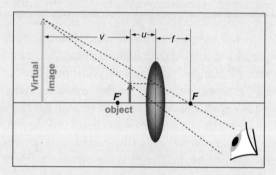

Figure D Glass, shown in gray, has a higher refractive index than air. Ray A is mainly refracted. The relationship $n_1\sin \theta_1 = n_2\sin \theta_2$ holds where n_1 is the refractive index of the glass and n_2 of air. The refraction of ray B is along the surface of the glass (θ_c is the critical angle). Ray C strikes the glass at an angle such that there is *total internal reflection*.

angles of reflection and refraction, and when $\theta = 0$ (see Figure 4.1.4). Consider the forward-scattered ($\theta = 0$) beam; it can be shown that an infinite array of scatterers produces a wave that is 90° out of phase with respect to the incident beam (see also Tutorial 7). This scattered wave combines with waves that are transmitted directly through the medium without interacting with it. The product is an emergent wave that is phase-shifted with respect to the incident beam (Figure 4.2.2). This phase shifted wave appears to have been retarded by passing through the slab, i.e. the slab changes the effective velocity of the light. (Note that the refractive index, n, used to describe optical phenomena, is related to the dielectric constant, ε, used to describe electrical phenomena (see Box 2.3) by the relationship $\varepsilon = n^2$.)

Refractive index depends on concentration

From the discussion above and in Chapter 4.1, it is clear that the extent of scattering will depend on α and the number and size of the scatterers. This means that the refractive index, n, depends linearly on solute concentration in dilute solutions. Changes in n change the phase of a beam passing through a sample (Figure 4.2.3). If the emerging beam is mixed with a reference beam, an interference pattern is obtained that reflects changes in n (see also Figure 3.2.4).

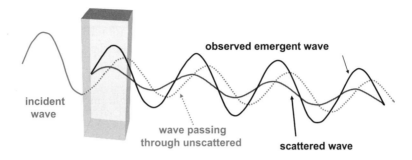

Figure 4.2.2 The combination of a scattered wave and a transmitted wave gives an emerging wave from a glass slab that appears to have had its velocity slowed down by the slab.

Optical devices designed to detect refractive index changes (Schlieren optics) are used routinely in analytical ultracentrifugation (Chapter 3.2; Figure 3.2.4) and in phase contrast microscopy (Chapter 7.1).

Dispersion

Dispersion describes the frequency dependence of the refractive index. As we sweep through a wide range of frequencies there will be several molecular processes that lead to absorption and dispersion

phenomena; these can be molecular rotations, vibrations, or electronic transitions (these will be discussed further in Section 5). Water, for example, does not absorb visible light significantly but it has vibrational resonances in the infrared and rotational resonances in the microwave spectrum. These molecular processes lead to both absorbance and refractive index variation of the sort shown in Figure 4.2.4. Insight into these frequency-dependent properties of n (and ε) can be obtained by considering the oscillating electrons of the scatterer to be analogous to a driven oscillator. As discussed in Tutorial 7, an oscillator produces absorption and dispersion phenomena, with a resonance condition when $\omega = \omega_0$, and a phase shift of 180° for the scattering below and above resonance. Vibrations and electronic excitation give typical Lorentzian absorption/dispersion curves

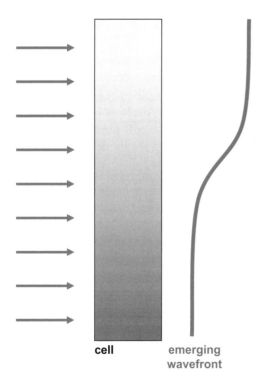

cell   **emerging wavefront**

Figure 4.2.3 Behavior of a plane wavefront passing through a solution with a concentration gradient that increases from top to bottom.

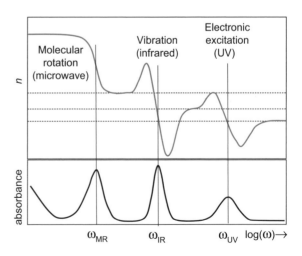

Figure 4.2.4 Three schematic absorbance regions for different molecular processes with the corresponding variation in refractive index, n (see Section 5 for a description of absorption).

(Tutorial 6; Figure 4.2.4). (Note that at lower frequencies other effects, including ion movement and heating, produce distorted curves in the microwave region.)

There is a general reduction in n with increasing applied frequency because some contributions, for example those arising from molecular reorientation, cannot respond fast enough to contribute to the polarizability and thus to the refractive index (or dielectric constant, ε).

In the case of visible light, the refractive index usually increases with decreasing λ because resonances are found in the UV region. Typical plots of the variation of refractive index with λ (and ω) are shown for various types of glass in Figure 4.2.5 (note the x-axes in Figures 4.2.4 and 4.2.5 run in opposite directions). This wavelength dependence of n is what causes the dispersion of white light by a prism (Box 4.4). The fact that different glasses have different dispersion properties is also important for the production of lenses with minimal chromatic aberrations (Chapter 7.1).

Dielectric spectroscopy

The refractive index (n) and dielectric constant (ε) are closely related ($\varepsilon = n^2$). Dielectric spectroscopy studies the way the dielectric constant (Box 2.3) of a medium varies as a function of frequency (in the range 10^2–10^{10} Hz). The steady-state value of ε and the rate at which the molecules respond to an applied electric field are related to the shape of the macromolecules, their dipolar character, and the different kinds of molecular process that occur at different frequencies (Figure 4.2.4). Dielectric spectroscopy can thus give insight into molecular motion of a system (see Problem 4.2.3).

Evanescent waves

Evanescent waves are formed when electromagnetic waves undergo total internal reflection at an interface. "Evanescent" means "tending to vanish" and the intensity of an evanescent wave decays exponentially with distance from the n_1/n_2 interface (Figure 4.2.6). Maxwell's equations, first described in 1862, can generally be used to explain the properties of electromagnetic radiation (Tutorial 8). With total internal reflection, there is an incident wave, a reflected wave, and no transmitted wave. Maxwell's equations require continuity at an interface, however, and this means that an additional, exponentially decaying wave has to be introduced to obtain a valid solution. This additional wave is the evanescent wave.

Evanescent waves have a decay rate, or "penetration depth", that is of the order of half a wavelength—about 100 nm for visible light (see Problem 4.2.2). The small penetration depth of the evanescent wave is valuable because the only physical effects induced are those that occur in the immediate vicinity of the interface. As discussed below, several powerful techniques use evanescent waves. These include ones

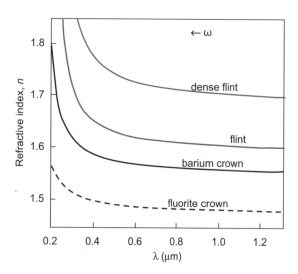

Figure 4.2.5 Typical refractive index dispersion curves for some glasses in the 200–1600 nm range.

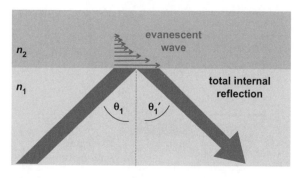

Figure 4.2.6 Total internal reflection in the n_1 medium produces an exponentially decaying evanescent wave in the n_2 medium.

with the abbreviations ATR, TIRF, and SPR, as will be described below. The first two of these are introduced very briefly here and will be discussed more in later chapters. SPR will be dealt with in more detail in this chapter.

Attenuated total reflectance (ATR)

As will be discussed in Chapter 5.2, infrared (IR) spectroscopy is a powerful tool for studying vibrations in molecules. A useful device in IR spectroscopy, one that exploits evanescent waves, is attenuated total reflectance infrared (ATR IR). A beam of IR light is passed through a crystal in such a way that it is reflected many times from the internal surface that is in contact with the sample. An evanescent wave is generated that extends into the sample, typically for a few microns at IR frequencies; the infrared beam is attenuated by the interactions between the evanescent wave and the sample near the surface (Figure 4.2.7). This method can be used with both liquid and solid samples. Examples of ATR IR will be given later (e.g. Figure 5.2.9). The method has a significant sensitivity advantage over conventional infrared detection systems.

Total internal reflectance fluorescence (TIRF)

Another powerful application of evanescent waves is found in the TIRF microscope, where fluorophores close to the coverslip surface are selectively excited. The big advantage of this excitation method is that it enables a visualization of surface regions only, such as a cell membrane (further discussion of this technique is given in Chapters 7.1 and 7.2; see also Problem 4.2.2).

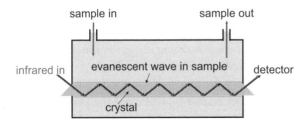

Figure 4.2.7 An ATR IR experiment. Multiple total internal reflections generate an evanescent wave that interacts with a sample at a surface.

Surface plasmon resonance

Introduction

Surface plasmon resonance (SPR) is a powerful tool for measuring molecular interactions between any pair of molecules. It does not need a labeled molecule and association and dissociation interactions can be measured in real time. SPR-based instruments are now in widespread use to define the kinetics and affinity of a wide variety of interactions between molecules. Initial developments by the Lundstrom group in Sweden 1983 led to the launch of the BIAcore SPR instrument in 1990. The utility and versatility of the method has led to rapid adoption in many laboratories and devices based on SPR methodology can now be bought from several companies.

A surface plasmon (SP) is an electron oscillation generated at a surface interface between a metal and a dielectric. Metals like gold or silver contain a large number of "free" electrons that can move and generate electromagnetic fields. Under the right conditions, an applied light wave can couple with these electrons and excite a wave in the metal. A plasmon resonance occurs when the E-wave in visible light (oscillating at $\sim 10^{15}$ Hz; see Tutorial 8) couples optimally with the oscillating electrons in the metal. Coupling of light photons and surface plasmons was demonstrated by Kretschmann and Otto in the late 1960s; they achieved it using a prism, total internal reflection, and a gold film (Figure 4.2.8). The photons and the plasmon wave also generate an evanescent wave, as shown in Figure 4.2.8.

There is a maximum loss, or reduction, in the reflected light intensity when plasmon resonance occurs. The resonance angle, θ_{spr}, is found by changing the angle of incidence of the light beam, giving a dip in a plot of intensity against angle. The plasmon resonance angle, θ_{spr}, is very sensitive to the refractive index of the medium near the metal surface because of interactions with the evanescent wave.

Experimental

SPR instruments can be configured in various ways, but most instruments use a Kretschmann configuration of the sort shown in Figures 4.2.8 and 4.2.9,

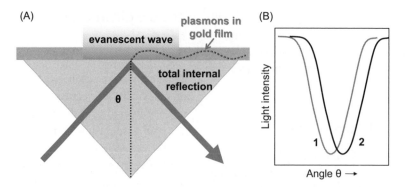

Figure 4.2.8 In (A), a prism is used to couple light, by total internal reflection, to surface plasmons in a gold film (dotted cyan line). This system can be tuned by changing the angle of incidence so that the plasmons resonate. (B) The position of the resonance dip changes (e.g. from 1 to 2) when changes in refractive index occur in the evanescent wave region.

where polarized light undergoes total internal reflectance at a surface covered with a thin gold film.

The sensor "chip" is housed in a flow cell that allows samples to be passed over the surface. One of the molecules (the ligand) is immobilized on the sensor surface, while the other (the analyte) is free in solution and is passed over the sensor. A variety of sophisticated flow cell systems are available; sample is injected through cells that can be opened and closed by a system of valves. Interactions may be analyzed in several flow cells at a time. Typical volumes of the flow cells are 20–60 nL, and a running buffer is passed

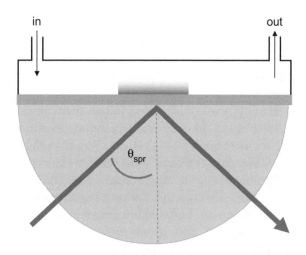

Figure 4.2.9 Illustration of SPR excitation and a flow system. A sensor chip with a gold coating is placed on a hemisphere of glass. SPR resonance occurs at θ_{spr} producing a dip in the intensity of the reflected light. Buffer and analyte can be passed over the sensor with a flow system and changes in θ_{spr} measured in real time.

through the cells at ~50 μL min^{-1}; very small amounts of sample (picograms) are required. Changes in dip position can be monitored rapidly, allowing real time measurements. The reflected light intensity is usually measured by a photodiode detector. The SPR signal, which is expressed in "response (or resonance) units", is a measure of mass changes (related to the refractive index) at the sensor chip surface. The signal is displayed in a graph called a "sensorgram" (see, for example, Figure 4.2.11).

Various prepared sensor surfaces (see below) are commercially available, to which a ligand can be attached. A closer view of the region of interest is shown in Figure 4.2.10 and a typical analysis cycle, with its sensorgram, is shown in Figure 4.2.11. The sensor surface, containing active ligands, is first exposed to a suitable buffer. Buffer, containing analyte, is then injected and if some analyte is captured by the ligand it changes the mass, and thus refractive index, near the surface. The accumulated mass of analyte can be obtained from the SPR response (ΔR) at this point, although non-specific binding may have also occurred at this stage. Buffer without analyte is then injected and non-specific binding components are washed off. Dissociation of the analyte is indicated by a decreasing response. Finally, a regeneration solution is injected to release any remaining analyte from the ligand. A new cycle can then begin with another analyte or a different analyte concentration.

SPR devices detect any mass change at the sensor surface, in the evanescent region, so they are **non-selective**. This means that any change in buffer

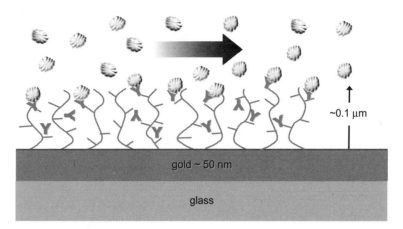

Figure 4.2.10 A close-up view of a region in Figure 4.2.9 where the evanescent wave and plasmon resonance are generated. Typically, a ligand, such as an antibody, here depicted by a Y, is covalently attached to a dextran matrix on the gold surface. Buffer containing the analyte (gray blob) is passed through a flow cell system. If analyte binds to Y then the mass (refractive index) will change near the surface and this can be detected in real time by a change in dip position in the sensorgram.

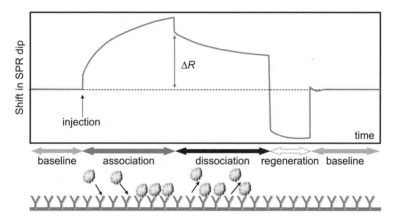

Figure 4.2.11 A typical "sensorgram" showing different stages of an analysis cycle (top). Initially, buffer is in contact with the sensor, giving a *baseline*; continuous injection of analyte solution then produces an *association* between ligand and analyte; injection of buffer alone then causes *dissociation*, with ΔR indicating the response due to the bound analyte. Any remaining bound analyte at the end of the dissociation period is removed by injection of a *regeneration* solution, followed by another cycle. Shown below is a schematic view of the association and dissociation stages. Note that shifts in θ_{spr} will be observed on injection of each of the different buffers. Corrections for these effects can be made by performing control experiments.

composition or concentration will produce detectable changes, as well as the effects of interest. Control experiments are thus very important and background responses must be subtracted from the sensorgram to obtain the actual binding response. It is also important to show that the immobilization procedure does not affect the interaction being studied. Specific experimental details of the matrix and coupling methods in use are described extensively in manufacturer's protocols and "further reading", but a few key experimental points are given below.

The surface chemistry

The gold surface can be modified, for example, by attaching a hydrated gel of carboxylated dextran. Such a surface provides a substrate for efficient covalent immobilization of ligands and a suitable environment for most interactions to occur. The thickness of the dextran layer (~100 nm) is similar to the depth of the evanescent wave. Other kinds of specialized surface (sensor chips) are commercially available, including ones designed to bind molecules with histidine-tags, and ones designed to provide a lipophilic environment. Immobilization to the surface can either be *direct*, by covalent coupling, or *indirect*, through capture by a covalently coupled molecule, as described below.

Direct coupling

In direct coupling, a ligand of interest is covalently attached to the sensor chip surface. There are three main types of chemistry used to couple a ligand to carboxymethyl groups on the chip, namely: amine (e.g. lysine groups on proteins), thiol (e.g. cysteine on proteins), or aldehyde (e.g. on carbohydrate). If the protein ligand to be immobilized has a free cysteine, ligand–thiol coupling is very convenient. Direct covalent immobilization can be used for any biomolecule, but as the attached molecule may have multiple copies of the functional group (e.g. lysines on the surface of a protein) the coupling can be heterogeneous. Direct coupling can also modify the ligand binding properties and displacement of the molecule from the chip (regeneration) can be difficult.

Indirect coupling

Here, the ligand is bound tightly to a **capturing molecule** that is covalently attached to the chip. A number of methods are available for ligand capture, including antibody fragments designed to bind the ligand and nickel-containing surfaces for proteins with histidine tags. (Recombinant proteins are often expressed with a specific binding tag attached to one end of the protein.) The **streptavidin/biotin interaction** is very useful for indirect coupling (Box 4.5). Although indirect coupling can only be used for molecules with a suitable binding site or tag, it has important advantages which usually make it the method of choice. The advantages include the fact that ligands are less likely to be inactivated; the ligand can be captured from an unpurified sample; all the molecules are immobilized in a consistent orientation; and regeneration of the chip surface is also relatively straightforward.

Concentration dependence and sensitivity

SPR detects refractive index changes, which are a measure of mass changes at the sensor surface. R_{max},

---

Box 4.5 The streptavidin/biotin complex

Streptavidin is a 52800-Da tetrameric protein found in the bacterium *Streptomyces avidinii*. It has an extraordinarily high affinity for biotin (also known as vitamin B7 and vitamin H). The dissociation constant (K_D) of the biotin–streptavidin complex is ~10^{-14} mol L^{-1}, making it one of the strongest non-covalent interactions known in nature. It is used extensively in molecular biology and bionanotechnology to link two molecules together or to attach a molecule to a chromatography support. The high affinity is believed to arise from shape complementarity between the binding pocket and biotin, and an extensive network of hydrogen bonds in the binding site. Harsh conditions may be needed to break the streptavidin–biotin interaction, but incubation above 70°C often breaks the interaction reversibly. Numerous variants of this system have been made for binding studies, including ones that are mono-valent rather than the natural tetra-valent state (see Figure A).

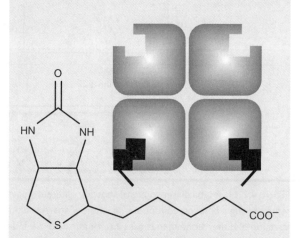

Figure A Representation of the streptavidin protein tetramer, occupied by two cyan biotin molecules. The chemical form of biotin is also shown.

---

in response units (RU), is the maximum binding capacity of a surface ligand for analyte. The formula $R_{max} = (M_{analyte}/M_{ligand}) \times R_L \times S$ can be used to relate R_{max} to the molecular weights (M) of the analyte and ligand; R_L is the immobilized ligand response (in RU) and S is the stoichiometry (the number of analyte binding sites on the ligand). For experimental purposes R_{max} should be around 100 RU. Note that the sensitivity of SPR instrument is higher for higher molecular weight analytes because the response (ΔR) is a measure of mass change (see Problem 4.2.4). A large number of small molecules may need to bind before the detection level is reached; small analytes are thus sometimes detected indirectly—for example by a competition assay where binding of a small molecule interferes with an interaction between larger molecules.

ΔR, which is a measure of the amount of bound analyte, can be translated into concentration by constructing a calibration curve. A typical concentration dependence experiment is shown in Figure 4.2.12. Concentration measurements and calibration curves are most reliable when an equilibrium plateau value is reached in the SPR measurements, as seen at higher concentrations in Figure 4.2.12 (50 and 100 nM).

Kinetics and thermodynamics

SPR experiments can provide information about both the kinetics and thermodynamics of interactions. In Chapter 2.1, we described how the reaction A + L $\rightleftharpoons$ AL is characterized by a dissociation constant $K_D = [A][L]/[AL]$, where [A], [L], and [AL] are the analyte, ligand, and complex con-

centrations, respectively. The association rate constant for this reaction is $k_1[A][L]$ and the dissociation rate constant is $k_{-1}[AL]$, thus $K_D = k_{-1}/k_1$. As association and dissociation rates can be observed directly (see Figures 4.2.11 and 4.2.12), K_D can be determined from their ratio. (Note that mass transfer limitations—an experimental situation in which the supply of analyte to the sensor chip surface is limiting—make it difficult to measure k_1 values faster than about $10^6 Ms^{-1}$.)

Even if kinetic measurements cannot be made easily, K_D can still be determined by analyzing the equilibrium plateau regions in plots of response units against time; the response is proportional to the amount of analyte bound to ligand. From Chapter 2.1, we know that the fraction of analyte bound ($\check{n}$) is $\check{n} = [A]/([A] + K_D)$. The concentration dependence of equilibrium responses can thus be used to determine K_D.

Integrins are large cell surface receptors that are essential for various functions, including cell migration. In a study of an integrin domain (the I domain), it was shown that changing specific amino acids could greatly enhance the affinity for another cell surface adhesion molecule, intercellular adhesion molecule-1 (ICAM-1). A combination of two activating mutations (F265S is a notation that means Phe 265 is changed to Ser) led to an increase of 200 000-fold in affinity. The affinity measurements were done with SPR, as shown in Table 4.2.1 and Figure 4.2.13.

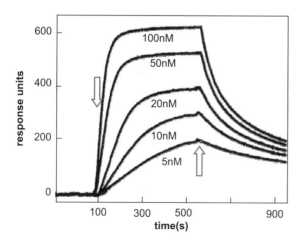

Figure 4.2.12 Typical concentration dependence of sensorgrams. The cyan arrows indicate the beginning of the association and dissociation stages in Figure 4.2.11.

Table 4.2.1 Kinetics and K_D values for I domains binding to ICAM-1

I domain	$10^{-3}k_{on}(M^{-1}s^{-1})$	$k_{off}(s^{-1})$	$K_D (\mu M)$
Wild type	3.1	4.6	1,500
F292A	9.5	0.19	20
F265S	11.8	0.0017	0.145
F265S/F292G	25.0	0.00015	0.006

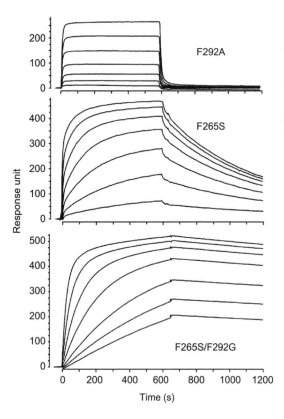

Figure 4.2.13 Binding of integrin I-domain mutants to ICAM-1 bound to a sensor chip. The signals obtained from a control experiment were subtracted from the sensorgrams shown. The I-domain mutants were injected in a series of twofold dilutions starting from 100 μM for F292A and 1 μM for F265S, and F265S/F292G (adapted with permission from *PNAS* 103, 5758–63 (2006)).

 ## Summary

1 Optical phenomena, including refraction and its wavelength dependence, can be explained by scattering from the molecules in the medium.

2 Refraction can be described by Snell's law ($n_1\sin\theta_1 = n_2\sin\theta_2$). Total internal reflectance occurs when the light ray is passing from a medium of higher to lower refractive index and the angle of incidence is greater than a "critical" angle.

3 Total internal reflectance leads to the generation of an evanescent wave, which can be used to monitor events near a surface. Infrared spectroscopy (ATR-FTIR) and microscopy (TIRF) exploit this wave for measurements.

4 Total internal reflectance at a metal surface generates surface plasmons, as well as an evanescent wave. This is used in the surface plasmon resonance (SPR) technique which is very powerful way of measuring molecular interactions in real time. SPR can measure on and off rates, as well as affinity.

 ## Further reading

Useful website

http://www.biacore.com/lifesciences/technology/index.html

Books

Feynman, R.P., Leighton, R.B., and Sands, M. *The Feynman Lectures on Physics*. Addison Wesley. Chapters 26 and 31 (Vol 1), 1963 (*a classic with an excellent description of refraction*).

Schasfoort, R.B.M., and Tudos, AJ. *Handbook of SPR*. Royal Society of Chemistry Publishing, 2008.

Reviews

Murphy, M., Jason-Moller, L., and Bruno, J.A. Using BIacore to measure the binding kinetics of an antibody–antigen interaction.*Curr. Protoc. Protein Sci.*19, 1–17 (2006).

Bonincontro, A., and Riseleo, G. Dielectric spectroscopy as a probe for the investigation of conformational properties of proteins. *Spectrochim. Acta A* 59, 2677–84 (2003).

 Problems

4.2.1 Snell's law can be derived in several ways, including the application of Fermat's principle of least time (of all possible paths, light will take the one that requires the shortest time). It can also be derived from a geometrical consideration of wave fronts travelling in the direction of the light beam and separated by the wavelength of the light. Derive Snell's law by considering the geometry of ABCD.

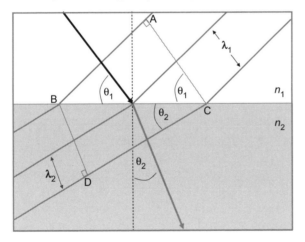

4.2.2 Calculate the critical angle θ_c when $n_1 = 1.52$, $n_2 = 1.33$. The intensity of an evanescent wave at a distance z from the surface can be represented approximately by $I_o\exp(-z/\delta)$ where δ, the penetration depth, is given by $\delta = \lambda_o/(4\pi n_2[\{\sin\theta_1/\sin\theta_c\}^2 - 1]^{1/2})$. What is the penetration depth for an incident angle, $\theta_1 = 65°$ and wavelength, $\lambda_o = 488$ nm?

4.2.3 The frequency dependence of the dielectric constant, ε, of protein solutions can be measured by exposing them in a capacitor to radiofrequencies. Plots of ε against frequency give dispersion curves (Figure 4.2.4) from which various parameters can be extracted, including relaxation frequencies associated with molecular tumbling. The hydrodynamic radius R_H (Box 4.2) and the electric dipole moment can be extracted from the "resonance" positions in the curves. Estimate R_H for lyzozyme at pH 5 and 9 if the observed relaxation frequencies at these values of pH were 8.9×10^6 Hz and 3.3×10^6 Hz. The relaxation frequency for a sphere is given by $\nu = kT/8\pi^2\eta R_H^3$ (see also Box 3.1). Assume that $\eta = 1cP = 10^{-3}$ kg m^{-1} s^{-1} and $T = 298$K. Comment on the results.

4.2.4 In SPR experiments, R_{max} (RU) is the maximum binding capacity of a surface ligand for analyte. $R_{max} = (M_{analyte}/M_{ligand}) \times R_L \times S$ relates R_{max} to the molecular weights (M) of the analyte and ligand; R_L is the immobilized ligand response (in RU), and S is the stoichiometry. For an $R_{max} \sim 100$ RU, how many pg mm^{-2} of ligand are required to obtain a suitable response for β2-microglobulin (11 kDa) binding to an antibody (150 kDa)? (Assume $S = 2$ and that 1 RU corresponds to the binding of approximately 1 pg protein mm^{-2}.)

4.2.5 In a screening program for inhibitors of HIV protease, the following kinetic constants were obtained for two potential drugs, acetyl pepstatin and indinavir, at two values of pH using SPR. Calculate the K_D values and comment of the results.

Inhibitor	pH	k_{on} (M^{-1} s^{-1})	k_{off} (s^{-1})
Acetyl pepstatin	7.4	2.90×10^4	2.11×10^{-1}
	5.1	8.20×10^4	6.21×10^{-1}
Indinavir	7.4	1.43×10^6	1.70×10^{-3}
	5.1	2.64×10^6	3.80×10^{-3}

4.3 Diffraction

Introduction

The scattering of electromagnetic radiation by molecules in solution was discussed in Chapter 4.1. As we will see in this chapter, scattering of short wavelength radiation by an *ordered* array of scatterers (e.g. crystals or fibers) can yield much more detailed information about molecules. X-rays, neutrons, and electrons have appropriate short wavelengths (<1nm) and can be used to derive atomic level resolution structural information. The most widely used diffraction method involves X-rays. In microscopy (Chapter 7.1), an image is formed by lenses, which focus the scattered light rays into a coherent, visible image. It is not yet possible to construct a suitable X-ray lens, so the image-formation stage has to be carried out by computer reconstruction procedures. In outline, therefore, most diffraction experiments can be considered to follow the general scheme shown in Figure 4.3.1.

Principles of diffraction

Diffraction arises from elastic scattering of radiation in directions other than the incident direction.

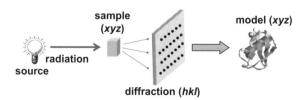

Figure 4.3.1 Schematic view of a diffraction experiment where the ordered array (crystal) has coordinates *xyz* while the diffraction pattern has coordinates *hkl*. X-rays, neutrons, and electrons are suitable radiation sources. The magnified image (model) has to be constructed from information derived from the diffraction pattern.

Diffracted waves can interact with each other and produce interference patterns which are most informative when the wavelength of the applied radiation is less than or equal to the dimensions of the diffracting object. In principle, the diffraction pattern from a single molecule contains the information required to generate the model in Figure 4.3.1 but scattering from large numbers of molecules in an ordered array is usually still required for practical reasons, such as adequate signal to noise ratio (but see Box 4.6).

A **crystal lattice** is a regularly repeated arrangement of molecules. The **unit cell** of a lattice is the smallest and simplest unit from which the three-dimensional (3D) periodic pattern can be produced by translation. The unit cell may be made up from several identical **asymmetric units** related by rotations. These ideas are illustrated for a simple cubic lattice in Figure 4.3.2. An ideal crystal is made from unit cells placed at every point on the lattice.

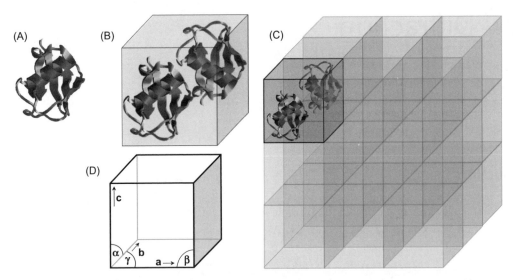

Figure 4.3.2 Illustration of (A) an asymmetric unit; (B) a unit cell, containing two symmetry related asymmetric units; (C) a lattice, shown here as a 3D array of cubes; (D) the unit cell has sides defined by vectors **a**, **b**, and **c** and angles α, β, and γ.

Symmetry relationships

In crystallography, most types of symmetry can be described in terms of an apparent *movement* of the object, such as a rotation—the movement is called a symmetry operation.

Symmetry operations include rotation, mirror reflection, inversion, and translation. Biological molecules are chiral, so possible symmetry operations are limited, for example rotations are allowed, but mirror planes are not. The complete collection of symmetry operations that apply to a particular crystal is known as its space group; there are 230 distinct space groups (65 with no mirror planes). They are described by letters and numbers, e.g. *P2* which has two-fold symmetry. A complete listing and description of space groups can be found, for example, at **http://xrayweb.chem.ou.edu/notes/symmetry.html**. Symmetry in the crystal lattice leads to *symmetry in the diffraction pattern*.

The object and the diffraction pattern have a reciprocal (Fourier transform) relationship

A crystal is a 3D array of atoms in real space with coordinate system *xyz*. Its diffraction pattern is a 3D array in reciprocal space with coordinate system *hkl* (Figure 4.3.1). Real space and reciprocal space are related by a Fourier transformation (FT) (Fourier transforms are explored in Tutorial 6).

The FT is a powerful tool that allows us to interconvert between *xyz* and *hkl* space. Examples are given in Figure 4.3.4. Note that scattering from objects close together results in a widely spaced diffraction pattern.

More examples of FTs of a single dot and rows of dots are shown in Figure 4.3.5; here they are shown in 1D to illustrate the reciprocal space intensities more clearly. A striking feature of Figure 4.3.5 is that the FT of the row of large dots can be reproduced by multiplying together the FT of the single large dot and the row of small dots; this is a practical illustration of the convolution theorem (Tutorial 6) which states that the convolution of two functions in real space is the multiplication of these two functions in reciprocal space (the FT domain). In Figure 4.3.5, the molecular information (dot size) is encoded in the intensity variation of the right hand diffraction

Figure 4.3.3 Illustration of two objects with two- and threefold symmetry, as depicted by an oval and a triangle, respectively.

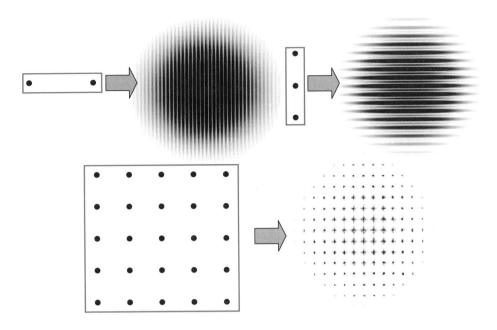

Figure 4.3.4 Examples of two-dimensional (2D) diffraction patterns. FT (filled gray arrows) of the dot patterns in the cyan rectangles (*xy*) produces the diffraction patterns in *hk* space. Note the reciprocal relationship between the dots and the separation and direction of the lines generated in the top two examples. In the bottom example, the grid in *xy* space is reproduced in *hk* reciprocal space.

pattern, while the line positions in the diffraction pattern arise from the lattice dimensions. This clearly demonstrates that the spot positions in diffraction patterns only tell us about the lattice properties and unit cells; the molecular information that we seek is defined by the *variation in the spot intensity* and **not** the *spot position*.

The Bragg equation and crystal planes

So far, we have discussed diffraction patterns in terms of Fourier transforms and the convolution theorem. As we saw in Chapter 4.1, scattering from irradiated atoms will occur at all angles but, in a crystal, only some angles give constructive interference and finite intensity; such positive diffraction spots

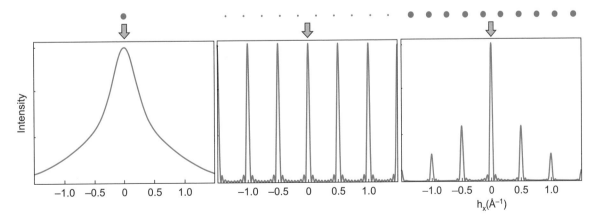

Figure 4.3.5 The cyan dots in *x* space are transformed to *h* space with a FT (gray arrows). On the left is a single large dot that gives a diffraction pattern that falls off with distance from the center. The row of very small dots gives a series of narrow lines (compare with top left in Figure 4.3.4). The row of large dots on the right gives a series of lines again but they fall off in intensity with distance from the center in the same way as the FT on the left.

are often called reflections. Can we predict the pattern of reflections in Figure 4.3.1 from knowledge of the crystal? At the beginning of the 20th century, three different ways of achieving this were proposed by Max von Laue (1912), Lawrence Bragg (1913), and Paul Ewald (1921). All of these constructions are useful but here we will only consider the Bragg method, which considers the diffracted beams as if they are reflected from planes passing through parallel points in a crystal lattice. It can be seen in Figure 4.3.6 that if X-rays, incident at angle θ, are scattered from adjacent crystal layers, separated by d, there is a path-length difference between scattering from adjacent layers that is $2p = 2d\sin\theta$. To get constructive interference between these two scattered waves, this path difference has to be an integral number of wavelengths ($n\lambda$). This leads to the famous Bragg equation $2\,d\sin\theta = n\lambda$.

There are many possible lattice planes in a lattice. Figure 4.3.7 shows a section though a crystal with some of the many possible lattice lines drawn on it. One line of each pair is drawn through the origin of a unit cell with axes **a** and **b** (written in bold because they are vectors). A set of lines crossing the **a** axis n times and the **b** axis m times, per unit cell, is given the label (n,m). For example, lines parallel to **b** are labeled (1,1); a (2,1) set of lines crosses **a** twice (at **a**/2 and **a** and **b** at 1).

Note the higher the indices, the closer together the lattice lines (smaller value of d). The lattice lines always intersect the unit cell at exact fractions of

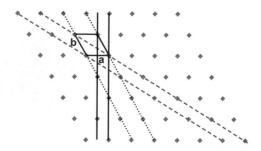

Figure 4.3.7 A section through a crystal showing the **a** and **b** axes of a unit cell. Lattice lines defined by (1,0) (dashed), (1,1) (dotted), and (2,1) (continuous) are drawn.

a and **b**. The situation in Figure 4.3.7 can be extended to three dimensions, as shown in Figure 4.3.8.

Any of the pairs of lines in Figure 4.3.7 or the planes in Figure 4.3.8 could act as the reflecting pair drawn in Figure 4.3.6. We can thus generalize the Bragg equation to $2d_{hkl}\sin\theta = n\lambda$. This formulation emphasizes the dependence of θ on d_{hkl} and shows that the reflections arising from any two lattice planes are related to the reflection coordinates (hkl), which are often called Miller indices.

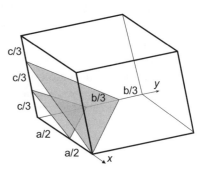

Figure 4.3.8 3D construction similar to the 2D version shown in Figure 4.3.7. This example shows two 2,3,3 planes.

Diffraction experiments

The sample

The usual samples in a diffraction experiment are fibers or crystals. Fibers often occur naturally, but some molecules can be oriented by applying shear or magnetic fields. Crystal growth is traditionally one of the most difficult and tedious stages in diffraction studies but robotic methods that can handle

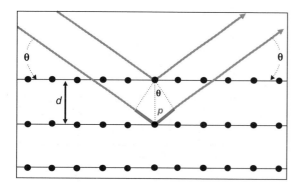

Figure 4.3.6 A beam of X-rays falling on two lattice planes separated by d. The angle of incidence and reflection is θ. The path length difference ($2p$) between the two beams is shown in cyan.

small sample volumes are greatly decreasing the tedium and increasing the chances of success. Before attempting crystallization, it is important that the macromolecule being studied is pure, homogeneous, and reasonably soluble (non-aggregated). Crystallization is then attempted by bringing the macromolecule up to, and through, the point of supersaturation (a situation where the solution contains more dissolved material than normally possible) in a controlled fashion. Two stages can often be identified in crystal formation: (i) nucleation where microcrystals form; and (ii) growth of a larger crystal around a nuclear seed. Figure 4.3.9 shows two ways of growing crystals. Small volumes (~200 nL) of protein in solution are allowed to equilibrate slowly with a reservoir containing various precipitants. The conditions are changed over a wide range by varying the salt concentration and environmental conditions (e.g. pH, temperature, ligands, or cofactors) of the protein solution, as well as the various precipitants in the reservoir (e.g. ammonium sulfate, polyethylene glycol, and organic solvents). Crystals must have dimensions of about 20 μm in two directions for diffraction studies, although increasingly small crystals are viable because of improvements in the collimation (precise parallel alignment) and intensities of available X-ray beams. Not all crystals give good diffraction data; even when diffraction is obtained, interpretation can be difficult because of artifacts such as crystal twinning, where unit cells in a crystal pack equally favorably in several orientations.

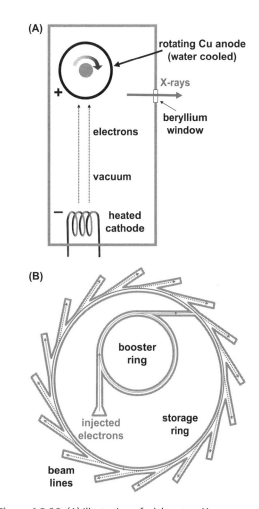

Figure 4.3.10 (A) Illustration of a laboratory X-ray generator; 0.154 nm wavelength X-rays are produced by bombarding a rotating copper anode with high-energy electrons.
(B) Illustration of a synchrotron with electrons (cyan) first accelerated, then contained in a storage ring. When electrons change direction, they emit radiation that enters "beamlines" at which experiments can be conducted.

Figure 4.3.9 Two methods of producing crystals. On the left is a "hanging drop" while a "sitting drop" is shown on the right. The protein solution is in cyan and the reservoirs in gray.

Sources of radiation

The radiation used for diffraction studies must have a wavelength that is of the same order of magnitude as the crystal lattice: X-rays, neutrons, and electrons come into this category (Tutorial 8).

X-rays can be produced in the laboratory or by synchrotrons. A common method of producing X-rays in the laboratory is to strike a target of a metal with electrons accelerated by a high voltage (Figure 4.3.10). A wavelength of 0.154 nm (1.54 Å) can be produced by a rotating copper target. Beryllium is used as a radiation window because it has a low absorption for X-rays due to its low atomic number (Table 4.1.1).

Synchrotrons provide increasingly important sources for X-ray crystallography. In these large devices, several hundred meters in diameter, high energy electrons travel in a circular orbit in a vacuum; the electrons are contained/stored in that orbit by the application of magnetic and electric fields. The electrons are injected into the storage ring, first using linear acceleration followed by a booster ring. The orbiting electrons emit intense synchrotron radiation in a tangential direction to the beam at points of curvature (where the electrons are accelerated). Synchrotrons produce intense radiation, covering a wide range of wavelengths, from about 1 μm to 0.1 nm. A crystal monochromator is used to select a wavelength appropriate for the experiment being conducted (see also Figure 5.6.4). The beams produced are also highly collimated (parallel).

Electron diffraction is usually carried out in an electron microscope, as discussed further in Chapter 7.1. For an electron microscope operating at 200 keV, the electron wavelength is around 2.5 pm (see Problem 7.1.1).

Neutrons can be produced by pulses of high-energy protons directed against a metal target. A thermal neutron produced in this way has a wavelength of around 0.2 nm. Neutron sources produce a relatively low intensity compared to the intensities of modern X-ray sources; large crystals are thus required and long measurement times are necessary to obtain good diffraction data. The number of experimental neutron sources dedicated to investigations of bio-macromolecules is also relatively limited. Neutrons do, however, have the significant advantage of detecting H-atoms much better than X-rays (see below).

Data collection

A typical X-ray set up, whether in the laboratory or at a synchrotron beamline, is shown in Figure 4.3.11. The crystal is mounted in the X-ray beam and rotated by a device called a goniometer. Rotation of the crystal about several angles is required in order to detect all possible reflections. The crystal is often scooped up in a small loop, sometimes made of nylon, and mounted in the goniometer. In the beam, the sample is kept cold with a stream of N_2 gas near 100 K. Low temperatures reduce the X-ray-induced damage to the crystals by immobilizing radicals produced by the ionizing radiation.

Detectors

X-ray diffraction experiments in structural biology involve measurement of thousands of reflections

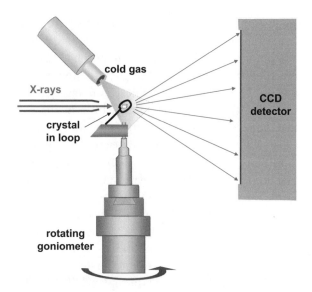

Figure 4.3.11 A typical experimental setup for crystallography with a collimated beam of X-rays, a cooled crystal in a loop, a rotating goniometer, and a detector.

(diffraction spots). Area detectors which collect an array of diffraction information at the same time are therefore used. There are two main types of area detector in current use: image plates (IP) and charge coupled devices (CCD). IP detectors employ a film-like image sensor comprised of phosphors that trap and store X-ray radiation energy stably until it is read out with a laser beam. This readout process is relatively slow but IP devices are versatile and are often used in laboratories. CCD detectors give faster read out times and are ideal for synchrotron use where higher throughput is required. A large phosphor screen (up to a thousand mm²) is again used to convert the incident X-rays into optical photons but these are now transmitted directly to the CCD with tapered optical fibers. In CCD detectors, light causes a build-up of charge in arrays of tiny photoelectric diodes (pixels). CCD detectors can also be used for electron crystallography (although photographic film is still preferred for high resolution electron microscopy images).

Neutron detectors are relatively inefficient and neutrons cannot be detected directly—they need to be first converted to charged particles (e.g. by the reaction: neutron + ³He → ¹H + ³H).

Note that all these detectors only measure the intensity of the reflections. As a wave is defined by both phase and amplitude, there is a loss of information that leads to the phase problem, as discussed below.

Interpretation of diffraction data

The aim is to interpret the diffraction patterns observed in terms of molecular structure. This is equivalent to the reconstruction procedures carried out by a lens in a microscope (Figure 4.1.1), something that cannot be done with most X-rays or neutrons. Before outlining the relatively complex reconstruction procedure, we briefly discuss some information that can be obtained directly from the diffraction pattern with a minimum of computation.

Direct interpretation

The unit cell dimensions can be obtained from the pattern of reflections observed (see section on Bragg equation and the 2D transforms in Figure 4.3.4). The unit cell dimensions are directly related to the inverse of the separation between the spots that are closest together in the diffraction pattern; this information can be used to obtain limits on the size and molecular weight of the diffracting molecule (see Problem 4.3.1).

Symmetry

Macromolecules often display symmetry. In hemoglobin, for example, two pairs of αβ subunits are related by a two-fold axis of rotation. Symmetry in the crystal is reflected in the diffraction pattern and information about the molecule and the asymmetric units can be deduced from that symmetry.

Diffraction patterns from fibers

Many biological molecules are long and occur naturally as fibers, or can be induced to line up in one direction. With such an oriented array, it is possible to obtain an informative diffraction pattern. The fiber is usually mounted perpendicular to the axis of the irradiating beam. Because the molecules in fibers are oriented randomly about the alignment direction, only regular repeating patterns that occur in the alignment direction appear in the diffraction pattern. The FT of a continuous helix with pitch p is

shown in Figure 4.3.12 (top). (See also the figure in Problem 4.3.11 for a cartoon of the pitch in a DNA helix.) Note the characteristic cross pattern and the separation of lines in the diffraction pattern, proportional to $1/p$. A better approximation to a real biological helix, such as DNA, is a discontinuous helix with a certain number of points (bases) per turn. The FT of a discontinuous helix is also shown in Figure 4.3.12 (lower panel). As expected from the convolution theorem, this is the product of the FT of the continuous helix and a set of planes made by the discontinuities. The planes generate new origins for the cross—if the discontinuities in the helix are a distance h apart then these cross origins will be $1/h$ apart.

Rosalind Franklin observed cross patterns of the sort shown in Figure 4.3.12 from moist DNA. The separation between the diffraction layer lines ($1/p$) corresponded to a helix pitch of 3.4 nm and the position of the repeated cross pattern suggested that there were around 10 bases per turn. Interpretation of this pattern by James Watson and Francis Crick led to their famous 1953 paper on the structure of DNA (see Problem 4.3.10).

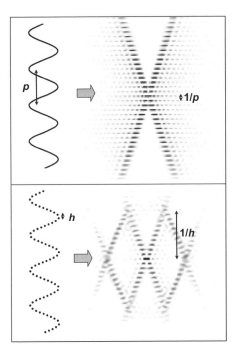

Figure 4.3.12 The FTs of a continuous helix (top) and a discontinuous helix (bottom). The pitch of the helix is p.

Determination of molecular structure

We now discuss in more detail the way in which molecular structure can be reconstructed from diffraction patterns (Figure 4.3.1) (look first at Tutorials 5 and 6).

Fourier synthesis, Fourier transformations, and the Patterson function

Let us first define some mathematical procedures useful in the reconstruction process. The crystal and the diffraction patterns are periodic systems and any such system can be simulated by the superposition of a set of sinusoidal waves (Tutorial 6). This operation is known as Fourier synthesis. Note that the representation of the square function by a summation of waves, as shown in Tutorial 6, improves with the addition of higher frequency components. In fact, a very good representation of the square function is generated if the series of waves is extended to about 60 terms. Another thing to note about wave summation is the importance of the phase of the various components.

The structure of a molecule can be synthesized by summing waves, but how do we determine the frequency, amplitude, and phase of the waves to be summed? The information can be derived from the diffraction pattern, and a convenient method of doing this is by Fourier transformation (FT). Summations, rather than integration, are appropriate for discrete diffraction reflections so that the equation that relates the real space domain $\rho(xyz)$ to the diffraction pattern $F(hkl)$ in the reciprocal space domain is $\rho(xyz) = K\Sigma_h\Sigma_k\Sigma_l F(hkl) \exp[-2\pi i(hx + ky + lz)]$, where K is a scaling factor that depends on experimental variables, such as the volume of the unit cell. $F(hkl)$ are known as structure factors; they are functions that describe both the amplitude and phase of a wave diffracted from crystal lattice planes characterized by the Miller indices hkl. We can write this equation in the shorthand form: $\rho(xyz) \Leftrightarrow F(hkl)$. The double-headed arrow reminds us that Fourier transforms have an inverse property that allows us to go in both directions.

There is a major problem when we come to perform the calculation $F(hkl) \Rightarrow \rho(xyz)$. The measurable parameter in a diffraction pattern is $I(hkl)$, the intensity of the reflection at position (hkl) not $F(hkl)$.

We can determine the magnitude of the structure factor $F(hkl)$ (proportional to $\{I(hkl)\}^{1/2}$), but the phase is unknown. How this phase problem can be solved will be discussed below.

The Patterson function, $P(uvw)$, is defined as the FT of $I(hkl)$ $\{I(hkl) \Rightarrow P(uvw)\}$. As $I(hkl)$ is a measurable quantity, $P(uvw)$ can be obtained directly. $P(uvw)$ is very useful but its interpretation is not straightforward. $P(uvw)$ can be visualized by placing, in turn, each atom of a molecule, on the origin (Figure 4.3.13). For large molecules, $P(uvw)$ gives very complex patterns that cannot be readily interpreted. Note, however, that "direct" methods, involving solution by the Patterson function, can be applied to small molecules and it can be used to locate heavy metal atoms (see below).

The phase problem and its solution

Phase information is lost because the experimental parameter in diffraction patterns is $I(hkl)$ rather than $F(hkl)$. In other words, **F(hkl)** is a vector with known amplitude but unknown direction. A convenient representation of a wave with a phase angle ϕ and amplitude a_o is $a_o\exp(i\phi)$. Thus, we can rewrite **F(hkl)** as $|F(hkl)|\exp(i\phi_{hkl})$. Another useful representation of a wave is as a vector in a complex plane with an amplitude $|a|$ and phase angle ϕ (Tutorial 5). As ϕ is unknown, **F(hkl)** will lie somewhere on a circle, as shown in Figure 4.3.14.

An advantage of this vector representation of a wave is that the mathematical problem of adding waves turns into the simple geometrical process of placing vectors nose to tail (Tutorial 5).

We now have the tools to discuss ways to solve the phase problem. The first method we will consider is multiple isomorphous replacement (MIR).

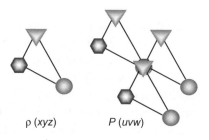

$\rho\ (xyz)$ $\qquad$ $P\ (uvw)$

Figure 4.3.13 Illustration of the Patterson function observed for a triangular molecule of three atoms. Each atom is placed, in turn, on the origin (when $u = v = w = 0$).

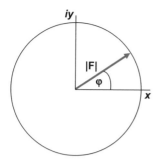

Figure 4.3.14 Representation of a structure factor, **F**, as an amplitude |F| and an unknown phase angle, φ.

This was the method used by Max Perutz and John Kendrew to solve the first protein structures—myoglobin and hemoglobin. The MIR method depends on attaching a strong scatterer—a metal ion, or cluster of atoms—to one or more specific sites on a crystal. It is important that the crystal should otherwise retain the same (*iso*) form (*morph*), i.e. the same molecular position and unit cell after the metal has bound.

For X-rays, derivatives of mercury, uranium, platinum, gold, and lead have been among the many "heavy atom" reagents used (remember that X-ray scattering is proportional to the atomic number of the scattering atom; see Chapter 4.1). Isomorphous crystals can be made by diffusing the heavy atom reagent into a preformed crystal or by crystallizing the protein in the presence of the reagent. The effect of suitable isomorphous replacement on the diffraction pattern is to change the intensities of the reflections from $I_P(hkl)$ to $I_{PH}(hkl)$ where P refers to protein alone and PH to protein plus heavy atom. The changes in intensity can be measured and a **Patterson difference map** can be obtained from the Fourier transform of $I_P(hkl) - I_{PH}(hkl)$ (Figure 4.3.15). This difference map can often be used to identify the positions of the heavy atoms, as it is relatively simple compared with a normal Patterson map, as long as the scattering from the metal introduced is strong enough. If the Patterson difference map can be interpreted, we then know both the *phase* and *amplitude* of $F_H(hkl)$.

If we denote the structure factor of the original crystal as F_P and that with the heavy atom as F_{PH}, we can perform what is known as a **Harker construction**. We have just seen that F_H can be obtained

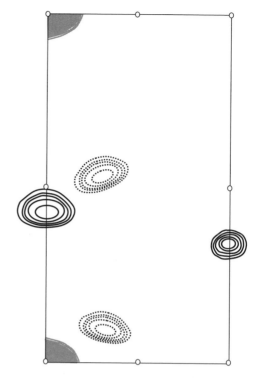

Figure 4.3.15 A Patterson difference map of a mercury derivative of hemoglobin. The dotted lines indicate negative density in the difference map.

from a Patterson difference map. F_H can thus be drawn as a defined vector in the diagram. We can also represent F_P as a circle of radius $|F_P|$ taking the end of F_H as the origin. F_{PH} can be represented by another circle of radius $|F_{PH}|$ with its center at the beginning of F_H. This construction takes account of the vector summation of waves, with $F_{PH} = F_P + F_H$ (Tutorial 5).

It can be seen from Figure 4.3.16 that there are two possible solutions—at A and B. To remove the ambiguity between A and B solutions, it is necessary to make at least one more, different, isomorphous replacement in order to obtain a "unique" (within experimental error) solution for the F_P phase angle—hence the description *multiple isomorphous replacement* (MIR).

Other solutions to the phase problem

MIR is the most general method for solving the phases of a structure when there is no other available information, but there are often severe practical difficulties in obtaining enough suitable derivatives.

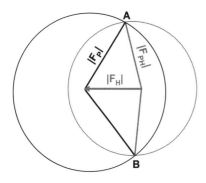

Figure 4.3.16 A Harker diagram, showing two possible solutions for the vector sum $\mathbf{F_{PH}} = \mathbf{F_P} + \mathbf{F_H}$.

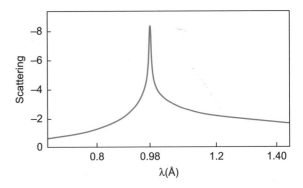

Figure 4.3.17 The scattering factor of selenium as a function of wavelength.

Sometimes it is possible to solve the phase with only one good heavy atom derivative (single isomorphous replacement) but additional information is then required. One way of gaining this extra information is to assume that regions where there is a large amount of mobile solvent (crystals contain 30–70% solvent) do not contribute to the electron density; this is known as solvent flattening. Sometimes extra information can also be gained from symmetry. In some viruses, for example, there are many similar subunits and averaging can be used to improve the phases.

The database of known structures continues to grow rapidly; this means that a similar structure to the one of interest can often be found in the protein databank (PDB; www.rcsb.org). Powerful ways of detecting similarities at the sequence level are available using database searches (Chapter 8.1). Molecular replacement is a method that can be used to find the phase when there is a known structure of a similar protein; one way to do this is to rotate and translate the known structure, comparing calculated Patterson maps of the known with that of the unknown until a solution is found.

Another powerful phasing method is to use anomalous scattering. We have seen that the phase of the scattering depends on the wavelength of the radiation and the "resonance" position of the scatterer (Chapter 4.1 and Tutorial 7).

Selenium resonates when $\lambda \sim 0.98\text{Å}$ (Figure 4.3.17), a wavelength that is readily available from synchrotron X-ray sources. Selenium can also be incorporated into proteins by including seleno-methionine in the growth medium of a bacterial expression system. With selenium (or sulfur—see Problem 4.3.9) in the sample, the phases can then be found using technique which is called multiple wavelength anomalous dispersion (MAD).

Consider the scattering from a pair of atoms, as shown in Figure 4.3.18. If the two atoms scatter with the same phase, then the path difference will be the same for the reflection at (hkl) as it is for $(-h-k-l)$, which is usually written as $(\bar{h}\bar{k}\bar{l})$; this is known as Friedel's law. If one of the atoms (e.g. the filled circle) is an anomalous scatterer (e.g. selenium scattering near resonance), the path difference for the (hkl) reflection will *not* be the same as for the $(\bar{h}\bar{k}\bar{l})$ reflection because the phase changes will interfere with the expected Bragg reflections from adjacent planes predicted by Figure 4.3.6. Difference maps calculated from $\Delta\mathbf{F} = |F(hkl)| - |F(\bar{h}\,\bar{k}\,\bar{l})|$ can then be used to locate the anomalous scatters in the unit cell

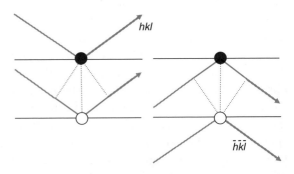

Figure 4.3.18 The path differences for scattering from two atoms (black and white) in a crystal are shown in cyan. Because the black atom scatters anomalously compared with the white, the *hkl* and $\bar{h}\bar{k}\bar{l}$ path differences will not be equal.

and thus calculate phases in a manner similar to that described in Figure 4.3.16.

Resolution

With estimates of phases, we can make a first attempt to calculate $\rho(xyz)$ from $\mathbf{F(hkl)}$. This will give some sort of electron density map. The quality of that map will, however, depend on the quality of the diffraction pattern and the phases. In Tutorial 6, we saw that good definition of a square wave requires high frequency components. In the same way, high frequency components are required for good resolution in diffraction experiments. High-frequency components in reciprocal space correspond to reflections at some distance from the center of the diffraction pattern. This is illustrated by the FT of the photograph of Max Perutz in Figure 4.3.19.

The better the resolution of the electron density map, the better the fit will be to any constructed model. With a map constructed from spacing corresponding to 5 Å (a 5 Å map), one can only see the overall shape of the molecule. At this resolution, strong features, such as α-helices, are observed as rods; at 4 Å it is possible to trace β-sheets; at 3 Å it is possible to discern the correct chain fold and see where many of the sidechains are, especially the more obvious ones such as those with an aromatic sidechain. At 1 Å resolution (rarely possible for a protein), subtle differences in conformation, and even some information about H atoms, can be obtained. Because of the importance of resolution in fitting the electron density, diffraction papers always quote the resolution of the experiment, e.g. "data collected to 2.5 Å resolution".

Determination of molecular structure from $\rho(xyz)$

The electron density obtained from the FT of $\mathbf{F(hkl)}$ is presented as a contour map of electron density;

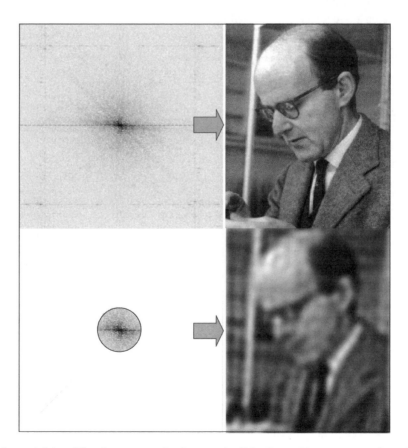

Figure 4.3.19 On the top left is a diffraction pattern of a photograph of Max Perutz. The reverse transform gives back quite a well resolved picture. If much of the diffraction pattern except the center is masked out, as shown below, the FT now gives a blurred picture.

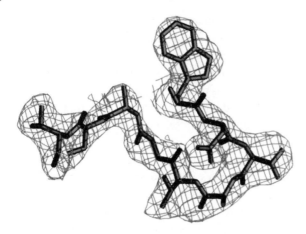

Figure 4.3.20 An electron density map, with some parts of a protein modeled into the density. This is a relatively high-resolution map with around 2 Å resolution. At lower resolution, the fits may not be so clear.

an example is given in Figure 4.3.20. Using a known sequence of amino acids (or nucleic acids), and assuming standard bond lengths and geometries, a model is built to match the density map as well as possible. This stage is done using computer visualization programs (Chapter 8.1).

R factors and refinement

Once a structure has been built into the density map, the expected diffraction pattern can be calculated from this structure using the FT $\rho_{model}(xyz) \Leftrightarrow F_{calc}(hkl)$. An estimate of how good the agreement is between the observed data (F_{obs}) and the calculated data (F_{calc}) is then given by $R = \dfrac{\sum(|F_{obs}| - |F_{calc}|)}{\sum|F_{obs}|}$, where the sum extends over all reflections. R is sometimes called the reliability value. An R value of around 0.2 is typically obtained for 2.5 Å resolution data—the lower the resolution the worse the R value. Errors in the structure can arise in a variety of ways—both in measurement of $|F_o|$ and in the phase angles. The errors in $|F_P|$ are usually around 5–10% while the errors in phase are higher.

It is possible, and important, to improve a calculated structure by minimizing differences between the calculated and observed diffraction patterns. This iterative process is called refinement. At each cycle, adjustments are made to the phase of the F factors and the coordinates of the model, subject to energy restraints, using molecular dynamics simulations (Chapter 8.1) or maximum likelihood—a method

where $\rho_{model}(xyz)$ is modified to optimize the statistical probability of generating $F_{calc}(hkl)$. Structures also often report an R_{free} value, which is obtained using a proportion of the diffraction data (5–10%) that was set aside and not used during structural refinement. R_{free} is generally 5–10% higher than R.

Temperature factors

We have seen that the X-ray scattering for individual atoms is proportional to the atomic number and that this falls off with the angular deviation of the diffracted or scattered beam from the incident beam (usually plotted as Q or $\sin(\theta/2)/\lambda$; see Chapter 4.1). Such a relationship is shown for carbon and nitrogen atoms in Figure 4.3.21.

If the scattering atom undergoes thermal vibration, then it is effectively bigger than a static atom and the curve will fall off faster with increasing θ. Curves of the kind shown in Figure 4.3.21 can be approximated by the relationship $F\exp(-2B\sin^2(\theta/2)/\lambda^2)$, where B is a measure of disorder, called the temperature factor. $B = 8\pi^2\hat{U}^2$, where $\hat{U}$ is the mean square amplitude of the atomic vibration. Large B values can arise from static disorder, caused by structural differences in different unit cells throughout the crystal, or dynamic disorder, which is real molecular motion. For a B factor of 15 Å^2, displacement of an atom from its equilibrium position is approximately 0.44 Å. A B factor higher than 60 Å^2 suggests that the observed atom is disordered in the crystal.

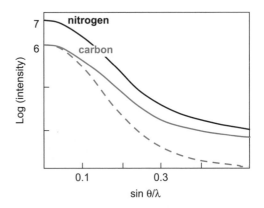

Figure 4.3.21 The scattering from an atom is proportional to the number of electrons (N = 7, C = 6) and falls off with increasing angle from the center of the diffraction pattern because of its finite size (see Chapter 4.1). Additional vibration causes an apparent increase in size so the scattering falls off faster with increasing angle (dashed cyan line).

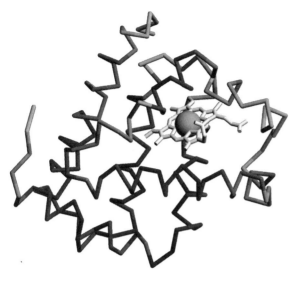

Figure 4.3.22 A representation of myoglobin (PDB:106M). Backbone regions with higher *B* factor are shown in a lighter gray.

B factor values for each atom are obtained from an analysis of the angular dependence of the diffraction pattern. Most X-ray structure coordinates deposited in the PDB have associated temperature factors that can be viewed using standard visualization programs (Chapter 8.1). An example is shown in Figure 4.3.22; this is myoglobin which, like most proteins, displays some regions of the structure with higher mobility than others. Relatively stable, low *B* factor, parts of the structure are shown in black.

Validation

There are frequently limitations in the structural data obtained and errors can arise, so it is important to carry out validation checks on structures produced. A number of such procedures are carried out by the PDB during the submission of coordinates. For proteins, one particularly useful check is to compare the backbone dihedral angles (ϕ and ψ) with those expected (Chapter 2.1). A number of programs are available to do this and an example is given in Figure 4.3.23. A good structure will have angles in the

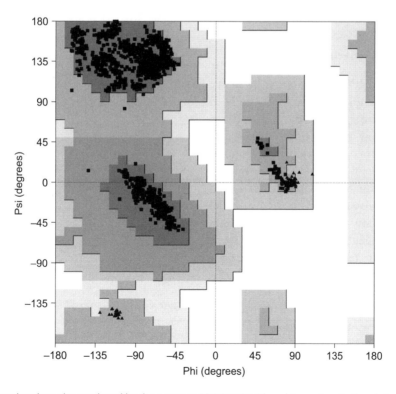

Figure 4.3.23 A Ramachandran plot produced by the program PROCHECK. The white regions in the plot are energetically unfavored and a structure should not have many residues in these unfavored regions. Dark gray regions are favored. It can be seen in this plot that most of the experimentally determined black points lie in the allowed regions.

allowed regions. Other indicators can also be used, such as the distribution of hydrophobic and hydrophilic amino acids and the population of sidechain rotamers.

Another validation consideration relates to the question "*Are the structures of macromolecules in the cell the same as in a crystal?*" This is a legitimate question as crystals are usually frozen for data collection and they contain very high concentrations of macromolecule and salt. The general answer, however, appears to be "*yes the structures are the same*". Some small changes can occur in regions where the macromolecules are in contact in the crystal, but there is good evidence for the validity of most structures determined: some enzymes retain their activity in the crystal form; solution state studies, using methods such as nuclear magnetic resonance (NMR), are also nearly always consistent with crystal structures. It should, however, also be remembered that the structural information obtained using these other methods is rarely as detailed as that from diffraction data.

Other crystallography techniques

Time resolved X-ray studies

It would often be valuable to see intermediates in a reaction pathway. Techniques have therefore been developed to visualize short-lived species. X-ray crystallography is now usually done at very low temperatures and this slows down reactions. Reactions can also be slowed down by using mutant proteins or substrate analogues. The start of a reaction in a crystal can also be synchronized by using caged substrates that are released by a pulse of light at $t = 0$. Fast data collection methods can also be used. For example, a technique called the Laue method speeds up data collection by using multiple, rather than single, wavelength irradiation. There are interpretation difficulties with the Laue method but it can be useful in situations where the structure is known and where only a few changes are happening in a reaction.

Neutron diffraction

The main advantage of using neutrons instead of X-rays in diffraction studies is the ability to observe

hydrogen atoms and to distinguish them from deuterium. (Note that for X-rays $\rho(xyz)$ represents electron density (electrons are the scatterers) while for neutrons we obtain nuclear density.)

Example Catalytic mechanism of serine proteases

Proteolytic enzymes, called serine proteases, have a serine, an aspartate, and a histidine at the active site. The ability of neutron diffraction to locate hydrogen atoms has allowed the protonation states of the catalytic site residues (Asp-102 and His-57) to be determined. Exchange of most of the ^1H for ^2H (D) in NH and OH groups in the enzyme was achieved by soaking crystals in D_2O before neutron diffraction data collection. Various H/D difference maps were used to identify the position of exchangeable hydrogens. In Figure 4.3.24, the difference map shows that a preferred position of the deuterium is on the His-57 ring rather than on the aspartate which had been an alternative proposal for the catalytic mechanism. This experiment, together with NMR experiments on the pKa of His-57 (Chapter 6.1), cleared up previous controversy about the role of the histidine and aspartate in the enzyme mechanism.

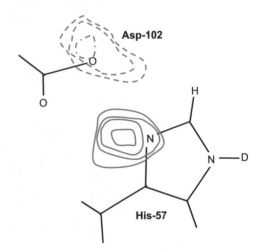

Figure 4.3.24 Neutron diffraction studies (adapted and redrawn with permission from *Biochemistry* 20, 6462–74 (1981)). Dotted lines signify negative density in the difference map.

Electron diffraction from 2D crystals

Electron microscopes can be adjusted to produce diffraction patterns, as well as images (Chapter 7.1). An early study of electron diffraction was by Henderson and Unwin who, in 1975, obtained 0.7 nm resolution electron density maps of bacteriorhodopsin. This was the first glimpse of the structure of a membrane protein. The bacteriorhodopsin was oriented in two-dimensional crystalline form in the membrane. Electron diffraction data were combined with electron microscope images collected at various tilt angles between the membrane and the electron beam. The amplitudes of the structure factors obtained from the diffraction pattern, combined with phase information from the images led to the structure determination. As will be described in more detail in Chapter 7.1, this approach has now been used to determine the structures of several important systems.

Achievements of crystallography

In February 2011, the protein databank (**www.rcsb.org**) contained over 60 000 structures determined by X-ray crystallography. This information has had a huge impact on our understanding of the molecules of life. Contributions include: information about enzyme specificity and mechanism; enzyme regulation by conformational changes (allostery); drug design; the folding, evolution, and prediction of protein structure; the way that proteins interact with DNA and RNA; the structure of membrane proteins, viruses, the ribosome, and so on—in other words almost all aspects of modern molecular biology! The importance of diffraction methods has also been recognized by the award of an unusually high number of Nobel Prizes, as shown in the following list:

1901 (Physics) W.C. Röntgen. *Discovery of X-rays.*
1914 (Physics) M. Von Laue. *Diffraction of X-rays by crystals.*
1915 (Physics) W.H. Bragg and W.L. Bragg. *Use of X-rays to determine crystal structure.*
1937 (Physics) C.J. Davisson and G. Thompson. *Diffraction of electrons by crystals.*
1946 (Chemistry) J.B. Sumner. *For his discovery that enzymes can be crystallized.*
1962 (Chemistry) J.C. Kendrew, M. Perutz. *For their studies of the structures of globular proteins.*
1962 (Physiology or Medicine) F. Crick, J. Watson, and M. Wilkins. *The helical structure of DNA.*

1964 (Chemistry) D. Hodgkin. *Structure of many biochemical substances including Vitamin B12.*
1982 (Chemistry) A. Klug. *Development of crystallographic electron microscopy and discovery of the structure of biologically important nucleic acid–protein complexes.*
1985 (Chemistry) H. Hauptman and J. Karle. *Development of direct methods for the determination of crystal structures.*
1988 (Chemistry) J. Deisenhofer, R. Huber, and H. Michel. *For the determination of the three-dimensional structure of a photosynthetic reaction center.*
1994 (Physics) C. Shull and N. Brockhouse. *Neutron diffraction.*
1997 (Chemistry) P.D. Boyer, J.E. Walker, and J.C. Skou. *Elucidation of the enzymatic mechanism underlying the synthesis of adenosine triphosphate (ATP) and discovery of an ion-transporting enzyme.*
2003 (Chemistry) R. MacKinnon. *Potassium channels.*
2006 (Chemistry) R.D. Kornberg. *Studies of the molecular basis of eukaryotic transcription.*
2009 (Chemistry) V. Ramakrishnan, T.A. Steitz, and A.E. Yonath. *Studies of the structure and function of the ribosome.*

It is not the intention of this book to repeat and expand on some of these exciting results which are well explained and presented in other textbooks. It is, however, worthwhile considering some of the practical aspects of some of these key diffraction results. Some extracts from three classic papers are given below, with some of the experimental details. Note some of the points made in these extracts. These include the quoted *resolution*, experimental difficulties, such as *twinning* of crystals, and how the *phase problem* was solved. You are also encouraged to download the coordinates of these structures from the PDB and to study them using a viewer such as Rastop, Chime, Pymol, or Swiss PDB viewer (all freely downloadable from the internet—see Chapter 8.1). The accession codes for the three examples are given below (warning—the ribosome structure file is very large!). Other Nobel Prize-winning structures are: 4tna.pdb (tRNA); 1bl8.pdb (a membrane spanning potassium channel); and 1nik.pdb (RNA polymerase II).

Photosynthetic reaction center: *J. Mol. Biol.* 180, 385–98 (1984); (1prc.pdb)

"X-ray analysis of three-dimensional crystals of the photosynthetic reaction center from the purple bacterium *Rhodopseudomonas viridis* led to an electron density distribution at 3 Å resolution calculated with phases from multiple isomorphous replacement. The protein subunits of the complex were identified. An atomic model of the prosthetic groups of the reaction center complex (4 bacterio-chlorophyll b; 2 bacterio-pheophytin b; 1 non-heme iron; 1 menaquinone, 4 heme groups) was built. The arrangement of the ring systems of the bacterio-chlorophyll b and bacteriopheophytin b molecules shows a local 2-fold rotation symmetry: two bacteriochlorophyll b form a closely associated, non-covalently linked dimer (special pair)."

F_1F_0 ATP synthase: *Science* 286, 1700–5 (1999); (1qo1.pdb)

"For data collection the crystals were harvested in a buffer containing 20% glycerol as cryo-protectant and then frozen rapidly in liquid nitrogen. The crystals are small (up to 150 μm in the largest dimension) and diffract X-rays weakly. They belong to the monoclinic space group $P2_1$ with unit cell dimensions $a = 135.9$ Å, $b = 1753$ Å, $c = 139.2$ Å, and $\beta = 91.6°$. Assuming one F_1c10^* complex with 453.2 kDa per asymmetric unit, the estimated solvent content is 66%. A data set was collected to 3.6 Å resolution at beamline IDO2B ($\lambda = 0.99$ Å) at the ESRF, Grenoble, France with a Mar Research Image plate detector (1600 pixel mode). Because of anisotropic diffraction and slight radiation damage, the final data set was restricted to 3.9 Å resolution."

*This structure was of the F_1 head with its 3 α and 3 β subunits plus 10 copies of the membrane-spanning c subunit.

Large ribosomal subunit: *Science* 289, 905–20 (2000); (1ffk.pdb)

"Several experimental approaches were used to extend the resolution of the electron density maps of the H. marismortui 50S ribosomal subunit from 5 to 2.4 Å. The twinning of crystals, which obstructed progress for many years, was eliminated by adjusting crystal stabilization conditions. The X-ray data used for high-resolution phasing were collected at the Brookhaven National Synchrotron Light Source. Osmium pentamine (132 sites) and iridium hexamine (84 sites) derivatives proved to be the most effective in producing isomorphous replacement and anomalous scattering phase information to 3.2 Å resolution."

➕ Summary

1 Informative diffraction patterns about molecules can be obtained when a wave is scattered by a periodic structure whose dimensions are comparable to the wavelength. X-rays, electrons, and neutrons have suitable wavelengths for detecting atomic detail in molecules.

2 The diffraction patterns can be interpreted directly to give information about the size of the unit cell (hence the molecular weight), the symmetry of the molecule, and, in the case of fibers, information about periodicity, e.g. pitch of a helix.

3 The determination of the complete structure of a molecule cannot be carried out from the diffraction pattern alone because it is necessary to recover lost phase information. The phases can be determined using the method of multiple isomorphous replacement (MIR), where heavy metals are incorporated into the diffracting crystals (the general procedures for crystallography using MIR are summarized in Figure 4.3.25). Other phasing methods are possible, including molecular replacement and anomalous scattering.

4 Once the electron (from X-rays and electrons) or nuclear (from neutrons) scattering density of a macromolecule has been calculated, a model can be built to match the density using a known sequence. The model can be can be refined by comparing calculated and observed diffraction patterns and incorporating energy minimization. Some of these procedures are summarized in Figure 4.3.25.

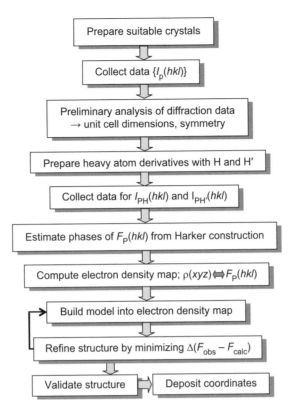

Figure 4.3.25 Summary of some of the procedures used in X-ray crystallography if MIR is used to solve the phase.

5 The biological information derived from diffraction experiments has been enormous. The 3D structures of over 60 000 different macromolecules and their complexes have been determined. This detailed information has greatly enhanced our understanding of the molecules of life. There is little evidence to suggest that the structures determined are significantly different in the living cell. Many Nobel Prizes have been won because of the structural insights gained from diffraction experiments; examples include the structure of DNA, hemoglobin, tRNA, the photosynthetic reaction center, F_1F_o ATP synthase, potassium channels, RNA polymerase II, and the ribosome.

6 While most information has been derived using X-ray crystallography, neutron and electron diffraction can yield valuable information. Neutrons, for example, are very useful for detecting the position of H-atoms.

Further reading

There are a number of excellent web-based diffraction tutorials and programs that allow you to perform instructive Fourier transforms and other exercises.

http://www.ysbl.york.ac.uk/~ cowtan/fourier/fourier.html

http://www.lks.physik.uni-erlangen.de/diffraction/index.html

http://www.ruppweb.org/Xray/101index.html

https://wasatch.biochem.utah.edu/chris/tutorial/index.html

http://www.bmsc.washington.edu/people/merritt/bc530/bragg/
http://escher.epfl.ch/fft/
http://www.jcrystal.com/products/ftlse/index.htm
http://www.brainflux.org/java/classes/FFT2DApplet.html

Synchrotron sites
http://www.esrf.eu/
http://www.diamond.ac.uk/

Books
Blow, D. *Outline of Crystallography for Biologists* Oxford University Press, 2002.
Drenth, J. *Principles of Protein X-ray Crystallography* (3rd edn) Springer, 2007.

Reviews
Petsko, G.A., and Ringe, D. Observation of unstable species in enzyme-catalyzed transformations using protein crystallography. *Curr. Opin. Chem. Biol.* 4, 89–94 (2000).
Blakeley, M.P., Cianci, M., Helliwell, J.R., and Rizkallah, P.J. Synchrotron and neutron techniques in biological crystallography. *Chem. Soc. Rev.* 33, 548–57 (2004).
Raunser, S., and Walz, T. Electron crystallography as a technique to study the structure of membrane proteins in a lipidic environment. *Annu. Rev. Biophys.* 38, 89–105 (2009).

 Problems

4.3.1 A crystal of a protein has unit-cell dimensions 7.02 nm × 4.23 nm × 8.54 nm. The density of the native crystal is measured as 1.29 g mL^{-1} and the water content is found to be 36%. Calculate the apparent molecular weight, assuming that the unit cell axes are orthogonal (at right angles).

4.3.2 Crystals of a protein of 55 000 Da have unit-cell dimensions $a = 6.65$ nm, $b = 8.75$ nm, and $c = 4.82$ nm. If the crystals have a density of 1.26 g mL^{-1} and are composed of approximately 50% water by weight, calculate the number of protein molecules in the unit cell (assume the cell axes are orthogonal). If there are eight asymmetric units in the unit cell, how many subunits might the protein have?

4.3.3 In the early determination of the structure of myoglobin by X-ray diffraction, 400 reflections were measured to obtain a resolution of 0.6 nm. Approximately how many reflections would be needed to obtain a resolution of 0.2 nm?

4.3.4 For the ideal diffraction experiment we want monochromatic, collimated, coherent radiation. Explain, with reference to Bragg's law, why these requirements are needed.

4.3.5 Calculate the wavelength of neutrons that emerge from a reactor at a temperature of about 100°C with a velocity of 2.8 × 10^3 m s^{-1} (mass of neutron = 1.67 × 10^{-27}kg).

4.3.6 Sometimes regions of a polypeptide chain give very weak electron-scattering density in X-ray studies of protein crystals, while other regions are clearly observed. Can you suggest a reason for this?

4.3.7 From X-ray crystallography studies, David Phillips and colleagues postulated that when a polysaccharide molecule binds to lysozyme, the cleavage takes place between binding sites labeled D and E. They used model building in this work and proposed that the sugar ring in site D was distorted from its normal position. Can you suggest why they could not observe this distortion directly?

4.3.8 Glycogen phosphorylase is involved in the breakdown of glycogen and is subject to a series of sophisticated control mechanisms, including phosphorylation and conformational changes. Crystals of phosphorylase b shatter in the presence of glucose-6-phosphate (G-6-P), a small molecule that activates the enzyme. Can you suggest why this might be?

4.3.9 The structure of crambin, a small hydrophobic protein, was determined in 1981 (*Nature* 290, 107) using the anomalous scattering of the sulfur in the three disulfide bonds in the protein. Sulfur has an absorbance band at 0.52 nm. The wavelength of the X-rays used in 1981 was 0.154 nm. Modern synchrotron beamlines are available that can be tuned over the range 0.07–0.21 nm. If $\lambda = 0.229$ nm and 0.071 nm were used to collect data, which would you expect to give the strongest anomalous scattering effect?

4.3.10 A real DNA diffraction pattern obtained from moist DNA is shown with some of its characteristics in terms of the parameters a, b, and c (see also Figure 4.3.12). Using the cartoon of the double stranded Watson Crick DNA structure shown on the right, describe the relationships between a, b, and c, and p, h, and θ.

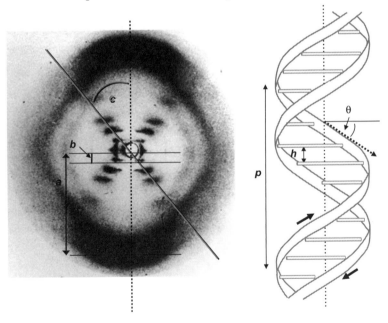

4.3.11 Chromosomes from higher organisms are composed of chromatin, which consists of protein and DNA. The protein part is largely histones, whose function is to fold DNA in an efficient manner. It is possible to degrade chromatin using an enzyme that attacks DNA thus producing a "nucleosome" particle consisting of about 146 base pairs of DNA and an octamer of histones (see also Figure 6.2.18). A low-resolution neutron diffraction map of nucleosome crystals is shown in the figure at two H_2O/D_2O ratios (top = 39% D_2O, bottom = 65% D_2O). What can you deduce about the structure of chromatin from these maps (see contrast matching section in chapter 4.1)?

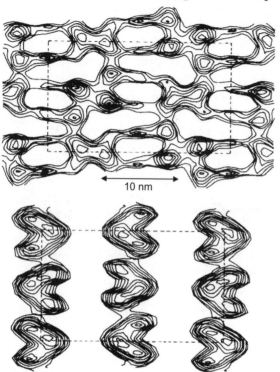

10 nm

5 Electronic and vibrational spectroscopy

"One important object of this original spectroscopic investigation of the light of the stars and other celestial bodies, namely to discover whether the same chemical elements as those of our earth are present throughout the universe, was most satisfactorily settled in the affirmative."

William Huggins, 1864

"The more I look back on my career, the more I feel that it has been a long history of frustration, disappointment, struggle and every kind of tribulation. But there have been a few gleams of success. It was poverty and the poor laboratories that gave me the determination to do the very best I could."

C.V. Raman in 1953, on the Silver Jubilee of his discovery of the Raman effect

"One of the most common requests I get is to explain why GFP and related FPs (fluorescent proteins) are significant. My usual answer has been that their genetically encoded fluorescent colors can make many key biochemical processes directly visible inside living cells and organisms. Using standard molecular biological tricks to connect host cell genes to FP genes, we can watch when and where those host cell genes get switched on and off, when the protein products are born, where they travel, with what other proteins they interact, and how long they survive."

Roger Tsien, Nobel Prize lecture, 2008

The interaction of electromagnetic radiation with matter gives rise to various phenomena. One of these, scattering, has already been dealt with in Section 4. We now deal with transitions between quantized energy levels. These experiments, called spectroscopy, come in two main classes: absorption (when incident radiation causes a change to a higher energy level) and emission (when energy is released on going from a higher energy level to a lower one). The colors we see around us arise largely from absorbance. Fluorescence emission, from a variety of "probes", has become one the most powerful tools available for studying and detecting molecules. In this section, we mainly deal with transitions between vibronic and electronic energy levels but aspects of Section 6, which discusses magnetic resonance, are closely related.

5.1 Introduction to absorption and emission spectra

Introduction

The development and application of spectroscopy has had an enormous influence on our understanding of the world; from microscopic atomic structure (nm) to the macroscopic (light year scale!) universe. William Wollaston discovered the existence of dark lines in the solar spectrum in 1801. A few years later, Joseph von Fraunhofer hypothesized that these dark lines were caused by an absence of certain wavelengths of light. By 1859, Gustav Kirchhoff managed to show that each pure substance produces a unique light spectrum; he, with Robert Bunsen, then went on to develop a technique for determining the chemical composition of matter using spectroscopic analysis. These and later spectroscopic investigations of light from stars by William Huggins and others showed that a common chemistry exists throughout the universe. Johann Balmer and Johannes Rydberg developed equations to describe observed spectra of atoms; these were largely rationalized by Niels Bohr's atomic model in 1913, but a universal explanation of the spectra of the elements had to wait until the development of quantum mechanics by Werner Heisenberg and Erwin Schrödinger in the 1920s.

This section deals with a variety of techniques that detect transitions between electronic and vibrational energy levels. We start with an introductory chapter that outlines some of the main features of absorption and emission spectroscopy; many of these features also apply to magnetic resonance techniques (Section 6).

Energy states

Some of the basic properties of matter were introduced in Chapter 2.1. Many of these can be explained by classical physics (e.g. Newton's laws) but to explain transitions between energy levels we need quantum mechanics (Tutorial 9).

Quantization

At the atomic level, not all energy states are possible. The ground state is the state of lowest energy, while those with higher energy are excited states. If two or more states of the molecule have the same numerical value of energy, they are said to be degenerate. In some cases, degeneracy can be removed by an external influence, such as an electric or magnetic field, thus leading to "split" energy levels. For example, electrons and some nuclei possess a property known as spin (Tutorial 9 and Section 6). This concept was introduced by Wolfgang Pauli to explain spectroscopic observations in the presence of a magnetic field. Particles with the property of spin, which can have values of $^1/_2$, 1, $^3/_2$, ..., interact with an applied magnetic field to give $2S + 1$ energy levels. The case of Mn^{2+}, which has an electron spin $S = ^1/_2$, and a nuclear spin $I = ^5/_2$, is illustrated in Figure 5.1.1.

Classification of energies

For most purposes, it is convenient to treat a molecule as if it has several distinct reservoirs of energy. The total energy, written in terms of energies that are most relevant here, is:

$$E_{total} = E_{translation} + E_{rotation} + E_{vibration} + E_{electronic} + E_{electron\ spin} + E_{nuclear\ spin}$$

Each E in this equation represents the energy indicated by its subscript. Translation and rotation arise from diffusion, which was discussed in Chapter 3.1; quantization of such diffusional energy levels is rarely observed. The other energies, however, give rise to informative spectra that will be discussed in more detail in this section and Section 6. The magnitude of

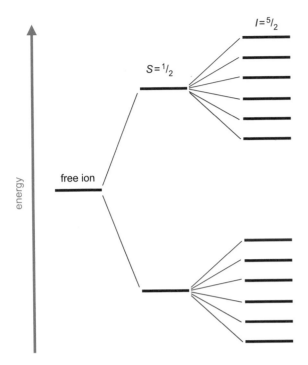

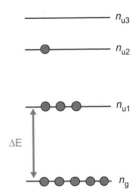

Figure 5.1.2 The population distribution among energy levels is governed by the Boltzmann distribution.

Figure 5.1.1 The energy levels of the Mn^{2+} ion in an applied field. This has an electron spin ($S = 1/2$), as well as a nuclear spin $I = 5/2$. The multiplicity arising from S is 2 and that from I is 6, giving a total of 12 energy levels.

$n_{u1}/n_g = \exp(-\Delta E/kT)$ where k is the Boltzmann constant (1.381×10^{-23} J K^{-1}). When ΔE refers to 1 mole, the term on the right becomes $\exp(-\Delta E/RT)$, where R is the gas constant ($R = 8.31$ J mol^{-1}). The very different energies associated with different processes, e.g. electronic and vibrational transitions, mean that the population distributions among energy levels are widely different for different molecular processes (see below).

these various E terms is very different. For example, the separation between energy levels associated with nuclear spin reorientation in a magnetic field is very small ($\sim 10^{-3}$ J mol^{-1}). The other separations are much larger: ~ 10 J mol^{-1} for molecular rotation, $\sim 2 \times 10^4$ J mol^{-1} for vibrations, and $\sim 2 \times 10^5$ J mol^{-1} for electronic energy levels. These energies correspond to very different regions of the electromagnetic spectrum, as discussed in Tutorial 8.

Absorption

Spectra

Energy is absorbed when a molecule is induced to move to a higher energy level. Absorption can be measured by plotting the amount of energy absorbed by the sample as a function of the kind of energy applied. The resulting plot is called a spectrum (Figure 5.1.3).

A spectrum consists of a series of "lines", "bands", "peaks", or "resonances", and it contains information about the intensity of the absorption, the width of the line, and the line position (the choice of energy scale depends on the technique used). Note that line a is "resolved", while b and c are unresolved (they overlap).

Energy level populations

In most practical studies, we deal with large numbers of molecules (but see Chapter 7.2). Thermal motion leads to the distribution of molecules among the energy levels available to them. Higher energy states are exponentially less likely to be populated than lower energy ones. The exact distribution depends on the temperature and the separation between energy levels (ΔE) (Figure 5.1.2). At a given temperature, the number of molecules in an upper state (n_{ui}, $i = 1, 2, 3....$) relative to the number in the ground state (n_g) is given by the **Boltzmann distribution law**:

Transition probabilities

Transitions between energy states occur when there is an *interaction* between the incident radiation and the molecule studied. We now briefly consider the

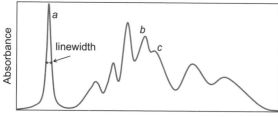

Figure 5.1.3 A spectrum, showing a resolved line, *a*, and several unresolved ones, including *b* and *c*. The horizontal axis can be frequency, wave number, energy, or any other convenient scale. The linewidth of a line is usually measured at its half height.

nature of this interaction and the likelihood of a transition occurring. A transition is probable when the applied energy (ΔE) matches the separation ($h\nu$) between the energy states of interest; ν is the frequency of the applied radiation and h is the Planck constant. The equation $\Delta E = h\nu$ is one representation of the **resonance condition** in a driven oscillator (Tutorial 7).

Another general rule applies to electronic and vibrational spectra: *there must be a net displacement of charge* in moving from one energy state to another.

The charge displacement that occurs during a transition between energy levels can be of two kinds:

1. A linear displacement—this corresponds to an **electric transition dipole** and is written as μ_e. (A dipole arises when there is a separation of charge (Tutorial 11).)

2. A rotation of charge—this corresponds to a **magnetic transition dipole**, μ_m. (A charge rotation generates a magnetic moment

perpendicular to the rotation (Tutorial 11 and Box 6.1).)

The applied electromagnetic radiation has an **E** component and an **M** component (Tutorial 8). The electric component **E** can interact with μ_e and the magnetic component M can interact with μ_m. The larger the **transition dipole moment**, the larger the probability of a transition will be (Box 5.1). A charge displacement generally creates a larger effect than charge rotation and the radius of rotation for an electronic or vibrational transition is very small, so μ_m is generally much less than μ_e (by a factor of $\sim 10^5$).

The fact that some transitions between energy states are much more likely than others leads to the idea of **selection rules**. Figure 5.1.4 illustrates the charge displacement rule for transitions between simple atomic orbitals; the spherical symmetry of *s* orbitals means there is no net charge displacement for a $1s \rightarrow 2s$ transition. Similarly, in infrared spectroscopy the N–N $\rightarrow$ N—N (symmetric stretch vibration) is "forbidden" but the O=C=O $\rightarrow$ O= $\cdots$ C=O (asymmetric stretch) is "allowed" because it involves a charge displacement. In fact, these rules are not very strictly obeyed because molecules have many possible interactions and distortions, and some important absorption spectra arise from what are "forbidden" transitions in a very symmetric environment, for example the *d–d* spectra of some metal ions (Tutorial 10).

In addition to the selection rules illustrated in Figure 5.1.4, others exist. One important one, relevant for our discussions of phosphorescence (Chapter 5.5) is related to the **spin properties** of the energy levels; this states that "allowed" transitions involve *no change in spin state* meaning, for example, that a transition

---

Box 5.1 Transition dipole moments and transition probability

An allowed transition between a ground state, g, and an excited state, e, has an associated dipole moment, defined as μ_{ge}; this can have both μ_e and μ_m components (see Tutorial 11 for definitions and discussion of dipoles and dipole moments). The value of μ_{ge} depends on the change in charge distribution between molecular orbitals that describe the g and e states. For the interaction with the electric transition dipole, μ_{ge} can

thus be calculated using the formula: $\mu_{ge} = \int \psi_g \mu_e \psi_e \, d\nu$, where ψ_g and ψ_e are the wave functions of the ground and excited states, and $\int d\nu$ indicates integration over space. The magnitude of μ_{ge} determines how the system will interact with an electromagnetic wave; transitions only occur if μ_{ge} is non-zero. The transition probability (the **dipole strength**) depends on the square of the magnitude of μ_{ge}: $D_{ge} = |\mu_{ge}|^2$.

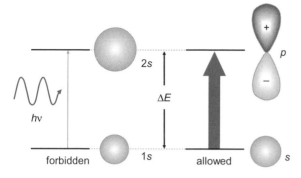

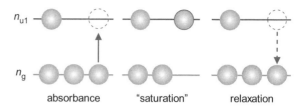

Figure 5.1.5 Absorption results in transitions to excited states. If the number of molecules in the upper of two levels equals the number in the lower level (saturation) absorption is undetectable and relaxation (an $n_{u1} \to n_g$ transition) is required before further absorption can be observed.

Figure 5.1.4 Illustration of the charge displacement selection rule. There is no net charge displacement in going between two symmetric s-type orbitals, thus the transition probability will be low and the transition is said to be "forbidden". There is, however, a charge displacement in a transition from an s-type to a p-type orbital; thus, this transition is "allowed".

between a spin $I = 1$ state to a spin $I = 1/2$ state is forbidden.

Absorption depends on the populations in different energy levels

We noted earlier that the distribution of molecules among energy levels depends on the energy differences between the levels. The relative populations of molecules in the different levels conform to the Boltzmann distribution law. Another factor to note is that when electromagnetic radiation is applied to a sample, it is just as likely to cause transitions from an excited state to a ground state as it is to cause transitions from a ground state to an excited one ($D_{ge} = D_{eg}$; Box 5.1). Consequently, for absorption to occur the ground state needs to be significantly more populated than the excited state.

The perturbation of populations by the applied electromagnetic radiation can be important. For example, the application of radiation can equalize the populations of the ground and excited state energy levels. It can also invert the populations (see discussion of the laser in Box 5.2 and the "180° pulse" in Chapter 6.1). If the populations are equalized, a situation known as saturation, (Figure 5.1.5), no signals are detected. Molecules can, however, achieve a Boltzmann distribution again after a process called relaxation. The processes that cause relaxation are

different for different kinds of spectroscopy (see also emission section below, Chapter 5.5, and Section 6).

The sensitivity of an experiment (the ability to detect a transition) also depends on the population difference between energy levels. The Boltzmann equation [$n_{u1}/n_g = \exp(-\Delta E/RT)$] allows us to calculate the expected population differences for different processes. For nuclear spin orientation with energies of $\Delta E \sim 10^{-3}$ J mol^{-1} this is $n_{u1}/n_g \sim 0.99999$ (i.e. *the populations are nearly equal*). In contrast, for electronic transitions $\Delta E \sim 2 \times 10^5$ J mol^{-1} and $n_{u1}/n_g \sim 10^{-21}$ (i.e. *essentially all the molecules are in the ground state*). This means that spectroscopy of electronic transitions is much more sensitive than nuclear spin resonance. (The energy involved in the transition (ΔE) is also important; higher energy transitions can generally be detected more easily than low energy ones.)

Absorption depends on concentration

The total absorption observed in a sample depends on the number of molecules in which transitions are induced. This means that absorption spectra can be used quantitatively. The effect of sample concentration on the spectral intensity is the basis of many applications of absorption and emission spectra. We will see in Chapter 5.3, for example, that the Beer–Lambert law is commonly used to measure concentration.

The width of an absorption line depends on the excited state lifetime

Spectral lines are not infinitely sharp. Various factors contribute to the broadening of a line. One is simple overlap of closely spaced lines (Figure 5.1.3). Another is the lifetime of the excited state. If the lifetime is short, the position of an excited state will be ill-defined and the line will be broad. If the lifetime is long, the position will be well-defined and the lines narrow. This is a demonstration of the Heisenberg uncertainty principle (Tutorial 9; Problem 5.1.2).

Absorption depends on the direction of the transition dipole moment

A transition dipole moment absorbs radiation by oscillating in sympathy with the applied radiation (Box 5.1). This oscillation has a certain direction associated with it, so only those components of the electromagnetic radiation that are in the same direction as the transition dipole moment will cause transitions (Figure 5.1.6). If the angle between the direction of the applied wave and the direction of the transition moment is θ, then the effective value of the transition moment is proportional to $\cos \theta$ and the transition probability is proportional to $\cos^2 \theta$.

The differential absorption of radiation polarized in two directions (e.g. $A_{\parallel}$ and $A_{\perp}$), as a function of frequency, is called dichroism. For plane-polarized light this is called linear dichroism (Chapters 5.2 and 5.3), while for circularly polarized light it is called circular dichroism (Chapter 5.4).

Emission

Emission of radiation can occur when a molecule changes from an excited energy state to a lower energy state. We noted above that absorption causes transitions to higher excited states and that a return to equilibrium involves relaxation. Emission can be observed during the relaxation process. There are three main ways that a molecule can go from an excited state to a lower one, as described below.

Spontaneous emission

The excited molecule can act as an oscillator, radiating its energy without any other interaction with its environment. The likelihood for spontaneous emission to occur can be calculated using wave mechanics. The probability that spontaneous rather

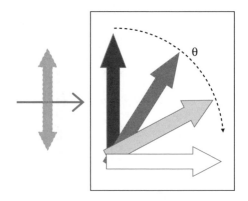

Figure 5.1.6 Absorption depends on the angle between the transition dipole moment and the polarization direction of the applied radiation, shown here as vertical. When $\theta = 0$ the absorption will be a maximum and a minimum when $\theta = 90°$.

than stimulated emission will occur at a frequency ν is proportional to ν^3. At radiofrequencies (e.g. $\nu \sim 5 \times 10^8$ Hz, corresponding to nuclear spin transitions), spontaneous emission is exceedingly unlikely whereas at optical frequencies (e.g. $\nu \sim 5 \times 10^{15}$ Hz, corresponding to an electronic transition) spontaneous emission dominates (see Problem 5.1.1).

Stimulated emission

As we stated in the absorption population section above, the application of electromagnetic radiation to a molecule is just as likely to lead to *emission* as absorption ($D_{ge} = D_{eg}$; Box 5.1). Such stimulated emission is very important in the laser (light amplification by stimulated emission of radiation; Box 5.2). A laser operates in systems where a non-equilibrium distribution of populations has been set up in an excited state by a pump. In such a system, applied radiation stimulates an emission cascade. As we will see in Section 7, stimulated emission is also used in some kinds of "super-resolution" microscopy.

Inter- and intramolecular de-excitation

The third way for a molecule to return from an excited state to a lower energy state is by encounters with other molecules and the generation of heat by collisions and vibrations. This type of decay is

Box 5.2 The laser

The laser, developed by Townes, Schwalow, and Gould in the 1950s and 1960s, is now an ubiquitous device that is in everyday use at home and in the laboratory. A populated excited state that can be stimulated to emit photons is required. More molecules should be in this excited state than in the lower states (population inversion). There are various energy level arrangements than can lead to this effect; the simplest is shown in Figure A. The process of supplying the energy required for the amplification is called pumping. The pump energy in Figure A is supplied by light (gray) with a higher energy than the stimulated emission (cyan).

A laser consists of a "gain medium" which is a material (e.g. crystals, glasses, gases, semiconductors) with energy levels of the sort illustrated in Figure A. The gain medium is placed inside a highly reflective resonant optical cavity which, in its simplest form, has two mirrors arranged such that light bounces back and forth through the gain medium, stimulating more emission as it passes (Figure B). As for a particle in a box (Tutorial 9), the only wavelengths that can be sustained in this cavity must satisfy the condition $n\lambda = 2L$, where n is an integer and L is the length of the cavity (all other wavelength destructively interfere). Multiple passes through the cavity are required to produce appropriate amplification so any divergent rays are also suppressed. The net result is a beam that emerges through a partially transparent mirror with high intensity and low divergence. The emitted laser light is also coherent (i.e. it all has the same frequency and phase). Laser light can be produced continuously or in pulsed mode (see Problem 5.1.5).

Figure A An illustration of the laser mechanism with a pump, a highly populated metastable state, and stimulating radiation.

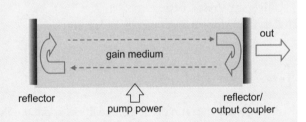

Figure B An illustration of a laser cavity.

referred to as "radiationless" because there is usually no detectable light emission. A special case of this type of de-excitation involves transfer of energy to another fluorophore (an acceptor) with matching energy levels. This resonance-energy transfer is important in photosynthesis, where sunlight absorbed by chlorophyll molecules is transferred via acceptors to the photoactive center, and in Förster fluorescence energy transfer, as will be discussed in Chapter 5.5.

Note Relaxation mechanisms in magnetic resonance arise from fluctuating magnetic fields at the appropriate frequencies. These will be discussed in Section 6.

 Summary

1 Energy is quantized and molecules have energies of various types:

$$E_{total} = E_{translation} + E_{rotation} + E_{vibration} + E_{electronic} + E_{electron\ spin} + E_{nuclear\ spin}$$

2 Absorption involves transitions to a higher energy level, induced by applied electromagnetic radiation.

3 Transitions between energy levels depend on the "transition dipole moment", a property related to the change in charge distribution that takes place when a transition occurs. Transition dipole moments are directional.

4 Absorption gives rise to a spectrum of lines on an energy scale.

5 At equilibrium, molecules are distributed among energy levels according to the Boltzmann law $[n_{ul}/n_g = \exp(-\Delta E/kT)]$. This equilibrium is achieved by relaxation processes. The population difference between energy levels strongly influences the sensitivity of the experiment.

6 The lifetime in an excited state influences the observed linewidth.

7 Depopulation of an excited state can arise from spontaneous, stimulated, or non-radiative transitions.

 Further reading

Books

Atkins, P., and de Paula, J. *Physical Chemistry for the Life Sciences* (2nd edn) Oxford University Press, 2010.

Hollas, J.M. *Modern Spectroscopy* Wiley, 2004.

 Problems

5.1.1 Calculate the ratio of the probability of spontaneous emission for an electronic transition with $\lambda = 300$ nm and a rotational transition with $\lambda = 30$ cm.

5.1.2 Calculate the ratio of molecules in an upper to a lower state at 298 K, when the energy separation is as follows (the associated molecular processes are shown in parentheses): (i) 1.2×10^{-2} J mol^{-1} (nuclear reorientation); (ii) 12 kJ mol^{-1} (a vibrational transition); (iii) $120 \times$ kJ mol^{-1} (an electronic transition).

5.1.3 Lifetime broadening can be estimated from $\delta E \sim h/2\pi\tau$, a formula derived from the Heisenberg uncertainty principle. Calculate the linewidth (in Hz and cm^{-1}) for a molecule that has a lifetime τ of (i) 10^{-13} s and (ii) 10^{-1} s. The separation between energy levels in NMR is about four orders of magnitude less than between rotational energy levels, yet transitions between rotational energy levels cannot be resolved in a solution while well-resolved NMR transitions are commonly observed. Using the information in the first part of the question, explain this observation.

5.1.4 A pulse laser delivers an average of power of 1 watt at $\lambda = 800$ nm. If its pulse rate is 8×10^7 s^{-1}, what is the energy in one pulse and how many photons does this correspond to (see Tutorial 8)? If the pulse width is 10 fs what is the power in the pulse? Will the output frequency be monochromatic for such a short pulse (see Tutorial 6)?

5.1.5 In photosynthesis, green plants harvest light by trapping energy from electromagnetic radiation and then use this energy to produce ATP. In fact, there are two processes that can deplete the excited state in the receptor: (i) ATP production and (ii) fluorescence. Upon addition of a molecule that acts as a weed killer, fluorescent intensity is observed to increase considerably. Why?

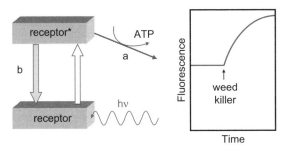

5.2 Infrared and Raman spectroscopy

Introduction

Molecules are said to undergo vibrational motion when their bonds stretch or bend. The energy of most molecular vibrations corresponds to the infrared (IR) region of the electromagnetic spectrum. Vibrations may be detected directly by measuring an IR spectrum or indirectly by Raman light scattering. The main biochemical applications of IR spectra include measurement of protein secondary structure, ionization states, and hydrogen–deuterium exchange. The intensity and position of IR bands is very sensitive to environment. The most useful region of the IR spectrum is in the range $\bar{v} \sim 700$–4000 cm^{-1} ($\lambda \sim 14$–$2.5 \text{ }\mu\text{m}$).

Physical basis of IR spectra

Using the Schrödinger equation (Tutorial 9), we can calculate the variation in energy of a diatomic molecule as a function of internuclear bond length; this curve, shown in cyan in Figure 5.2.1, is often called a Morse curve (compare the Lennard-Jones potential for *non-covalent* interactions discussed in Chapter 2.1; see also Figure 5.2.2). At the minimum point, the attractive and repulsive forces in the molecule are in balance. This molecular potential energy curve can be well approximated by a parabola, especially near the equilibrium position r_e (Figure 5.2.1).

The parabola drawn in Figure 5.2.1 has the same form as the potential energy of a weight on a spring ($1/2kr^2$; Tutorial 3). The vibrations of the bond can, therefore, be analyzed to a first approximation, in terms of harmonic oscillations of a weight on a spring (Tutorial 7). The vibrational frequency, $\bar{v}_{vib}$, is then given by, $\bar{v}_{vib} = (k/\mu)^{1/2}/2\pi$, where k is the force constant of the bond; μ is the reduced mass of the molecule AB, which is related to the atomic masses of A and B by $1/\mu = 1/M_A + 1/M_B$.

Isotopic substitution is a useful tool in IR spectroscopy because the vibration frequencies change. Consider an OH group: the reduced mass is $1/\mu_{OH} = 1/16 + 1/1$; therefore $\mu_{OH} = 16/17$. If the group is changed to OD (readily done by changing solvent to D_2O) we have $1/\mu_{OD} = 1/16 + 1/2$ and $\mu_{OD} = 16/9$. Assuming the force constant for the OH and OD bonds is the same, the frequency change given by the ratio $\bar{v}_{vib}(OH)/\bar{v}_{vib}(OD) = \sqrt{\mu_{OD}/\mu_{OH}} = \sqrt{9/17} \approx 0.73$ (Note the reduced masses here are relative masses; to obtain the absolute mass we multiply by the atomic mass constant—see Appendix A1.3 and Problem 5.2.4.)

Vibrational energy levels and transitions

Quantum mechanics tells us that only certain values of energy are allowed. These values are represented by the horizontal lines on the molecular energy curve shown in Figure 5.2.2. As the energy increases, we enter a non-parabolic region where the vibrations can no longer be treated by the simple harmonic motion model described by Figure 5.2.1 (they are then called anharmonic). The spacing between such vibrational energy levels becomes smaller as the energy increases. The dominant observed transitions are from the ground state to the first excited state. Transitions to higher levels are usually weak. As mentioned in Chapter 5.1, energy level transitions must have an associated *transition dipole moment* and selection rules apply.

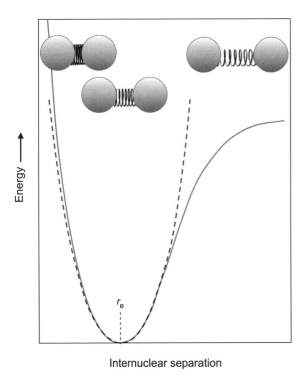

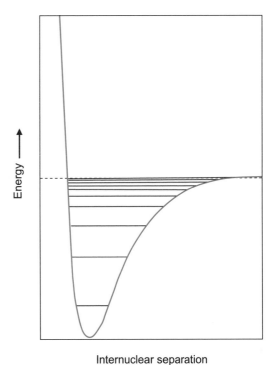

Figure 5.2.1 A molecular potential energy curve for a diatomic molecule. The equilibrium bond length of the molecule is r_e. The parabola (dotted line) represents the potential energy of a simple harmonic oscillation about r_e.

Figure 5.2.2 The potential energy curve of a diatomic molecule in its ground *electronic* state. The horizontal lines represent possible *vibrational* energy levels. The uppermost vibrational level corresponds to the "dissociation limit".

Measurement of IR spectra

An IR spectrum usually consists of a plot of absorption as a function of wavenumber (cm^{-1}). Suitable IR radiation sources include a heated wire or rod of silicon carbide. Glass absorbs IR so the sample cells are made of materials NaCl, KBr, CaF, or LiF. Water has a high absorbance in the IR region so measurement is restricted to selected regions, or "windows", of the spectrum. The solvent can be changed to D_2O which has different windows than those of H_2O but the solvent background absorption is often significant, even in the window regions. An advantage of Raman scattering spectroscopy, is that measurements can be carried out in aqueous solution (see later).

It is possible to use a spectrometer similar to the UV spectrometer described in Chapter 5.3, but most measurements are now carried out using the Fourier transform IR (FTIR) method (Figure 5.2.3). In an FTIR spectrometer, a dichroic (half-reflecting) mirror splits the beam from the IR source into two. The two beams are reflected at two mirrors, one movable and one fixed; the beams are then recombined at a detector. The amplitude of the recombined signals will depend on the frequency and the distance the mirror moves. This produces an interferogram (Figure 5.2.3) which is the Fourier transform of the spectrum observed using conventional IR methods.

FTIR has several important advantages over conventional methods. One is the efficient and rapid collection of digital data which allows many scans to be collected; this improves the signal-to-noise ratio because noise adds up as the square root of the number of scans, whereas coherent signals add linearly. The stability and dynamic range of modern FTIR instruments also helps in the collection of IR spectra from aqueous solutions and in the production of good quality difference spectra.

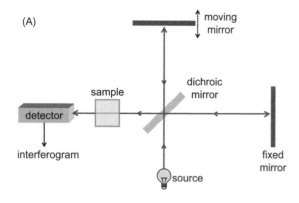

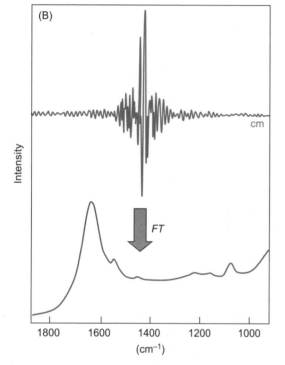

Figure 5.2.3 (A) Illustration of a Fourier transform infrared (FTIR) spectrometer (see text). (B) An interferogram and an IR spectrum produced by a Fourier transform.

Attenuated total reflection IR is a modification of FTIR (ATR-FTIR) that gives improved sensitivity

Note Fourier transformation (FT) methods have the recurring theme of an inverse relationship; in FTIR, the distance the mirror moves (cm) transforms to wavenumber (cm⁻¹), in NMR a time (s) to hertz (s⁻¹) conversion is used, and in X-ray crystallography we transform from real space (xyz) to reciprocal space (hkl); various FT relationships are described in Tutorial 6.

compared with normal transmission-IR. An **evanescent wave** is generated at a surface of the sample and this reduces the problem of strong attenuation of the IR signal in highly absorbing media, such as aqueous solutions. This method was discussed and illustrated in Chapter 4.2, Figure 4.2.7. ATR-FTIR is very sensitive and can be used to analyze trace amounts of materials, e.g. in forensic science. Examples of applications using ATR-FTIR are given in Figures 5.2.6 and 5.2.9.

Two-dimensional methods

The source in Figure 5.2.3 can be short, femtosecond infrared laser pulses. Application of a series of these pulses can produce a "two-dimensional" (2D) infrared spectrum. In its simplest form, two pulses with a waiting time of a few picoseconds between them are applied. As will be discussed in more detail under NMR (Chapter 6.1), this kind of experiment can lead to a 2D spectrum which gives improved resolution, as well as additional information about coupling between different vibrational modes compared with a normal spectrum. Time resolution of femtoseconds is possible. This short time scale, combined with the ability to obtain detailed information about molecular structure from IR spectra, means that 2D IR methods can probe very rapid structural fluctuations in proteins and other molecules.

IR spectra and applications

Spectra of polyatomic molecules

In a diatomic molecule there is only one mode of vibration—the stretching of the bond. In polyatomic molecules, bonds can stretch and bend. To describe the possible motions of a molecule of N atoms, we might expect to need $3N$ coordinates (x, y, and z for each atom). Of these, however, three coordinates are needed to specify the position of the center of mass and three rotational angles are required to define the orientation of a non-linear molecule. $3N - 6$ possible vibrations are thus generally found in a non-linear molecule. A linear molecule only requires two rotation angles to define its orientation so it has $3N - 5$ possible vibrations. Large polyatomic molecules are therefore expected to have very complex vibrational spectra. Fortunately, many vibration modes can be

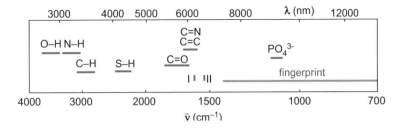

Figure 5.2.4 Characteristic IR absorption bands. The positions marked I and II refer to important bands associated with amide groups, e.g. in proteins. Scales (non-linear) are indicated for both wavenumbers (cm^{-1}) and wavelength (nm).

localized because they involve displacement of just two bonded atoms with little interaction with other vibrations; characteristic frequencies can thus be assigned to certain vibrations. The general regions of the IR spectrum, where different vibrational bands are observed, are summarized in Figure 5.2.4.

IR spectra are complex in the 1450 to 700 cm^{-1} region but, because characteristic patterns are found there, it is often called the fingerprint region. Absorption bands in the 4000 to 1450 cm^{-1} region are usually caused by stretching vibrations of diatomic units, and this is sometimes called the group frequency region. Analysis of the characteristic group frequency region forms the basis of most applications of IR spectroscopy to biological systems.

> **Example** Peptide bond vibrations
>
> Two characteristic bands found in the infrared spectra of proteins and polypeptides are called amide I and amide II. These arise from vibrations in the planar peptide group (Figure 5.2.5). The amide I band (~1645 cm^{-1}) predominantly arises from C=O stretching, while the amide II band (~1550 cm^{-1}) arises mainly from the out-of-phase combination of NH in-plane bending and CN stretching vibrations (see also Figure 5.2.6).

Solvent and H-bonding effects

Associated with each vibrational energy level are many closely spaced rotational energy levels that arise from molecular rotation; this usually leads to rather broad vibrational bands. The solvent may also cause shifts in the vibrational frequencies because specific interactions with the molecule can lower or raise some energy levels resulting in a shift in the frequency of absorption, an effect also observed for electronic transitions (this point is also discussed in Chapter 5.3).

H-bonding

The presence of an H-bond can change the vibrational frequency. For example, both the C=O and N–H groups in a peptide bond (Figure 5.2.5) are involved in H-bonding in protein secondary structure (α-helix and β-sheet). Most H-bonds formed between C=O and H–N groups in a protein (Chapter 2.1) are approximately linear; formation of an H-bond thus makes it easier to stretch in the C=O direction, causing the amide I band to move to a lower wavenumber by an amount (~20–30 cm^{-1}) that depends on the strength of the H-bond. In contrast, bending motions become harder to perform on H-bonded groups and associated bending bands tend to move to higher wavenumber.

Protein secondary structure

One common application of IR spectroscopy is the analysis of protein secondary structure. The locations of both the amide I and amide II bands are sensitive to the secondary structure content of a protein, partly from H-bonding effects. Figure 5.2.6 shows typical IR protein spectra in the informative region. Table 5.2.1 summarizes the positions of these bands.

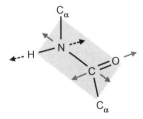

Figure 5.2.5 Three stretch vibration directions in a peptide bond (NH, black; C=O, cyan; and C–N, gray).

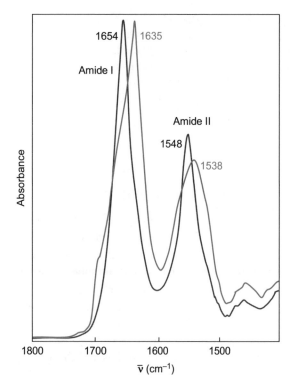

Figure 5.2.6 FTIR spectra of myoglobin, a predominantly α-helical protein (in gray) and concanavalin A, a predominantly β-sheet protein (in cyan) (adapted from *Trends Biochem. Sci.* 17, 328–33 (1992)).

A collection of protein IR spectra is available at **http://www.unco.edu/nhs/chemistry/faculty/dong/ irdata.htm.**

Hydrogen deuterium exchange

ATR-FTIR spectra of a small protein recorded at room temperature in H_2O and after a long incubation

Table 5.2.1 Assignment of amide I band positions to secondary structure; β-sheets have a strong band around 1633 cm⁻¹ and a weaker band around 1684 cm⁻¹ (see Figure 5.2.5) (data from *Biochim. Biophys. Acta* 1767, 1073–101 (2007))

Secondary structure	Average band position in H_2O (cm^{-1})	Average band position in D_2O (cm^{-1})
α-helix	1654	1652
β-sheet	1633	1630
	1684	1679
Turns	1672	1671
Disordered	1654	1645

in D_2O are shown in Figure 5.2.7. Significant differences between the H_2O and D_2O spectra are observed; this allows the rate of exchange to be studied by monitoring spectra as a function of time after the solvent is changed from H_2O to D_2O.

The amide I band generally shifts to lower wavenumber by a few cm⁻¹ upon H/D exchange; large changes in intensity of the amide II bands are also observed. The rate of H/D exchange is an indication of the degree of exposure to solvent and protein flexibility. Different proteins have very different exchange properties. It can be seen from the rate curve in Figure 5.2.7 that ~75% of hydrogens in this protein exchange rapidly with solvent. The remaining 25% have much slower solvent exchange kinetics, presumably because they are involved in H-bonds in the core of the protein.

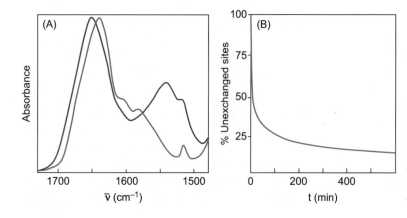

Figure 5.2.7 (A) ATR-FTIR spectra of a protonated (gray) and a deuterated protein (cyan). (B) The overall rate of H/D exchange when solvent is changed from H_2O to D_2O (adapted with permission from *Biochemistry* 41, 15267–76 (2002)).

Example Bacteriorhodopsin

Bacteriorhodopsin (bR) is a protein found in the purple membrane of a salt-loving bacterium; it is a popular model system for studying light-driven proton pumping. It converts solar energy into a pH gradient across the cell membrane that can be used to drive ATP synthesis. bR undergoes a light-induced cycle of changes for every proton it pumps out of the cell. The bR photocycle has been well characterized using fast UV/visible spectral methods and the major intermediates have been identified (see Figure 5.2.8 and Chapter 5.3).

An example of an FTIR difference spectrum, generated by subtracting spectra of bR obtained in two states of the photocycle, is shown in Figure 5.2.9. This difference technique has been developed in sophisticated ways to define the role of specific carboxylic acid sidechains in proton pumping by bR.

The difference spectra in Figure 5.2.10 were obtained using ATR-FTIR, a sensitive detection method that was introduced in Chapter 4.2. Analysis of the pH dependent spectra allowed the transient pK_a changes of single amino acid sidechains of bacteriorhodopsin embedded in the membrane to be measured. Asp 96 has a high pK_a (>12) in the ground state (bR_{570}) but this was found to drop to 7.1 (Figure 5.2.10B) during the lifetime of the N intermediate,

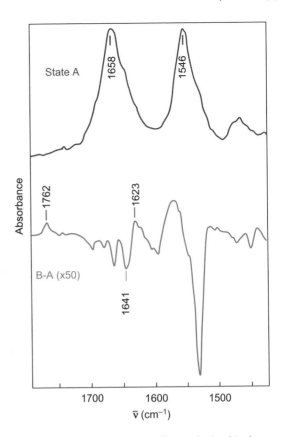

Figure 5.2.9 The FTIR spectrum of bR_{570} obtained in the dark at −20°C (state A, top). Application of light produces state B (M_{412}). The light-minus-dark difference IR spectrum is shown on a 50 × expanded scale (bottom) (adapted with permission from *J. Bioenerg. Biomembr.* 24, 147–67 (1992)).

suggesting that Asp96 plays a key role as an internal proton donor in the pumping process.

Oriented samples

The directional property associated with the transition dipole moment (Chapter 5.1; Figure 5.1.5) influences the observed spectra. Vibrations with transition dipoles parallel to the electric vector of the radiation will show preferential absorption. In oriented samples, this directional property can be detected using IR light, polarized along a particular direction. With oriented samples, linear dichroism measurements can give information on molecular conformation. In a protein α-helix, the C=O groups are aligned along the long axis of the helix; the transition dipole moments are thus aligned preferentially along the helix axis and this is reflected in the absorption spectra (see Problem 5.2.4).

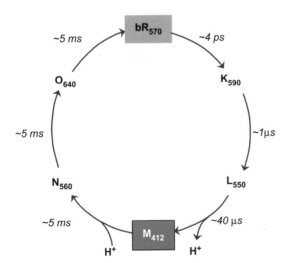

Figure 5.2.8 The photocycle of bacteriorhodopsin. The subscripts refer to the wavelengths (in nm) of the maximum absorptions of the various intermediates K, L, M, N, O, etc.

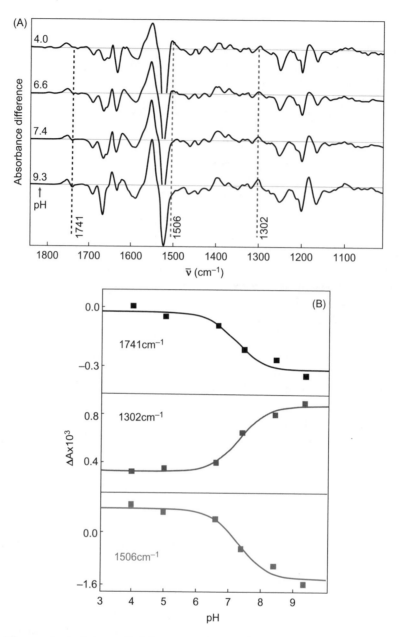

Figure 5.2.10 (A) IR difference spectra between the bR$_{270}$ state and a state characteristic of the N, O states, obtained ~7 ms after the exciting laser flash (observed at pH values 4.0, 6.0, 7.4, and 9.3). (B) Plots of the intensity of vibrational bands, at positions identified by the broken lines in A, as a function of pH. Continuous lines are fits to the Henderson–Hasselbalch equation (Chapter 2.1). Fitting the data at 1741 cm^{-1} (assigned to the C=O stretch of Asp96) shows that the transient pK$_a$ for this group is 7.1 (adapted with permission from *PNAS* 96, 5498–503 (1999)).

Example Amyloid fibrils

Amyloid fibrils are insoluble aggregates that form *in vivo*; these are associated with diseases such as Alzheimer's and transmissible spongiform encephalopathies. They result from the self-assembly of partially unfolded proteins. Whatever the structure of the original precursor protein, the predominant secondary structure found in these fibrils is β-sheet. The presence of amyloid is often identified using fluorescent dyes, circular dichroism, or FTIR. A peptide fragment [21NFLNCYVSGFH31] from a protein called β$_2$-microglobulin forms amyloid fibrils after incubation for some hours at pH 7.5, 37°C. Its FTIR spectrum with fitted components is shown in Figure 5.2.11.

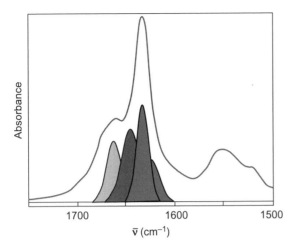

Figure 5.2.11 FTIR spectrum of the amyloid fibril formed from the peptide [21]NFLNCYVSGFH[31]. The component bands correspond to β-sheet (cyan), disordered (dark gray), and "non-α-non-β" (light gray), respectively (adapted with permission from *Biochim. Biophys. Acta* 1753, 100–7 (2005)).

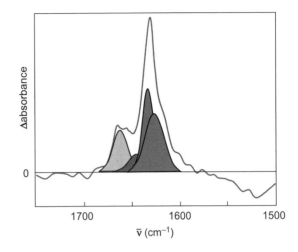

Figure 5.2.12 IR linear dichroism of peptide [21]NFLNCYVSGFH[31] and decomposed bands (same color code as Figure 5.2.11); the spectrum was obtained by collecting spectra with parallel and perpendicular polarized light and taking a difference spectrum.

IR linear dichroism spectra of the same peptide were also collected using samples oriented by centrifugation. The measurements were made using an IR microscope so as to focus solely on the oriented regions of the sample. From this experiment, the direction of the transition dipole moment of each absorption band could be determined.

Analysis of the decomposed bands in Figures 5.2.11 and 5.2.12 confirmed that the predominant structure of the amyloid formed is β-sheet. The direction of the transition dipole moment for each kind of absorption band could also be determined and the results were interpreted in terms of two kinds of β-sheet, twisted or bent with respect to each other by about 30°.

Raman scattering

Raman spectroscopy is another way of obtaining information about molecular vibrations. In IR spectroscopy, this information is obtained from the direct absorption of the incident electromagnetic waves. In contrast, the information in Raman spectroscopy is obtained from changes in the frequency of scattered light. Raman spectroscopy can be performed in water, and detectors in the UV/visible region are superior to those in the IR region, so the method is better for some applications. The intensity of a Raman spectrum is, however, usually very weak, although powerful enhancement methods can be used. One enhancement method is resonance Raman spectroscopy, where the frequency of the applied radiation corresponds to an electronic absorption band of a chromophore in the molecule; some vibrations associated with that chromophore can be then be dramatically enhanced.

Physical basis of Raman spectroscopy

We saw in Chapter 4.1 that scattering arises when electrons in a molecule oscillate in sympathy with an applied electromagnetic wave. The extent of oscillation depends on the polarizability of the molecule. Most of the oscillating electrons scatter light at the same frequency as the incident beam (Rayleigh scattering). However, a vibration can occur while the electrons oscillate, causing the oscillation frequency and the scattering to change; this vibration-sensitive effect is called Raman scattering.

Figure 5.2.13 gives a schematic representation of the energy processes involved in Raman scattering. The applied electromagnetic radiation raises the system to a higher energy state from which there can be three main outcomes: (i) Rayleigh scattering, occurring at essentially the same frequency as the applied

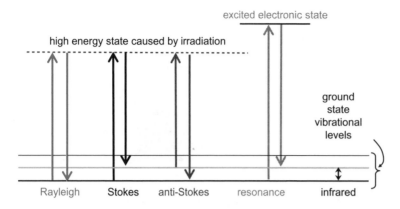

Figure 5.2.13 Processes leading to Raman scattering. The horizontal lines represent different energy levels. The measurement made by IR spectroscopy is shown by the small double headed arrow on the right.

radiation; (ii) lower energy light arising from a change to a higher vibrational state ("Stokes" lines), or (iii) higher energy light arising from a change to a lower vibrational state (anti-Stokes lines). Rayleigh scattering is much stronger (>10^6-fold) than Stokes scattering which is stronger than anti-Stokes scattering. A special case is the resonance Raman condition, where the high energy state is tuned to be close to an excited electronic state.

As discussed in Chapter 4.1, the intensity of Rayleigh scattering is related to the size of the scatterer (R_G; see Box4.1), as well as the wavelength of the scattered light and α, the molecular polarizability. Raman scattering has the same dependence, so higher sensitivity is obtained for larger molecules with high polarizability.

Raman spectra have different selection rules from IR. It was pointed out in Chapter 5.1 that "symmetric stretch" IR bands are forbidden because there is no charge displacement in going from the ground to the excited *vibrational* state. However, there can be a change in *electronic* distribution during such transitions, so some IR-forbidden vibrations can be Raman active.

Resonance Raman

When the frequency of the exciting light is near, or matches, the absorption of a chromophore, the polarizability becomes relatively large. Vibrations associated with the chromophore are thus resonance enhanced. The observed enhancement near resonance can be at least 1000-fold, so that vibrational frequencies of a chromophore can be probed *selectively* with high sensitivity. [Note fluorescence (Chapter 5.5) also involves an excited state but, unlike Raman, fluorescence involves complete absorption to an excited state, followed by a return to a lower state after a certain lifetime.]

Measurement of Raman spectra

A simple experimental apparatus for measuring Raman spectra is illustrated in Figure 5.2.14. The sample is irradiated with laser light and the scattered light is analyzed in a spectrometer. The more intense Stokes lines are usually measured. For resonance Raman, suitable monochromatic laser light sources are required to irradiate the absorbance region of the chromophore being studied. (Note that detection of the resonance Raman signal can be hindered by fluorescence signals that can overwhelm the Raman signal.)

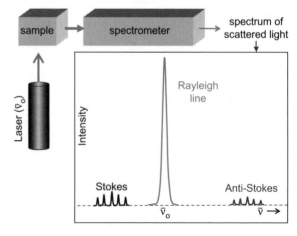

Figure 5.2.14 An experimental set-up for measuring Raman spectra. Note the output, consisting of a relatively strong Rayleigh line plus weaker Stokes and even weaker anti-Stokes lines arising from vibrational transitions.

Surface-enhanced Raman scattering (SERS)

If Raman-active molecules are adsorbed on the surface of some metals, surface plasmons (Chapter 4.2) can be generated that efficiently couple the incoming laser light to the molecules, thereby amplifying the Raman signal (compare attenuated total reflectance methods). One such technique is known as surface-enhanced Raman scattering (SERS). When combined with the resonance Raman effect, surface-enhanced resonance Raman scattering (SERRS) has high enough sensitivity to detect single molecules (compare fluorescence methods in Chapters 5.1 and 7.2) An example of the SERRS method is given at the end of this chapter (see Figure 5.2.17).

Applications of Raman spectroscopy

Raman scattering is often used to provide a "molecular fingerprint" of the sample. Water scatters weakly and no longer obscures the spectral range, as it does in IR. The 200–2000 cm^{-1} region is accessible to Raman in either D_2O or H_2O solvent. In this range, there are many low-frequency vibrations that are sensitive to sample composition and macromolecule conformation. These include the S–S stretching vibration in cystine bridges, which gives a strong Raman line at about 675 cm^{-1}; several vibration bands from aromatic sidechains and nucleic acids are also highly polarizable and thus give strong scattering and relatively intense Raman lines. High sensitivity techniques like SERRS have been incorporated in assays for the detection of proteins, DNA, and other molecules of interest, for example to detect disease-specific biomarkers and to discriminate between different pathogens.

The amide I band of proteins can be studied either by IR or Raman but the amide II band, which is intense in IR, is very weak in Raman. There is also an amide III band, however, and this can only be effectively studied by Raman scattering in aqueous solution. The amide III band (~1235 cm^{-1}) is strong for an α-helix and relatively weak (1260–1295 cm^{-1}) for a β-sheet. Bands that are sensitive to conformation can also be found in the Raman spectra of nucleic acid-protein complexes.

Resonance Raman spectroscopy can probe vibrations of sites associated directly with a chromophore,

such as heme, retinal, carotenoids, and chlorophylls. A resonance Raman scattering study of heme groups in oxy- and deoxyhemoglobin is shown in Figure 5.2.15. The UV/visible absorbance bands of $\pi - \pi^*$ transitions of a typical heme group are shown in the top spectra (see Chapter 5.3 for a description of heme UV/visible spectra). Irradiation at 514 nm in the β-band results in a significant increase in molecular polarizability which is reflected in a strong enhancement of vibrations associated with the heme chromophore. Resonance Raman spectra are thus a sensitive probe of heme environment. Note that the sulfate ion in the spectrum shown in Figure 5.2.15 is 1000-fold more concentrated than heme but gives a similar intensity spectrum; unlike heme, its spectrum is not resonance enhanced.

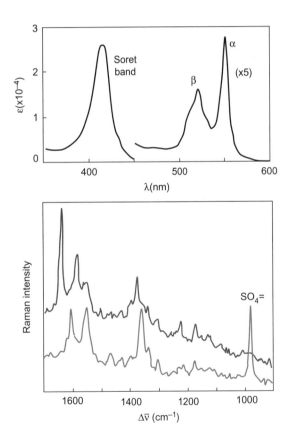

Figure 5.2.15 Resonance Raman spectra of oxy- (gray) and deoxyhemoglobin (cyan), concentration ~0.4mM; 0.4M sulfate ion is also present in the deoxy sample (the band ~981 cm^{-1}: bottom spectrum). The excitation wavelength was 514 nm, which corresponds to the β-band in the UV/visible spectrum (top) (data derived with permission from *Proc. Roy. Soc. A* 345, 89–105 (1975)).

Example Green fluorescent protein

Green fluorescent protein (GFP) forms a chromophore from its constituent amino acids (Chapter 5.5). Time-resolved absorption and fluorescence measurements of GFP in H_2O and D_2O have shown that the chromophore has at least two interconvertible protonation states, commonly referred to as states A and B, with absorption maxima of 400 and 475 nm respectively; the A form dominates. A close analogue of the GFP chromophore is 4-hydroxybenzylidene-2,3-dimethylimidazolinone, HBDI. Resonance Raman spectra of GFP in both H_2O and D_2O (Figure 5.2.16) were collected under various conditions and compared with HBDI spectra. The results indicated that the main absorption band of GFP at 400 nm represents the neutral chromophore (A form) and the minor band at 475 nm represents the anionic B form. Significant differences were also observed between the resonance Raman spectra of GFP and HBDI, presumably due to the influence of the protein on the structure of the chromophore's excited state.

As mentioned previously, an experimental problem with resonance Raman arises because of strong fluorescent emission from the excited state. This can lead to difficulties in detecting Raman spectra; this was the case for the spectra shown in Figure 5.2.16

(see also Problem 5.2.7). GFP protonation states have, however, also been studied by SERRS at the single molecule level as shown by the following example.

Example Surface-enhanced resonance Raman scattering spectra of single GFPs

Raman spectra can be obtained from individual molecules adsorbed on metallic nanoparticles, usually silver or gold. The spectra obtained arise from a surface enhanced resonance Raman scattering (SERRS) effect (see measurement section). SERRS spectra of a single enhanced green fluorescent protein (EGFP) molecule adsorbed on a silver colloid and excited at 488 nm are shown in Figure 5.2.17. An obvious feature in this time series is the frequency jump after ~15 s from 1524 (gray) to 1562 cm^{-1} (cyan); these bands can be assigned and the frequency jump interpreted in terms of a conversion of the chromophore from the deprotonated to the protonated form. Other time series from single molecules show similar results, indicating rapid reversible interconversion between the A and B forms.

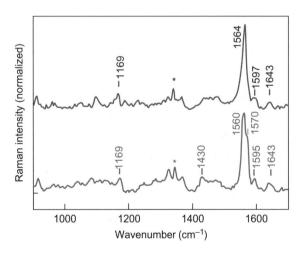

Figure 5.2.16 Resonance Raman spectrum of GFP in aqueous solution with UV excitation at 354.7nm. Top spectrum (gray) in H_2O; bottom spectrum (cyan) in D_2O. The asterisks indicate artifacts seen in the absence of GFP. These spectra were carefully compared with the spectra of a GFP chromophore analogue (HBDI) (adapted from *J. Phys. Chem. B* 105, 5316–22 (2001)).

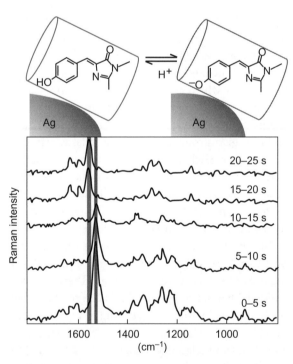

Figure 5.2.17 SERRS spectra of single GFP molecules in a silver colloid. The GFP chromophore, part of which is shown in the top cartoon, can exist in protonated and deprotonated form (see also Figure 5.5.5) (adapted with permission from *J. Am. Chem. Soc.* 125, 8446–7 (2003)).

Summary

1 Infrared (IR) spectra give information about molecular vibrations. The most common detection instruments use a moving mirror to generate an interferogram, followed by a Fourier transformation (FTIR).

2 The commonly studied IR region of the spectrum is 1400–4000 cm^{-1}. The main experimental parameter is $\bar{v}_{max}$ (cm^{-1}), the position of the maximum of the absorption band.

3 For a vibration to be IR active it must have an associated transition dipole moment.

4 Insight into molecular vibrations can be obtained by considering an oscillating spring. The vibrational frequency $\bar{v}_{vib}$ of a bond can be approximately described by $\bar{v}_{vib} = (k/\mu)^{1/2}/2\pi$, where k is a force constant and μ is the reduced mass; μ can be changed by isotope substitution.

5 The number of modes of vibration is $3N - 6$ for a molecule consisting of N atoms ($3N - 5$ if the molecule is linear). For a macromolecule, there are very many vibrational transitions but many of these can be assigned to particular bonds or groupings, such as the amide I and amide II bands in proteins. Isotope substitution is a useful assignment tool.

6 IR absorption by solvent water is a technical problem because it leaves only certain "windows" where studies can be done.

7 Raman spectra arise from light scattering experiments when a change in vibrational level occurs during the scattering process. Raman spectra can be obtained more readily from water than IR spectra.

8 Resonance Raman spectra result when the wavelength of the exciting light falls within an electronic absorption band of a chromophore in the molecule, thus giving a large sensitivity enhancement. Other sensitivity enhancement procedures are possible, such as surface-enhanced Raman scattering (SERS).

9 In biological applications, IR and Raman spectra are mainly used to: identify molecules; give information about protein conformation; measure ligand binding to macromolecules; define ionization states; probe hydrogen bonds and hydrogen–deuterium exchange.

Further reading

Useful website
http://www.infochembio.ethz.ch/links/en/spectrosc_ir.html

Reviews
Harris, P.I., and Chapman, D. Does Fourier-transform infrared spectroscopy provide useful information on protein structures? *Trends Biochem. Sci.* 17, 328–33 (1992).

Barth, A. Infrared spectroscopy of proteins. *Biochim, Biophys, Acta* 1767, 1073–101 (2007).

Hiramatsu, H., and Kitagawa, T. FT-IR approaches on amyloid fibril structure. *Biochim. Biophys. Acta* 1753, 100–7 (2005).

Hunt, N.T. 2D-IR spectroscopy: ultrafast insights into biomolecule structure and function. *Chem. Soc. Rev.* 38, 1837–48 (2009).

Hogiu, S., Weeks, T., and Huse, T. Chemical analysis *in vivo* and *in vitro* by Raman spectroscopy-from single cells to humans. *Curr. Opin. Biotechnol.* 20, 63–73 (2009).

Dafforn, T.R., and Rodger, A. Linear dichroism of biomolecules: which way is up? *Curr. Opin. Struct. Biol.* 14, 541–6 (2004).

 ## Problems

5.2.1 CO_2 is a linear molecule and therefore has $3N - 5 = 3 \times 3 - 5 = 4$ expected vibrational bands. In CO_2 IR spectra, two (degenerate) bending vibrations are observed at 667 cm^{-1} and one asymmetric stretch is observed at 2349 cm^{-1}. A further (symmetric stretch) band is predicted at 1537 cm^{-1} but is not observed. Comment on these results.

5.2.2 When CO_2 binds to carbonic anhydrase, a zinc-containing enzyme that catalyzes CO_2 hydration, no change in the 2349 cm^{-1} asymmetric stretch band was observed. Comment on this result.

5.2.3 Alanine has a band at 1308 cm^{-1}, assigned to −CH deformation. In deuteroalanine this band is absent, but a new band appears at 960 cm^{-1}. Why?

5.2.4 Calculate the vibrational frequency of $^{12}C-^{16}O$; assume the bond force constant is 1850 Nm^{-1} and the atomic mass unit is 1.66×10^{-27} kg.

5.2.5 CO binds covalently to the heme group in myoglobin (Mb). When it is bound, the heme is planar and its Fe atom lies in the heme plane (state a) (see also Problem 5.1.3). The bond between Fe and CO can be broken by light. The figure illustrates myoglobin, irradiated with a light pulse to displace the bound CO atom. The rate of rebinding can be monitored by FTIR.

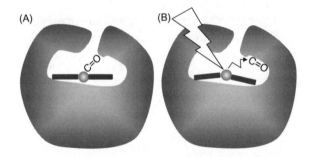

After photodissociation, the heme group buckles and the Fe moves out of the heme plane (state B). At low temperatures, CO ultimately rebinds and the system returns to state A. The stretching frequency of free CO is 2140 cm^{-1} for the isotope combination $^{12}C^{16}O$. When bound to Mb, this changes to 1945 cm^{-1}.

(i)   What would the stretching frequencies be for $^{13}C^{16}O$, when free and bound?

(ii) X-ray crystallography and other experiments suggest that the CO becomes trapped in a docking site near the heme after photo-dissociation. Using pulsed time-resolved FTIR, two rate constants were observed for this relocation between the docking and iron bound sites; one with rate constant 0.2ps, the other with 0.5 ps. The absorption direction for the IR bands also appeared to be different for the two decay processes. Can you suggest an explanation for these observations?

5.2.6 An IR dichroic study of bacteriorhodopsin (bR) in oriented dried purple membranes showed that the amide I band is polarized perpendicular to the membrane bilayer. Using information in Figure 5.2.8, what does this suggest about the conformation of bR in the membrane?

5.2.7 "Applications of resonance Raman techniques are limited by chromophore fluorescence". Give an explanation for this statement.

5.2.8 It has been suggested that the conversion of inter-chain –SH groups of lens proteins to S–S bonds leads to less soluble, higher molecular weight products whose presence may help to account for some of the properties of aging in an eye lens with cataract. Raman scattering can be used to study the intact living lens. The Raman scattering spectra in the –SH and S–S vibrational regions of an intact rat lens at ages 28 days and 7 months are shown below. The signal at 2580 cm^{-1} arises from the –SH stretching vibration. Interpret the changes in these bands. Would any other information help your conclusion?

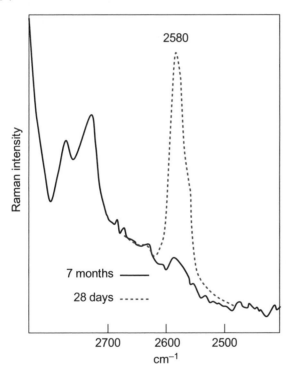

5.3 Ultraviolet/visible spectroscopy

Introduction

Absorption in the UV (~200–400 nm) and visible (~400–750 nm) ranges arises from electronic transitions between energy states in a **chromophore** (Gr: "color bringer"). This absorption can give rise to color, vision, and the conversion of sunlight into energy by plants. Color can also arise from selective reflection (iridescence—e.g. from a peacock's tail) but here we are mainly concerned with color arising from selective absorption. (Note that color then is usually determined by light that is *not* absorbed, e.g. copper sulfate solution appears blue because it absorbs in the red). Chromophores that absorb in accessible regions of the spectrum include aromatic amino acids, nucleic acid bases, NADH, porphyrins (e.g. chlorophyll and heme), carotenoids (e.g. beta-carotene), melanin, bilirubin, and some transition metal ions.

Most UV/visible spectroscopy experiments are relatively simple and inexpensive to perform, yet they can be very informative. The most general application is to measure concentrations, e.g. of proteins, NADH, and nucleic acids. Concentration time courses can be measured very rapidly (< ps) using pulse laser techniques. In most cases, there is a direct and simple relationship between the number of molecules present and the observed absorption (Beer's law). In some cases, however, the absorption significantly depends on the chromophore *environment*. Spectral changes can then be interpreted in terms of biological events, such as the unfolding of DNA.

Measurement of electronic spectra

A spectrum in the UV/visible region is usually presented as a plot of absorbance versus the wavelength of the applied irradiation, expressed in nanometers. It is obtained by measuring the amount of light absorbed by a sample.

The intensity of light falls off exponentially as it passes through an absorbing sample (Figure 5.3.1). The absorbance, or optical density, A, of a compound, is thus defined by the **Beer–Lambert law** $A = \log_{10} I_o/I_t = \varepsilon c l$, where I_o is the intensity of the incident radiation, I_t is the transmitted intensity ($I_o - I_t$ is the radiation absorbed), c is the concentration of the compound, l is the path length of the light through the cell, and ε is the absorption coefficient. The units used for ε frequently depend on the units used for c and l, but a convenient unit is $L\ mol^{-1}\ cm^{-1}$. Spectra are characterized by the values of λ_{max} (the positions of maximum absorbance) and the corresponding value of ε (ε_{max}).

A UV/visible spectrometer (or spectrophotometer)

An outline of a double-beam **UV/visible spectrometer** is shown in Figure 5.3.2. A "source" produces a wide range of wavelengths; typically a deuterium arc lamp

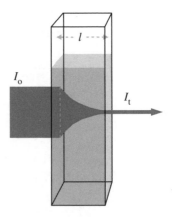

Figure 5.3.1 Light (cyan) absorbed in a sample cell of width *l*. There is an exponential decay in signal as the light passes through the cell.

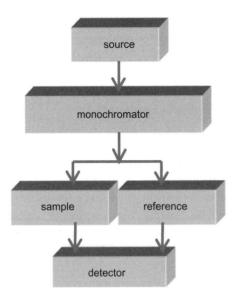

Figure 5.3.2 Schematic view of a double-beam UV/visible spectrometer.

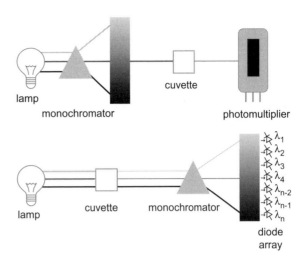

Figure 5.3.3 The top illustration shows normal spectrum collection, where the absorbance at one wavelength is measured; the monochromator is then adjusted to pass another wavelength and another measurement made. The bottom picture illustrates the collection of a whole spectrum at once using an array of photodiodes, each detecting absorbance at a different wavelength.

and a heated tungsten filament are used to produce UV and visible light, although light emitting diodes are increasingly employed. The wavelengths are separated by a monochromator, usually a grating made from ruled glass with grooves of the same dimensions as the wavelength of light to be dispersed. The beam produced is then split into two, one beam passing through the sample, the other acting as a reference beam.

Samples are typically placed in a transparent cell, known as a cuvette. The detectors are often a photomultiplier—vacuum tubes that are extremely sensitive to light; photodiodes are also efficient low cost light detectors although they have lower sensitivity than photomultipliers. Routine instruments only extend down to about a wavelength of 185 nm because absorption by oxygen and many solvents is considerable in the far UV region ($\lambda < 200$ nm).

For the rapid detection of signals from a range of wavelengths, diode array spectrometers of the sort illustrated in Figure 5.3.3 are often used. Examples of the use of such devices will also be given later when rapid spectral data collection is discussed.

Electronic energy levels and transitions

Some of the basics of absorption spectra were introduced in Chapter 5.1. You are reminded that

electronic energy levels are associated with different molecular orbitals (Tutorial 10). When an electron undergoes a transition, it is transferred from one molecular orbital to another. Transitions between electronic energy levels can be induced by electromagnetic radiation if they have an electric transition dipole moment (Box 5.1). Symmetrical transitions, such as between a 1s and 2s orbital, are forbidden.

The π bonding orbital, the π^* anti-bonding orbital, and the n non-bonding (lone-pair) orbital are involved in many of the electronic transitions in biological systems (Tutorial 10). The π–π^* transition, which results in an unsymmetrical displacement of charge, is an allowed transition. The n–π^* transition is generally weaker and can be thought of as occurring in two stages: a rotation and a displacement of charge. The rotation part interacts only with the magnetic component of the radiation. This interaction with the magnetic component is much weaker than the interaction with the electric component, although it is very important for optical activity (Chapter 5.4).

For transition metal complexes (e.g. involving copper or cobalt) most of the observed electronic transitions can be understood by considering the d-orbitals of the central atom (Tutorial 10). In the presence of ligands, the d-orbital levels split in a way that depends on the stereochemical arrangement of the ligands around the metal. d–d transitions are generally "forbidden", but strict symmetry selection rules only apply to isolated molecular orbitals and many of these "forbidden" transitions are, in fact, observed.

Time scale of an electronic transition and the Franck–Condon rule

Some of the properties of transitions to excited states can be understood by considering Figure 5.3.4 which shows Morse curves for the ground and excited electronic states. As explained in Chapter 5.2, such curves represent the energy of a diatomic molecule as a function of internuclear separation; horizontal lines represent allowed vibrational energy levels. The minimum position of the ground state curve (r_e) is the equilibrium bond-length of the two nuclei in the

diatomic molecule. This equilibrium position, r_e, tends to be greater in the excited state because the excited electron orbitals are further from the nuclei. Essentially, all the molecules will initially be in the lowest vibrational level of the ground electronic state (see Boltzmann law, Chapter 5.1).

Another factor we have to consider for transitions to excited states is time scales. The resonance condition for a transition between energy states is $E = h\nu$. This involves a charge displacement induced by the applied radiation; the time taken for this displacement will be related to $1/\nu$. A typical electronic transition with a wavelength $\lambda = 420$ nm has a frequency $\nu = c/\lambda = 7.14 \times 10^{14}$ Hz (Tutorial 8); so the transition will take $\sim 1.4 \times 10^{-15}$ s. In contrast, a vibration with $\bar{\nu} = 1000$ cm^{-1} has a transition time of $\sim 33 \times 10^{-15}$ s ($\bar{\nu} = 1/\lambda = \nu/c$; $\nu = 3 \times 10^{13}$ s^{-1}). In other words, the electronic transition is 20-fold faster than the vibrational one. These relative time scales provide the basis for the Franck–Condon rule, which states that nuclei (associated with vibrations) are essentially static during an electronic transition. An electronic transition in which the nuclei do not move can be represented by a vertical line in Figure 5.3.4. Depending on the relative displacement of the ground and excited state curves, this vertical line will meet the upper curve at vibrational levels that are usually higher than the lowest vibrational level of the excited state. There will be a whole series of such lines at positions that depend on transient interactions with solvent and other environmental factors; these environmental shifts usually lead to broad, unresolved bands in solution.

Absorption properties of some key chromophores

There are two main classes of spectra observed in biological systems: those from organic molecules, often involving transitions to π^* orbitals, and those from metal ions (e.g. Cu), involving d–d transitions. The three main types of organic chromophore are: (i) peptide bonds and aromatic amino acids in proteins; (ii) purine and pyrimidine bases in nucleic acids; and (iii) highly conjugated (double bonded) molecules, such as heme and chlorophyll.

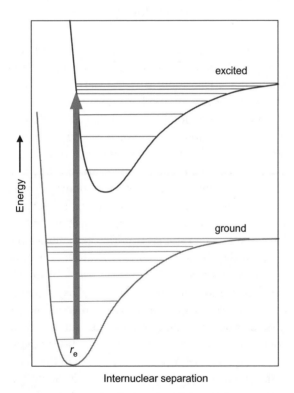

Figure 5.3.4 Potential energy curves for the ground and first excited electronic states, with ground state equilibrium position r_e. The horizontal lines represent the allowed vibrational energy levels and the vertical cyan line represents absorption.

Peptide bonds and amino acids

The spectra observable from peptide bonds occur in the far UV (~200 nm). These mainly arise from a π–π^* transition which is centered at 190 nm with a minor contribution from an n–π^* transition centered at 220 nm. A number of amino acid sidechains, including histidine, also have transitions around 210 nm, but these are usually masked by the relatively strong peptide bond absorption. A useful wavelength range for proteins is above 230 nm, where there are significant absorptions from the aromatic sidechains of phenylalanine, tyrosine, and tryptophan and cystine (S–S bonds) (see Figure 5.3.5 and Table 5.3.1).

It is possible to estimate the molar absorption coefficient of a protein from knowledge of its amino acid composition. A convenient way to determine protein concentration is to compute the absorption coefficient of a native protein from the molar absorption coefficients of the constituent tyrosines, tryptophans, and cystines at a given wavelength, using an equation of the form: $\varepsilon_{prot} = n_{trp} \times \varepsilon_{trp} + n_{tyr} \times \varepsilon_{tyr} + n_{cystine} \times \varepsilon_{cystine}$, where n_{trp} is the number of tryptophans, etc. This method is used by the ExPASy Proteomics Server (**http://www.expasy.ch/tools/protparam-doc.html**) to calculate the absorption coefficient of any submitted protein, assuming the following values at 280 nm: $\varepsilon_{trp} = 5500$, $\varepsilon_{tyr} = 1490$, and $\varepsilon_{cystine} = 125$ in units of L mol^{-1} cm^{-1} (Table 5.3.1).

Table 5.3.1 Absorption maxima (λ_{max}) and approximate absorption coefficients (ε_{max}) for selected amino acids and for a typical protein (bovine serine albumin, BSA) at neutral pH

Sidechain	λ_{max} (nm)	ε_{max} (L mol^{-1} cm^{-1})
Tryptophan	280	5500
Tyrosine	274	1400
Phenylalanine	257	200
Cystine	250	300
BSA	280	40 900

Local environmental and intramolecular effects can change the absorbance (see below); one way round this problem is to measure the absorbance of a protein denatured in 6M GdHCl. ε_{prot} can then be calculated from the sum of the absorbances of model compounds in 6M GdHCl.

Purine and pyrimidine bases in nucleic acids

The absorption of nucleic acids arises mainly from the constituent purine and pyrimidine bases in the region 200 and 300 nm and results from n–π^* and π–π^* transitions. Some representative spectral values of purine and pyrimidine bases and derivatives are listed in Table 5.3.2. The absorption of

Table 5.3.2 Absorption maxima (λ_{max}) and absorption coefficients (ε_{max}) for selected nucleic acid bases and their derivatives at neutral pH

Base	λ_{max} (nm)	ε_{max} (L mol^{-1} cm^{-1})
Adenine	261	13 400
NADH	340	6220
NAD$^+$	259	16 900
Guanine	275	8100
Cytosine	267	6100
Uracil	260	9500
Thymine	264	7900

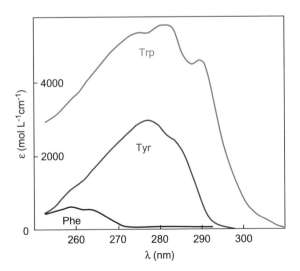

Figure 5.3.5 Near-UV absorbance spectra of Trp, Tyr, and Phe sidechains.

oligonucleotides can also be calculated using online servers (e.g. **http://www.scripps.edu/researchservices/ corefac/biopolymercalc2.html**).

A_{260} measurements can be used quantitatively: for double-stranded DNA, A_{260} is 0.020 µg mL^{-1} cm^{-1}, and it is 0.027 µg mL^{-1} cm^{-1} for single-stranded DNA and RNA.

Nucleic acid samples are often contaminated with other molecules (e.g. proteins). The ratio of the absorbance at 260 and 280 nm ($A_{260/280}$) can be used to assess purity. For pure DNA, $A_{260/280}$ is ~1.8, and for pure RNA $A_{260/280}$ is ~2. As proteins absorb at 280 nm, changes in these ratios can indicate contamination. For example, 10% protein contamination would typically increase the $A_{260/280}$ ratio for DNA to ~1.98. (Note the $A_{260/280}$ ratio is more sensitive to DNA contamination of protein solutions than protein contamination of DNA because the A_{260} and A_{280} values are higher for DNA than protein.)

Highly conjugated systems

In large organic molecules that are highly conjugated (have many alternating single and double bonds), there is extensive electron delocalization which lowers the energy required for a transition (see Tutorial 9) and results in spectra in an accessible region. For example, the porphyrin ring system has important absorption bands. One example is the spectrum from heme, as shown in Figure 5.3.6. The most intense band, around 420 nm, is called the Soret band, after its discoverer, and several weaker absorptions (Q bands) at higher wavelengths (450–700 nm).

Note that the π–π* transitions dominate the porphyrin spectra and the d–d transitions in iron play a minor role. Variations in the peripheral substituents of the porphyrin ring can cause spectral changes (see also Figure 5.3.7), as can a change of metal atom.

Chlorophyll, another example of a conjugated system, also has a porphyrin ring with a coordinated Mg ion rather than Fe. It absorbs sunlight for use in photosynthesis. There are two main types of chlorophyll, named a and b, that differ only in the composition of a sidechain (in a it is –CH$_3$, in b it is CHO), yet there is a relatively large spectral change between a and b, illustrating the sensitivity of absorbance spectra to chemistry. The two kinds of chlorophyll complement each other in absorbing sunlight. It can be seen from the spectra in Figure 5.3.7 that plants obtain their energy from the blue and red parts of the spectrum, with very little light absorbed in the green 500–600 nm region. Plants appear green because of this non-absorbing window.

Transition metal spectra

The spectra of transition metal complexes can be interpreted in terms of $d \rightarrow d$ transitions (Tutorial 10). The intensities and band positions depend on the metal, the ligand, and the arrangement of the ligands around the metal. $d \rightarrow d$ transitions are forbidden in a symmetrical environment but become increasingly intense when there is a breakdown in symmetry. For example $[Co(H_2O)_6]^{2+}$, which is octahedral, has absorption bands at 513 nm and appears

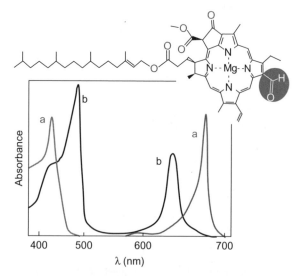

Figure 5.3.7 The formula of chlorophyll b is shown; chlorophyll a has a CH$_3$ group in the cyan position. The spectra of the two molecules are also shown.

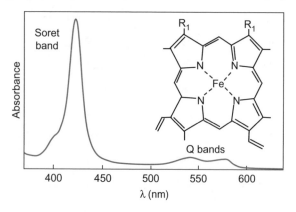

Figure 5.3.6 UV/visible spectrum of ferrous heme bound to carbon monoxide showing Soret and Q bands.

pale pink; in contrast $[Co(Cl)_4]^{2-}$ has lower symmetry with intense absorbance bands at 667 nm and appears blue.

The difference in energy between the ground and first excited states (and hence the position of the absorption bands) in *d*-orbitals depends on the "ligand field splitting", Δ, between lower and upper *d*-energy levels (Tutorial 10). The more strongly the ligand binds, the lower the wavelength of the absorbance will be because Δ is larger (higher energy). The ligands may be placed in a series according to their binding, e.g. $I^- < Br^- < Cl^- < SCN^- < F^- < OH^- < H_2O < NO_2^- < CN^- \approx CO$.

Charge-transfer spectra

Many transition metal complexes exhibit intense absorption bands, often referred to as **charge-transfer spectra**; these can be considered as arising from the removal of an electron from one atom and its "transfer" to another. A well-known example is Fe(III) thiocyanate, $Fe(SCN)_3^{2+}$, which is intensely red because an electron from the thiocyanate is transferred to the iron. Another example is intensely purple permanganate, MnO_4^-. The position of charge-transfer bands depends on both metal ion and ligand and on the relative ease of oxidation or reduction of these species; many metal-containing electron-transfer proteins show strong charge-transfer bands.

Applications of UV/visible spectra

The most universal use of absorption spectra is the measurement of **concentration** (see discussion of key chromophores above). A colored ligand is often exploited in enzyme assays; a well-known example is the use NADH because of its convenient absorbance band at 340 nm (Table 5.3.2).

Because the absorption bands of organic biological chromophores are broad, the spectra of individual chromophores are rarely resolved; their electronic spectra therefore do not often give direct information about structure (see, however, the carbonic anhydrase example below). The spectra are, however, **sensitive to environment**. The electron distribution in the excited and ground states is different, e.g. the electrons in a π^* orbital are generally more exposed than in a π orbital (Tutorial 10). Differential effects of environment (e.g. solvent) on ground and

excited states can change the separation between the energy levels and the spectra. The excited state is usually more sensitive to environment than the ground state and interactions are particularly likely when the solvent is polar. Interactions with other transition dipoles can also affect the spectra (Tutorial 11 and see below). An increase in the energy gap causes a blue shift in the absorption band, while a decrease causes a red shift.

Solvent perturbation

In solution, the molecules in each electronic state will be surrounded by a solvent "cage" that can perturb the spectra. One effect of these solvent interactions is to cause line broadening because of diffusion induced fluctuations. The environment of different chromophores can also be probed using **solvent perturbation** methods, where spectra from a sample placed in two slightly different solvent systems are compared. The spectral shifts are very small, so difference techniques are used to detect them, as shown in Figure 5.3.8.

An example of a solvent perturbation experiment is shown in Figure 5.3.9. The number of exposed chromophores in a protein can be estimated using this kind of experiment as only those chromophores near the surface can sense a change in solvent

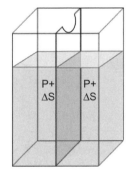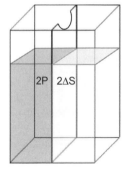

Figure 5.3.8 Tandem cells used to obtain solvent perturbation spectra. On the left the protein is exposed to solvent perturbation in both halves of the cuvette (e.g. ΔS = 20% ethylene glycol). On the right, we have the same number of molecules in the optical path length but no solvent perturbation, because P and ΔS are separated. These two tandem cells can be placed in a double beam spectrophotometer (Figure 5.3.2) as reference and sample. The semicircular gap at the top of the cuvette allows ready mixing of the samples for control experiments.

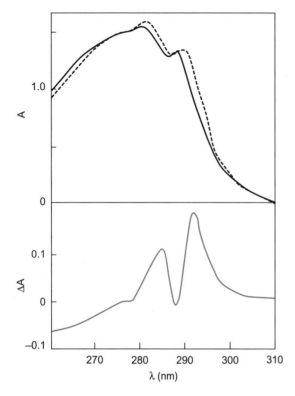

Figure 5.3.9 An example of solvent perturbation spectroscopy. The absorbance spectra are of tryptophan in water (solid line) and water + 20% dimethyl sulfoxide (dotted line). The difference spectrum below is shown in cyan.

composition. (It is assumed that the protein is not denatured by the solvent change.)

Spectral perturbation by protein environment

In human photoreceptors, or cones, there are large numbers of proteins called opsins—transmembrane bundles of 7 α-helices surrounding the chromophore, 11-*cis*-retinal. There are three different kinds of receptor in humans with distinct absorption spectra (blue, green, and red), yet they are all opsins and all use the same chromophore. The absorption differences between the blue, green, and red receptors arise because of *different* amino acid sidechain interactions in the vicinity of the chromophore (Figure 5.3.10).

Spectra of proteins containing transition metal ions

The electronic spectra of transition metals can provide structural information about the ligands and their geometry; usually this is achieved by comparisons with model compounds. Metals with no suitable absorbance spectrum can often be replaced with ones that do, e.g. zinc with cobalt (see example below), or magnesium with manganese.

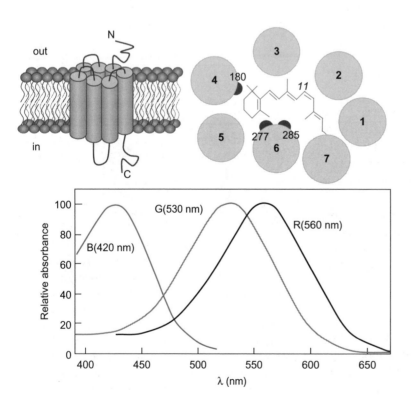

Figure 5.3.10 A representation of the 7 α-helix protein, opsin, and its chromophore 11-*cis*-retinal. There are three main classes of opsin that give blue, green, and red receptors with maximal absorption at 420, 530, and 560 nm, respectively (see spectra in lower diagram). Variation of certain amino acids in the sequence causes the changes in absorbance position. The blue receptor has many changes, but red/green sensitivity is mainly brought about by changes in the illustrated residues, namely 277, 285, and 180.

Example Carbonic anhydrase spectra

The enzyme carbonic anhydrase hydrates CO_2 rapidly and efficiently. It is a zinc-containing enzyme but the enzyme still works well when the zinc is substituted by cobalt. The cobalt spectra are very sensitive to environment, and pH titrations detect a group ionizing with a $pK_a \sim 6.8$ (inset Figure 5.3.11). This has been interpreted as an ionizing water molecule coordinated to the Zn/Co; the large reduction in water pK_a (from ~14 to 6.8) arises because of the enzyme environment and the acidic effect of the metal ion.

Box 5.3 Isosbestic points

The spectra in Figure 5.3.11 illustrate what are known as isosbestic points. If two species with different spectra (e.g. A and B) are in equilibrium, then the observed spectrum at a particular wavelength will be $S(\lambda) = \varepsilon_{A\lambda}[A] + \varepsilon_{B\lambda}[B]$, where [ ] means concentration and $\varepsilon_{A\lambda}$ is the extinction coefficient of A at wavelength λ. If $\varepsilon_{B\lambda} = \varepsilon_{B\lambda}$ at some wavelengths then the same absorbance will be observed for all ratios of [A]/[B]. Isosbestic points are useful for measuring the absolute concentration of a chromophore in a sample.

Properties associated with the direction of the transition dipole moment

As we have seen in Chapter 5.1, transition dipole moments are directional. Three important absorption properties that arise from this directional property are: (i) linear dichroism, (ii) transition coupling, and (iii) hypochromism (and hyperchromism).

Linear dichroism of oriented samples

The differential absorption of parallel- and perpendicular-polarized light by an oriented sample is known as linear dichroism (LD = $A_\parallel - A_\perp$). The transition dipole associated with a given transition (e.g. $\pi \to \pi^*$) has a definite orientation, which causes selective absorption of light polarized along this orientation (see Figure 5.1.5 and IR dichroism discussion in Chapter 5.2). If molecules are, or can be induced to be, oriented, linear dichroism can give useful information additional to that of normal absorption spectra. LD has been applied to various naturally occurring ordered arrays, including membranes, protein fibers, and DNA strands, and new sample alignment methods are being

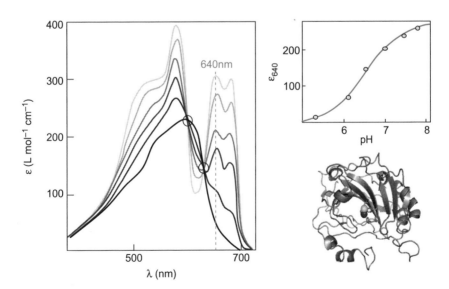

Figure 5.3.11 The Co^{2+} spectra of carbonic anhydrase as a function of pH. The enzyme structure (PDB: 1CA2) is shown with the natural zinc ion (cyan sphere). Top right gives a plot of ε_{640} as a function of pH (corresponding to the dotted line on the left). This experiment shows that there is a group with a pK_a ~6.8 closely associated with the metal ion. "Isosbestic" points are marked with circles (see Box 5.3).

developed, including orientation by shear flow, electric fields, and gel squeezing.

Example Linear dichroism of DNA

DNA can be oriented by flow and LD has been applied quantitatively to lengths of 800 bases or longer. The π–π^* transitions of DNA are all polarized in the plane of the bases, so that when the bases are assembled approximately perpendicular to the helix axis in B-form DNA the transitions are all polarized perpendicular to the helix axis. If the B-DNA bases were all oriented at exactly 90° to the helix axis, the LD of B-DNA would simply be the inverse of the absorbance spectrum. In practice, this is not quite true since the bases are slightly tilted. LD measurements have been used to show that the bases of B-DNA in solution lie at an average angle ~80° to the helix axis (see further reading).

Coupling between transitions

A transition between two energy levels involves a transition dipole moment (Box 5.1). If two chromophores with their associated dipole moments are in close proximity, the resulting spectrum is different from when they are far apart. As discussed in Tutorial 11, two shifted bands arise when two dipoles interact. This effect is sometimes known as exciton coupling. Figure 5.3.12 shows the difference between the monomer and dimer spectra of bacteriochlorophyll. In the photosynthetic reaction center where two bacteriochlorophyll molecules (the special pair—see Figure 5.3.19) are stacked together in a specific way, the dimer spectrum is significantly perturbed.

Hypochromism in DNA

A special case of transition coupling applies to DNA spectra. The term hypochromism (less color) means that the absorption intensity of a sample is less than the sum of its constituent parts; hyperchromism (more color) arises when the absorption is more than expected. Hypochromism is common in the spectra of nucleic acids with the absorbance of double-helical DNA being significantly less than that of the free bases or that of denatured DNA. The effect is widely used to monitor nucleic acid conformational changes.

A qualitative way of explaining hypochromism/hyperchromism effects is to assume that when a transition occurs, the transition dipole moment μ_{ge} (Box 5.1) induces instantaneous dipoles moments in neighboring chromophores. Interactions between μ_{ge} and these induced dipoles can reduce or enhance the transition, depending on the structural arrangement of the chromophores. Hypochromism arises when there is a parallel arrangement of dipoles (see Figure 5.3.13); here, the direction of the induced dipoles is opposite to that of the inducing dipole and this leads to a net reduction in μ_{ge} and a decrease in absorption. If the dipoles form a head-to-tail arrangement, the net effect will appear as an increase in μ_{ge} and an increase in absorption (hyperchromism).

Illustrations of hypochromism are given in Figures 5.3.14 and 5.3.15; these examples all show single stranded DNA having stronger absorbance than double stranded DNA. This can be used to monitor temperature-induced DNA unfolding (Figure 5.3.15) and the sensitivity of the DNA melting temperature to GC base content in DNA (see also Problem 5.3.6.).

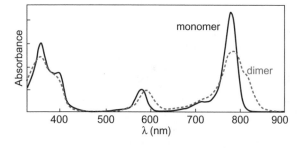

Figure 5.3.12 Absorption spectra of bacteriochlorophyll in carbon tetrachloride for a monomer and dimer (adapted with permission from *J. Am. Chem. Soc.* 88, 2681–8 (1966)). Compare also with the spectra in Figure 5.3.7, collected under different conditions.

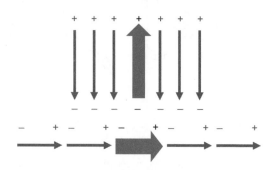

Figure 5.3.13 Arrangements of transition dipoles that accounts for hypochromism (top) and hyperchromism (bottom).

Monitoring rapid reactions

Many biological reactions are rapid and UV/visible spectroscopy is widely used to measure them. A method for monitoring rapid reactions is the "stopped flow" method (Figure 5.3.16). Such devices can measure reactions after rapid mixing, with a "dead" time that is less than 1 ms. This apparatus and variants such as "quench flow" and "continuous flow" measure non-equilibrium reaction rates.

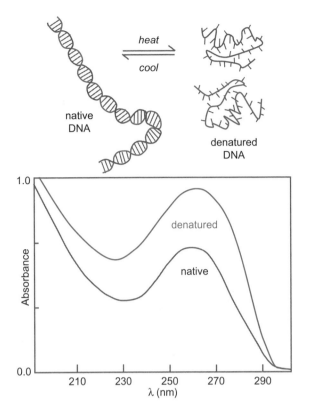

Figure 5.3.14 Illustration of hypochromism in DNA where the absorbance of native DNA is about 40% less than that of denatured DNA.

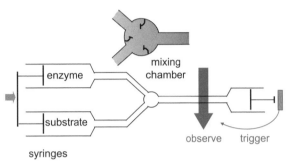

Figure 5.3.16 Schematic illustration of a stop flow apparatus, showing an expanded view of the mixing chamber with three baffles to aid mixing.

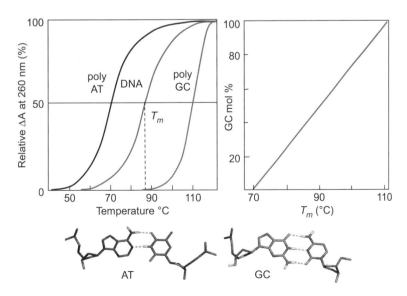

Figure 5.3.15 An application of hypochromism. The melting temperature, T_m, of DNA can be readily monitored by measuring the point at which there is 50% denaturation. GC base pairs are more resistant to unfolding than AT pairs because of three rather than two H-bonds. This gives a good correlation between GC content and T_m (right).

Example Lactate dehydrogenase

A simple example of a stop flow experiment is shown in Figure 5.3.17. This shows the change in absorbance observed at 340 nm, after 1 mmol L^{-1} pyruvate was added to a sample containing 2.5 µmol L^{-1} lactate dehydrogenase bound to stoichiometric amounts of NADH. The enzyme is a tetramer so the bound NADH concentration is 10 µmol L^{-1}. The overall reaction is pyruvate + NADH → lactate + NAD+ and the change in A_{340} is caused by the oxidation of NADH to NAD+. The time course is an exponential curve, suggesting a first order process with the four subunits behaving identically.

Example Light-driven reactions

Sophisticated methods, involving laser pulses and special detectors, have been developed to study very rapid light-driven reactions. **Bacteriorhodopsin** is a model for the study of light-driven proton pumping. The protein has already been discussed in Chapter 5.2 and its "photocycle" was shown in Figure 5.2.8. It has seven transmembrane helices

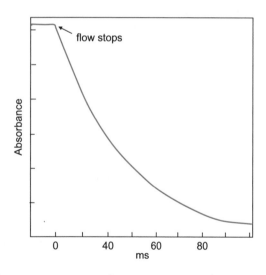

Figure 5.3.17 Time course observed in a stop flow apparatus with the A_{340} measured as a function of time after mixing pyruvate with the NADH bound to the enzyme lactate dehydrogenase.

that surround a retinal chromophore covalently bound to lysine 216. Photoisomerization of the all-*trans* retinal to the 13-*cis* configuration initiates the vectorial translocation of a proton to the extracellular side, followed by reprotonation from the cytoplasm. There are several conformational states in the pumping process that can be resolved by optical spectroscopy (Figure 5.3.18). By using pulsed lasers and diode array detectors of the kind shown in Figure 5.3.3 it has been possible to measure the interconversion rate between these states with picosecond resolution.

Example Photosynthetic reaction center (PRC)

The PRC in plants and bacteria captures light energy for photosynthesis using special chromophores. Light energy can be absorbed directly by the PRC, or passed to them from surrounding "antenna" chromophores. The bacterial PRC has been an important model for understanding the structure and chemistry of this light-capturing process. The crystal structure determined by Robert Huber and colleagues in 1982 (Chapter 4.3) revealed how four hemes, four bacteriochlorophyll b (BChl-b), two bacteriopheophytin b (φ), two quinones (Q$_A$ and Q$_B$), and a ferrous ion are oriented in the protein (Figure 5.3.19) which has two homologous subunits (L and M). Optical spectroscopy, using laser flashes as short as 1 fs and rapid detection devices, of the sort illustrated in Figure 5.3.3, has been used to monitor the appearance and disappearance (bleaching) of PRC absorption bands as a function of time. These measurements showed that the primary event is absorption of a photon by the "special pair" of BChl-b molecules. Photons arrive at the special pair very rapidly via four heme groups (not shown). The excited state created in the special pair has a very short lifetime and an electron is transferred to φ$_A$ within 3 ps leaving a positive charge on the special pair. Some 200 ps later the electron has migrated to quinone A and within a further 100 µs the Q$_A$ radical transfers its electron to Q$_B$. The non-heme Fe(II) is not reduced in this process and its role is not entirely clear.

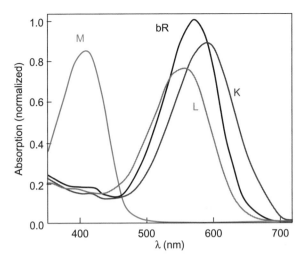

Figure 5.3.18 The respective normalized absorption curves for the K, L, and M states of bacteriorhodopsin in the photocycle—see Figure 5.2.8 (adapted with permission from *PNAS* 96, 4413–9 (1999)).

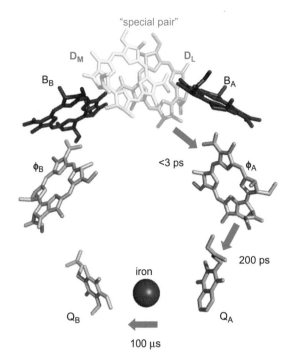

Figure 5.3.19 Diagram of the arrangement of chromophores in the bacterial PCR; from PDB coordinates 1PRC with many sidechains removed. The cyan arrows show the pathway of electrons after leaving the special pair and their lifetimes, as determined by pulsed laser spectroscopy.

➕ Summary

1. Electronic spectra involve transitions between different electronic energy states. The observable regions are ~200–400 nm (ultraviolet) and ~400–750 nm (visible).
2. The term chromophore describes the group giving rise to the electronic transition. Examples of chromophores in the accessible UV/visible range are aromatic amino acids, nucleic acid bases, NADH, hemes, polyunsaturated centers, and some transition metal ions.
3. Two parameters characterize an absorption band: the position of the maximum (λ_{max}) and the absorption coefficient (ε). The relationship between absorption coefficient, A, and concentration, c, is given by the Beer–Lambert law, $A = \log_{10}(I_o/I_t) = \varepsilon cl$.
4. The probability of a transition occurring depends on a displacement of charge and conservation of spin. The extent of charge displacement is related to the electric transition dipole moment. Selection "rules" exist but are not rigorously obeyed.
5. Different time scales for an electronic transition ($\sim 10^{-15}$ s) and molecular rearrangement ($\sim 10^{-13}$ s) give rise to the Franck–Condon principle (absorption is "straight up" in Figure 5.3.4).

6 The electronic energy levels can be described by molecular orbitals. For organic molecules with unsaturated bonds, the molecular orbitals involved in observable transitions are predominantly π, π^*, and n.

7 Transition metal ions give spectra that are characteristic of the nature and stereochemistry of the coordinated ligands. The spectra can mostly be explained by transitions involving the d-electrons of the metal ion. Charge transfer between groups can give intense absorption bands.

8 Important biological applications of UV/visible spectroscopy are measurements of concentrations and the binding of ligands to macromolecules.

9 The sensitivity of electronic spectra to the solvent environment of the chromophore leads to shifts in λ_{max} and ε. This is the basis of solvent perturbation spectra and fine tuning of chromophore absorption by protein environment, e.g. in photoreceptors in the eye.

10 Neighboring electric transition dipole moments can interact. This leads to hypochromism in nucleic acid spectra.

11 Rapid reactions, including photocycles, can be monitored by devices such as "stop flow" and very short laser pulses.

 Further reading

Books

Van Holde, K.E., Johnson, W.C., and Ho, P.S, *Principles of Physical Biochemistry* (2nd edn) Prentice Hall, 2005.

Reviews

Berera, R., van Grondelle, R., Kennis, J.T. Ultrafast transient absorption spectroscopy: principles and application to photosynthetic systems. *Photosynth. Res.* 101, 105–18 (2009).

Rodger, A. How to study DNA and proteins by linear dichroism spectroscopy. *Sci. Prog.* 91, 377–96 (2008).

Sundström, V. Femtobiology. *Annu. Rev. Phys. Chem.* 59, 53–77 (2008).

Pace, C.N., Vajdos, F., Fee, L., Grimsley, G., and Gray, T. How to measure and predict the molar absorption coefficient of a protein. *Protein Sci.* 4, 2411–23 (1995).

 Problems

5.3.1 If 20.8% of the 340 nm radiation incident on a given solution of NADH is transmitted and if the absorption coefficient of NADH at 340 nm is 6220 L mol^{-1} cm^{-1}, what is the concentration of NADH in the solution? (Assume the path length is 1 cm).

5.3.2 Calculate the absorbance of (i) a 1 mmol L^{-1} solution and (ii) a 1 µmol L^{-1} solution of NADH in a cell of path length 1 cm. Comment on the percentage of transmitted radiation in these two cases.

5.3.3 The absorption spectra of dinitrophenyl (Dnp) ligands undergo a blue shift when the ligands are transferred from water to a less polar solvent, but when they are bound to a hydrophobic anti-Dnp-antibody binding site, a red shift is observed in the spectrum. Why?

5.3.4 Lysozyme has six tryptophan and three tyrosine residues. A "model" mixture of the esters of these two amino acids in water had a maximum absorbance at 281 nm. The difference spectrum between water and a 20% ethylene glycol solution of this mixture had a maximum absorbance at 292 nm. The ratio of the absorbance in the difference spectra ($\Delta\varepsilon_{292}$) to that in water (ε_{281}) was 0.042. When lysozyme was used, this ratio was 0.034. What can you deduce from these results?

5.3.5 The tyrosine residues of an antibody fragment (molecular weight 25 kDa) were reacted with tetranitromethane. Nitrotyrosine has an absorption band at 428 nm, which, in this protein, has an absorption coefficient of 4100 L mol^{-1}cm^{-1} at pH 10. At this pH, the absorbance is 0.154 for a 4×10^{-5} mol L^{-1} solution of protein in a cell with a path length of 1 cm. Determine the number of tyrosines modified and the pK_a value(s) from the following data.

A_{428}	0.065	0.067	0.069	0.079	0.084	0.117	0.126	0.134	0.142	0.156
pH	5.47	5.86	6.09	6.41	6.54	7.26	7.42	7.69	7.92	8.5

5.3.6 Explain the observation that the denaturation-induced increase in intensity in the absorption spectrum of viral DNA is much smaller than the increase for the replicative form of this DNA that is found in host bacteria.

5.3.7 The main 260 nm $\pi - \pi^*$ absorption of nucleic acid bases has a transition dipole in the plane of the bases. The absorption of polarized light in a sample of oriented DNA is polarized mainly perpendicular to the long axis. What can you deduce about the orientations of the base pairs?

5.3.8 β-Lactamases are bacterial enzymes that catalyze the hydrolysis of β-lactam antibiotics, making them ineffective as inhibitors of bacterial cell wall synthetic enzymes. Rapid-scan and stopped-flow UV/visible studies of the hydrolysis of the antibiotic nitrocefin by a β-lactamase identified three species: the substrate (S: nitrocefin) with λ_{max} = 390 nm; the product (P: hydrolyzed nitrocefin) with λ_{max} = 485 nm; and an intermediate (I) with λ_{max} = 665 nm. Comment on the results shown in the figure.

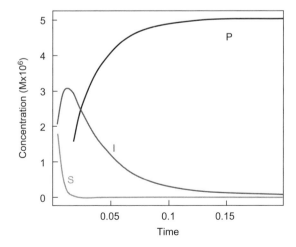

5.4 Optical activity

Introduction

A molecule is optically active if it interacts differently with left- and right-circularly polarized light. This differential interaction gives rise to two related phenomena known as optical rotatory dispersion (ORD) and circular dichroism (CD). Optical activity measurements in the 180–350 nm wavelength range are sensitive probes of molecular conformation in proteins and DNA; these are the main focus of this chapter. Other kinds of optical activity measurements can also be informative, including those where a magnetic field is applied to the sample.

The phenomenon

Circularly polarized light arises from the sum of two E waves which are at right angles to each other *and* phase shifted. This is illustrated in Figure 5.4.1 which shows the generation of left (L) and right (R) circularly polarized light by phase shifted E waves.

It follows from Figure 5.4.1 that plane-polarized light is equivalent to two circularly polarized beams rotating in opposite directions (the two perpendicular E waves—gray and dashed—cancel because they are 180° out of phase). This effect is illustrated another way in Figure 5.4.2, where the sum of the circulating L and R waves gives a plane polarized wave (cyan). When L and R come together at the top and bottom of the circle they will produce a cyan line twice their length but they will cancel when they are right angles to that position.

An understanding of both CD and ORD arises from consideration of the differential effects a sample can have on the two circularly polarized beams, L and R.

Optical rotatory dispersion (ORD) arises from the difference in refractive indices of the L and R beams in a sample ($n_R - n_L$). (The term "dispersion" is used to denote the frequency dependence of the refractive index.) As refractive index is related to the velocity of light in a medium (Chapter 4.2), finite $n_R - n_L$ means that the L and R beams will travel at different rates through the sample. The result of differential velocity of L and R through the sample is to rotate the plane wave formed from the sum of L and R waves by

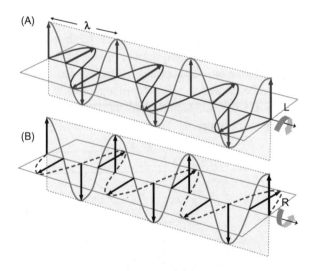

Figure 5.4.1 L and R circularly polarized light can be generated by summing two perpendicular E waves. The cyan wave is constant in A and B, but the perpendicular wave is either advanced or retarded in phase by 90° (λ/4). Consider the points where the cyan wave crosses the z-axis. The gray waves in A and the dashed wave in B produce a finite E wave at these points, as indicated by the gray and black arrows. It can be seen that the gray arrows rotate counterclockwise in the direction of propagation (L), while the black arrows rotate clockwise (R).

Figure 5.4.2 An illustration of two circularly polarized beams of equal amplitude, where the E vectors are rotating in a right-handed (R) and left handed (L) sense. The vector sum of these two components is a plane-polarized wave (cyan). Three representative positions of L and R vectors and their resultants are shown (this is zero when L and R point in opposite directions—broken dashed lines).

an amount that will depend on $n_R - n_L$ (Figures 5.4.2 and 5.4.3).

The rotation is defined as $[\alpha]_\lambda = \alpha_{obs}/cl$, where $[\alpha]_\lambda$ is the specific rotation at wavelength λ, α_{obs} is the observed rotation angle, c is the sample concentration, and l is the sample path length.

Circular dichroism (CD) arises from differential absorption of the L and R beams in a sample. This can be expressed as $\Delta A = A_L - A_R = \Delta\varepsilon cl$, where A_L and A_R are the absorbances of the L and R beams and $\Delta\varepsilon = \varepsilon_L - \varepsilon_R$ is the differential absorption coefficient, $\Delta\varepsilon$ has values in the range 10^{-3}–10^{-6} L mol^{-1} cm^{-1} that are much smaller than typical values of ε (see Table 5.3.2). The parameter molar ellipticity $[\theta]$

is often used to report CD measurements. $[\theta]$ and $\Delta\varepsilon$ can be interconverted using the relationship $[\theta] = 3298\Delta\varepsilon$ (degree cm^2 L mol^{-1}). (Sometimes, mean residue ellipticity is reported for normalization purposes; for proteins this is related to $[\theta]$ divided by the number of amino acids in the protein.)

The physical basis of optical activity

Transitions from the ground state to the excited state involve transition dipole moments (Box 5.1; Tutorial 11). Transitions can have both an electric, μ_e, and a magnetic, μ_m, transition dipole moment (Figure 5.4.4). Optical activity requires both μ_e and μ_m components, which results in a helical displacement of charge (Figure 5.4.4). The magnitude of the optically active transition is proportional to the vector product of μ_e and μ_m.

The required helical displacement of charge may arise directly from the chemical properties of the chromophore, e.g. if it involves a chiral center; the optical activity is then *intrinsic*. Optical activity can, however, also arise from *induced* effects due to the local environment of the chromophore, e.g. DNA bases in a helix.

Measurement

ORD measurements determine the angle, α, through which a sample rotates an incident plane-polarized wave (Figure 5.4.3). ORD measurements are still performed on some systems like polysaccharides but have largely been replaced by CD and will not be discussed further here.

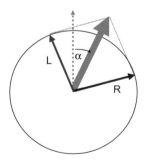

Figure 5.4.3 Optical rotation arises from different values of refractive index for the L and R beams. Assuming no differential absorption, the resultant of L and R remains plane polarized but the plane will be rotated by an angle α that depends on their differential velocity through the sample (proportional to $n_R - n_L$).

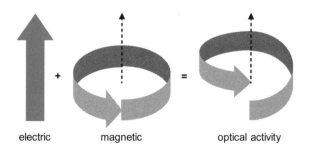

electric   magnetic   optical activity

Figure 5.4.4 A helical displacement of charge on going from the ground to excited state is required for optical activity. It arises from both μ_e and μ_m transition dipole moments.

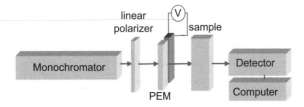

Figure 5.4.5 Schematic diagram of a CD spectrometer. A key component is the photoelastic modulator (PEM) which generates L and R components by applying a high frequency voltage (V) modulation.

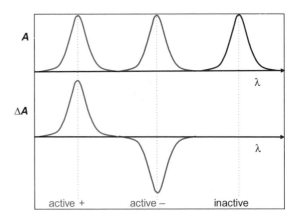

Figure 5.4.6 The frequency dependence of three bands with absorbance (*A*) and different kinds of CD (Δ*A*) bands.

CD could be detected by a double beam instrument with separate L and R paths, but it is more practical to use a device called a photoelastic modulator (PEM; Figure 5.4.5). As explained in Figure 5.4.1, switching the phase of one of two E waves can generate either an L or an R beam. PEM devices can rapidly change the phase of a light in response to an applied voltage. A high frequency modulation of the beam, switching between L and R, can thus be generated and used to measure the differential absorption ($A_R - A_L$) with high sensitivity. The light source can be produced in the laboratory or by a synchrotron source (see Chapter 4.3). Synchrotrons have much higher intensity than conventional sources at short wavelengths and can be used to record CD spectra down to 160 nm. At such short wavelengths there are, however, technical difficulties because of absorption by buffers and oxygen. Careful sample preparation and purging with nitrogen are then important.

The frequency dependence of CD

A molecule may have several absorption bands: some optically active, some inactive. A schematic frequency dependence for a molecule with three resolved absorbance bands, one with Δ*A* positive, one with Δ*A* negative, and one optically inactive, is shown in Figure 5.4.6.

Applications of CD spectra

The main applications of CD arise from its sensitivity to different conformational states in macromolecules. Absorption bands arising from peptides generally have *intrinsic* optical activity because the geometry around the C_α carbon in amino acids, other than glycine, is chiral. Many other biological chromophores do not have *intrinsic* optical activity but *induced* optical activity is often found; for example, heme absorption transitions in nucleic acids and aromatic amino-acid sidechains, and the *d*–*d* absorption bands in metalloproteins. In polysaccharides, both the nature of the individual sugars and the linkages between them are important in determining the optical activity; however, carbohydrates generally absorb only in the far ultraviolet and they are technically difficult to study by CD.

CD spectra of proteins

For proteins, the main chromophores of interest are the peptide bond, the aromatic sidechains, and some prosthetic groups. Most peptide bond spectra are *intrinsically* optically active. There are two main observed peptide bond transitions: π–π* and *n*–π*, as discussed in Tutorial 10 and Chapter 5.3. The spectral region dominated by the peptide backbone (170–250 nm) is of great interest because it can be used to estimate the secondary structure of proteins.

As illustrated in Figure 5.4.7, the CD spectra in the 170–250 nm region are very different for proteins with different structure. Standard curves for different kinds of secondary structure have been constructed—see Figure 5.4.8. One way to match the observed and calculated CD spectra is to assume that the observed spectrum is a linear combination of various types of secondary structure—a procedure sometimes known as **multicomponent analysis**. The fractions of secondary structure (α, β, etc.) are varied until the sum gives

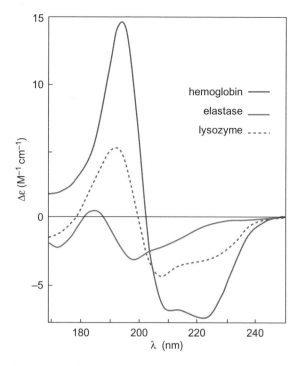

Figure 5.4.7 The CD spectra of three representative proteins: hemoglobin, elastase, and lysozyme (redrawn and adapted with permission from *Proteins* 7, 205–14 (1990)).

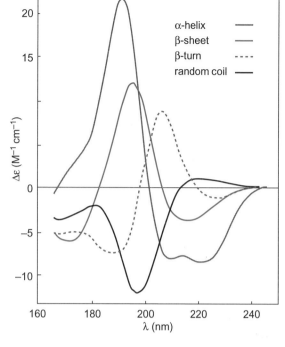

Figure 5.4.8 The CD spectra of representative secondary structure (redrawn and adapted with permission from *Proteins* 7, 205–14 (1990)).

a best fit to the experimental curve over the entire wavelength region.

A problem with multicomponent analysis is that the reference spectra for the different secondary structure types are not perfect. To try to get around this problem, methods have been developed which analyze the experimental CD spectra using databases of reference protein CD spectra. Servers have been set up to analyze experimental data in terms of fractions of secondary structure and proteins with known structure (see further reading). CD is particularly reliable in predicting α-helix content in a protein (~97% accuracy when compared with known X-ray structures), but less good for β-sheet (~75%) and β-turns (~50%). Successful secondary structure analysis using these techniques requires the CD spectra to be recorded at low wavelength (down to ~170 nm); accurate knowledge of the protein concentration is also important.

The *induced* optical activity in aromatic side-chains, observed in the 260–320 nm range, also gives useful information. It is sometimes possible to obtain residue-specific information from this region in contrast to the overall conformational information

about the protein backbone given by the 170–250 nm region. An example is given for lysozyme in Figure 5.4.9. It was possible, using specific esterification of residues and comparison of different lysozymes, to ascribe most of the negative CD band at 305 nm to Trp 108 and the pH dependence to the ionization of a nearby group, Glu 35, i.e. very specific structural information.

CD Spectra of DNA

Nucleic acid bases are optically inactive, but they can have *induced* activity, e.g. in a DNA helix. If the DNA bases are near each other in space, the transition dipole moments interact, which causes a splitting of the absorption bands (see Tutorial 11). When a CD band is split in this way the resulting absorption bands turn out to have CD spectra with opposite sign (see also Figure 5.4.12). DNA spectra normally show a positive band around 270 nm and a negative band around 240 nm. Any change in the angle between adjacent chromophores alters the transition probability, the position of the absorption bands, and the

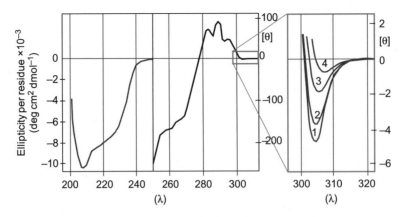

Figure 5.4.9 CD spectra of lysozyme. Three different spectral regions are shown. The first (200–250 nm) provides information about secondary structure. The next region, 250–310 nm, arises from induced activity in aromatic sidechain absorbance. Note this region gives weaker spectra and the – scale is expanded by ~40-fold compared to the 200–250 nm region. Finally, the region above 300 nm (cyan box), expanded again by 40-fold, shows spectra obtained at different values of pH (numbers 1–4 correspond to pH values of 7.2, 6.0, 5.0, and 3.5, respectively) (adapted with permission from *J. Biochem.* 76, 671–83 (1974)).

rotational strengths (Tutorial 11). The induced CD spectra of DNA are thus very sensitive to the overall structure (Figure 5.4.10).

An example illustrating the difference between the A and B forms of DNA and a DNA fragment that binds a transcription factor is shown in Figure 5.4.10.

Protein DNA complexes

The CD spectra arising from DNA and protein can often be resolved, making CD useful in studies of protein/DNA complexes.

Example Jun/Fos DNA interactions

Jun and Fos are transcription factors containing a basic DNA binding domain and a "leucine zipper" region with interacting α-helices that form a Jun/Fos heterodimer (see Figure 5.4.11). When formed, the heterodimer interacts with a region of DNA called AP-1 to regulate transcription of the gene. CD spectra of the sort shown in Figure 5.4.11 were used to explore the interactions between Jun and Fos, and with AP-1. Several features can be readily deduced from the spectra, for example the heterodimer has higher helix content than Fos on its own, and the addition of AP-1

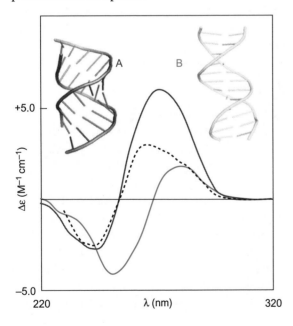

Figure 5.4.10 CD spectra of DNA. The gray line shows the spectrum from DNA in the A conformation (PDB: 2D47), while the cyan is from the hydrated B-form (PDB: 1BNA). The dotted line shows the CD spectrum of a DNA sequence that binds to the transcription factor TFIIIA; this led to the conclusion that this region of DNA is distorted even before TFIIIA binds (see *EMBO J.* 8, 1809–17 (1989)).

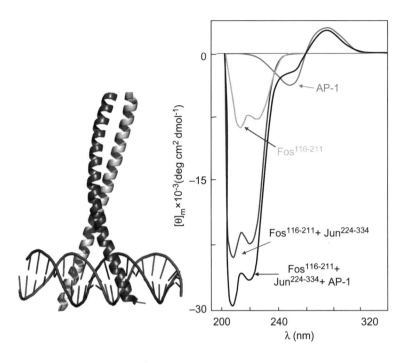

Figure 5.4.11 An image of the X-ray structure of a Jun/Fos complex with DNA in cyan, Jun in light gray, and Fos in dark gray (PDB: 1Fos). On the right are four CD spectra: AP-1 alone (cyan); a Fos fragment (116-211) on its own; Fos$^{116-211}$ + Jun$^{224-334}$; and Fos$^{116-211}$ + Jun$^{224-334}$ with AP-1 (adapted and redrawn with permission from *Nature* 347, 572–5 (1990)).

further increases the helical content. The DNA region of AP-1 (cyan) does not seem to be much perturbed by binding to protein.

Magnetic circular dichroism

In 1845, Michael Faraday observed that optical activity could be induced by the application of a magnetic field. Experimentally, magnetic circular dichroism (MCD) is very similar to CD, except that in MCD a strong magnetic field is applied to the sample in the same direction as the light is propagated. There are three distinct effects that an applied magnetic field can induce on electronic transitions—these are designated as the A, B, and C terms. The A term arises when there is a degenerate excited state in the absence of an applied field. In the presence of a magnetic field, B_o, the energy levels split and the resulting absorption band has two optically active components (Figure 5.4.12). The B-term is due to field-induced mixing of states, and the C-term is due to field-induced changes in the ground state.

MCD has been applied to metalloproteins to give information about the oxidation and spin state of the metal and the effects of inhibitors and ligation on the metal (see Problem 5.4.5). Magnetic fields can also be applied to induce optical activity in IR (Chapter 5.2) or X-ray absorption edge spectra (see Chapter 5.7).

It should also be noted that direct optical activity measurements of fluorescence (Chapter 5.5), Raman (Chapter 5.2), and X-ray absorption spectra (Chapter 5.6) can also give useful information, even in the absence of an applied magnetic field.

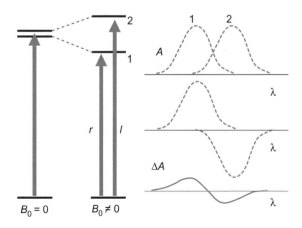

Figure 5.4.12 Illustration of the A term effect in MCD. An applied magnetic field splits a degenerate excited state. Each of the two resulting absorption bands is optically active, but with opposite sign (middle spectra). The net result is a derivative-like spectrum (solid cyan line on right).

Summary

1 A molecule is optically active if it interacts differently with left- and right-circularly polarized light. This interaction can be detected either as a differential change in velocity of the two beams through the sample—optical rotatory dispersion (ORD)—or as a differential absorption of each beam—circular dichroism (CD). CD is the main technique in current use.

2 CD spectra are characterized by $\Delta\varepsilon$ (the differential absorption coefficients of the two beams) or the molar ellipticity $[\theta]$. CD bands can be positive or negative.

3 Peptide groups are intrinsically optically active but many other biological chromophores are not. Optical activity *induced* by interactions with the local environment is very important, e.g. in aromatic amino acid sidechains and the bases of DNA.

4 One of the main applications of CD spectra arises from their sensitivity to protein secondary structure. Other uses include the detection of conformational changes and measurement of ligand binding.

5 Optical activity can be induced by the application of a magnetic field (MCD).

 Further reading

Useful websites

http://www.enzim.hu/~szia/cddemo/edemo0.htm (a useful animated tutorial on circular polarized waves and CD).

http://dichroweb.cryst.bbk.ac.uk/html/home.shtml (an online server for protein CD; DichroWeb incorporates five popular and effective algorithms to calculate protein secondary structure content).

Reviews

Johnson, W.C. Protein secondary structure and circular dichroism: a practical guide. *Proteins* 7, 205–14 (1990).

Greenfield, N.J. Using circular dichroism spectra to estimate protein secondary structure. *Nat. Protocols* 1, 2876–90 (2006).

 Problems

5.4.1 In three preparations of a purified amino acid, the UV absorbance spectra were all found to be the same. However, the CD spectra of the three samples showed that one was optically inactive; the other two samples were optically active but had opposite sign. Explain.

5.4.2 Glycophorin A spans the red blood cell membrane; its primary structure is known. This, with other information, suggests that the structure is as indicated in the figure with: a heavily glycosylated extracellular domain (residues 1–65); a very hydrophobic region (70–90); and a hydrophilic intracellular region (92–131). The CD spectra of various fragments are shown in the figure. The CD spectrum of the hydrophobic peptide 60–90 had to be measured in detergent for solubility reasons, but the CD spectra of the other peptides here were essentially unchanged by the same amount of detergent. Comment on the significance of these spectra.

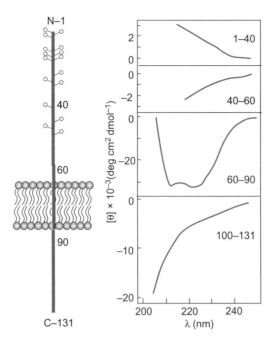

5.4.3 Some encephalopathy diseases are associated with an accumulation of an abnormal form (PrP^{Sc}) of the normal prion protein (PrP^{C}). A fragment of the protein that contains a single disulfide bond that can be reduced was studied by CD. The figure shows the CD spectra of a fragment of the human PrP protein in oxidized and reduced form (solid gray and dashed cyan lines). Comment on these spectra and suggest their possible significance.

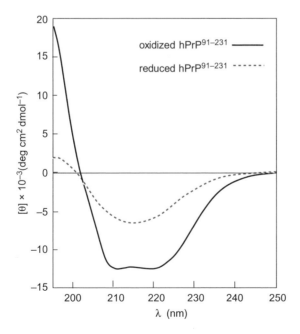

5.4.4 Like many spectroscopic techniques, CD spectra can be used to measure kinetics with rapid reaction devices such as stop flow (see Chapter 5.3). A particular advantage of CD is that measurements in the "far UV" (190–250 nm) are sensitive to backbone conformation, while the "near UV" (260–310 nm) is sensitive to aromatic sidechain conformation. The figure shows CD kinetic results for α-lactalbumin in the far (gray)

and near UV (cyan) regions at two different Ca^{2+} concentrations. (Protein refolding was monitored after rapid dilution from 6M GuHCl.) Comment on the results.

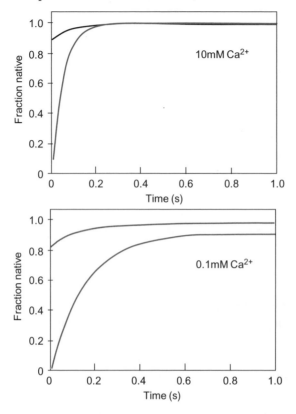

5.4.5 There are a number of di-cobalt enzymes, but there is some confusion about their ligand geometry. Crystallography of a phosphodiesterase from one bacterium indicated that it had two 6-coordinate metal binding sites, while that from another species indicated two 5-coordinate binding sites. The figure shows the MCD spectra of (a) a model compound with two Co(II) ions, one made to have a 5-coordinate ligand geometry and the other 6-coordinate. (b) and (c) show the MCD spectra of two enzymes containing di-cobalt centers. The bands arise from $d–d$ transitions. The bands marked with gray lines at around 500 nm and 580 nm are characteristic of spectra with 6- and 5-coordination, respectively. Suggest possible ligand geometries of enzymes b and c.

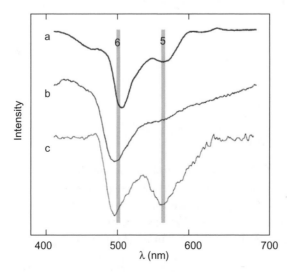

5.5 Fluorescence

Introduction

Fluorescence occurs when light is emitted from an excited state. A molecular group that fluoresces is called a fluorophore. Naturally occurring fluorophores in biology include tryptophan, the Y-base of t-RNA, NADH, and chlorophyll. A wide range of synthetic fluorescent probes, dyes, and proteins can be added to an experimental system to probe particular molecules or environments. The lifetime of the fluorophore's excited state depends on competition between the radiative (fluorescent) emission and "radiationless" processes that transfer the excitation energy to the surroundings. The result is that fluorescence is very sensitive to environment (e.g. to the presence of acceptors and quenchers) and molecular motion. Fluorescence probes are widely used as sensitive indicators, e.g. in DNA sequencing, immunofluorescence (Box 5.4), cell sorting, and microscopy (Chapter 7.1).

Physical basis of fluorescence

In general, a fluorophore will be in its ground electronic state and its lowest vibrational level (Chapters 5.1 and 5.2). Absorption of the appropriate energy usually results in excitation into upper vibrational levels of the first singlet excited state. As discussed in Chapter 5.3, the spectral transitions between different electronic states are governed by the Franck–Condon rule (straight up/straight down in Figure 5.5.1) because the nuclear framework is constant during the relatively rapid transitions between energy states. Different vibrational levels in the excited state will receive the transition depending on the relative positions of the ground- and excited-state energy curves. After molecules have reached the excited state, those in higher vibrational levels lose their excess energy

(usually as heat) and make their way to the lowest vibrational level. Fluorescence then occurs via spontaneous emission to the ground state, usually at a wavelength longer than the exciting light.

Lifetime and quantum yield

The lifetime is a bulk property that is a measure of how long the molecules exist in the excited state. Typical fluorescent lifetimes are $\sim 10^{-9}$ s; this is not

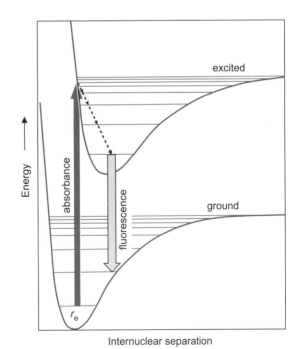

Figure 5.5.1 Fluorescence energy diagram. Absorption (cyan arrow) takes place vertically, according to the Franck–Condon rule. This transition is likely to reach a higher vibrational level in the excited state, followed by relaxation to the lowest vibrational level. Fluorescence then usually occurs by spontaneous emission to a higher vibrational level in the ground state.

the same as the time taken for a transition, which is $\sim 10^{-15}$ s $(1/\nu)$. If fluorescence is the only means of depopulating the excited state, then the fluorescence rate can be described by a first-order rate constant, k_f; the inverse of k_f is a relaxation time, τ_f. Depopulation of the excited state can also occur by radiationless processes (Figure 5.5.2) that decrease the observed fluorescence intensity and the lifetime (Chapter 5.1), with no detectable light emission. We can describe these radiationless processes by a rate constant k_r. The observed fluorescence intensity will depend on the relative rates of the competing processes, k_f and k_r. The overall depopulation rate will be $k_f + k_r$ and the lifetime will be $\tau = 1/(k_f + k_r)$.

The quantum yield (ϕ_f) is the fraction of excited molecules that emit fluorescence. From the above discussion about the balance between k_r and k_f processes, we can see that $\phi_f = k_f/(k_f + k_r)$. It follows that ϕ_f can also be written as τ/τ_f. Absolute values of ϕ_f are difficult to measure experimentally because instrumental correction factors have to be known. In practice, ϕ_f can be obtained by comparison with a standard sample for which the quantum yield is known. Measurements of fluorescence intensity changes are then usually sufficient to obtain relative values of quantum yield. The quantum yield is sensitive to the immediate surroundings of the fluorophore and any added quenching agent (see below).

Measurement

Fluorescence is characterized by both an excitation spectrum and an emission spectrum. The emission spectrum, usually measured at right angles to the excitation (see Figure 5.5.3), occurs at longer wavelengths than the excitation spectrum.

At very low concentration of fluorophore with low absorbance, the emission of light will be relatively uniform from the front to the back of the sample cuvette (see Figure 5.3.1). At high absorbance, however, much more fluorescence will be generated at the front of the cell than the back (this is called the inner filter effect). Accurate fluorescence measurements therefore require low absorbance so that the light intensity does not drop off strongly across the cuvette. Fluorescence emission can be strongly temperature dependent, so it is also important to

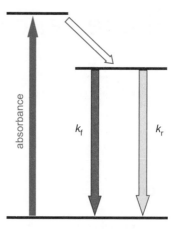

Figure 5.5.2 Depopulation of an excited state by k_f and k_r processes.

carefully control the temperature in experiments. Irradiation of fluorophores can also cause photobleaching, where there is photochemical destruction of the fluorophore (see also Chapter 7.1).

If a molecule contains more than one copy of the same fluorophore (e.g. tryptophan residues in a protein), the environments may be different, but the individual emission spectra will not usually be resolved. If the fluorophores have different

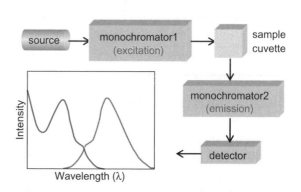

Figure 5.5.3 A fluorescence spectrometer. An emission spectrum is obtained by scanning monochromator 2 (M2) with monochromator 1 (M1) at a fixed wavelength. An excitation spectrum is obtained by scanning M1 with M2 fixed. The inset shows absorption (cyan) and emission (gray) spectra.

Note Because transitions probable in absorption are also probable in emission (Box 5.1), and because the nuclear framework, and hence the vibrations, are also similar in ground and excited states, there is often a mirror-image relationship between the excitation and emission spectra (see Problem 5.5.1).

Table 5.5.1 Characteristics of some intrinsic fluorophores

Fluorophore	Absorption: λ_{max} (nm)	Fluorescence: λ_{max} (nm)
Trp	280	348
Tyr	274	303
Phe	257	282
Y-base	320	460

fluorescence lifetimes, however, then the decay of the fluorescence gives another chance to resolve them. This can be done by exciting the sample with a short pulse of light (<1 ns) and monitoring the emission as a function of time. The observed decay will be the sum of exponential terms from the different fluorophores. Analysis of these decay rates can often resolve individual fluorophore lifetimes (see also the discussion of FLIM in Chapter 7.1).

Fluorophores

The main fluorophores used in biochemistry can be classified as *intrinsic* or *synthetic/extrinsic*. Intrinsic fluorophores include the aromatic amino acids, flavins, vitamin A, chlorophyll, and NADH (see Table 5.5.1). Nucleic acids do not have appreciable fluorescence, with the exception of the Y-base in t-RNA. Trp usually has a higher quantum yield than other amino acids and its fluorescence completely dominates the emission spectrum of an unlabeled

protein, partly because of efficient energy transfer from Tyr to Trp (but see Problem 5.5.7).

Many of the applications of fluorescence involve the addition of an extrinsic fluorescent probe to the system. These probes may be bound, either covalently or non-covalently, and much ingenuity has gone into designing site-specific probes. The structures of some common extrinsic probes are shown in Figure 5.5.4. In addition to these, a number of bright, relatively stable dyes with a range of emission wavelengths (e.g. Alexa fluor, Cy3, Cy5, etc.), are now available. Transition metal ions are not usually fluorescent but some lanthanide ions do have observable fluorescence (e.g. terbium). These ions can often be used to probe the roles of non-fluorescent ions such as calcium.

Fluorescent proteins

The introduction of genetically encoded proteins, such as green fluorescent protein (GFP), has been a major advance in cell biology. GFP, first found in the jellyfish *Aequorea victoria*, is now one of the most exploited proteins in biochemistry and cell biology. It autocatalytically generates an internal fluorophore (Figure 5.5.5), thus giving a very valuable probe with a wide range of applications, including the determination of protein localization in intact cells. By making mutants that change the environment of the GFP fluorophores (see also Chapter 5.3, Figure 5.3.10), it has been possible to generate fluorescent proteins with a wide range of emission colors. For example, cyan fluorescent protein (CFP) and yellow fluorescent protein (YFP) are both derived from GFP. Enhanced GFP (EGFP) is a mutant of GFP that is more stable than wild type. A new generation of fluorescent proteins is derived from corals. The discovery and development of GFP and related

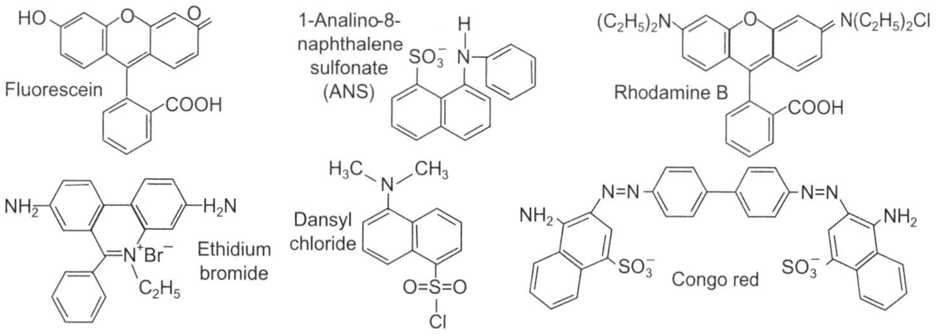

Figure 5.5.4 Some extrinsic fluorophores.

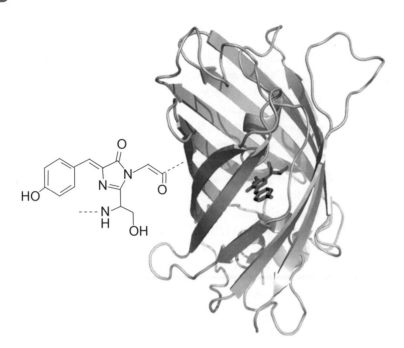

Figure 5.5.5 The green fluorescent protein. The chemical structure of the fluorophore that is formed from constituent Ser, Tyr, and Gly residues in the protein is shown on the left. The crystal structure is also shown (PDB: 1EMA) with the fluorophore in cyan in the core of the protein.

molecules led to the award of a Nobel Prize to Chalfie, Shimomura, and Tsien in 2008.

While applications using GFP and other fluorescent proteins have been extraordinarily informative, GFP is a rather large addition to a target protein. An alternative is to use membrane-permeable fluorophores, such as biarsenical reagents that bind a specific amino-acid sequence inserted into the protein of interest (see Figure 5.5.6). A typical recognition sequence contains four cyteines (e.g. CysCysProGlyCysCys); this requires a smaller alteration to the target protein than the insertion of GFP. Another advantage of these biarsenical reagents is that their fluorescence is quenched until they are attached to the tetracysteine tag.

Quantum dots

These are "nanoparticles" (2–10 nm in diameter) made of a semiconductor material. Because of their small size (~50 atoms wide) quantum dots display unique optical properties including intense fluorescence. The wavelength of their emission depends on the dot size; quantum dots can thus be "tuned" to emit any color of light desired, with small dots emitting in the blue and larger ones in the red. They are exceptionally bright, photostable (not susceptible to degradation from the incoming radiation), and have narrow emission spectra compared to fluorescent proteins and dyes. Quantum dots have great potential as fluorescent probes, but they can be hard to manipulate because their surface properties are not yet very well understood.

Environmental effects on fluorescence

As discussed in Chapter 5.3, a molecule has different electron distributions in its excited and ground states and these lead to different interaction possibilities in the two states. This means that electronic transitions are sensitive to solvent and other environmental factors. Fluorescence is even more sensitive to environment than absorption because additional factors, such as the radiationless depopulation pathway, can have a large influence. This sensitivity of fluorescence to environment can be used in a number of ways. One is to estimate environmental polarity. In general, the excited state will be more polar (Box 2.3) than the ground state; interactions between the excited electrons and a polar environment thus tend to decrease the energy of the excited state and shift the emission spectrum toward the red.

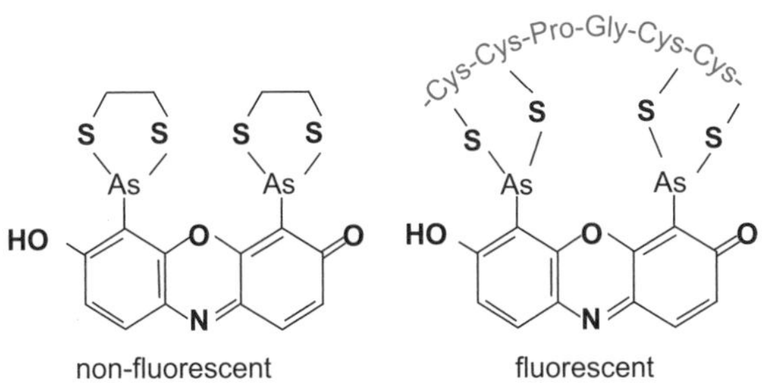

non-fluorescent   fluorescent

Figure 5.5.6 A biarsenical dye system for specific labeling of proteins containing a CCPGCC motif. The one shown, called ReAsh, has a fluorescence maximum in the red (609 nm). Other derivatives are available with different emission colors. These reagents do not fluoresce until covalently attached to the protein CCPGCC motif.

Example Environmental effects on the fluorescent spectra of l-anilino-8-naphthalene sulfonate (ANS)

Figure 5.5.7 shows a strong blue shift of the ANS emission spectrum as the probe environment becomes less polar (bound to a protein rather than exposed to solvent). The spectra in Figure 5.5.7 also illustrate the large changes in quantum yield that can occur in different environments. In the absence of protein, the ANS emission is almost completely quenched; this means that the radiationless pathway (k_r) dominates. When bound to protein the k_r pathway is much reduced and strong ANS fluorescence is observed.

Example Thermal shift assay/ligand binding

As shown by the ANS example above, many fluorophores only fluoresce when bound to a macromolecule. Some fluorophores also bind selectively to the **unfolded state** of a protein. Fluorescence is then observed to increase greatly when the fraction of unfolded species increases, e.g. with increasing temperature. This approach (Figure 5.5.8) can be used with small quantities of protein in thermal cycling devices. The method is especially useful as binding of a ligand to a protein often increases the thermal stability by an amount proportional to the affinity. Increases in melting temperature can thus be used to screen for ligands and new drugs (see also Chapter 3.7 and Problem 3.7.4).

Quenching

Addition of oxygen or ions, such as I^- or Cs^+, often leads to an enhanced radiationless rate, k_r, and

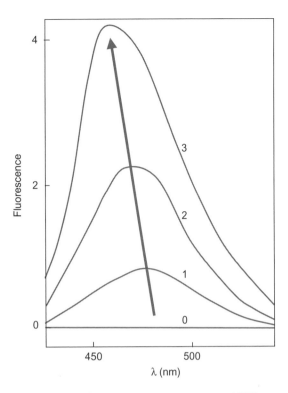

Figure 5.5.7 The fluorescence emission spectra of ANS at increasing concentrations of the protein bovine serum albumin (BSA). The cyan arrow indicates the shift of λ_{max} to the blue, as well as a large increase in emission intensity. Note that fluorescence is generally measured in arbitrary units.

therefore a reduction in quantum yield. This **quenching** of fluorescence mainly arises from collisions between quencher and fluorophore. Analysis of the concentration dependence of the quenching can give information about the dynamics of the collision process and the accessibility of a quencher to a fluorophore.

Quenching can be *quantified* in a simple way as follows: the quantum yield, defined earlier, is $\phi_f = k_f/(k_f + k_r)$. In the presence of quencher this becomes $\phi_{fQ} = k_f/(k_f + k_r + k[Q])$, where [Q] is the concentration of quencher and k is the rate of quenching (a first-order process is assumed because [Q] > concentration of fluorophore). It follows from these equations that $\phi_f/\phi_{fQ} = 1 + k[Q]\tau$, where τ is the lifetime in absence of quencher. This equation is often written in terms of the observed fluorescent intensities in the presence (I) and absence (I_0) of quencher. The expression $I_0/I = 1 + K[Q]$ is called the **Stern–Volmer equation**, where K is a quenching constant. A plot of $[I_0/I - 1]$ versus [Q] gives a straight line of slope K.

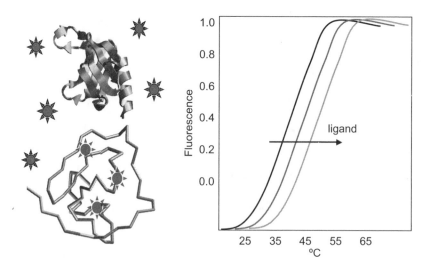

Figure 5.5.8 Thermal shift assay: a dye is chosen that does not fluoresce in water or in the presence of a globular protein (gray stars) but it does fluoresce when it binds to the unfolded protein (cyan stars). This can be used as a sensitive and convenient probe of thermal unfolding; rapid measurements of the influence of different ligands on thermal stability can be made.

The quenching rate (k) depends on the probability of a collision between fluorophore and quencher. This probability depends on diffusion rates, size of colliders, and concentration. It can be shown that $k = 4\pi aDN_A \times 10^{-3}$, where D is the sum of the diffusion coefficients of quencher and fluorophore, a is the sum of their molecular radii, and N_A is Avogadro's number. (For example, for oxygen quenching of tryptophan fluorescence in solution, $a \sim 0.4$ nm, $D_{O_2} = 2.6 \times 10^{-9}$ cm^2 s^{-1} and $D_{trp} = 0.66 \times 10^{-9}$ cm^2 s^{-1} which gives $k \sim 1 \times 10^{10}$ M^{-1} s^{-1}.)

Example   Accessibility of Trp to O$_2$

The tryptophan fluorescence of aldolase was monitored in the presence of increasing concentrations of oxygen, up to 0.13M—corresponding to an oxygen pressure of 100 atmospheres. (This concentration of O$_2$ had negligible effect on aldolase catalysis.) A Stern–Volmer plot is shown in Figure 5.5.9. The linearity of the plot suggests that the various tryptophans in aldolase do not have significantly different properties and that all are accessible to oxygen quenching. This early experiment demonstrated that rapid structural fluctuations on the nanosecond time scale permit ready diffusion of oxygen through proteins (i.e. some protein properties are "liquid" like).

Example   Quantitative real time PCR (qRT-PCR)

qRT-PCR is an extremely sensitive method for detecting and quantifying the amounts of DNA, or RNA, in a sample; it is sensitive enough to measure material from a single cell. A key component is a quenched fluorophore that "lights up" when quencher and fluorophore are released from a probe by enzyme action (Figure 5.5.10). The **polymerase chain reaction** (PCR) amplifies DNA by cycles of heating and cooling to expose the two DNA strands and allow enzymatic replication of a DNA template. Probe oligonucleotides with a fluorescent reporter dye attached to one end and a quencher at the other end are used to monitor the reaction. These probes are designed to hybridize to an internal region of a PCR product (see also Box 5.5). In the initial state, the close proximity of the fluorophore and quencher results in no observed fluorescence. During the PCR, the 5′-nuclease activity of the polymerase cleaves the probe, thus decoupling the fluorophores and quencher so that fluorescence increases in each thermal cycle of the PCR reaction. The observed fluorescence intensity is then directly proportional to the amount of probe-specific DNA in the sample.

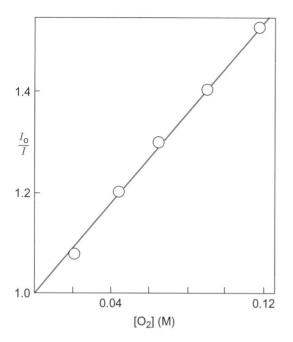

Figure 5.5.9 A Stern–Volmer plot of the quenching of tryptophan in aldolase by O_2 (adapted with permission from *Biochemistry* 12, 4171–9 (1973)).

Fluorescence anisotropy

Measurements of **fluorescence polarization** can give information about motion and macromolecular interactions. If fluorophores are excited with plane polarized light and the fluorescence is observed through analyzing polarizers, it is found that the fluorescence is also polarized. The reason for this is **photoselection** (Figure 5.5.11); only those fluorophores with components of their transition dipole moments oriented along the direction of polarization are excited. The excitation depends on the angle θ between the plane of polarization of the incident light and the transition dipole moment of the transition. The probability of absorption is proportional to $\cos^2\theta$. The incoming light thus only excites a subset of molecules in a random array (see also Figure 5.1.5 and discussions of linear dichroism in Chapters 5.2 and 5.3).

The **fluorescence anisotropy** is defined as $A = (I_\parallel - I_\perp)/(I_\parallel + 2I_\perp)$, where $I_\parallel$ and $I_\perp$ are the fluorescence intensities polarized parallel and perpendicular to the direction of the exciting beam. ($I_\parallel + 2I_\perp$ is the total emitted light parallel to the incident axis and the two directions at right angles.) If the molecules rotate during the time between absorption and emission, the observed A value will be less than A_0, the value for the "rigid" state, i.e. the motion causes depolarization. The reduction in polarization depends on the degree of motion (slow motion → little depolarization; fast motion → strong depolarization).

The anisotropy, A, is thus a direct measure of the molecular rotation in solution. For isotropic motion, the relationship between A and A_0 is given by the

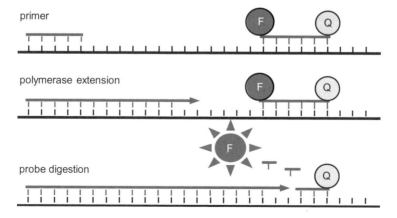

Figure 5.5.10 Illustration of real time quantitative PCR. In the intact probe, which binds to its target DNA, the fluorophore F and quencher Q are close together and all the fluorescence is quenched. When the probe is cleaved by the polymerase, however, the quencher and fluorophores drift apart and the fluorophores emits light.

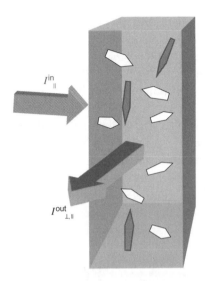

Figure 5.5.11 Illustration of photoselection and fluorescence polarization. The only molecules excited will be those with their transition dipole in the appropriate orientation (e.g. the cyan molecules). In turn, these will be the only molecules to emit so the output light will also be polarized.

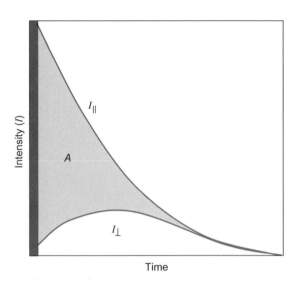

Figure 5.5.12 Measurement of $I_\parallel$ and $I_\perp$ as a function of time after a short excitation pulse. The thick gray line near $t = 0$ is an experimental "dead" time of the apparatus that occurs after the intense excitation pulse. The shaded gray area between the $I_\parallel$ and $I_\perp$ lines is related to A.

Perrin equation, $1/A = 1/A_o\{1 + (\tau_f)/(\tau_r)\}$, where τ_f is the fluorescence lifetime and τ_r is the rotational correlation time of the molecule (see Box 4.3 for a definition of correlation time). The value of A_o can be estimated by altering the viscosity η (e.g. by addition of glycerol) or the temperature ($\tau_r \propto \eta/T$) and extrapolating to the "rigid" state.

This kind of measurement, using constant illumination, gives information about the average motion of the system. Alternatively, time-resolved fluorescence anisotropy experiments (Figure 5.5.12) can be done. Here, fluorescence is excited with sub-nanosecond pulses of polarized light and the intensity and polarization of the resulting fluorescence ($I_\parallel$ and $I_\perp$) are measured as a function of time after the pulse (Figure 5.5.12). The response time of the instrumentation is short, ~50 ps, but finite so the measurements have to take account of this "dead time" effect.

After the pulse, there is an initial difference in $I_\parallel$ and $I_\perp$ intensities because of photoselection; $I_\parallel$ then decays due to both motional and lifetime effects. However, $I_\perp$ initially increases because of motion before it also decays due to the finite fluorescence lifetime. In the simple case of isotropic motion, $A(t)$ decays exponentially: $A(t) = A_o \exp(-t/\tau_r)$. A plot of ln A versus time is then linear, with a slope of $-1/\tau_r$,

the inverse of the correlation time. The pulse method is more informative than the steady-state method as τ_f (or A_o) needs not be known to calculate τ_r. Also, if there is more than one motion contributing to the decay curve, plots of log $A(t)$ versus t will be nonlinear. The curves can then be analyzed in terms of two or more exponentials to give the rate constants and amplitudes of each motion as long as the rate constants are significantly different. This is illustrated in the following example.

Example Analysis of motion in antibodies

Figure 5.5.13 shows plots of time-resolved fluorescent anisotropy for dansyl-lysine bound to an antibody (IgG) and antibody fragments. The dansyl-lysine fluorophore is held rigidly in the binding site. Antibodies are Y-shaped molecules that bind two fluorophores, one at the end of each arm. The decay for the small (F_{ab}) fragment with only one bound fluorophore is linear, showing motion represented by an isotropic rotational correlation time, τ_r, of ~25 ns. Figure 5.5.13 shows, however, that the decay of A is more complex for the larger antibody fragments. The decay for the intact antibody can, in fact, be analyzed in terms of two rotational times: 26 ns and 100 ns. This was interpreted as overall tumbling of antibody (100 ns) plus "wagging" of the two arms (26 ns).

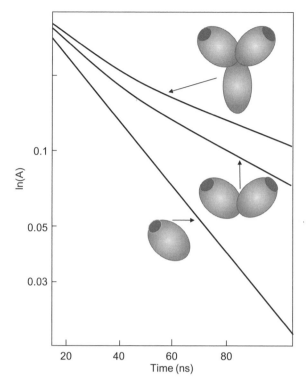

Figure 5.5.13 Time-resolved fluorescence depolarization of DNP (cyan) bound to an intact antibody (top) and two different antibody fragments (data derived with permission from *Biochemistry* 20, 6842–52 (1981)).

Measurement of fluorescence anisotropy is also a powerful way of looking at complex formation. In general, a macromolecule will rotate more slowly in solution when it is a part of a macromolecular complex than when it is alone. Higher fluorescence anisotropy is thus observed after complex formation, due to a slower rotation rate, as illustrated in the following example.

Example CAP/RNA polymerase interactions

The catabolite gene activator protein (CAP) brings RNA polymerase close to DNA thus stimulating transcription of nearby genes. CAP/RNA polymerase interactions were analyzed using fluorescence anisotropy, as summarized in Figure 5.5.14. A double-stranded DNA fragment containing the CAP DNA binding site was fluorescently labeled. This binds tightly to CAP which, in turn, binds to RNA polymerase. The molecular weight of the CAP/DNA complex is around 73 kDa while the CAP/DNA/RNA polymerase complex is around 520 kDa; a large change in rotational correlation time, τ_r, is thus expected on complex formation. It was shown that CAP interacts with RNA polymerase (with $K_D \sim 2.8 \times 10^{-7}$M), whereas a mutant of CAP, defective in its ability to activate transcription, interacts much more weakly.

Förster resonance energy transfer (FRET)

As well as quenching by collisions, fluorescence can be reduced by the transfer of excitation energy to other fluorophores (acceptors). **Förster (or fluorescence) resonance energy transfer (FRET)** occurs when the excited singlet state of a donor is transferred to the excited singlet state of an acceptor. In this transfer of energy, the donor returns from the excited state to the ground state and the acceptor is simultaneously excited from its ground to its excited state. The energy separations in each case must match, i.e. be in **resonance**. Figure 5.5.15 shows excitation and

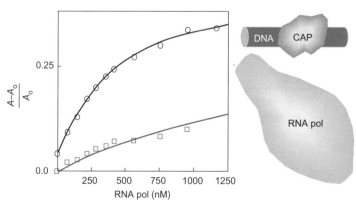

Figure 5.5.14 Fluorescence anisotropy of fluorescent DNA is shown for the CAP/DNA/RNA pol interaction (circles). As more and more of the larger molecular weight complex forms with increasing concentration of RNA pol the fluorescence anisotropy increases in a way that depends on K_D (see Chapter 2.1). The squares show a control titration of DNA and RNA pol (data derived with permission from *Nature* 364, 548–9 (1993)).

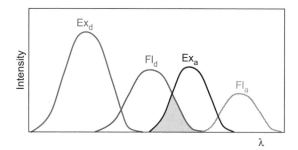

Figure 5.5.15 Resonance-energy transfer between donor and acceptor. Ex_d and Fl_d are the excitation and emission spectra of the donor; Ex_a and Fl_a are the corresponding acceptor spectra. The shaded area represents the region of overlap between donor fluorescence and acceptor absorption. Under suitable condition irradiation of Ex_d can result in the appearance of Fl_a.

emission spectra of a donor (Ex_d, Fl_d) and an acceptor (Ex_a, Fl_a). The original excitation energy (Ex_d) can reappear as acceptor emission (Fl_a), or be dissipated by non-radiative processes. The observed fluorescence from the donor (Fl_d) and the acceptor (Fl_a) depend on the spectral overlap between Fl_d and Ex_a (the shaded area in Figure 5.5.15). The efficiency (E_T) of depopulation by FRET can be described by $E_T = 1 - Fl_d/Fl_d^o$ where Fl_d^o is the fluorescence intensity of the donor in absence of acceptor. (E_T can also be written in terms of lifetimes or quantum yield, e.g. $E_T = 1 - \tau_D/\tau_0$ where τ_D and τ_0 are the depopulation rates of F_{ld} with and without acceptor). Förster explained FRET in terms of a dipole–dipole interaction between the donor and acceptor pair. The energy of a dipole–dipole interaction depends on $1/R^3$ where R is the intermolecular distance (Tutorial 11). The rate of energy transfer is proportional to the square of this interaction and hence to $1/R^6$. If we define R_o as the distance at which the energy transfer is 50% efficient, i.e. when $E_T = 1/2$, then $E_T = R_o^6/(R^6 + R_o^6)$; this is known as the Förster equation.

A model system to study the dependence of FRET on the separation of donor (naphthyl) and dansyl (acceptor) was developed using a variable number of proline residues between them (Figure 5.5.16). The proline residues form a relatively rigid helix, whose dimensions are known, so the distances between the naphthyl and dansyl groups can be estimated. The measured efficiencies of transfer fit well with the Förster equation (see, however, Figure 7.2.10 in Chapter 7.2).

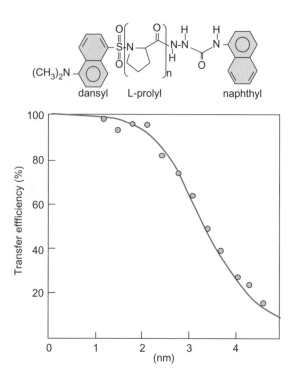

Figure 5.5.16 Efficiency of energy transfer between a napthyl and a dansyl group as a function of distance. The distance was varied by synthesizing the molecule shown at the top, with a variable number of prolines (adapted with permission from *PNAS* 58, 719–26 (1967)).

The main factors that influence the efficiency of resonance-energy transfer are: (i) the degree of spectral overlap between Fl_d and Ex_a (shaded area in Figure 5.5.15); (ii) the quantum yield of the donor; and (iii) R, the distance between donor and acceptor.

Example Conformational changes in calmodulin

Calmodulin is a calcium-binding protein that is expressed in all eukaryotic cells. It binds to and regulates a number of different protein targets. It has two domains that each bind two Ca^{2+} ions; the domains are separated by a flexible helical region (Figure

---

Note Dipole–dipole interactions have an angular dependence $(3\cos^2\theta - 1)$, as well as a distance dependence (Tutorial 11). The angle, θ, between the electric transition dipole moments of donor and acceptor and relative motion of donor and acceptor will usually be uncertain. Most calculations of distance assume an angular dependence term appropriate for a rapidly rotating isotropic system where $(3\cos^2\theta - 1)^2 \approx 0.67$; the error introduced by this assumption is usually relatively small.

5.5.17). The hypothesis that calmodulin undergoes a large conformational change when it binds target peptides was tested using FRET. By attaching *N*-1-pyrenylmaleimide to Cys27 in one of the two domains and nitrating a tyrosine (Y139; nitrated tyrosine is fluorescent) in the other, the average donor/acceptor distance was measured by FRET. In the Ca^{2+} form, without peptide, a distance of 2.26 nm was obtained, while that distance was found to be 1.95 nm when the calmodulin was bound to a kinase-derived peptide. This experiment thus confirmed that there is a conformational change induced in calmodulin by peptide binding (see also Problem 5.5.8).

Example   FRET and GFP

As mentioned above, it is possible to tune the fluorescence of green fluorescent protein (GFP) by site-directed mutagenesis. A simple example of FRET between two variants of GFP is shown in Figure 5.5.18. BFP (blue fluorescent protein) was generated from GFP by the mutations Ser66H (serine 66 changed to histidine) and T145F (see Tutorial 1 for amino acid nomenclature). A DNA construct was made that linked a GFP mutant (S65C) to BFP with a 25-residue linker sequence that could be cleaved by trypsin. This

construct could be expressed and purified in an expression system. Excited at 368 nm, the intact GFP/BFP dimer emits bright green light ~500 nm; this is equivalent to Fl_a in Figure 5.5.16 while the band ~440 nm corresponds to Fl_d. When the sample was exposed to trypsin, the FRET between the proteins was gradually abolished because the trypsin cleavage allows the proteins to diffuse apart, increasing R and reducing E_T.

Note that fluorescent proteins like GFP are large (as large as R_0 in the Förster equation), so FRET involving fluorescent proteins is mainly used qualitatively.

Bioluminescence resonance energy transfer (BRET)

Figure 5.5.18 illustrates FRET between fluorescent proteins. For some types of experiment it is an advantage to avoid the application of external radiation. Some organisms emit light (**bioluminescence**) via the following chemical reaction: luciferin + O_2 → oxyluciferin + light. This reaction is catalyzed by the enzyme luciferase. In **bioluminescence resonance energy transfer (BRET)**, the donor fluorophore in a

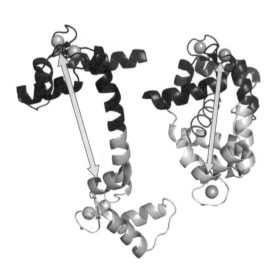

Figure 5.5.17 Structures of calmodulin with (PDB:1CDL) and without (PDB:3CLN) a bound target peptide (cyan helix in the structure on right). The arrows show the separation between the residues with the fluorescent labels (Cys27 in the dark gray domain and Y139 in the light gray domain). These distances were observed to change in FRET experiments (see *Biochemistry* 35, 6815–27 (1996) for more details).

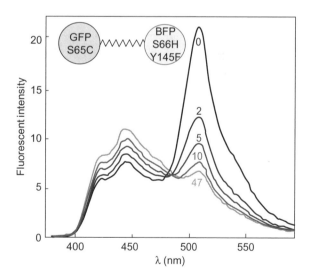

Figure 5.5.18 Fluorescence emission spectra observed from a linked GFP/BFP dimer. As cleavage of the linker peptide by trypsin progresses (0, 2, 5, 10, and 47 min), the green peak (~500 nm) diminishes while the cyan peak (~440 nm) increases. There was no further change in the spectrum after 47 min (adapted with permission from *Curr. Biol.* 6, 178–82 (1996)).

FRET pair is replaced by luciferase-induced biolumi-nescence. BRET can be used to detect protein–protein interactions by fusing one protein to the donor and the other to the acceptor. Addition of coelentera-zine, the natural substrate of luciferase from the sea pansy *Renilla reniformis*, leads to the emission of blue light (480 nm). Energy transfer occurs if the emission spectrum of the donor overlaps the excita-tion spectrum of the acceptor (Figure 5.5.15). If the two fusion proteins do not interact, only blue light is emitted upon substrate addition. If the two fusion proteins interact, BRET occurs and an additional light signal corresponding to acceptor reemission is detected (535 nm if yellow fluorescence protein, YFP, is the acceptor).

Exploitation of fluorescence sensitivity

As mentioned above, fluorescence can be used as a very sensitive detection system because light with essentially no background can be measured. Two examples where this sensitivity of fluorescence is exploited are given in Boxes 5.4 and 5.5. These methods involve combining

---

Box 5.5 Fluorescence *in situ* hybridization (FISH)

This technique is used to detect and localize specific DNA sequences on chromosomes. Single-stranded DNA probes with an attached fluorescent label bind to a complemen-tary strand of DNA (Figure A) in a sample (this could be a frozen permeabilized cell or a purified chromosome prep-aration). Fluorescence microscopy (Chapter 7.1) can be used to find the location of the probe. FISH can also be used to detect and localize specific mRNAs within tissue samples.

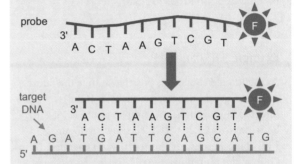

Figure A Illustration of a fluorescently labeled DNA fragment binding a complementary strand of DNA.

Box 5.4 Immunofluorescence

Antibodies labeled with fluorescent dyes can be used to visualize biomolecules of interest. Immunofluorescent-labeled tissue sections can be studied using fluorescence confocal microscopy (Chapter 7.1). Usually, two sets of anti-bodies are used (Figure A); a primary antibody binds the antigen of interest and a secondary dye-coupled antibody binds the primary antibody. Different primary antibodies that recognize a variety of antigens can all be recognized by a single dye-coupled antibody as they share a common "constant" region. Cells and tissue are frequently permeabi-lized with detergent before antibody labeling.

In some cases, primary antibodies are directly labeled with fluorophores. Direct labeling decreases the number of steps in the staining procedure and can reduce the problem of high background because of non-specific labeling. Many uses of fluorescent antibodies have been superseded by the development of recombinant proteins like green fluorescent protein (GFP). Use of such "tagged" proteins allows much better localization of proteins in live cells.

Figure A Illustration of immunofluorescence. A primary antibody (cyan) recognizes a receptor, e.g. on the cell surface. This primary antibody is recognized in turn by a fluorescently labeled antibody (gray).

fluorophores with specific binding reagents, namely antibodies (**immunofluorescence**) and DNA fragments (*in situ* hybridization).

In addition to immunofluorescence and FISH, many other important techniques exploit the ability to detect fluorescence at very low concentrations. As we will see later in Section 7, fluorescence is widely used in microscopy and it can even detect single molecules. Other techniques that use fluorescent labels include DNA sequencing and fluorescence activated cell sorting (FACS) which is briefly described below.

Fluorescence-activated cell sorting (FACS)

This technique (Figure 5.5.19) can sort a heterogeneous mixture of biological cells into two or more containers, one cell at a time, based upon the fluorescent characteristics of each cell. Thus, if cell A has a cell surface receptor with a bound fluorescent antibody while cell B does not, the A and B populations can be separated. **Fluorescence-activated cell sorting** (FACS) provides fast, and quantitative, separation. Cells in a rapidly flowing stream of liquid pass through a vibrating mechanism that causes the stream to break up into individual droplets with one cell per droplet. These pass through a detector which measures the fluorescence of each cell. An electrical charging ring puts a charge on the droplet depending on the result of the fluorescence intensity measurement. The charged droplets then fall through an electrostatic deflection system that diverts droplets into containers based upon their charge. (Although antibodies are commonly used, FACS can be applied to any population of cells with specific fluorescent labels, e.g. GFP.)

Phosphorescence

As well as fluorescence, another emission process, called **phosphorescence**, can occur. Phosphorescence is useful in some circumstances because it has a longer lifetime than fluorescence. Phosphorescence generally arises from emission from a spin 1 **triplet state** (T^0 in Figure 5.5.20) rather than the spin 0 first **singlet state** (S^1), which is the main fluorescence emitter. The phosphorescent lifetime is relatively long because the transition from the triplet state back to the ground state ($T^0 \rightarrow S^0$) is **spin forbidden**. Consider the inset in Figure 5.5.20, showing the spin orientations in the S^0, S^1, and T^0 states. The singlet ground state will usually have its electrons paired up in an antiparallel fashion in one level with a net spin $I = 0$ and a degeneracy of $2I + 1 = 1$, i.e. it is a singlet (S^0) (see Chapter 5.1). When an electron is excited to the first excited state it will at first retain its orientation, the net spin will remain zero, and the excited state will be another singlet (S^1). Once in the excited state, however, the spins are no longer required to be antiparallel by the Pauli exclusion principle and it is possible, via a mechanism called **intersystem crossing**, for the spin to reorient and go into a parallel, triplet state where the net spin is $I = 1$ (with $2I + 1 = 3$ energy levels in a magnetic field, hence the name triplet).

After passing through vibrational levels, the forbidden $T^0 \rightarrow S^0$ phosphorescent transition can take place with a relatively low probability (long lifetime). The emission wavelength is red shifted with respect

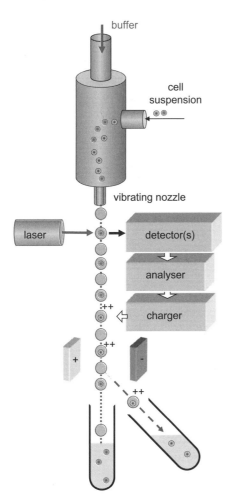

Figure 5.5.19 A fluorescence-activated cell sorter.

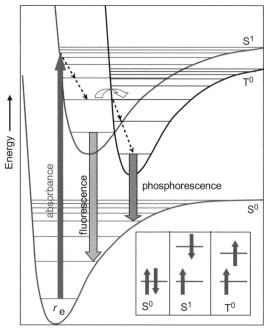

Figure 5.5.20 Illustration of phosphorescence energy levels and spin states. Absorbance occurs from S^0 to a higher vibrational level in S^1. Energy is lost in the vibrational levels and most will go to the lowest level and fluoresce. There is also a finite probability of transfer, via intersystem crossing (open cyan arrow), to the triplet state. Again, there will be loss of energy through the vibrational levels of the triplet state to the lowest energy level. From here, return to the ground state ($T^0 \rightarrow S^0$) is spin forbidden and the electron has to reorient before this can occur. The inset shows the spin orientations in the different singlet and triplet states.

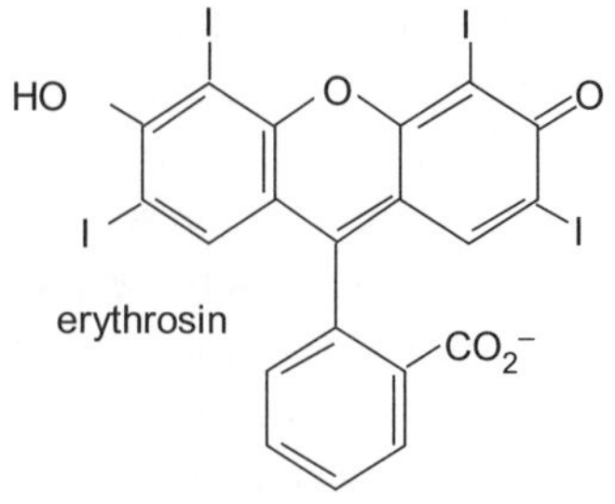

Figure 5.5.21 The structure of the phosphorescent probe erythrosin.

to fluorescence because the T^0 state has a lower energy than the S^1 state. The lifetime of phosphorescence is relatively long (typically 10^{-4} s). This makes it vulnerable to quenching processes such as collisions with the solvent. Some members of the fluorescein family substituted with the heavy atoms Br or I (Figure 5.5.21) are used as probes, as these halogens enhance the amount of intersystem crossing to the triplet state.

Phosphorescent probes, of the kind shown in Figure 5.5.21, have been widely used to study the rotational motion of membrane bound receptors and large macromolecular complexes. Typically, an erythrosine derivative is attached via an SH group to the protein of interest, phosphorescence anisotropy is then measured after an excitation pulse (similar to the experiment in Figure 5.5.11 but on a slower time scale—µs rather than ns—see, e.g. *Biochim. Biophys. Acta* 1464, 242–50 (2000)).

⊕ Summary

1 Fluorescence is the emission of radiation that occurs when a molecule in an excited electronic state returns to the ground state. It involves excitation; this takes place in $\sim 10^{-15}$ s, followed by emission after a finite lifetime of $\sim 10^{-9}$ s in the excited state.

2 The measurable parameters in fluorescence are the excitation and emission spectra, as well as the intensity, lifetime, quantum yield, and polarization. The emission spectrum is red shifted with respect to the excitation spectrum.

3 The term fluorophore describes the molecular group that gives rise to fluorescence. The naturally occurring fluorophores in biology include tryptophan, the Y-base of t-RNA, NADH, and chlorophyll. A wide range of synthetic dyes and fluorescent proteins (e.g. GFP) are used as specific *extrinsic* fluorescent probes.

4 A physical picture of fluorescence comes from considering the energy levels involved in electronic transitions, using "Morse" energy curves and the Franck–Condon principle. There are various processes by which the excited state can lose energy—only one of which is fluorescence.

5 Fluorescence is very sensitive to environment, e.g. polarity. The quantum yield can be influenced by quenching agents such as oxygen. Quenching measurements can give information about the accessibility of the fluorophore.

6 Molecular events on the time scale of the fluorescence lifetime and binding can be investigated by analyzing fluorescence anisotropy, either by static or time resolved experiments.

7 Distances between fluorophores can be obtained from Förster resonance energy transfer (FRET).

8 Because it is highly sensitive, fluorescence is widely used analytically, e.g. in immunofluorescence assays, DNA sequencing and fluorescence activated cell sorting.

9 Phosphorescence is the emission of radiation from a triplet state to the ground state. It has a longer lifetime than fluorescence and can be used to measure relatively slow motions.

Further reading

Book
Lakowitz, J.R. *Principles of Fluorescence* (3rd edn) Springer, 2006.

Reviews
Tsien, R.Y. Constructing and exploiting the fluorescent protein paintbox (Nobel Lecture). *Angew. Chem. Int. Ed. Engl.* 48, 5612–26 (2009).

Jovin, T.M., and Vaz, W.L. Rotational and translational diffusion in membranes measured by fluorescence and phosphorescence methods. *Methods Enzymol.* 172, 471–513 (1989).

Hawe, A., Sutter, M., Jiskoot, W. Extrinsic fluorescent dyes as tools for protein characterization. *Pharm. Res.* 25, 1487–99 (2008).

Jameson, D.M., and Croney, J.C. Fluorescence polarization: past, present, and future. *Comb. Chem. High Throughput Screening* 6, 167–73 (2003).

McCombs, J.E., and Palmer, A.E. Measuring calcium dynamics in living cells with genetically encodable calcium indicators. *Methods* 46, 152–9 (2008).

Ciruela, F. Fluorescence-based methods in the study of protein–protein interactions in living cells. *Curr. Opin. Biotechnol.* 19, 338–43 (2008).

Problems

5.5.1 Draw an energy level diagram to explain the mirror-image relationship between excitation and emission spectra.

5.5.2 When an antigen labeled with a fluorescent group bound to its antibody, the polarization of the fluorescence greatly increased. Why?

5.5.3 Thermolysin is a proteolytic enzyme ($M = 37.5$ kDa) that binds four calcium ions which stabilize it. Two of these calcium ions are close together and can be substituted by a single terbium ion (Tb^{3+}). The active site of the enzyme also contains a zinc ion, which is essential for activity, but this can be replaced effectively by Co^{2+}. The fluorescence of Tb-thermolysin in the presence of zinc and cobalt is shown in the figure. The fluorescence of terbium is partially quenched when Co^{2+} replaces zinc at the active site because of energy transfer. Given that $R_0 = 1.63$ nm for the Tb donor and Co acceptor, calculate the distance between the two metal sites.

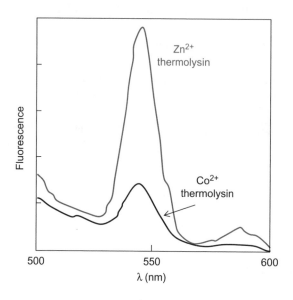

5.5.4 In a solution of chlorophyll, the measured fluorescence lifetime is 7 ns. However, in a photosynthetic unit, the lifetime is estimated to be about 0.1 ns. Can you suggest a reason for this difference? What are the relative fluorescence yields of chlorophyll in solution and in the photosynthetic unit? Assume the radiative lifetime of chlorophyll in the absence of any quenching is 25 ns.

5.5.5 Tryptophan fluorescence quenching is widely used to monitor the binding between 2,4-dinitrophenyl (Dnp) ligands and specific antibodies. There is usually significant overlap of the absorption spectrum of the DNP derivative (e.g. ε-Dnp lysine) with the emission spectra of the antibody; this results in highly efficient energy transfer. The figure shows the effect on the intrinsic fluorescence of an antibody when two different DNP-derivatives are added. One of the DNP derivatives binds strongly, the other weakly, to the antibody. The fluorescence quenching of a control solution containing tryptophan (at the same concentration as the protein) is also shown. Explain the data. Is it necessary to correct for the data of the blank quenching in determining the binding constant? Does this limit the utility of the method?

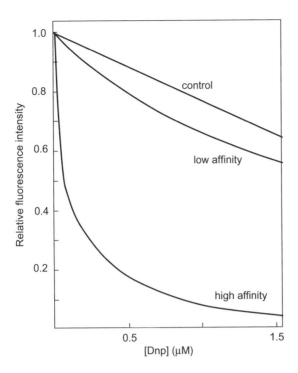

5.5.6 The kinetics of carboxypeptidase cleavage of the peptide dansyl-Gly-L-Phe were followed by monitoring enzyme tryptophan fluorescence and dansyl (DNS) fluorescence as a function of time after rapid mixing using stopped-flow methods. The structure of DNS, the kinetics and the fluorescence spectra of tryptophan and DNS, together with the absorption spectra of DNS and Co^{2+} in the enzyme are shown in the figure. (i) Explain why the fluorescence of tryptophan increases while that of dansyl decreases. (ii) When the zinc ion is replaced by cobalt the enzyme is still active, but no dansyl fluorescence is observed. Suggest a reason for this and what information might be obtained from this observation.

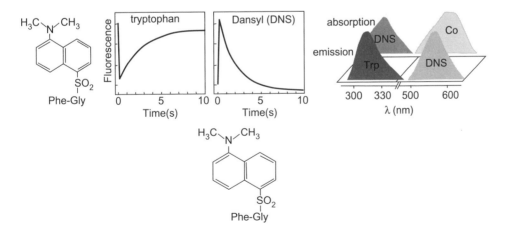

5.5.7 The 3-phosphoglycerate kinase of yeast contains two Trp and 8 Tyr per molecule. It has an unusual fluorescence emission spectrum with a maximum at 308 nm when excited at 280 nm. When denatured in 4M guanidine hydrochloride, the protein showed two emission peaks, one at 345 nm and the other at 303 nm. Try to interpret these results.

5.5.8 A number of ingenious fluorescent biosensors have been developed to detect and measure particular ligands. There has, for example, been much interest in detecting intracellular calcium and there are commercially available kits for measuring intracellular calcium concentrations by fluorescence. Roger Tsiens's group developed a reagent with a BFP and a GFP donor/acceptor pair tethered together with a peptide that binds calmodulin. The figure shows the fluorescence emission from the BFP/GFP sensor and the sensitivity to calcium in the presence of calmodulin. Can you explain this sensitivity? (Hint: study Figures 5.5.16 and 5.5.17.)

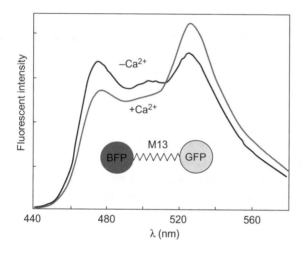

5.6 X-ray spectroscopy

Introduction

This chapter gives a brief description of techniques that involve absorption and emission of high-energy X-rays. These X-ray spectra, which arise from transitions involving core electron energy levels, give information about metal ion environments and the elemental composition of a sample. We have previously discussed UV/visible spectroscopy (Chapter 5.3) which probes transitions between outer electronic energy levels (levels occupied by electrons that determine a molecule's chemistry). In contrast, X-ray photons, which have energy >10^3 times higher than light, cause transitions between inner electronic levels of atoms. The incident high energy X-rays can also eject electrons completely from the core, a process called photo-dissociation.

A typical **X-ray absorption spectroscopy** (XAS) spectrum obtained by irradiating a molecule with X-rays is shown in Figure 5.6.1. The spectrum can be divided into "edge" and "EXAFS" regions. We will see below that these two regions arise from different physical phenomena. Analysis of the edge region can give some information about oxidation states, the nature of ligands, and the geometry around metals. EXAFS can give atomic resolution information about the coordination geometry around a metal in a metalloprotein and can be particularly informative when a cluster of metals is present. XAS-related phenomena have been known since the 1930s, but they did not become a practical experimental tool until synchrotron radiation sources became widely available. XAS does not require ordered samples and gives structural information about the local environment that is complementary to information available from other techniques, such as crystallography (Chapter 4.3) and EPR (Chapter 6.2).

Various names are found in the literature for X-ray induced phenomena. Analysis of the edge region in Figure 5.6.1 is sometimes referred to as XANES (X-ray absorption near-edge structure) or NEXAFS (near-edge X-ray absorption fine structure). EXAFS (extended X-ray absorption fine structure) applies to the spectral region beyond the edge (Figure 5.6.1). XAFS (X-ray absorption fine structure) is a term applied to experiments that investigate both the edge and the EXAFS region. **X-ray fluorescence** (XRF) is a related technique that has powerful analytical uses (see below).

Theory

Incident X-ray photons result in transitions between the atomic energy levels of the absorbing atom. Typical X-rays have energies ranging from ~500 eV to 500 keV, corresponding to wavelengths of ~2.5–0.025 nm. Absorption occurs when the energy of the incident photon is greater than or equal to the ionization energy of the core electrons. The absorbing electron can either be raised to an unfilled electron energy level or ejected into the "continuum".

The most commonly studied edge region corresponds to K transitions, where electrons come from the lowest $1s$ energy level. At lower X-ray energies, ionization of $2p$ or $2s$ electrons can give rise to what

---

Note Electron volts (eV) are often used to describe energy in XAS. As explained in Tutorial 8, an electron volt is the amount of kinetic energy gained by a single unbound electron when it accelerates through an electric potential difference of one volt; it corresponds to an energy of 1.602×10^{-19} J. 1 eV is equivalent to a wavelength of ~1.24 μm. In the X-ray region, $\lambda = 1$ nm corresponds to 1240 eV.

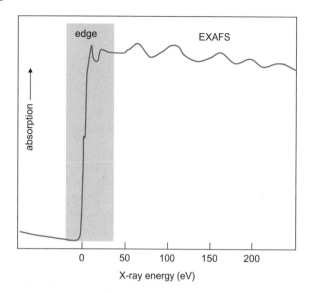

Figure 5.6.1 A typical XAS spectrum showing the edge and EXAFS regions.

Table 5.6.1 Energies associated with K and L_3 transitions for some elements. The energies underlined are accessible to X-rays from synchrotrons

Element	Energy (keV)	
	K	L_3
Mn	6.5	0.6
Fe	7.1	0.7
Ni	8.3	0.8
Cu	9.0	0.9
Mo	20.0	2.5
Tb	52.0	7.5
W	69.5	10.2
Hg	83.1	12.3

are called L_3, L_2, and L_1 absorption edges (see also Tutorial 10 and Figure 5.6.2). The energies required to induce these transitions in some elements are summarized in Table 5.6.1.

X-ray absorption causes a vacancy in one of the inner electron shells. One way of filling the vacancy is by X-ray fluorescence. This occurs when a higher energy electron drops to a lower level, emitting an X-ray of well-defined energy. This is analogous to the fluorescence discussed in Chapter 5.5, but here the emission spectrum is characteristic of the atom involved and can be used to identify and quantify the elements in a sample (see discussion below). An L shell electron dropping into the K level gives a K_α fluorescence line, while a transition from M to K gives a K_β line and so on (see Figure 5.6.3).

Measurement

XAS requires a tunable X-ray source and high-quality X-ray detectors. Synchrotrons provide a wide range of X-ray wavelengths, and monochromators made from silicon can select a particular energy using Bragg diffraction (Chapter 4.3); these filters can have energy resolution of ~1 eV at 10 keV. The simplest experiments are carried out in transmission mode (Figure 5.6.4).

A beam of monochromatic X-rays is passed through the sample and, in transmission mode, the absorbance is monitored as a function of wavelength (electron volts). The incident and transmitted X-ray fluxes are measured with sensitive X-ray detectors. X-ray absorption can be described by an absorption coefficient, μ, and Beer's law: $I = I_0 e^{-\mu t}$, where I_0 is the X-ray intensity incident on a sample, t is the sample thickness, and I is the intensity transmitted through

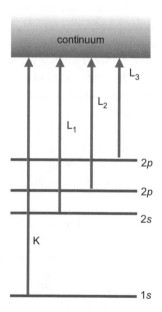

Figure 5.6.2 Core electron energy levels (K and L) that are susceptible to transitions induced by X-ray photons.

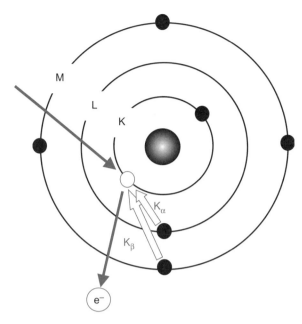

Figure 5.6.3 Schematic illustration of X-ray fluorescence (white arrows) following the ejection of a core electron by incident X-rays (solid cyan arrows). The vacancy is filled by K_α or K_β fluorescence.

the sample. μ depends on the sample density ρ, the atomic number Z, the atomic mass A, and the incident X-ray energy E. Fluorescent emission from the sample can also be detected, as shown in Figure 5.6.4.

Edge spectra

For a given element, K edge energies depend slightly ($\pm$ a few eV) on the chemical environment of the element. For example, higher oxidation state metals have higher positive charge, making it slightly more difficult to photo-dissociate a $1s$ electron, thus shifting the K edge to higher energy. Shifts of 1–2 eV per oxidation state are typical for first-row transition metals. Sometimes, especially when there is no previous structural information about the metal site

being examined, XAS can provide qualitative information about the kinds of ligands holding a particular metal in a metalloprotein. Edge analysis can also indicate the molecular geometry or the arrangement of ligands around the metal. Comparison of the edge spectrum before and after addition of an inhibitor or a reductant can give information about the role of a metal site in enzyme catalysis.

EXAFS

Emitted photoelectrons can be considered as spherical waves radiating from the absorbing atom. The theoretical basis of EXAFS arises from modeling the interference between this radiating photoelectron wave and waves backscattered from the surrounding atoms (Figure 5.6.5).

EXAFS data are processed by subtracting a smoothed edge spectrum to emphasize the oscillations. The X-ray energy scale is also usually converted to a k (wavenumber) scale where $k = \{(2m(E - E_0)/\hbar^2\}^{1/2}$; E_0 is the absorption edge energy, m is the electron mass, and $\hbar$ is the reduced Planck constant (see Appendix A1.3). The oscillations in the EXAFS region are often further enhanced by multiplying by k^3 (Figure 5.6.6). Distinctive features of the oscillations are more readily detected by taking a Fourier transform (FT) of $\chi(k)$ to give a plot that relates to distance, R (see Chapters 4.3, 5.2, and 6.1 for discussions of Fourier transforms).

The different frequencies apparent in the oscillations in $\chi(k)$ correspond to different near-neighbor coordination shells which can be described and modeled. Methods for analyzing EXAFS data have greatly improved in recent decades and online computer programs are available that fit the data obtained to a model for the coordination geometry around a metal in a protein (see, e.g., **http://cars9.uchicago.edu/feffit/**).

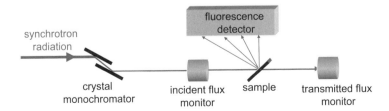

synchrotron radiation

crystal monochromator

incident flux monitor

sample

transmitted flux monitor

fluorescence detector

Figure 5.6.4 Schematic view of an XAS experiment. It is possible to operate either in transmission or fluorescent mode.

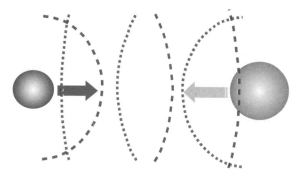

Figure 5.6.5 Illustration of a wave radiating out from the cyan center. This generates (gray) backscattered waves from neighboring atoms resulting in an interference pattern generated by mixing of cyan and gray waves.

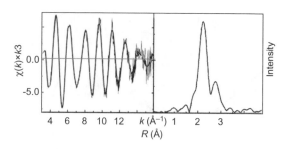

Figure 5.6.6 Illustration of a typical EXAFS spectrum after correction and multiplication by k^3. The FT of $\chi(k)$ is shown on the right. [Note that angstroms (1 Å = 10^{-10} nm) are used to measure distances here.]

Example Formate dehydrogenase H (FDHH)

FDHH is an enzyme in the respiratory pathway of *E. coli* that is active when the bacteria are grown anaerobically on glucose. This molybdenum enzyme is unusual in that it contains a selenocysteine residue (SeCys140) which is essential for full activity and which has been shown by EPR (Chapter 6.2) to interact with the molybdenum ion at the active site. Molybdenum enzymes have diverse functions; they catalyze a variety of two-electron redox reactions with the molybdenum cycling between the Mo(VI) and Mo(IV) oxidation states (Chapter 6.2).

Selenium and molybdenum K-edge X-ray absorption spectroscopic data were collected using synchrotron radiation (Figure 5.6.7). Samples of oxidized, dithionite-reduced, wild-type and mutant *E. coli* FDHH, at a concentration of ~2.0mM, were studied by XAFS in the frozen state at 10 K. Figure 5.6.7 shows the selenium and molybdenum edge spectra of oxidized and reduced states. Figure 5.6.8 shows EXAFS spectra of the oxidized state and a mutant form of the enzyme where the SeCys at residue 140 was replaced by a cysteine. Note that the band assigned to the Mo–Se distance disappears in the Fourier transform of the mutant spectrum.

A crystal structure of the oxidized protein at 2.9 Å resolution indicated an unusual molybdenum active site with four Mo–S ligands at 2.4 Å, one Mo–OH at 2.1 Å, and a coordinated selenocysteine, with a Mo–Se distance of 2.6 Å. EXAFS at the molybdenum edge largely confirmed these observations but the selenium EXAFS (not shown in Figure 5.6.8) also indicated the presence of a Se–S distance at 2.19 Å. This had been missed in the crystallographic analysis, which had suggested a closest Se–S contact of 2.9 Å.

EXAFS thus indicated, for the first time, that the active site of *E. coli* FDHH contains a novel seleno-sulfide ligand to molybdenum, as shown in Figure 5.6.9. The seleno-sulfide bond is believed to be important for understanding the catalytic mechanism of FDHH as it is a potentially redox active ligand.

Analytical uses of X-ray emission

X-ray fluorescence spectroscopy (XRF) is widely used to measure the elemental composition of materials. As this method is fast and nondestructive to the sample, it is used for analysis in diverse fields.

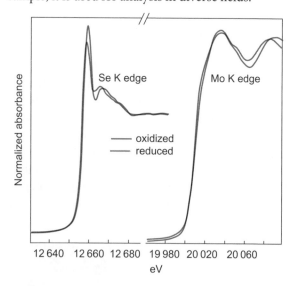

Figure 5.6.7 The selenium and molybdenum K edge spectra for FDHH in the oxidized, Mo(VI), and reduced, Mo(IV), states (adapted and redrawn with permission from *J. Am. Chem. Soc.* 120, 1267–73 (1998)).

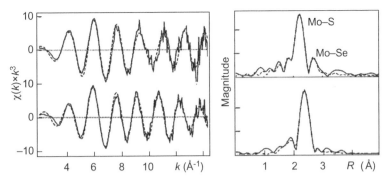

Figure 5.6.8 The molybdenum K edge EXAFS spectra of wild-type FDHH in the oxidized state (above) and a mutant enzyme (below) where the rare SeCys residue was replaced by Cys. Views on the right are the FTs of the EXAFS. Peaks assigned to Mo–S and Mo–Se distances are indicated. The dotted lines correspond to theoretical fits (adapted and redrawn with permission from *J. Am. Chem. Soc.* 120, 1267–73 (1998)).

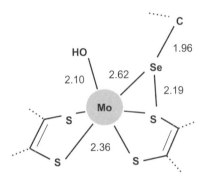

Figure 5.6.9 A structure of the active site of FDHH consistent with X-ray and EXAFS data. Bond distances are shown in Å units.

XRF can be produced not only by incident X-rays but also by other excitation sources, including electrons, alpha particles or protons. Synchrotron X-rays provide a suitable source for XRF experiments but a beam of high energy (MeV) protons, produced by an accelerator (a device that produces a stream of charged particles using applied electric and magnetic fields), has distinct advantages. In a technique called **particle-induced X-ray emission** (PIXE), an incident proton beam is applied to a sample to eject inner core electrons and produce XRF. Very fine irradiation beams can be made, allowing a very small area of a sample to be irradiated. Sensitive detection is also possible because the background X-ray levels are much lower than when X-ray excitation is used to produce the XRF.

One useful application is to measure the composition of specific elements in a protein sample (e.g. Zn, Cu, sulfur, etc.). This is a demanding experiment, however, as the sensitivity depends on the mass fraction of the element in the protein; typically one might be trying to detect a metal atom of mass around 50 Da in a molecule of around 100 kDa. When proteins are analyzed with PIXE, every element heavier than fluorine can, however, be observed simultaneously. This means that the sulfur in the amino acids methionine and cysteine can be used as an internal standard to indicate the amount of protein present if, as is usually the case, the amino-acid composition is known. Concentrations of any elements of interest can then be measured relative to that sulfur signal. The measured ratio can be used to calculate the stoichiometry of, say zinc to protein, with an accuracy of about ±10% (see also Problem 5.6.2).

⊕ Summary

1 X-rays have a high energy that corresponds to transitions between inner electron energy levels. Irradiation of atoms by X-rays can eject electrons to give an "edge" spectrum. An electron wave is also produced that can interact with neighboring atoms to give **extended X-ray absorption fine structure** (EXAFS). The edge spectrum can be used to identify ligands around a metal and EXAFS can be used to measure distances to these ligands.

2 After the ejection of inner electrons by applied radiation; outer electrons fill the vacancies producing X-ray fluorescence (XRF) in the process. XRF is a powerful analytical tool for measuring the elemental composition of a sample.

 ## Further reading

Useful website

http://gbxafs.iit.edu/training/tutorials.html (XAFS tutorials)

Reviews

Bunker, G., Dimakis, N., and Khelashvili, G. New methods for EXAFS analysis in structural genomics. *J. Synchrotron Radiat.* 12, 53–6 (2005).

Garman, E.F., and Grime, G.W. Elemental analysis of proteins by microPIXE. *Prog. Biophys. Mol. Biol.* 89, 173–205 (2005).

 ## Problems

5.6.1 The figure shows the K edge XAFS spectra of rubredoxin, which is an Fe–S protein with one Fe atom tetrahedrally coordinated to four cysteine thiol groups. It is as an electron carrier, shuttling between the Fe(II) and Fe(III) forms. The cyan and gray spectra correspond to the Fe(II) and Fe(III) oxidation states. Can you predict which is which?

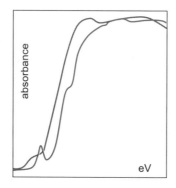

5.6.2 Provided a protein has not been post-translationally phosphorylated, phosphorus is unique to nucleic acids. Sulfur, which occurs in the amino acids Met and Cys, is unique to proteins, so the phosphorus to sulfur ratio can give information about DNA/protein stoichiometry in a complex. In a study of a protein domain that forms a complex with a 15 nucleotide long DNA, microPIXE analysis of three different crystals found that crystals I and II contained single stranded DNA (measured by phosphorus), while crystal III had little, or no, phosphorus. The measured ratio of phosphorus to sulfur in crystals I and II was 4.4 ± 0.3 and 4.7 ± 0.4. If the synthetic oligonucleotide had 14 phosphates and the protein contained one sulfur atom, estimate the stoichiometry of protein domain:DNA in the crystal.

6 Magnetic resonance

Spin is a fundamental property of elementary particles and atomic nuclei. Wolfgang Pauli was the first to propose the concept of spin. He worked out a mathematical theory in 1927 that predicted "the Pauli exclusion principle" which requires a pair of electrons in the same energy level to have opposite spin. Kronig, Uhlenbeck, and Goudsmit later suggested that Pauli's concept could be interpreted in terms of particles spinning around their own axis. Spin was also an integral part of Paul Dirac's relativistic quantum mechanics developed in 1928.

A rotating charge generates a magnetic field; electrons and nuclei with the property of spin thus have an associated magnetic moment. Absorption spectra can arise by inducing reorientation of magnetic moments in a magnetic field with applied radiation, as first demonstrated by Isador Rabi in 1938, using molecular-beam detection. The principles and applications of techniques that detect *nuclear* and *electron* magnetic reorientation (resonance) in an applied magnetic field are discussed in the following two chapters.

6.1 Nuclear magnetic resonance

Introduction

Nuclear magnetic resonance (NMR) is a technique that detects the reorientation of nuclear spins in an applied magnetic field, B_o (Box 6.1). The phenomenon was first demonstrated independently by the Bloch and Purcell groups in 1945. It is extraordinarily versatile and has become a valued tool in many fields: in chemistry for analyzing and identifying small molecules; in molecular biology for giving information about the structure and interactions of macromolecules; in the pharmaceutical industry for screening drug binding and analyzing their metabolic products; in medicine for giving detailed anatomical images and detecting molecular markers of disease; and in neuroscience for giving information about how the brain responds to different stimuli. The reasons for its power and versatility arise from its ability to detect changes in chemical structure and environment and the fact that magnetic waves readily penetrate membranes and tissues without damaging them. The main disadvantage is the relatively weak signals compared to other forms of spectroscopy; this insensitivity is a consequence of the small energy separation between the energy levels in the nucleus (Chapter 5.1).

The NMR phenomenon

Spin

Nuclei in many isotopes have the property of spin. Nuclei also have an associated charge, rotation of which produces a magnetic field (Box 6.1). A nucleus with finite spin thus has a magnetic moment, μ_N, that can interact with an applied static magnetic field, B_o. As discussed in Chapter 5.1, this interaction with B_o leads to a nucleus with spin I having $2I + 1$ possible energy levels.

Box 6.1 Magnetism

Magnetic fields are measured in Tesla; the Earth's magnetic field strength is around 5×10^{-5} T. A magnetic field can be generated by a moving charge. For example, a field $B_o = \mu_o IN/r$ is generated at the center of a solenoid of wire of N turns, radius r, and current I, where μ_o is a constant called the permittivity of free space (see Figure A and Appendix A1.3 for values of constants). One thousand turns of radius 0.1 m carrying 1.5 amps gives a field in the solenoid: $B_o \sim 10^{-2}$ T. The magnitude of the field generated also depends on the *material within the solenoid*. B_o is the field generated in a vacuum; for other materials, $B = B_o + B_M$ where BM is an additional field set up by the material. B_M is negative for a diamagnetic material and positive for a paramagnetic material. Ferromagnetic materials (like iron) preserve their magnetism when B_o is removed but paramagnetic materials do not.

A magnetic field or current loop has an associated magnetic dipole moment, μ (see Tutorial 11). A magnetic moment can be considered as having two magnetic poles separated by some distance r; the magnetic moment will be related to pole strength $\times r$. The material-induced field, B_M, can also be written as $\mu_o M$, where M is the magnetization, which is defined as the magnetic moment per unit volume. Diamagnetic materials have a very small induced magnetic moment in a direction that opposes B_o. The effect is created by B_o-induced circulation of electrons.

The magnetic susceptibility is another measure of the response of a material to an applied field. It is defined as $\chi = B_M/B_o$ which is a number that is positive or negative, depending on whether the material is paramagnetic or diamagnetic.

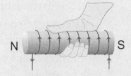

Figure A Generation of a magnetic field by a solenoid. The right hand grip rule predicts that if the current is in the direction of the fingers, a north pole is in the direction of the thumb.

Precession

The nuclear magnetic moments, μ_N, do not just behave like a compass needle; μ_N has quantum properties which means that it can only take up certain allowed angles in the applied field; the "spin" of μ_N also means that it has the property of **angular momentum** (Figure 6.1.1; Tutorial 3). An important consequence of angular momentum, **L**, is **precession** which occurs when a twisting force (a torque, **T**) is applied (see Figure 6.1.2). Because nuclear spin moments cannot align perfectly along B_o, there is a permanent moment, or **torque**, trying to align the μ_N with B_o. An illustration of the concept "torque + angular momentum = precession" is given by a spinning gyroscope; a weight applied to one side results in precession rather than tipping over (an animation of a precessing gyroscope with a weight attached to one side can be found at **http://www.youtube.com/watch?v=xeTTSW9UH0I&feature=related**). The vector cross product (Appendix A2.2) relationship between **T**, **L**, and precession, **P**, results in precession of μ_N around B_o (Figure 6.1.2; Tutorial 3).

The resonance frequency

The frequency of the induced precession is $\omega_o = \gamma B_o$, where γ, the **magnetogyric ratio**, is a proportionality constant and ω_o is the rotation frequency (often called the Larmor frequency). γ is different for different nuclei, resulting in a wide range of frequencies where NMR signals are detected (Table 6.1.1 and see below).

Many important isotopes found in biology (e.g. ^1H, ^{13}C, ^{15}N, and ^{31}P) have nuclei with spin $I = 1/2$ with only two energy levels. These are very convenient observation nuclei because the resulting spectra and their interpretation are relatively simple. When $I > 1/2$, e.g. ^2H and ^{23}Na nuclei, the spins can also interact with their environment via something called a quadrupole moment which makes spectra more complex (see discussion below and Figure 6.1.36).

Magnetization and phase coherence

When $I = 1/2$ the two energy states are occupied by magnetic moments μ_N aligned along and against B_o, as shown in Figure 6.1.3.

In a real sample, we have a very large number of spins. As a result, simple classical physics can be used to describe the properties of the bulk nuclear magnetization, which is a vector with magnitude **M**. The induced component along the B_o direction, M_z, is proportional to the population difference $(n_u - n_g)$ between the two energy levels in Figure 6.1.3. The many precessing magnetic moments in the sample can also be visualized by considering them to have a common point of origin, as shown in Figure 6.1.4.

As shown in Figure 6.1.4, the μ_N vectors are randomly distributed around the z-axis at equilibrium (there is no **phase coherence**); this means that the

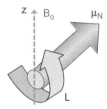

Figure 6.1.1 A nuclear magnetic moment μ_N does not align perfectly along an applied magnetic field, B_o. μ_N also has property called angular momentum, **L**. The B_o direction is defined to be along the z-axis.

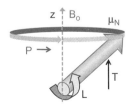

Figure 6.1.2 A combination of torque (**T**) and angular momentum (**L**) results in precession (**P**) of μ_N around the B_o direction.

Table 6.1.1 Spin, resonant frequency, and natural abundance of some isotopes. This table means, for example, that ^{13}C signals are detected around 126 MHz when $B_o = 11.74$ T

Nucleus	Spin	Frequency at 11.74 T (MHz)	Natural abundance (%)
^1H	½	500.0	99.98
^2H	1	76.75	1.5×10^{-2}
^{13}C	½	125.72	1.1
^{14}N	1	36.12	99.63
^{15}N	−½	50.66	0.37
^{19}F	½	470.39	100
^{23}Na	3/2	132.26	100
^{31}P	½	202.40	100

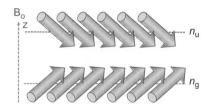

Figure 6.1.3 Spin ½ nuclei, interacting with an applied field B_o, have two possible energy states. These are occupied by magnetic dipoles, μ_N, with populations n_u and n_g in the upper and ground states. As a result of the Boltzmann distribution law, there will be a slight excess population in the ground state, usually aligned along B_o.

net magnetization component in the x–y plane, M_{xy}, is zero. Phase coherence in the x–y plane can, however, be induced by applying another field, B_1, rotating around the z-direction, as explained in the next section.

Measurement

The B_o field

A typical experimental setup for an NMR experiment is shown in Figure 6.1.5. A constant, strong,

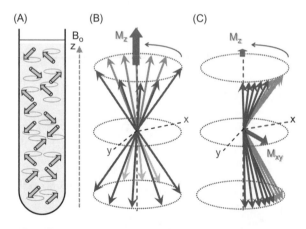

Figure 6.1.4 A representation of precessing magnetic moments, μ_N. (A) The equilibrium situation in an NMR sample with many magnetic dipoles, μ_N; some are aligned along the field, some against. (B) The μ_N can all be depicted as having the same origin. The net magnetization in the $B_o(z)$ direction is M_z, but $M_{xy} = 0$ at equilibrium because the μ_N are randomly distributed around the z-axis (there is no phase coherence). In (C), finite M_{xy} has been induced in the x–y plane because there is some phase coherence.

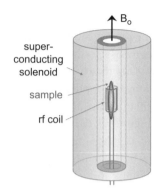

Figure 6.1.5 In an NMR experiment, a sample is placed in a static B_o field generated by a superconducting solenoid. It is also placed in a coil to which a radiofrequency (rf) B_1 field is applied (see Figure 6.1.6).

and very homogeneous B_o field (2–23T) is applied to the sample. This B_o field is generated by a current in a solenoidal coil (Box 6.1) made of "superconducting" wire. Superconductivity occurs for some metals and alloys (e.g. niobium/tin) at very low temperatures. This means that there is zero resistance and a current in the solenoid flows for ever, as long as the circuit can be closed and the conductor is kept cold enough; usually this is done by bathing the solenoid in liquid helium at 4 K or less. The field has to be very homogeneous across the sample and sets of "shim" coils are used to make fine adjustments.

The B_1 field

In addition to the static B_o field, we need a rotating B_1 field to induce phase coherence in the x–y plane (see Figure 6.1.4). To induce M_{xy} signals efficiently, B_1 has to be at right angles to B_o in the x–y plane. B_1 can be generated by a solenoid, but for convenience of sample access it is usually generated by a resonator or coil, shaped as shown Figure 6.1.6. The coil is tuned to the resonant frequency of the signal to be detected, for example ^{1}H signals resonate around 500 MHz if B_o is 11.7 T (see Table 6.1.1).

The applied radiofrequency (rf) B_1 field is an oscillating sine wave; this is equivalent to two rotating fields going in opposite directions (see also Figures 6.1.7A and 5.4.2). Only one of these fields rotates in the right way to induce phase coherence in the precessing spins (Figure 6.1.4).

To understand the effect of the rotating field, it is helpful to change to a new reference frame, $x'y'$

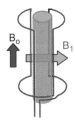

Figure 6.1.6 An rf coil generates a $\mathbf{B_1}$ field at right angles to $\mathbf{B_o}$. The same rf coil acts as a receiver for signals generated by $\mathbf{M_{xy}}$ magnetization in the sample.

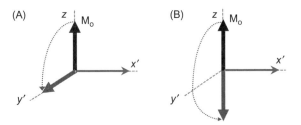

Figure 6.1.8 (A) A $\mathbf{B_1}$ field (cyan arrow) applied along the rotating axis x' causes precession of the equilibrium magnetization, $\mathbf{M_o}$, around x'. If it is applied for time t, $\mathbf{M_o}$ is rotated to the y' axis (this is a 90° pulse). (B) If $\mathbf{B_1}$ is applied for time $2t$, $\mathbf{M_o}$ is inverted, by rotation, to the $-z$ axis (a 180° pulse).

(Figure 6.1.7B) that rotates at the same frequency as $\mathbf{B_1}$. $\mathbf{B_1}$ then appears to be static in this new reference frame.

NMR pulses and transients

Modern NMR spectrometers operate in the **transient mode**, where a $\mathbf{B_1}$ rf **pulse** is applied to the sample; this results in a transient response, usually detected by the same coil that applied the $\mathbf{B_1}$ pulse (Figure 6.1.6). As we will see, the application of various pulse sequences to NMR samples can be used in sophisticated ways to generate diverse information about the system under study. The effect of different kinds of pulses on the bulk equilibrium magnetization, $\mathbf{M_o}$, is illustrated in Figure 6.1.8. Like the individual spin moments, $\mathbf{\mu_N}$, $\mathbf{M_o}$ has angular momentum; application of the rotating $\mathbf{B_1}$ field along x' generates a torque around the rotating x' axis and induces precession of $\mathbf{M_o}$ at right angles to that axis (Figure 6.1.2). The rotating $\mathbf{B_1}$ field thus causes $\mathbf{M_o}$ to rotate around x' with a precession frequency of $\omega = \gamma B_1$, as shown in Figure 6.1.8. As ω is an angular frequency, we can see from Figure 6.1.8A that if $\mathbf{B_1}$ is applied for time $t = \pi/(2(\gamma B_1))$ it generates $\mathbf{M_o} = \mathbf{M_x}$; this is called a 90° pulse, while a 180° pulse, generates $\mathbf{M_o} = -\mathbf{M_o}$.

The **transient signal** induced in the coil by spins that have been perturbed by the $\mathbf{B_1}$ pulse is often called a **free induction decay**. It can readily be translated into a frequency-dependent spectrum by Fourier transformation (FT; Figure 6.1.9; Tutorial 6), even if it arises from many chemical groups with

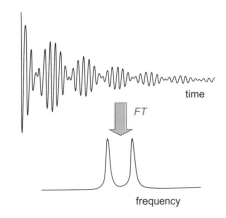

Figure 6.1.9 An illustration of Fourier transform NMR, where a transient $\mathbf{M_{xy}}$ signal is converted into a frequency-dependent spectrum by a Fourier transformation (FT). The illustration is for two "lines" with equal intensities and decay rates but different frequencies.

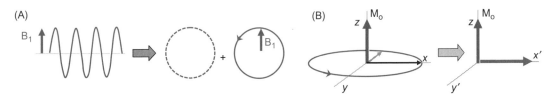

Figure 6.1.7 Generation of $\mathbf{B_1}$ fields and rotation of magnetization. (A) An applied sine wave is equivalent to two fields in the x–y plane rotating in opposite sense. (B) One of these fields is ineffective because it rotates in the wrong direction; the other can be considered as static in a new rotating coordinate system, $x'y'$.

different values of ω and decay rates. The FT method is equivalent to a piano tuner striking all the notes at once rather than one at a time, so it provides very efficient data collection. We will learn about what the spectrum generated tells us in the next section.

The spectral parameters in NMR

An NMR spectrum is characterized by a series of "lines" or "resonances", e.g. the two-line spectrum in Figure 6.1.9 and the three-line spectrum of ethanol, shown in Figure 6.1.10. There are five parameters that define such lines: (i) the intensity (I) or, more rigorously, the area of the line; (ii) the chemical shift (δ), which defines the line position on a frequency scale; (iii) the multiplet structure, which is related to a spin–spin coupling constant (J); (iv) a relaxation time (T_2), related to the linewidth and the decay rate of the $\mathbf{M_{xy}}$ magnetization; (v) another relaxation time (T_1), related to the lifetime of a spin in an energy level and the decay of $\mathbf{M_z}$ magnetization. These five parameters, together with another, called "the nuclear Overhauser effect", are discussed, in turn, below.

Intensity

If a system of nuclear spins is at equilibrium, then the amplitudes of the transient signals after a pulse are directly proportional to the number (concentration) of nuclei in the sample. Instrumental variables are such that estimates of absolute concentration are not very reliable, but relative concentrations can be obtained quite accurately from measurement of resonance intensity. For example, in the ethanol spectrum depicted in Figure 6.1.10, the integrated intensities of the lines arising from the –OH, –CH$_2$–, and –CH$_3$ groups are found to be in the ratio 1:2:3, i.e. directly proportional to the number of protons in the groups.

Chemical shift

Nuclei are surrounded by electrons; the applied field, $\mathbf{B_o}$, induces currents in the electron clouds that reduce the effective field experienced at the nucleus. These shielding fields induced by the circulating electrons are directly proportional to $\mathbf{B_o}$. Thus, we can write $\mathbf{B_{eff}} = \mathbf{B_o} - \sigma$, where σ is a shielding constant that depends on the nature of the electrons around the nucleus.

A normalized frequency scale is used to measure chemical shifts with respect to a signal from a reference compound, usually tetramethylsilane, $Si(CH_3)_4$, or a derivative thereof. The chemical shift scale used in NMR is $\delta = (\nu - \nu_{reference})/\nu_o$; this has dimensionless units of parts per million (ppm); $(\nu - \nu_{reference})$ is the frequency difference between the line of interest

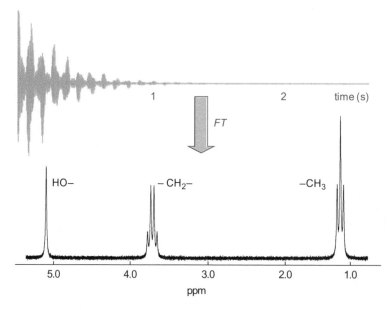

Figure 6.1.10 Spectrum of ethanol and its transient signal (in cyan) before Fourier transformation (FT).

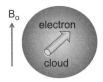

Figure 6.1.11 The nuclear magnetic moment is shielded from B_o by circulating electrons.

and the reference line, in Hz; ν_o is the applied frequency, in megahertz (MHz); ν_o can be calculated from Table 6.1.1 for any value of B_o. This ppm scale has the advantage that it is independent of B_o.

There are two main classes of shift that are important for our purposes; as illustrated in Figure 6.1.12 these are: (i) an *intrinsic*, or *primary*, shift, which is characteristic of a particular chemical group; and (ii) an *induced*, or *secondary*, shift arising from the influence, through space, of environmental effects (mainly from neighboring magnetic moments).

The *intrinsic* shifts of different groups, which depend on the local chemical bonding arrangement of nuclei, can vary widely. Methyl protons are well shielded and appear to the right (high field) in NMR spectra, while aromatic amide protons are less well shielded and appear to the left (lower field). Intrinsic shifts are also very different for different nuclei (^{1}H, ^{15}N, ^{13}C, etc.). ^{1}H spectra have a range of around 15 ppm, while ^{13}C spectra have a much larger shift range (>200 ppm).

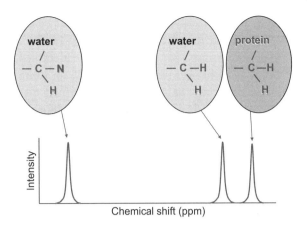

Figure 6.1.12 Chemical shift is influenced by bonding and environment. The two resonances in an aqueous environment have different shifts because of different chemical structure (CNH, CH_2), while two resonances with the same chemistry (CH_2) have different shifts because of different environments (water and protein).

Figure 6.1.13 shows a high field spectrum of the enzyme lysozyme (14.5 kDa); this spectrum of the native, folded state arises from a combination of *intrinsic* and *induced* shifts. Induced shifts are very important in macromolecule NMR and the spectrum of a folded globular protein is very different from the spectrum of an unfolded protein. The local environmental shifts allow the resonances of groups with the same chemistry to be resolved and assigned,

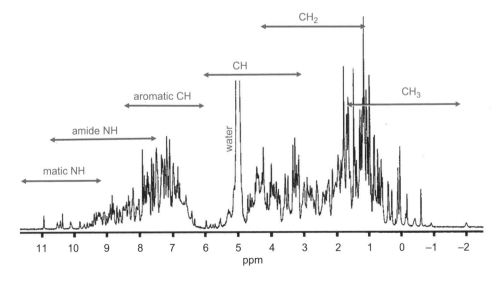

Figure 6.1.13 ^{1}H spectrum of the enzyme lysozyme, at 950 MHz. The shifts from different chemical groupings and their approximate range are shown. If the protein is heated to a high temperature it unfolds and the environmental shifts that cause a spread in one type of group (e.g. alanine methyl resonances) are then very small, giving a much simpler spectrum.

even in a large macromolecule; this is the key factor that allows macromolecular structure to be determined and site-specific interactions to be identified.

Some *induced shifts can be quantified*; for example aromatic rings cause shifts because their electrons are essentially delocalized and a field opposing an applied $\mathbf{B_o}$ field is produced by the circulating electrons. If $\mathbf{B_o}$ is applied in a direction perpendicular to the plane of the ring, a field that opposes $\mathbf{B_o}$ is produced by the circulating electrons. This is a dipole which generates a field proportional to $(1 - 3\cos^2\theta)/r^3$ (see Tutorial 11 and Problem 6.1.2). Induced shifts can thus sometimes be used to give structural information, although this is not yet a reliable procedure because of other unknown shift contributions.

Spin–spin coupling and multiplet structure

High-resolution NMR lines can exhibit additional structure that arises from weak interactions between magnetic nuclei (see, for example, the CH_3 resonance in Figure 6.1.10 which has three components). These spin–spin coupling interactions are communicated between the nuclei by the electrons in a chemical bond. The size of the interaction is defined by a **spin-spin coupling constant** (J) in units of Hz. J is, to first order, independent of the applied field, $\mathbf{B_o}$, but its magnitude depends on both the nature of the bond and the number of bonds involved. Consider, for example, a hypothetical molecule consisting of two spin ½ nuclei and a shared electron cloud (Figure 6.1.14). Each nuclear moment can be oriented in one of two ways. These two orientations cause slightly different electronic distributions, which result in small shielding effects at the other nucleus; the result is that each of the nuclear resonances is split in two, giving a pair of doublets, separated by the splitting constant J (Hz).

The spectrum of a nucleus coupled to two equivalent nuclei is a triplet with intensities in the ratio 1:2:1 because the following orientations are possible: $\uparrow\uparrow$, $\uparrow\downarrow$, $\downarrow\uparrow$, and $\downarrow\downarrow$, two of which ($\uparrow\downarrow$, $\downarrow\uparrow$) are equivalent. Similarly, three equivalent nuclei give rise to a quartet with intensity ratios 1:3:3:1 ($1 \times \uparrow\uparrow\uparrow$, $3 \times \uparrow\uparrow\downarrow$, $3 \times \uparrow\downarrow\downarrow$, $1 \times \downarrow\downarrow\downarrow$), and so on. Note the triplet structure of the $-CH_3$ group in ethanol in Figure 6.1.10 which arises from J coupling to the neighboring $-CH_2-$ group.

There is a relationship named after Martin Karplus who developed the theory that relates bond angles to the observed value of J. As we will see later in the protein structure determination section, this can give useful structural information; the dihedral angle, ϕ, is related to the three-bond coupling constant, $J_{NH\alpha CH}$, as illustrated in Figure 6.1.15.

The T_2 relaxation time and linewidth

The relaxation parameters T_2 and T_1 give a variety of useful information about a molecule, including molecular motion and distances to neighbors. These uses will be discussed later. T_2 is the time constant of the decay of $\mathbf{M_{xy}}$ components (see the section on measurement). The transient decay, $f(t)$, induced by a $\mathbf{B_1}$ pulse and the corresponding frequency-dependent lineshape, $F(\omega)$, are a Fourier transform pair (Tutorial 6), where: $f(t) = \cos(\omega_o)\exp(-t/T_2)$ $\Leftrightarrow$ $F(\omega) = CT_2/(1 + (\omega - \omega_o)^2 T_2^2)$. The lineshape, described by $F(\omega)$, is known as a Lorentzian; C is a constant and the width of the line at half its height (the linewidth) is $1/\pi T_2 = (\Delta\nu_{1/2})$ (Figure 6.1.16).

Inhomogeneity of $\mathbf{B_o}$, as well as the intrinsic T_2, affects the observed linewidth because a spread of $\mathbf{B_o}$ produces a spread of frequencies (Figure 6.1.17); this results in the experimentally observed

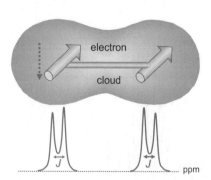

Figure 6.1.14 A diatomic molecule with two nuclei sharing electrons. The nuclear dipoles can communicate their orientation to each other via the electron cloud. The cyan nuclear dipole is shown causing a slight polarization (dotted arrow) of the electron cloud in its vicinity. This results in two doublets with splitting J Hz.

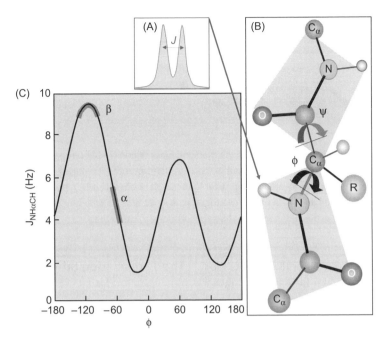

Figure 6.1.15 (A) In ¹H spectra of a protein, the amide NH resonances are split into doublets by spin coupling to the C_α proton three bonds away. (B) The NH resonance that is split is shown as a cyan sphere. J splitting of the doublet varies between approximately 2 and 10 Hz depending on the angle φ. (C) A plot of the Karplus equation which relates φ to J. The values of φ in proteins give J values of ~5 Hz in α-helices and ~9 Hz in β-strands (see also Figure 6.1.43).

value, T_2^*, being shorter than the intrinsic or "real" T_2 value.

There are, however, methods that can be used to reduce the effect of experimental inhomogeneities. One way is to use a "spin echo" technique with the pulse sequence $90°_x–\tau–180°_y–\tau–$, as illustrated in Figure 6.1.18. The 90° pulse along x' creates $\mathbf{M}_{xy}$ magnetization along y'. The cyan and black signals in Figure 6.1.17 arise from spins that are in slightly different values of $\mathbf{B}_o$; they thus precess with different

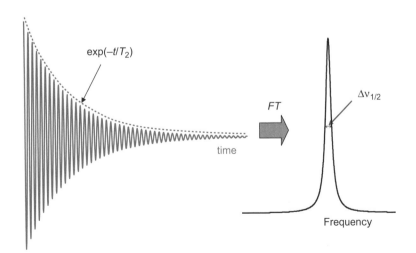

Figure 6.1.16 The decay rate of $\mathbf{M}_{xy}$ in the x-y plane is related to the linewidth. The faster the decay, the broader the line.

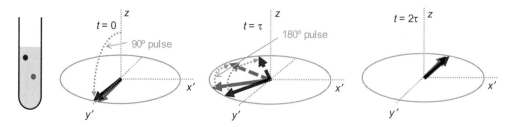

Figure 6.1.17 An inhomogeneous $\mathbf{B_o}$ field will result in the cyan and black regions of the sample having different resonant frequencies. The cyan and black components then fan out in the x'–y' plane after the 90° pulse because they have different precession frequencies; here the cyan signals are rotating faster than the black ones. The 180° pulse at time τ flips the positions so that black is now ahead of cyan. Cyan and black then arrive together at $-y'$ at time 2τ. Different effects can be obtained by applying pulses along different directions, e.g. the 180° pulse could be applied along y' rather than x'.

frequencies in the x'–y' plane. After they precess for a time τ the cyan signal, shown here to "run" faster than the black signal, is ahead. A 180° pulse, applied along the x' axis, flips these signals so that the black spins are now ahead of the cyan ones. As the cyan signals will continue to run faster than black, cyan catches up with black at time 2τ, thus generating an "echo". The intensity of the echo signals, at time 2τ, plotted as a function of τ, gives a measure of the real T_2 value that is relatively insensitive to experimental $\mathbf{B_o}$ inhomogeneities.

The T_1 relaxation time

T_1 is the time constant that describes the recovery of $\mathbf{M_z}$, the magnetization along the $\mathbf{B_o}$ field direction, after an applied perturbation. A 180° pulse can be used to invert $\mathbf{M_o}$ (Figure 6.1.8B) at $t = 0$. Figure 6.1.19 illustrates an experiment where a 180°–τ–90° pulse sequence has been applied to a sample with three lines; the recovering magnetization is monitored with a series of 90° pulses at increasing τ values. The three lines have different T_1 values,

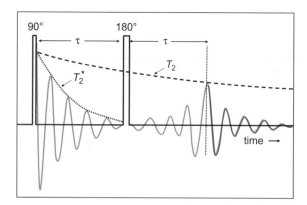

Figure 6.1.18 A spin echo 90°–τ–180°–τ pulse sequence. The dashed line represents the intrinsic T_2 decay; the faster experimental transient decay seen after the 90° pulse (T_2^*) arises because of $\mathbf{B_o}$ inhomogeneity, as well as T_2. The 180° pulse generates an echo at time 2τ because it flips the cyan and black signals in Figure 6.1.17. The cyan part of the echo in Figure 6.1.18 can be Fourier transformed to give a series of normal spectra at different values of τ. The intrinsic T_2 value can then be extracted from this decay series.

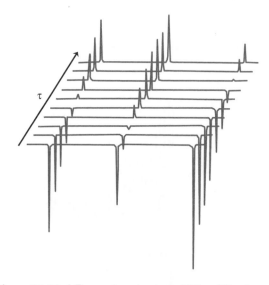

Figure 6.1.19 A T_1 experiment, using a 180°–τ–90° pulse sequence. Plots of intensity against τ (usually) give exponential decays for each line, each with a characteristic time constant T_1.

suggesting that they have different magnetic environments.

What causes relaxation?

Let us now consider some mechanisms that can induce T_1 and T_2 relaxation. Unlike some forms of spectroscopy (e.g. UV/visible/fluorescence) the probability of spontaneous transitions between energy states is negligible in NMR (see Chapter 5.1; Problem 5.1.1). This means that transitions between nuclear energy levels only arise from fluctuating magnetic fields that have the right frequency components—these are equivalent to the applied $\mathbf{B_1}$ fields discussed above but occur within the sample itself. For nuclei with $I = 1/2$, the most important sources of local fields are usually interactions between magnetic dipoles and we focus on those here. (Note there is another relaxation mechanism that we will introduce later called **chemical shift anisotropy** (CSA); this arises from variations in chemical shift that occur as the molecule takes up different directions in $\mathbf{B_o}$.)

Figure 6.1.20 shows how a magnetic moment, μ_y, experiences a magnetic field via a dipolar interaction with a neighboring moment, μ_x. Brownian motion causes the x–y molecule to tumble randomly in solution. The separation between x and y is constant but the angle the vector $\mathbf{r}$ makes with $\mathbf{B_o}$ varies as the molecule tumbles (remember the magnetic dipoles can only point along or against $\mathbf{B_o}$). The net result is that μ_y experiences a fluctuating field from μ_x (Figure 6.1.21) and μ_x experiences a similar fluctuating field, induced by μ_y.

The local magnetic fields ($\mathbf{B_{loc}}$) generated by the random molecular motion contain a spectrum of frequencies. As described in Box 4.3, random motions in solution can be characterized by a correlation function with a correlation time τ_c. Just as we can analyze the frequency content of a transient NMR signal by a Fourier transform (Figure 6.1.9) we can analyze the frequency content of the motion-generated local magnetic field fluctuations. As illustrated in Figure 6.1.21, this analysis shows that fast molecular motion generates a wide range of frequencies with relatively low amplitude while slow motion generates a much more

Figure 6.1.20 Dipole–dipole interactions between two nuclei, a distance r apart in a diatomic molecule. The form of the dipole–dipole interaction between μ_x and μ_y is $(1 - 3\cos^2\theta)/r^3$ (Tutorial 11); this relationship is illustrated by the magnetic lines of force drawn around μ_x. Note, for example, the change in arrow direction as θ goes from 0°, when $1 - 3\cos^2\theta = -2$, to 90°, when $1 - 3\cos^2\theta = 1$. As the molecule x–y tumbles randomly in solution it causes fluctuations in θ so that the field generated at μ_y by μ_x also fluctuates. (Note that the quantum angle the dipole makes with respect to $\mathbf{B_o}$, as shown in Figure 6.1.1, has been neglected here because it averages out.)

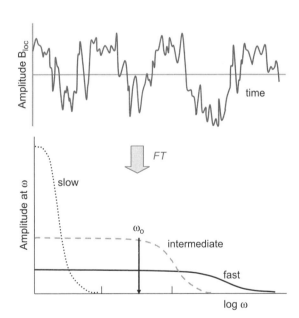

Figure 6.1.21 Brownian motion of a magnetic moment generates random local field variation at a neighbor (top). The frequency dependence of $\mathbf{B_{loc}}$ can be found by FT; this is shown here for three cases: long τ_c, slow tumbling, dotted line; short τ_c, fast tumbling, solid line; and an intermediate case (gray dashed line). Note: the frequency scale here is logarithmic so the Lorentzian lineshape shown in Figure 6.1.16 looks different but the forms are in fact the same.

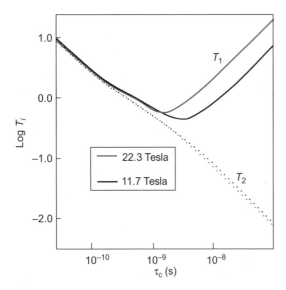

Figure 6.1.22 Dependence of T_1 and T_2 on τ_c at two field strengths. The curves are calculated for a ^{15}N nucleus bonded to a ^{1}H 0.102 nm away (typical for an amide group in a protein).

Example Carbon relaxation in a lipid bilayer

The ^{13}C T_1 values observed for a ^{13}C labeled lipid in a bilayer are shown in Figure 6.1.23. Note that the T_1 values for the CH$_2$ groups vary between 0.1 and 1.8 s. The ^{13}C dipolar relaxation in these CH$_2$ groups will be completely dominated by the two ^{1}H atoms attached to the carbon at a fixed bond-length distance away. The observed variation in T_1 values for these CH$_2$ groups must therefore arise from differing flexibility along the chain (different values of τ_c). For ^{13}C relaxation in a lipid we are in a region of the T_1 vs. τ_c plot (Figure 6.1.22) where T_1 decreases directly with increasing τ_c. The results in Figure 6.1.23 thus show that the lipid chain is much more flexible in the middle of the bilayer than near the glycerol backbone. Note that the long T_1 value for the C=O carbon in Figure 6.1.23 is not due to increased flexibility; the C=O carbon does not have two attached protons, so the effective value of r in the $f(\omega,\tau_c)/r^6$ relaxation term is large.

limited range of frequencies, but these have higher amplitude.

The probability that fluctuating fields will induce a transition between energy levels depends on their frequency and amplitude. One frequency that causes a transition is the Larmor frequency, ω_o (see Table 6.1.1); it can be seen from Figure 6.1.21 that as τ_c increases, the intensity of fluctuations at ω_o will first increase and then decrease. This means that T_1 will go through a minimum value that will depend on ω_o, which depends on the applied field, $\mathbf{B_o}$ (see also Problem 6.1.7).

The reason we are interested in relaxation is that it gives information about motion and distances. The magnitude of the local fields in Figure 6.1.21 depends on the square of the amplitude of the fluctuation, which depends on $1/r^3$ (r is the separation between dipoles). The net result is that $1/T_1$ is proportional to $nf(\omega,\tau_c)/r^6$, where n is the number of magnetic dipoles around the nucleus of interest and $f(\omega,\tau_c)$ is a function that depends on ω and τ_c. An example of the dependence of relaxation times T_1 and T_2 on τ_c is shown in Figure 6.1.22. For short τ_c, T_1 and T_2 are often equal, while $T_2 < T_1$ for larger τ_c. Typical macromolecules have τ_c values of ns so that $T_2 < T_1$ and T_1 is frequency dependent.

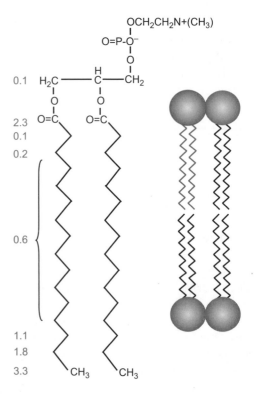

Figure 6.1.23 ^{13}C T_1 values (in cyan) for lipids in a lipid bilayer.

The nuclear Overhauser effect

As discussed above, a dominant relaxation mechanism is dipolar interactions with other spin magnetic moments. In such a case, irradiation of one set of spins by a selective $\mathbf{B_1}$ field can cause intensity changes in the resonances of nearby spins. For example, in the case of two nuclear-spin dipoles, A and B (Figure 6.1.24) saturation (equalization of the populations—see Chapter 5.1) of A will cause a change in intensity at B (Figure 6.1.24) because the relaxation effect at B arising from A is changed. The resulting fractional change in intensity of B is called the nuclear Overhauser effect (NOE) named after Al Overhauser who first observed an equivalent effect in 1953.

NOEs can be positive or negative. As, like T_1, they are caused by dipole–dipole induced relaxation their value depends on ω and τ_c and the distance to neighboring dipoles. Figure 6.1.25 shows the frequency dependence of the ^{15}N NOE observed on irradiation of an attached proton 0.102 nm away (see also Figure 6.1.22 and Problems 6.1.7 and 6.1.8). Note that a value of –3 means that the resonance is inverted and three times more intense than without irradiation.

As we will see later, multiple NOEs in a complex system like a protein can be more conveniently detected by two dimensional methods rather than by selective irradiation.

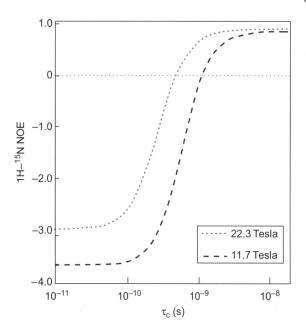

Figure 6.1.25 The field and τ_c dependence of the ^{15}N NOE observed on irradiation of the H in an amide NH group (data for figure kindly provided by J. Boyd).

Chemical exchange

A molecule in solution can exist in many different environments. Examples are a ligand that exists both free in solution and bound to a protein, or a phosphate group that exists in protonated and unprotonated forms. In both of these examples, the molecule can exchange between the two environments during the measuring time which typically takes tens of milliseconds. Such exchange processes have important consequences for the appearance of the NMR spectra and useful information can be extracted about the exchange rates from the behavior of the resonances. (Exchange also explains something that may have puzzled you about the spectrum of ethanol in Figure 6.1.10. Why is the OH resonance a single line rather than split into a triplet by the CH_2 group? The answer is that fast exchange of the proton of the OH group with solvent water averages out the spin–spin coupling.)

Consider the situation, where A and B are interconverting molecular forms: $A \underset{k_b}{\overset{k_a}{\rightleftharpoons}} B$. The spectral parameters of A are δ_A, J_A, and $1/T_{2A}$, with a similar set for B. If the interconversion between A and B forms is slow (k_a and k_b small) then the resonances of A and B are observed separately and independently

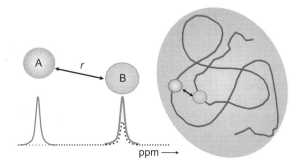

Figure 6.1.24 Illustration of an NOE between A and B. Selective irradiation (saturation) of A causes a change in the intensity of B (dotted gray line). In a protein (right) with many ^1H signals, selective NOEs can be seen between any two adjacent protons provided they are close enough ($r \leq 0.5$ nm).

of one another. If the exchange is fast, however, then a weighted mean of the spectral parameters is observed. A simulation of the observed effects for A and B, separated by $\Delta\omega = 2\pi(\delta_A - \delta_B) = 2\pi\Delta\nu$, with exchange rate $k = k_a + k_b$, is shown in Figure 6.1.26. There are two extremes in Figure 6.1.26: one, when $k \ll \Delta\omega$, has two observed resonances, and is called the *slow* exchange limit; the other, when $k \gg \Delta\omega$, has one observed resonance and is called the *fast* exchange limit. The regime between these limits, where line broadening is observed is called *intermediate* exchange. Note that we have only considered the case where A and B are equally populated with $k_a = k_b$. In the non-equal case, the observed chemical shift becomes $\delta_{obs} = f_A\delta_A + f_B\delta_B$ in the fast limit, where f_A and f_B are the fractions of the A and B species present in solution. As we will see, this equation is very useful in a variety of situations, including the determination of a pK_a value or analysis of ligand binding.

Paramagnetic centers

The magnetic moment of an unpaired electron is much greater than that of a ^1H nucleus (658-fold larger—Chapter 6.2); thus, the presence of a paramagnetic center in a solution can have a large effect on the observed NMR spectra. These effects can be

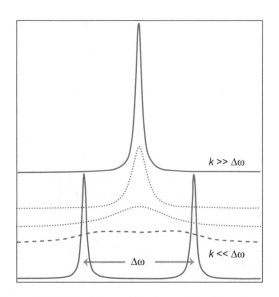

Figure 6.1.26 Simulation of the observed lineshapes for the interconversion of A and B (shift separation $\Delta\omega$) at increasing rate, k, going from lower to upper spectra.

interpreted to give information about molecular motion, distance from the paramagnetic center to the observed nucleus, and binding of a paramagnetic ion to a macromolecule. They can also be used to enhance contrast in imaging experiments (see below). Useful paramagnetic centers for NMR studies include spin labels, transition metals and lanthanide ions.

The interaction between a paramagnetic center (S) and a nucleus (I) can give rise to observed shifts or changes in relaxation time. These changes arise from two kinds of mechanism: (i) a **dipolar interaction** between I and S and (ii) a **delocalization** of the S electron which changes the effective field experienced by I. Mechanism (ii) is often called a scalar or "contact" mechanism and it has similarities to the through-bond effect that causes spin–spin coupling J.

The most useful paramagnetic effects usually arise from the paramagnetic **dipolar mechanism** which can cause significant, interpretable shifts and relaxation. The shifts are similar to those mentioned for ring currents, with NMR effects that depend on r and θ (refer to Figure 6.1.20 where one of the dipoles is a paramagnetic center). As we saw above, relaxation is induced by fluctuating fields, characterized by a correlation time τ_c. The dipolar relaxation mechanism arising from paramagnetic centers has a similar form to that described in Figure 6.1.21 but the values of B_{loc} induced by the electron magnetic moment are much larger. The relaxation time of the S spin (τ_s) is also important as it can contribute to B_{loc} fluctuations in a similar way to random motion.

Different paramagnetic metal ions have very different magnetic properties. τ_s, for example, is very different for different paramagnetic centers; the two lanthanide ions Gd(III) and Eu(III) are both paramagnetic, but for Gd(III), τ_s is about 10^{-9} s, while for Eu(III) it is 10^{-13} s. The slower field fluctuations generated by the Gd(III) τ_s are much more effective than those for the fast fluctuations generated by the Eu(III) τ_s (Figure 6.1.21); Gd(III) is thus an effective relaxation (or broadening) probe, while Eu(III) is a poor relaxation agent. Paramagnetic centers also induce dipolar chemical shifts and Eu(III) turns out to be a very good shift reagent. Similarly, the transition metal ion, Mn(II), is an effective relaxation agent while Ni(II) and Co(II) can act as shift agents.

Applications of NMR

Now that we have described some of the basics, we are in a position to consider some illustrative examples of how NMR can be applied to biological problems. Four main classes of experiment will be considered briefly here: (i) magnetic resonance imaging (MRI); (ii) magnetic resonance spectroscopy (MRS) of tissues and body fluids; (iii) studies of membranes by solid state NMR; and (iv) protein structure determination. This is a limited subset and the reader is referred to the "Further reading", websites, and problems at the end of the chapter for further information and insight.

Magnetic resonance imaging (MRI)

MRI has become a standard imaging technique in most hospitals and a variation of the method, functional MRI (fMRI), has also become important in neuroscience. The method, first demonstrated in 1973, led to a Nobel Prize for Paul Lauterbur and Peter Mansfield in 2003. An image of water protons in the specimen is obtained by collecting spectra while different field gradients are applied to the sample (Figure 6.1.27) in different directions. These field gradients "label" the position of sample components in space. This allows an image to be reconstructed, much as microscopy images are reconstructed using tomography (Chapter 7.1). Consider the illustration in Figure 6.1.27 which has a simple sample made of two capillary tubes, one cyan one gray. Application of strong field gradient in the "1" direction will give two, well separated, lines from the two tubes, while a gradient along direction "2" will only produce one line. Intermediate directions of gradient, e.g. direction "3", will give intermediate line separations. The gradients thus encode structural information that can be reconstructed to generate an image.

Contrast in MRI

To generate a good image, it is very important to have intensity variation in signals from different parts of a sample in MRI, just as in microscopy. One way to obtain contrast is to exploit different T_1 and T_2 values in a sample. These can either be intrinsic differences or differences induced by an added paramagnetic agent, such as Gd(III). As illustrated in Figure 6.1.28, the water in the "gray matter" in the brain relaxes faster than the relatively mobile fluid the brain is bathed in. A T_1 pulse sequence can exploit these intrinsic differences to generate contrast. T_2 differences can also be highlighted with the spin echo sequence shown in Figure 6.1.18.

Functional magnetic resonance imaging (fMRI)

Functional magnetic resonance imaging (fMRI) has become an important tool in neuroscience. Two MR images are acquired in rapid succession (≤1 s apart)

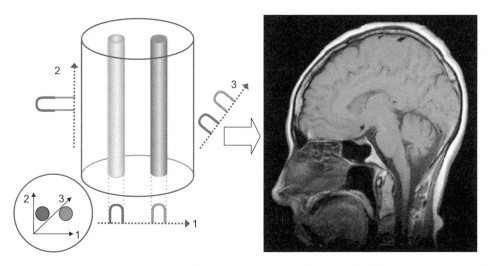

Figure 6.1.27 A typical MRI image of a human head is shown on the right. The principle of the method is illustrated on the left. Three field gradients are applied to a sample of two tubes in different directions. A gradient along the "1" direction gives separated lines from the cyan and gray tubes. Application along the "2" direction, however, gives superimposed lines.

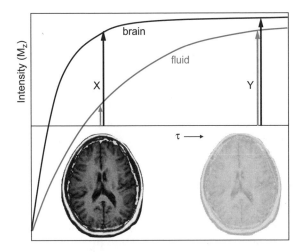

Figure 6.1.28 Different intrinsic T_1 values can be exploited to obtain contrast in MRI images. If the τ value in a 180°–τ–90° sequence (Figure 6.1.19) is set to correspond to point A, then good contrast is obtained between brain and fluid. If it is set to point B, then the contrast is relatively poor (note that the different arrow lengths at A and B indicate the relative intensities that will be observed).

while the subject performs some specified brain task, e.g. recite poetry. Small differences in signal intensity occur in "activated" regions of the brain. The difference signals are then superimposed on a high resolution image of the same region (this requires a longer time—several minutes—to collect). The observed signal changes in fMRI mainly arise from small differences in the oxygenation state of the blood (deoxyhemoglobin is paramagnetic so its presence causes extra relaxation; oxyhemoglobin is diamagnetic so it does not affect relaxation). These oxygenation differences occur in specific regions of the brain while the task is performed. This is therefore a powerful, non-invasive method for studying brain function with a resolution of around 1 mm.

Magnetic resonance spectroscopy of tissues

Most metabolites of interest have relatively low concentration in tissues (~1mM or less) and this leads to detection problems with an insensitive method like NMR. Tissues are also relatively inhomogeneous, so that local variation in magnetic susceptibility (Box 6.1) and binding to macromolecular networks leads to relatively broad spectral lines being observed. In spite of these difficulties, studies of molecules in

tissues and cells, usually called **magnetic resonance spectroscopy** (MRS), has been widely applied. The main advantage of MRS is that it can study molecules *in situ*, non-invasively. Many different nuclei can also be used to observe the metabolites or ions under study (^{1}H, ^{13}C, ^{23}Na, ^{31}P, etc.). We now consider some simple examples involving ^{31}P and ^{1}H nuclei illustrating advantages and disadvantages of each nucleus type.

^{31}P studies

^{31}P is a very suitable nucleus for MRS; there are no significant background signals from solvents; the isotope is 100% abundant, and it has a sensitivity of ~0.07 compared with ^{1}H (we define ^{1}H sensitivity as 1; see also Problem 6.1.4). ^{31}P studies have been widely used to study, non-invasively, muscle and heart energetics and metabolism in healthy and disease states, as well as the kinetics of enzymes including creatine kinase.

The inorganic phosphate (P) signals in Figure 6.1.29 are of particular interest. Phosphate ionizes with a pK_a value around 6.9. The HPO_4^{2-} species resonates around 3.4 ppm and the $H_2PO_4^-$ species around 5.7 ppm (using PCr as a 0 ppm reference). As shown in Figure 6.1.30, the observed resonance moves continuously between these two extremes as the pH is varied. This is because the ionizing proton exchanges rapidly and an average chemical shift position is observed ($\delta_{obs} = f_A\delta_A + f_B\delta_B$, as discussed in the chemical exchange section). P_{in} and P_{out} signals are thus

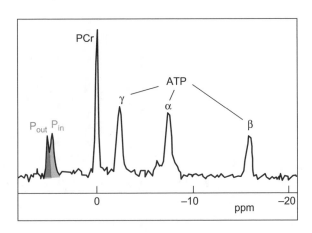

Figure 6.1.29 ^{31}P NMR spectra from an isolated, beating rat heart. Note the signals from ATP, phosphocreatine (set to 0 ppm), and inorganic phosphate (P).

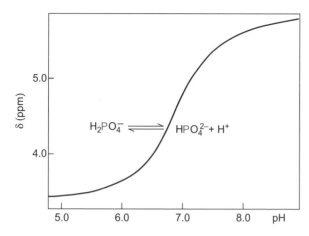

Figure 6.1.30 The resonance position of the inorganic phosphate signal as a function of pH; the equilibrium between the ionizing species present at low and high pH are also shown.

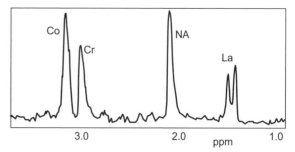

Figure 6.1.31 Typical ^{1}H brain spectra obtained using a long τ value (270 ms) in a spin echo sequence. Peaks from choline (Co), creatine (Cr), N-acetyl aspartate (NA), and lactate (La) are indicated.

seen in the spectrum in Figure 6.1.29 because the intracellular and extracellular pH values are slightly different, with different resulting shift values. It also follows that the pH can be determined from δ_{obs} using a calibration curve of the sort shown in Figure 6.1.30.

^{1}H studies

The ^{1}H has a high relative sensitivity and it has a natural abundance that is essentially 100% (see Table 6.1.1 and Problem 6.1.3). It would thus appear to be the most obvious nucleus to use to observe metabolites but low concentration metabolite signals have to be detected against a very strong background signal from water protons. Mammalian tissue has around 70% water content which means that the ^{1}H concentration is around 80M ($0.7 \times 55 \times 2$). This high concentration makes MRI, which mainly detects water, a practical method. To detect relatively weak metabolites using ^{1}H, however, one has to use spectroscopic "tricks" to discriminate between metabolites and water, as well other strong signals from lipids and macromolecules. One such "trick" exploits the fact that water and macromolecules have short effective T_2 values; if the spin echo sequence shown in Figure 6.1.18 is applied, the water and macromolecule signals decay away rapidly, while the echoes from small metabolites decay relatively, slowly (Figure 6.1.31). The small molecules then observed can often be used as markers of disease states.

Localization

It is often important to detect signals from a defined region of a tissue, e.g. a tumor. The methods used to do this can be very simple, e.g. a "surface" coil, which is merely a flat solenoid placed above the region of interest. Alternatively, a specified region can be selected by applying a field gradient, followed by a rf pulse that only excites spins that resonate within the field range $\mathbf{B_o} + \Delta\mathbf{B_x}$ (Figure 6.1.32). The gradient approach can be extended to 3D, and various interesting acronyms are given to the selection methods used; these include "PRESS" and "STEAM".

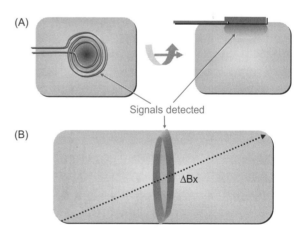

Figure 6.1.32 Illustration of simple methods used to localize a signal in a tissue. (A) shows how a flat coil can be used to detect a region immediately below the surface. (B) shows how localization can be achieved using an applied field gradient together with rf detection pulses tuned to excite and detect only signals with the field value $\mathbf{B_o} + \Delta\mathbf{B_x}$. Other regions of the sample, where this resonance condition does not apply, are not observed.

Signals from body fluids and biopsies

Studies of body fluids (e.g. urine, plasma) do not have the line broadening problem that arises from magnetic susceptibility effects in inhomogeneous tissue. High resolution, high sensitivity spectra can be obtained and some purification/concentration can be carried out on the sample before the NMR analysis. Very informative spectra can thus be obtained; metabolic profiles of disease states can be generated and drug metabolism products can be identified. Untreated biopsy material does not give good spectra because of the inhomogeneous nature of the samples, but this can be improved by magic angle spinning (see below).

⋮   **Example**   Toxicity response to a cancer drug

⋮   In patients undergoing chemotherapy, optimum drug
⋮   dosage depends on the drug toxicity and different
⋮   patients respond differently to this toxicity. The study
⋮   illustrated in Figure 6.1.33 suggests that NMR meta-
⋮   bolic profiles can identify patient sub-populations that
⋮   are more susceptible to adverse symptoms than others.

Solid state NMR

Up to now we have discussed situations where the molecules tumble relatively rapidly in solution. This means that the fluctuating local fields that induce relaxation (Figure 6.1.21) are not observed in the spectrum because they are averaged out as in the "fast exchange" situation illustrated in Figure 6.1.26. In "solids", the molecules are relatively static so that dipolar and other effects are not averaged out. Consequently, the observed spectra from solids are generally very broad.

A "powder" spectrum

Any NMR parameter that depends on the angle the molecule makes with the applied field direction will give rise to a spectrum that is a sum over all these angles, if the tumbling is slow. We saw, for example, that the interaction between two dipoles depends on the angle, θ, they make with respect to $\mathbf{B_o}$ (Tutorial 11, Figure 6.1.20). Another mechanism that causes a $\mathbf{B_o}$-dependent effect is called chemical shift anisotropy

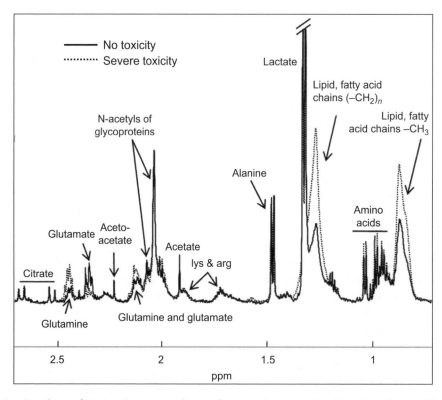

Figure 6.1.33 Overlay of mean ^{1}H spin echo spectra of serum from a patient group that experienced no toxicity and another that experienced severe toxicity to treatment with the anti-colorectal cancer drug capecitabine. Various identified metabolites are indicated (adapted with permission from *Clin. Cancer Res.* 17, 3019–28 (2011)).

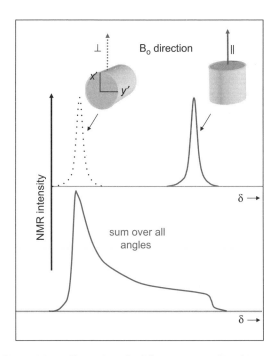

Figure 6.1.34 Illustration of solid state spectra. Top shows the spectra obtained at two fixed angles (⊥ and ∥ to **B₀**). The cyan spectrum below represents the sum over all angles. **B₀** is more likely to be ⊥ than ∥ to an axially symmetric molecule, hence the spectrum is asymmetric.

(CSA) which arises from asymmetry in the electron cloud around a nucleus (e.g. ^{31}P; see Figure 6.1.35) and leads to different chemical shifts being observed for different **B₀** directions (Figure 6.1.11).

To reproduce a solid spectrum, a sum over all angles, we need to consider the probability that a molecule will have a particular value of θ in **B₀**. For example, if a molecule has axial symmetry it is more likely that the dipoles will be perpendicular (⊥) to **B₀** than parallel (∥) to **B₀** because all the angles in the x'–y' plane are equivalent. The sum over all nuclei then gives an asymmetric "powder" spectrum, as shown in Figure 6.1.34. While CSA gives rise to a spread for one resonance and powder spectra of the sort shown in Figure 6.1.34, a dipolar interaction between two similar dipoles gives resonances for each angle; their positions are dominated by the dipolar $(1 - 3\cos^2\theta)$ term with maximum and minimum values of ∓ 2 and ± 1. We will return to the solid state spectrum arising from a dipolar interaction in Chapter 6.2 (see Figure 6.2.15).

^{31}P spectra of lipid membranes

Figure 6.1.35 illustrates the kind of NMR spectra observed from membranes. Spectra of the phosphate in the lipid head group are shown for three different membrane preparations. For small, sonicated vesicles, there is fast tumbling and a narrow line is observed because the spectra are averaged over all angles. For large unsonicated liposomes, however, a broad asymmetric powder pattern is seen because molecular tumbling is much slower.

In the case shown in Figure 6.1.35, the main spread in shifts seen in the liposomes arises from chemical shift anisotropy (CSA) and the sum over all angles leads to the sort of line shape seen in Figure 6.1.34. In the third type of sample, lipid multilayers are oriented on stacked glass plates. This allows spectra to be collected at a specified angle, θ, between the bilayer and the magnetic field. It can be seen that both the shift position and the linewidth change as θ is changed. The shift effect arises from CSA but the

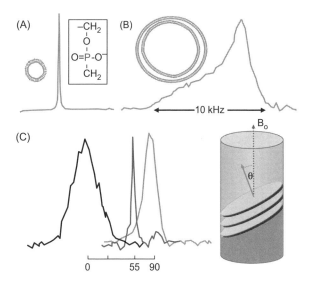

Figure 6.1.35 ^{31}P spectra of three different preparations of phosphatidyl choline membrane bilayers: (A) small vesicles with narrow averaged spectrum; an inset is also shown with the phosphate head group. (B) The spectrum obtained from aqueous liposomes. (C) Multilayers coated on glass plates; spectra obtained at three different orientation angles of the plates with respect to **B₀** are shown. Note, in (C) the line (cyan) is especially narrow when θ ~ 55°; this is the angle when $1 - 3\cos^2\theta = 0$ (see below) where the dipolar interaction between phosphate and the –CH₂– group is near zero.

linewidth changes because of dipolar broadening, caused by an interaction between the ^{31}P nucleus and the adjacent protons in the CH_2 group (see inset in Figure 6.1.35). This dipolar broadening has an angular dependence described by $1 - 3\cos^2\theta$; this is zero when $\cos\theta = \sqrt{1/3}$, i.e. when $\theta = 54.74°$. This angle is known as the "magic" angle (see also below).

^{2}H spectra of lipid membranes

Deuterium (^{2}H) spectra of membranes are very sensitive to motion and environment, e.g. the spectrum changes significantly with temperature and the presence of membrane proteins. Spectra of deuterated lipids are thus often used to probe both lipid and membrane protein properties. The deuteron, ^{2}H, has spin $I = 1$ which means that there are $2I + 1 = 3$ energy levels when B_0 is applied. These three energy levels are labeled by their quantum numbers 0 and ± 1 in Figure 6.1.36; the observed "allowed" transitions are those between -1 to 0 and 0 to 1. If there were no interaction, other than with B_0, then these -1 to 0 and 0 to 1 transitions would coincide. But an $I = 1$ nucleus also has a property called a **quadrupole moment**, Q, which means that it can interact with local electric fields in the sample. Q gives rise to a splitting of the energy levels even when $B_0 = 0$, and this results in two lines being observed and the position of the two lines is different for each angle that the molecule makes with B_0.

When the two ^{2}H lines are summed over all angles we get spectra like the one shown in Figure 6.1.36; this is like two back-to-back superimposed powder spectra of the sort shown in Figure 6.1.34, each one arising from one of the two 0 to ± 1 transitions. ^{2}H spectra of the sort shown in Figure 6.1.36 are obtained from lipids in a bilayer. The spectral shape, e.g. the separation between the two prominent peaks, is very sensitive to lipid motion and environment. Incorporation of ^{2}H at specific positions in a lipid chain shows the mobility gradient along that chain in a similar way to ^{13}C relaxation (Figure 6.1.23).

Magic angle spinning (MAS) and decoupling

The membrane example in Figure 6.1.35(C) showed narrow lines in a sample oriented at a particular angle to B_0; this "magic" angle occurs when $1 - 3\cos^2\theta = 0$ or $\theta_m = 54.74°$. In 1958, Raymond Andrew showed that if the sample is spun rapidly at θ_m with respect to B_0 (inset in Figure 6.1.37), then narrow lines can be obtained because the angular dependent variations in the spectra are averaged out by the spinning.

Another spectroscopic trick for producing narrower lines is **decoupling**. In Figures 6.1.37 or 6.1.34, the effects of dipolar broadening from ^{1}H neighbors could have been removed by applying a strong rf field at the ^{1}H frequency while the ^{13}C or ^{31}P signals are being detected. The applied decoupling field "stirs" the ^{1}H spins in such a way as to average out their broadening effects. J coupling and other effects can also be removed with this kind of decoupling irradiation.

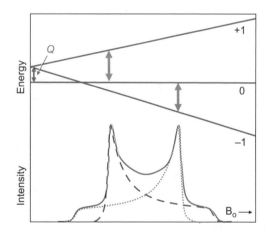

Figure 6.1.36 The top picture shows the three energy levels associated with ^{2}H. Two transitions, between the 0 and ± 1 levels, are shown as double-headed black and gray arrows. The degeneracy of the three levels is removed at zero field because of the quadrupole splitting, Q (cyan arrow). The lower solid state spectrum (cyan) is typical of that observed from a ^{2}H sample. This results from the sum of the two lines over all angles; it is equivalent to two powder patterns, one from each of the two lines.

Protein structure determination

In 2002, Kurt Wüthrich was awarded a Nobel Prize for his contributions to the determination of macromolecular structure by NMR in the solution state. The first NMR spectrum of a protein was obtained in 1957, but many technical improvements had to be made before useful information could be extracted from such spectra. A measure of the successful

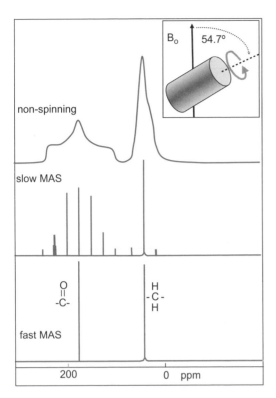

Figure 6.1.37 ^{13}C NMR of enriched glycine in a powder. Without spinning the spectrum is very broad (top), but the magic angle spinning (MAS) leads to a simple two-line spectrum (bottom) (adapted with permission from *Biol. Chem.* 390, 815–34 (2009)).

current status of the method, however, is that over 8700 NMR structures of macromolecules had been deposited in the protein database by the end of 2010 (PDB; **www.rcsb.org**). Compared to X-ray crystallography (Chapter 4.3), NMR is still relatively laborious and is restricted to the study of fairly small systems. A major advantage, however, is that macromolecular structures can be obtained in the solution state and crystals are not required. As many proteins do not crystallize, especially those containing unstructured regions, NMR may often be the only atomic resolution structural tool available to study them.

Technical difficulties associated with all spectroscopic methods include **sensitivity** (how good is the signal to noise ratio at a given concentration?); **resolution** (can we resolve individual chemical groups?); and **assignment** (can we assign a resolved resonance to a particular chemical group?). These three technical difficulties are particularly pertinent

to NMR. Consider the spectrum of lysozyme shown in Figure 6.1.13. This spectrum, clearly rich in information, was obtained at 950 MHz using a concentration of 500 μM in a sample volume of 300 μL; in contrast, the first protein spectrum, obtained in 1957, was obtained at only 40 MHz. This large increase in applied **B$_0$** field (and corresponding frequency) gives much better *resolution*, as the chemical shift, δ, is proportional to **B$_0$**, while linewidth is relatively field independent. Higher **B$_0$** also gives larger energy level separation, a larger population difference, and better *sensitivity*. While increases in **B$_0$** have been very important, two further technical improvements were required before NMR could be applied in a systematic way to study macromolecules. One was the use of multidimensional NMR methods (Figure 6.1.38), the other was the use of isotope labeling.

Multidimensional methods in NMR

Richard Ernst was awarded the Nobel Prize in 1991 for his contributions to the development of pulse and multidimensional NMR methods. Figure 6.1.9 shows an example of a one-dimensional (1D) spectrum, obtained by Fourier transformation of a transient signal. This approach can be extended by collecting a series of transients with different time dimensions. Figure 6.1.38 illustrates the generation of a two-dimensional (2D) spectrum of ethanol by a pair of 90° pulses, separated by time period, t_1, which is increased incrementally. The t_1 period is called the evolution time and data acquisition is carried out during a period t_2. As explained in the caption to Figure 6.1.38, this stepped, two-pulse, sequence produces a 2D spectrum with peaks along a diagonal, as well as informative off-diagonal "cross"-peaks (see Figure 6.1.38). The major advantages of 2D spectra are improved spectral resolution and easier identification of correlations between peaks via the cross-peaks (see also below).

Once the idea of 2D spectra is established, one can imagine extending this to other dimensions. Figure 6.1.39 shows how **3D spectra** can be obtained by adding another evolution period between pulses. The increased spread in 3D compared with 2D further improves the ability to resolve and characterize complex NMR spectra. The *F1*, *F2*, and *F3* axes in 3D spectra often represent different isotopes, e.g. ^{1}H,

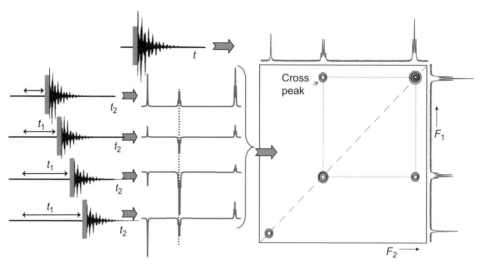

Figure 6.1.38 Two-dimensional spectra: at the top is shown the same 1D ethanol spectrum as in Figure 6.1.10, transformed from a time to a frequency dimension with a Fourier transform (gray arrow). Below right are four ethanol spectra collected at different separation times (t_1) between two 90° pulses; at each t_1 value the t_2 dimension is transformed into four F_2 spectra (cyan). In practice, the t_1 increments are small and hundreds of t_1 values may be used. The effect of t_1 variation is to modulate the intensity of the spectra in F_2; this is because the M_{xy} magnetization fans out in the x-y plane during t_1 in a way that depends on chemical shift (and/or B_o—see Figure 6.1.17). The dotted line on the middle quartet of the four cyan spectra emphasizes the modulation; a second FT along the direction of this dotted line (the t_1 dimension) will give another version of the spectrum but this time along F_1 rather than F_2. Combination of the F_1 and F_2 dimensions gives a 2D spectrum with the three ethanol signals spread out along the diagonal shown with a dashed gray line. The peaks in the 2D spectrum on the right are shown as a plot of contours with the same intensity. This particular sequence of two 90° pulses (called COSY—see below) gives off-diagonal cross-peaks arising from J-coupling between the –CH$_2$– and –CH$_3$ groups.

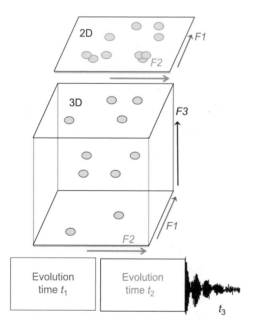

Figure 6.1.39 Representation of 3D spectra with two evolution times, t_1, t_2, and a detection period, t_3. The corresponding frequency axes are $F1$, $F2$, and $F3$.

^{15}N, and ^{13}C. For such heteronuclear experiments the usual detection frequency is ^{1}H; this is done on an inner rf coil (Figure 6.1.6), while pulses of ^{15}N and ^{13}C frequencies are applied via a second coil placed outside the ^{1}H coil (see Problem 6.1.11).

Through-bond and through-space cross-peaks

The presence of cross-peaks in multi-dimensional spectra indicates that pairs of resonances are **correlated** in some way. We have seen above that two nuclei can communicate with each other, either via a **shared bond** (observed as J-coupling) or via a through-space dipolar interaction (observed as a NOE). The cross-peaks indicate whether these different kinds of correlations exist and these in turn tell us about the molecule being studied.

Different pulse sequences are used to select NOE-type or J-type correlations (Figure 6.1.40). 2D COSY (**co**rrelated **s**pectroscop**y**) spectra detect through-bond correlations; for example the sequence used to generate Figure 6.1.38 is a COSY spectrum and the –CH$_2$– and –CH$_3$ groups, which are correlated by J

(A)

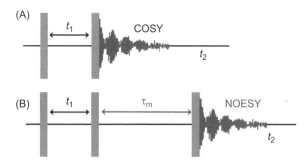

Figure 6.1.40 Different correlations can be observed with different pulse sequences. (A) The COSY sequence (90°–t_1–90°) detects through-bond (J) correlations while (B) the NOESY sequence (90°–t_1–90°–τ_m–90°) detects through-space (NOE) correlations. The fixed τ_m evolution period allows dipolar NOE correlations to develop.

coupling, have cross-peaks between them. It is also possible to observe NOE cross-peaks using a different pulse sequence. 2D NOESY spectra arise when an extra 90° pulse and a fixed delay, τ_m, are inserted in the sequence (Figure 6.1.40).

Figure 6.1.41 illustrates the COSY and NOESY spectra expected from the amino acid tyrosine, and shows the structural characteristics of tyrosine that can be deduced from these spectra. NOESY detects more interactions than COSY and, crucially, it can

detect dipolar interactions from ^{1}H nuclei that are not only within the tyrosine sidechain but also to other amino acids in the protein.

The size limit in NMR

Our ability to detect through-bond correlations depends on the linewidth compared to the values of the coupling constants, J. In ^{1}H spectra of small proteins, J is large enough for **through-bond correlations** to be detected for atoms up to three bonds apart, using the COSY sequence. The NOESY experiment can also detect **through-space interactions** up to about 0.5 nm apart.

With increasing molecular weight, molecules tumble more slowly in solution (longer τ_c; Figure 6.1.22); this results in faster T_2 relaxation and increased linewidth. For example, the linewidth ($\Delta\nu_{1/2}$) of ^{1}H spectra of a 10 kDa globular protein is ~5 Hz at 35°C; for 20 kDa and 50 kDa proteins these linewidths are ~10 Hz and ~25 Hz, respectively. Observation of through-bond cross-peaks (COSY) becomes increasingly difficult when the linewidth becomes similar to the coupling constant, J. In addition, the number of peaks and spectral overlap increases with molecular weight (we will return to the size problem below).

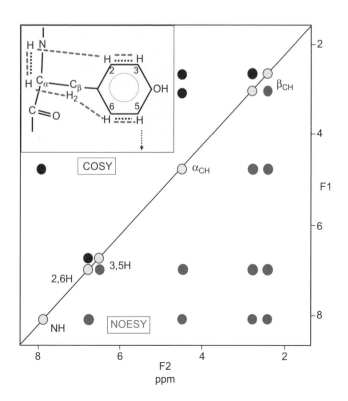

Figure 6.1.41 An illustration of observed COSY and NOESY cross-peaks for tyrosine. COSY cross-peaks are shown in gray, while NOESY peaks are in cyan. The inset shows the corresponding interactions in the molecule as gray and cyan dotted lines. (Note: the spectra are normally symmetrical about the diagonal but here the region above the diagonal is shown with the expected COSY cross-peaks while the region below the diagonal shows the expected NOESY peaks.)

¹H assignment and protein secondary structure

How do we assign observed resonances to a particular group in a protein? In early studies, assignment was carried out with non-labeled proteins, using ¹H spectra alone. Some amino-acid types have unique COSY peak patterns, e.g. Gly, Ala, Val, Ile. In addition, immediately neighboring amino acids could be identified using NOEs from the NH of amino acid i to the NH (NN) and CαH (αN) of amino acid $i - 1$ (Figure 6.1.42). Once the neighbors are identified (e.g. the Gly neighbors, Ala and Val in the sequence Ala–Gly–Val) the sequence fragment can be compared with the known protein sequence. It is then often possible to make unique assignments to many residues.

It also turns out that the unique structural properties of α-helices and β-sheets give rise to patterns of ¹H NN and αN NOES that are *characteristic for different kinds of secondary structure*. As shown in Figure 6.1.43, strong NN NOEs are observed in α-helices and strong αN NOEs are observed in β-strands (see also Problem 6.1.9).

Isotope labeling

For proteins larger than about 100 amino acids, the ¹H-only assignment methodology described above becomes increasingly difficult as the linewidths and spectral overlap increase. With ¹H only spectra, NOEs have to be used to link amino acids because the four-bond coupling constants between amino acids are smaller than the linewidths of most macromolecules. A solution to such problems is to use isotope labeling. Common isotope substitutions are: (i) ¹⁵N for ¹⁴N (¹⁴N is spin 1 with broad, generally undetectable, signals); (ii) ¹³C for ¹²C (¹²C is NMR "silent" as it has no spin); and (iii) ²H for ¹H which simplifies spectra and gives better relaxation properties due to

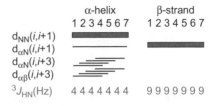

Figure 6.1.43 The pattern of NN and αN NOEs seen in different α and β secondary structure for seven consecutive residues. The thickness of the lines indicates the strength of the NOE. Also shown are the αN $(i,i + 3)$ and αβ $(i,i + 3)$ patterns seen in helices and the values of observed NH coupling constants are shown in cyan (see also Figure 6.1.15).

the fact that the ²H nucleus has a much smaller magnetic dipole than ¹H (see relaxation mechanism section). As most proteins can be produced in *E. coli* or yeast expression systems, these substitutions can be done by manipulating the growth media. ¹⁵N substitution is inexpensive and straightforward and is usually the first to be done.

The introduction of ¹⁵N and ¹³C gives new detection possibilities (e.g. 3D NMR; Figure 6.1.39) and gives rise to a new set of observable coupling constants. As shown in Figure 6.1.44, some of these are quite large. Crucially, the C=O–NH coupling constant between neighboring amino acids is ~15 Hz, which is large enough to be observed in many proteins. Systematic through-bond assignment procedures can thus be applied to a ¹³C, ¹⁵N labeled sample, using 3D detection methods of the kind shown in Figure 6.1.39. This means that 3D, through-bond procedures applied to ¹⁵N, ¹³C labeled samples is now the method of choice for resonance assignment.

Other solutions to the size problem

As illustrated above, the use of isotope labels extends the range of macromolecules that can be studied by

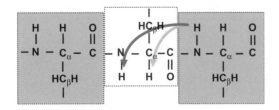

Figure 6.1.42 Three amino acids showing NN and αN connections as can be observed as ¹H-¹H NOE cross-peaks in ¹H 2D spectra.

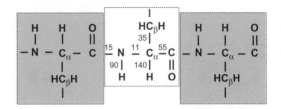

Figure 6.1.44 In a ¹³C, ¹⁵N labeled sample, direct through-bond coupling between amino acids is detectable (~15 Hz between ¹³C=O and ¹⁵NH). Other coupling constants shown are: ¹⁵N-¹H ~ 90 Hz; ¹³Cα-¹H ~140 Hz; etc.

NMR, although routine structure determination of very large systems by NMR is still not very practical. Solutions to the size problem are, of course, being sought and, in favorable cases, studies of very large systems (~900 kDa) have given novel information (see Journal Club). One useful pulse sequence that can extend the accessible molecular weight range is called "TROSY" (transverse relaxation optimized spectroscopy). Consider the NH doublet spectrum observed for the NH resonance in proteins (Figure 6.1.15). It turns out that one of the resonances in this doublet is broader than the other. This is because chemical shift anisotropy (CSA) and dipolar relaxation mechanisms add for one component while they subtract for the other resonance in the doublet. By tuning the TROSY experiment to detect the narrower signal, the accessible molecular weight range can be considerably extended. TROSY works best at high magnetic fields, where the CSA effect is larger, and in ^{2}H enriched proteins where the dipolar effect is reduced, thus optimizing the subtraction effect involving these two relaxation mechanisms.

Yet another way of extending the size limit is to use more sophisticated isotope labeling patterns. For example, Lewis Kay's group has grown proteins in ^{2}H$_2$O media containing 2-keto-3-d_2-1,2,3,4-^{13}C-butyrate (Figure 6.1.45). When this labeled protein is dissolved in ^{1}H$_2$O, there are only two classes of observable protonated groups: the ^{13}CH$_3$ groups from the labeled metabolite in the growth medium

and –NH groups whose protons are exchangeable with the ^{1}H$_2$O solvent. This means we have much fewer resonances than in an unlabeled protein and these are narrower because of the deuteration. This allows us to deal with much larger proteins than would otherwise be possible.

Experimental restraints in structure determination
NMR spectra of macromolecules give us various kinds of information about their structure. The *structure*-related data derived from the spectra are often termed "experimental restraints". Some of these restraints are briefly discussed below.

^{1}H–^{1}H NOEs: The most powerful and numerous restraints are ^{1}H–^{1}H NOEs, detected in NOESY spectra as cross-peaks between specific assigned pairs of ^{1}H resonances. In a large protein there may be thousands of NOESY cross-peaks, each one giving specific pairwise distance information.

Coupling constants: Observed J values can help us define some dihedral angles. This is illustrated, for example, in Figure 6.1.15.

Chemical shifts: The observed *pattern of chemical shifts* is very informative. The database of known structures, with known chemical shifts, continues to grow; application of this knowledge, together with structure-prediction algorithms (see Chapter 8.1) means that shift patterns can now be used to predict 3D structures of small proteins, even in the absence of any other restraints (see "Further reading").

Residual dipolar coupling: As mentioned above, dipole–dipole interactions are usually averaged out in solution because of rapid molecular tumbling. It is, however, possible to reintroduce some of the lost dipolar interaction information if a small amount of sample alignment is induced in a solution-state sample. Weak alignment (~10^{-3} of the full effect) can be introduced by manipulating the NMR sample in various ways; one way is to add lipid bicelles which line up in $\mathbf{B_o}$, as illustrated in Figure 6.1.46, another is to use compressed gels. With some alignment, small residual dipolar couplings (RDCs) can be observed, and these give information about the direction of specific groups with respect to $\mathbf{B_o}$. The relatively long-range order information given by RDCs can be incorporated in structure calculations and this complements the short-range information (≤0.5 nm) given by NOEs.

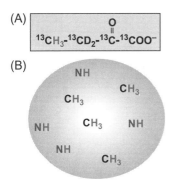

Figure 6.1.45 Labeling strategy for large systems. (A) By growing protein in media containing 2-keto-3-d_2-1,2,3,4-^{13}C-butyrate and ^{2}H$_2$O but then suspending the expressed protein in ^{1}H$_2$O it is possible to obtain a protein sample (B) where the only protonated, and thus observable, groups are CH$_3$ and NH.

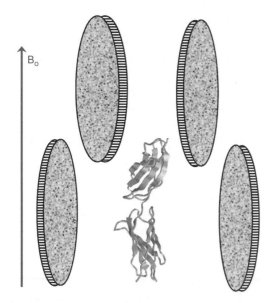

B_0

Figure 6.1.46 Introduction of weak alignment of a protein by lipid bicelles that align in **B_0**.

Solvent H-exchange: The protons in NH and OH groups can *exchange with solvent* and this exchange can be readily observed in NMR. For example, if the solvent is changed from ^1H_2O to ^2H_2O, resolved ^1H peak intensities can be monitored as a function of time as they disappear upon exchange. If a particular assigned ^1H is relatively slow to exchange with solvent *and* it has a pattern of NOEs that suggests that it is in a defined region of secondary structure then it

may be possible to assign it as an *H-bonded* ^1H. This H-bond can then be incorporated in the structure calculations. This procedure is, however, usually only applied at the refinement stage of the structure calculation (see below).

Calculation of structures

Calculation of structures is usually done by incorporating the various observed NMR restraints discussed above into a molecular dynamics (MD) simulation protocol (Chapter 8.1). MD simulations calculate the position and velocity of protein atoms in a series of small steps (~10^{-15} s) using Newton's equations of motion. NMR-derived experimental restraints can be added to the standard forces fields used to treat other interactions in the MD simulations. For example, an energy term can be added to take account of observed NOEs. Starting with an extended structure, and assuming standard information about amino acids, such as planarity of the peptide bond, restrained MD simulations produce families of structures that are consistent with the experimental restraints. Several hundred structures may be produced and structures of a subset that has the lowest energy are superimposed (Figure 6.1.47).

Refinement and validation of structures

Figure 6.1.48 summarizes the protein structure determination procedures described so far. At the "compute a 3D model" stage, a number of experimental

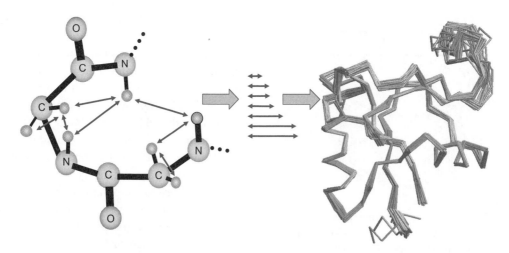

Figure 6.1.47 Illustration of structure calculation procedure in NMR. On the left, a large number of ^1H–^1H NOEs ($\leftrightarrow$) are observed in a protein of known sequence. These are included in a MD simulation (Chapter 8.1) that folds up the protein in a way that fits the experimental data. A number of such calculated structures are then superimposed and compared (right).

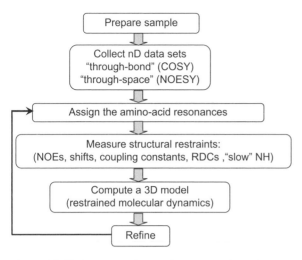

Figure 6.1.48 Summary of a protein structure determination protocol using NMR.

structure by looking for consistency with experimental restraints and any deviation from expected protein dihedral angles (the Ramachandran plot)—see also Chapter 4.3.

Motion and rmsd

It may be noted that the family (ensemble) of calculated structures in Figure 6.1.47 has some regions of structure that superimpose better than others. One way of analyzing the family is to use a "**root mean square deviation**" (rmsd). The rmsd for any atom is the square root of the sum of squares of distances between that atom in all models in a computed ensemble and the average position for that atom in the ensemble. A family of structures, showing its rmsd per residue, is shown in Figure 6.1.49.

There are two ways in which ill-defined structural regions (high rmsd) could arise. One could be poor quality experimental restraints; another possibility is that some regions are genuinely more mobile than others. One good way of distinguishing these two possibilities is to use the ^1H–^{15}N heteronuclear NOE experiment described in Figure 6.1.21. This experiment is very sensitive to protein flexibility. As seen in

ambiguities and mis-assignments are likely to be detected and resolved, so a cycle of iterative corrections are made as indicated on the flow diagram. Other refinement procedures can be added, such as the addition of H-bonds and solvent water in the MD simulations. It is also important to validate the

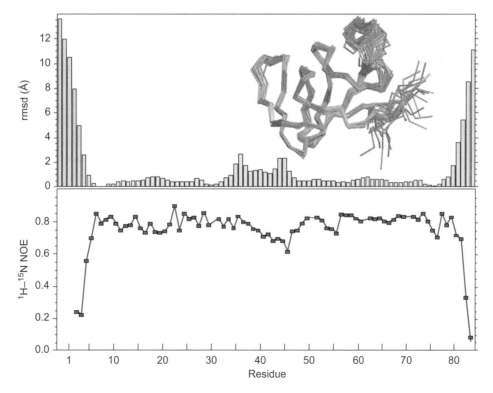

Figure 6.1.49 A family of structures for an SH3 domain with the rmsd plotted per residue (top). The bottom plot is of the ^1H–^{15}N NOE. Note the inverse correlation between the two plots.

(A)

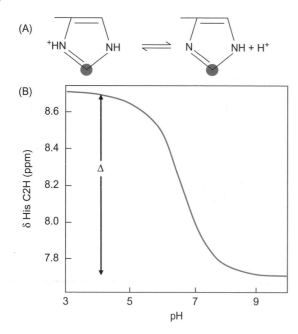

(B)

Figure 6.1.50 Titration of (A) His C2 resonance in a histidine in the catalytic triad of a serine protease, showing (B) that the $pK_a \sim 6.7$.

the bottom plot of Figure 6.1.49 there is a good correlation between low NOE and high rmsd, suggesting that the variation found in the calculated protein family arises here from motion rather than poor experimental restraints.

Other structures

While we have only discussed proteins here, it is important to realize that NMR has made significant contributions to the determination of the structure of RNA and a wide range of complexes. Novel isotopic labeling schemes and restraints derived from residual dipolar couplings now make it possible to carry out NMR structure determination of RNAs up

to 30 kDa (see "Further reading"). NMR structures of integral membrane proteins are also beginning to appear, several done with solid state NMR methods.

Other NMR applications

In an attempt to illustrate some more aspects of the versatility and range of the NMR method that have not been covered, two further examples are given below and more are in the problem section.

Protein ionization states

There was early controversy about the ionization states of the key aspartate and histidine residues in the catalytic triad of serine proteases; the argument was concerned with whether the aspartate accepts a proton during the charge relay or merely acts to orient the His sidechain. NMR was used to measure the pK_a of His 57 in α-lytic protease (an enzyme chosen because it only has one histidine). Rapid exchange of the H^+ between water and the histidine sidechain means that the C2 His resonance (marked in cyan in Figure 6.1.50A) moves continuously between protonated and deprotonated resonance positions (see also Figure 6.1.30). This observed pK_a value cleared up the controversy by showing that the His sidechain has a normal pK_a value. (A pK_a value of ~6.7 can be readily calculated from the shift vs. pH plot (the pK_a is when pH = $\Delta/2$).)

Ligand binding and 2D ^{15}N–^1H spectra

A very useful 2D spectrum in solution-state NMR studies of proteins is one where ^1H resonances are spread along one axis and ^{15}N along the other. This can be obtained with a pulse sequence of the sort shown in Figure 6.1.51A; it is called a **heteronuclear single**

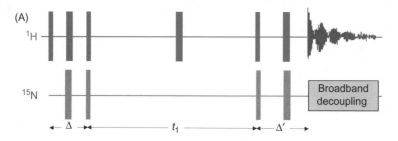

Figure 6.1.51 (A) A pulse sequence to generate a ^{15}N–^1H HSQC spectrum. Narrow and wide bars correspond to 90° and 180° rf pulses, respectively, applied to either ^1H (cyan) or ^{15}N (gray) nuclei. (B) A HSQC spectrum of a folded protein domain, isolated from a ^{15}N-labeled extracellular matrix protein fibronectin.

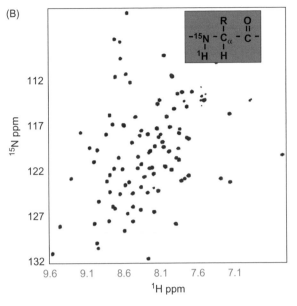

Figure 6.1.51 (*Continued*)

quantum coherence (HSQC) spectrum. The HSQC spectrum of a protein gives a "fingerprint" of the NH groups in a protein, i.e. a peak is obtained from every amino acid except proline. This spectrum can be used to indicate whether a protein is properly folded. For example the well-dispersed protein spectrum shown in Figure 6.1.51B is characteristic of a folded globular protein. An HSQC spectrum is also very sensitive to ligand binding and we can measure the affinity of weak interactions, as well as map the interaction site where ligand and protein make contact.

An example of a titration experiment using ^1H–^{15}N HSQC spectra is shown in Figure 6.1.52. In this case, a peptide, derived from an integrin cytoplasmic tail, was added to a domain from a protein called talin. Several resonances are greatly perturbed by the addition of integrin tail while others are not (for further details see *Cell* 128, 171–82 (2007)).

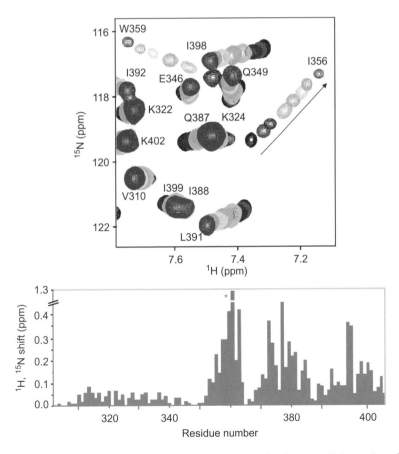

Figure 6.1.52 The top picture shows a series of overlaid ^1H–^1N spectra of a talin domain with increasing additions of an integrin tail fragment. Some resonances, e.g. the one from I356, traced with an arrow, show a large perturbation while others do not. The lower picture maps the observed shifts (weighted mean of ^1H and ^{15}N) for each residue. I356 is marked with an asterisk. Observed perturbations can be mapped on a known structure. A plot of shift perturbation against concentration also gives an estimate of the affinity of the interaction.

 **Summary**

1 Nuclear magnetic resonance (NMR) is a technique that detects nuclear-spin reorientation in an applied magnetic field. The method is very versatile, being sensitive to environment and molecular structure. Relatively concentrated samples are, however, required because the signals produced are often weak.

2 The parameters describing NMR lines are chemical shift(s), spin–spin coupling (J), the intensity of the signal, two relaxation times (T_1 and T_2), and the nuclear Overhauser effect. T_2 is inversely related to the linewidth.

3 Measurement in NMR is achieved by application of a pulse of radiofrequency to the sample placed in a strong static magnetic field. A transient signal is detected from the nuclear spins which is Fourier transformed to give an NMR spectrum. The decay rate of the transient is inversely related to the linewidth.

4 Dipolar interactions with neighboring spins are very important in NMR. These can give rise to induced shifts and relaxation. The dipolar interaction depends on distance ($1/r^3$), allowing structural information to be deduced. In addition, Brownian motion of the molecules gives rise to a fluctuating dipolar field that causes transitions between energy levels and relaxation. Relaxation times can be interpreted in terms of molecular motion. Strong dipolar interactions can be induced by paramagnetic ions.

5 If a chemical group exchanges between different environments, the resulting spectra may be broadened, or the observed parameters may be an average of all the different environments.

6 The applications of NMR in biology are diverse. They include: imaging; measurement of the concentration and isotopic composition of molecules in intact tissue and body fluids; the structure of small macromolecules (proteins and nucleic acids); molecular dynamics information about a variety of systems, including macromolecules and membranes; ligand binding and ionization states.

 Further reading

Online book and websites

http://www.biophysics.org/Portals/1/PDFs/Education/james.pdf

Hornack, J. *Basics of NMR* (online book with animations; **http://www.cis.rit.edu/htbooks/nmr/**)

http://www.ismrm.org/mr_sites.htm (online site list)

http://opa.faseb.org/pdf/mri.pdf (downloadable .pdf introduction)

Books

Freeman, R. *Magnetic Resonance in Chemistry and Medicine* Oxford University Press, 2003.

Keeler, J. *Understanding NMR Spectroscopy* Wiley, 2005.

Gadian, D. *NMR and its Application to Living Systems* (2nd edn) Oxford University Press, 1995.

Jezzard, P., Matthews, P.M., and Smith, S.M. *Functional MRI: An Introduction to Methods* Oxford University Press, 2003.

Cavanagh, J., Fairbrother, W.J., Palmer, A.G., and Skelton, N.J. *Protein NMR Spectroscopy: Principles and Practice* (2nd edn) Academic Press, 2007.

Reviews

Betz, M., Saxena, K., and Schwalbe, H. Biomolecular NMR: a chaperone to drug discovery. *Curr. Opin. Chem. Biol.* 10, 219–25 (2006).

Latham, M.P., Brown, D.J., McCallum, S.A., and Pardi, A. NMR methods for studying the structure and dynamics of RNA. *ChemBioChem* 6, 1492–505 (2005).

Sprangers, R., Velyvis, A., and Kay, L.E. Solution NMR of supramolecular complexes: providing new insights into function. *Nat. Methods* 4, 697–703 (2007).

Shen, Y., Lange, O., Delaglio F., Rossi, P., Aramini J.M., Liu, G., et al. Consistent blind protein structure generation from NMR chemical shift data. *Proc. Natl. Acad. Sci. U.S.A.* 105, 4685–90 (2008).

Watts, A. Solid-state NMR in drug design and discovery for membrane-embedded targets. *Nat. Rev. Drug Discovery* 4, 555–68 (2005).

Hellmich, U.A., and Glaubitz, C. NMR and EPR studies of membrane transporters *Biol. Chem.* 390, 815–34 (2009).

Steitz, T.A., and Shulman, R.G. Crystallographic and NMR studies of the serine proteases. *Annu. Rev. Biophys. Bioeng.* 11, 419–44 (1982).

 ## Problems

6.1.1 In ^{1}H NMR experiments at 400 MHz, the 90° pulse was measured to be 10 μs. What are the magnitudes of the $\mathbf{B_0}$ and $\mathbf{B_1}$ fields? (Hint: see Table 6.1.1 and Figure 6.1.8.)

6.1.2 Aromatic rings give rise to ring-current shifts because the electrons are essentially delocalized. The circulating electrons produce a field that opposes the applied field $\mathbf{B_0}$. The field produced can be considered as a dipole with magnitude $\mu = \mathbf{B_0} e^2 a^2 / 4m_e$ where e and m_e are the charge and mass of the electron; a is the ring radius which is ~0.14 nm for a benzene ring (the figure shows the geometry used for this discussion of ring current effects; the vector $\mathbf{r}$ makes an angle θ with the perpendicular to the ring and an angle θ' with the $\mathbf{B_0}$ direction).

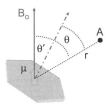

From Tutorial 11 we know that the field at point A in the figure is proportional to $\mu(1 - 3\cos^2\theta')/r^3$. In solution, however, the molecule tumbles and the field at A is averaged over all θ'. The induced dipole μ will also change with θ as the molecule tumbles because the induced current will change with θ. If $\mu_\perp$ and $\mu_\parallel$ are the dipoles induced when $\mathbf{B_0}$ is $\perp$ and $\parallel$ to the ring, it can be shown that the net ring current induced shift is $\mathbf{B_{ring}} = (\mu_o/4\pi)(\mu_\perp - \mu_\parallel)(1 - 3\cos^2\theta)/r^3$. Use this information to calculate the shift (i) on a group 0.3 nm above the ring and (ii) on one of the hydrogens attached to the ring (values of constants are given in Appendix A1.3).

6.1.3 The relative signal to noise (S:N) ratio of spin 1/2 nuclei is ~γ^3. In data accumulation, the S:N ratio improves as the square root of the time taken. Calculate the relative times required to acquire ^{1}H, ^{13}C, and ^{31}P spectra of a solution of ATP with the same S:N ratio.

6.1.4 The diagonal of a ^{1}H 2D COSY spectrum of leucine is shown in the figure (the structure of leucine is also shown). Mark, above the diagonal, the expected positions of the off-diagonal COSY cross-peaks.

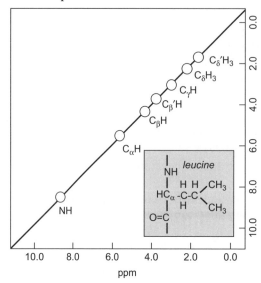

6.1.5 The A and B resonances in the exchange process $A \xrightleftharpoons[k_a]{} B$ have the following parameter: $\delta_A = 820$ Hz, $\delta_B = 740$ Hz, $1/T_{2A} = 3$ s^{-1}, and $1/T_{2B} = 2$ s^{-1}. What would the observed shift be when $k_a = 10^4$ s^{-1} and $k_b = 5 \times 10^3$ s^{-1}? Assuming the relaxation in the slow exchange regime is given by $1/T_{2obs} = 1/T_2 + k$, what would the observed linewidth be when $k_a = 3$ s^{-1} and $k_b = 1.5$ s^{-1}?

6.1.6 In the slow exchange regime, two resonances are observed for the exchanging species A and B. Irradiation of A is observed to cause a reduction in the intensity of the B resonance. This is not due to a NOE effect, but is a result of interconversion of A and B and a resulting transfer of saturation from A to B. The fractional change in the intensity of the B resonance can be shown to be given by $1/T_{1B}/(1/T_{1B} + k_b)$ (see Problem 6.1.5 for nomenclature). In ^{31}P spectra of intact muscle, ATP and phosphocreatine are readily observed. Phosphocreatine regenerates the ATP used up during muscle contraction by the following reaction:

$$\text{MgADP}^- + \text{phosphocreatine}^{2-} + \text{H}^+ \leftrightarrow \text{MgATP}^- + \text{creatine}$$

In an experiment on intact muscle, selective irradiation of ATP (the γ resonance) produced an 18% decrease in the phosphocreatine resonance, while irradiation of the phosphocreatine resonance gave a 30% decrease in the ATP resonance. The T_1 values for phosphocreatine and ATP were observed to be 3.2 s and 1.5 s and the concentration ratio of phosphocreatine to ATP was 6. Calculate the relative flux rates of ATP→ phosphocreatine and phosphocreatine → ATP.

6.1.7 When two spins interact via a dipolar mechanism they create a four-level system as shown in the figure (four energy levels also arise when two nuclei are linked by J-coupling—see Problem 6.1.11).

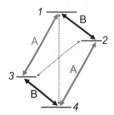

At equilibrium, the populations (P) in the four energy levels in the figure are dictated by the Boltzmann distribution with P4 > P3 > P2 > P1. These populations can be disturbed by pulses and/or by "saturation" where, for example, selective irradiation of the 3–4 transition would make P3 = P4. After such a perturbation, the populations relax back to equilibrium. There are several relaxation routes they can take, with different probabilities. The probabilities corresponding to the A and B transitions are $W_A = W_{31} = W_{42}$; $W_B = W_{43} = W_{21}$. There are two other possible relaxation routes, corresponding to what are known as double quantum (W_{14}) and zero quantum (W_{32}) transitions. It can be shown by considering the various values of W and the populations that the NOE obtained when the A spins are saturated (i.e. P3 = P4 and P1 = P2) is $\eta = (I - I_0)/I_0 = \gamma_A/\gamma_B \{(W_{14} - W_{32})/(W_{14} + W_{32} + 2W_B)\}$. As we saw in the text (e.g. Figure 6.1.21), the transition probabilities depend on τ_c and ω, as well as r, the separation between A and B. In fact, it can be shown that $W_{23} = C^2 \tau_c/\{20[1 + (\omega_A - \omega_B)^2\tau_c^2]\}$; $W_{14} = 3C^2 \tau_c/\{10[1 + (\omega_A + \omega_B)^2\tau_c^2]\}$; $W_B = 3C^2\tau_c/\{40[1 + \omega_B^2\tau_c^2]\}$; and $W_A =3C^2\tau_c/\{40[1 + \omega_A^2\tau_c^2]\}$, where $C = \mu_0\gamma_A\gamma_B\hbar/4\pi r^3$, assuming that the only relaxation process is between A and B. Note that η can be positive or negative because of the term $W_{14} - W_{12}$. Using these formulae, calculate the NOE expected for a ^{1}H–^{1}H pair and a ^{13}C–^{1}H pair in the limit of rapid tumbling ($\omega\tau_c \ll 1$).

6.1.8 A 217-residue fragment of bovine prion protein bPrP(23-230) was studied by NMR. The C-terminal domain, bPrP(121-230), was found to have a well-defined globular structure. Explain the ^{1}H–^{15}N NOE pattern observed for the backbone amide groups measured in a 1mM solution of bPrP(23-230), as shown in the figure.

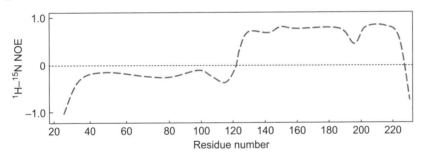

6.1.9 The amino acid sequence of a fragment of hen lysozyme is shown below. Filled, half-filled, and empty circles indicate amides with slow, intermediate, and rapid exchange rates with solvent, respectively. A summary of d_{NN} and $d_{\alpha N}$ NOE connectivities obtained from a ^{1}H 2D NOESY spectrum is also shown. Comment on what information this diagram gives about the secondary structure in lysozyme.

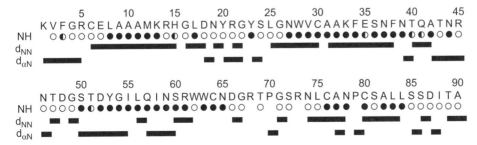

6.1.10 Triose phosphate isomerase (TIM) is a glycolytic enzyme with a flexible loop (residues 166-176) at the active site that is believed to be important in product release. A mutant form of yeast TIM (W90Y, W157F) was labeled at one remaining Trp (W168)

with 5-fluorotryptophan. ^{19}F NMR spectra are shown in the figure. Spectra are shown at two temperatures and different concentrations of the TIM substrate analogue (SA) glycerol-3P. The shift between free and bound states at 25°C was ~340 Hz. Comment on the results.

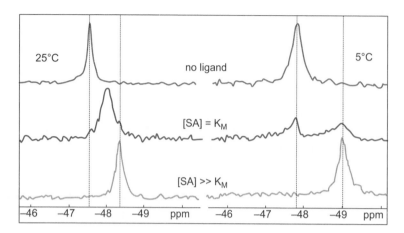

6.1.11 There are numerous powerful pulse sequences in modern NMR that are used to select specific kinds of information from a sample. One basic building block used in many of them is the INEPT sequence (insensitive nuclei enhanced by polarization transfer). Signals from a relatively weak nucleus (X; e.g. ^{15}N or ^{13}C) are detected via a proton signal that it is coupled to it via a coupling constant J. The basic INEPT sequence can be described by:

Proton: $90°_x$–τ–$180°_y$ –τ–$90°_y$

X: $----180°--90°$ acquisition

(the x and y subscripts give the direction along which the ^{1}H **B$_1$** fields are applied—see Figure 6.1.18). In a coupled NH system there are two ^{15}N transitions and two ^{1}H transitions, so a four energy level diagram can be drawn for the coupled NH group, as shown in the figure.

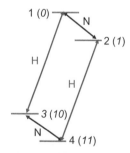

The transitions associated with ^{15}N are about 10× lower frequency than ^{1}H (Table 6.1.1). This results in an approximate energy distribution of populations among the four energy levels, as indicated by the numbers in brackets. Explain how the INEPT sequence gives a signal enhancement. (Hint: the interval τ is chosen to be $1/4J$.)

6.2 Electron paramagnetic resonance

Introduction

Electron paramagnetic resonance (EPR) was first observed by Yevgeny Zavoisky in 1944. Also known as electron spin resonance (ESR), the method detects unpaired electrons. Unpaired electrons are relatively rare in biological systems but they occur in important processes—for example, free radicals in photosynthesis. EPR has also been used to study the structure and function of many paramagnetic metalloproteins. Stable nitroxide free radicals (**spin labels**) added to a system are widely used as extrinsic probes of molecular dynamics and structure. Other uses of EPR include dating of fossils and measurement of exposure to radiation.

The electron has a magnetic moment, μ_e, by virtue of its spin ($S = 1/2$). It has two allowed orientations in an applied magnetic field ($\mathbf{B_o}$) and transitions between these orientations can be induced if oscillating electromagnetic radiation, with appropriate frequency, ν, is applied perpendicular to $\mathbf{B_o}$. Typical applied values of $\mathbf{B_o}$ lead to absorption, observed in the microwave frequency range. (Note: many of the concepts of EPR are very similar to those in NMR (see Chapter 6.1) and these are not all repeated here.)

Measurement

As noted above, EPR studies measure absorption in the microwave region of the electromagnetic spectrum. The microwave region is defined by "bands" named, for historical reasons, by letters like X and Q. Table 6.2.1 lists some of these bands, together with their approximate frequencies (in GHz; 10^9 Hz), their wavelengths (λ in cm), and typical $\mathbf{B_o}$ fields (in Tesla) that cause an electron to resonate at that frequency.

The X band is the most commonly used band in EPR studies but higher frequencies are being

Table 6.2.1 Microwave bands

	S	X	Q	W
ν (GHz)	~ 3	~ 9	~ 35	~ 95
λ (cm)	10	3	0.85	0.3
$\mathbf{B_o}$ (T)	0.1	0.33	1.25	3.5

increasingly applied because of the improved resolution and sensitivity obtained. Microwaves can be conducted along hollow metal tubes, called waveguides, whose dimensions are related to the wavelengths of the conducted microwaves. At X-band frequencies, the required applied field (~0.3 T) is low enough to be readily obtained with an electromagnet. A resonator (or "cavity") contains the sample; the detection system usually has the cavity in one arm of a balanced network that involves a "bridge" or "circulator" (Figure 6.2.1). Microwave sources are

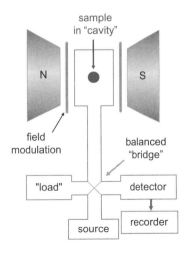

Figure 6.2.1 Representation of an EPR spectrometer operating with an iron magnet; also shown are a "cavity", containing the sample, field modulation coils, and a balanced "bridge" system that is set up to detect absorbance changes in the cavity arm.

given names like klystron and gyrotron, depending on the method of generation. At very high frequencies (e.g. 180 GHz) microwaves are directed onto the sample via a reflector system.

Most EPR instruments operate in what is known as **continuous-wave mode** where the field is **modulated** as it is swept through the resonance line. It can be seen from Figure 6.2.2 that if the modulation is less than the linewidth of the absorption line, the signal detected appears as the **derivative** of the absorption line. Samples can be aqueous or solid. Water absorbs microwaves so aqueous samples change the resonant properties of the cavity and optimization of the sample shape and position are important. The sensitivity of the method is high (compared to NMR) because of the much larger magnetogyric ratio of electrons. For samples with relatively narrow lines, such as spin labels, a concentration of $1\mu M$ or less is usually sufficient; with high fields and low temperatures, it is possible to detect as few as 10^7 spins.

Pulsed EPR

There are many advantages in applying pulses, as is done in NMR, rather than sweeping the field; data acquisition in the time domain is faster and a range of flexible and powerful new techniques become available, such as 2D spectroscopy. EPR spectra are relatively broad so these advantages are hard to implement technically (it is easier to cover the

spectral width with a pulse in NMR than EPR). Pulsed EPR is, however, becoming much more widely available; a short (<20 ns), intense (>300 W) microwave pulse is applied to the sample, followed by measurement of the generated transient microwave signals. A frequency spectrum is then obtained by Fourier transformation. An example of a pulsed EPR experiment, one that can be used to measure distances between paramagnetic centers up to around 8 nm apart, will be described in more detail below.

Spectral parameters

An EPR signal is characterized by four main parameters: (i) intensity; (ii) linewidth; (iii) g-value; and (iv) multiplet structure. These parameters are now briefly discussed in turn.

Intensity

The integrated area of the **absorption** signal is generally proportional to the concentration of the unpaired spins in the sample. (EPR spectra are usually displayed as a derivative rather than an absorption spectrum—see Figure 6.2.2). A number of factors can, however, alter the observed spectral intensity; these include "quenching" of the signal by interactions between spins at high concentrations and "saturation" of resonances (i.e. the population difference between energy levels is reduced by the applied radiation—see Figure 5.1.4).

g-value

Resonance positions are characterized by a *g-value*; this is a measure of the local magnetic field experienced by the electron (compare chemical shift in NMR). The resonance condition in EPR is defined as $h\nu = g_e\beta_e\mathbf{B_o}$, where g is the g-factor, which, for a free electron, is 2.0023; β_e (sometimes denoted μ_B) is the Bohr magneton $(0.927 \times 10^{-23}$ J T^{-1}). This resonance condition can also be written in terms of the "Larmor" precession frequency $\omega = 2\pi\nu = \gamma_e\mathbf{B_o}$, where $\gamma_e = g\beta_e/\hbar$ is the magnetogyric ratio of the electron $(1.76 \times 10^{11}$ s^{-1} T$^{-1})$. The g-value for a molecule containing unpaired electrons is characteristic

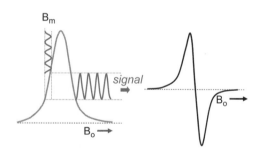

Figure 6.2.2 An EPR absorption line is shown in gray on the left. The applied field, **B$_o$**, modulated by an oscillating field, **B$_m$**, is swept through the resonance line at constant applied microwave frequency. The signal reaches a maximum as it goes up the side of the resonance; the output is then zero at the maximum of the absorption peak and then reaches another maximum with the opposite phase on the other side of the resonance. The resulting observed "derivative" spectrum is shown on the right.

of the electronic environment. A wide range of *g*-values is observed in transition metal ion spectra but the *g*-value of many free radicals is close to 2.0.

Linewidths and relaxation times

As with NMR, the width of EPR lines at their half height is inversely related to a relaxation time ($1/\pi T_2$). The relaxation mechanism can, however, be treated, simplistically, as arising from two contributions. The first is related to T_1, which is a measure of the recovery rate of the spin magnetization along the $\mathbf{B_o}$ direction, after a perturbation from equilibrium; the faster the relaxation rate, the broader the line. As with NMR, the basic mechanism causing T_1 relaxation involves fluctuating magnetic fields experienced by the unpaired electron. (Note that short T_1 also implies short T_2 and broad lines.)

In most transition metal complexes, the linewidth is determined by the T_1 contribution. T_1 is often so short that the metal ion spectra are not detectable at room temperature. However, T_1 increases considerably with decreasing temperature, so such systems are often studied at very low temperatures (<77 K) in order to obtain acceptably narrow lines.

The second major contribution to the linewidth arises because local magnetic field effects can have a range of values. This mechanism is known as inhomogeneous broadening; it is apparent when the molecules are either static or tumbling relatively slowly (see anisotropy section below). This broadening can be reduced or removed entirely by averaging that occurs when molecules tumble rapidly in solution. When T_1 is relatively long, as occurs in many free radical spectra, this inhomogeneous broadening mechanism dominates.

Multiplet structure

In NMR, we saw that resonances can be split into multiplets by interactions with other nuclei. In EPR, there are two main types of interaction that give rise to multiplet splitting of the resonance lines—hyperfine and zero-field splitting. These splitting are very informative features of EPR spectra, especially the hyperfine splitting.

Hyperfine splitting arises from the interaction of the electron spins (*S*) with nuclear spins (*I*). This

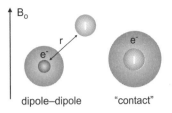

Figure 6.2.3 Representation of two kinds of hyperfine interaction between an electron and a nucleus: a dipolar term and an isotropic term.

interaction is of two types: (i) a through-space dipolar interaction between electron and nucleus; this is directional and *anisotropic* (i.e. it depends on the angle the electron nuclear vector, **r**, makes with $\mathbf{B_o}$ in Figure 6.2.3; see also Tutorial 11 and next section); (ii) an interaction that mainly results from delocalization of the unpaired electron on the nucleus; this "Fermi contact" interaction is *isotropic* and we will refer to this interaction as the isotropic hyperfine interaction here.

Consider first the information available from the isotropic hyperfine interaction. This is frequently the dominant effect seen in spectra of free radicals or metal ions tumbling rapidly in solution because the dipolar interaction is averaged to zero. The contact term is independent of $\mathbf{B_o}$ and is characterized by a hyperfine splitting constant (compare *J* in NMR). An illustration of splitting arising from an interaction with one $I = 1/2$ nucleus is shown in Figure 6.2.4.

Isotropic interactions between an electron and a nuclear spin, *I*, give $2I + 1$ lines of equal intensity—see for example the Mn^{2+} example in Figure 6.2.5.

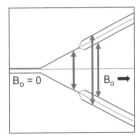

Figure 6.2.4 Isotropic hyperfine splitting. The effect of an applied field on energy levels is shown. For illustration, the hyperfine interaction only appears at higher field to show the effect of a simple interaction between an electron and one, $I = 1/2$ nucleus. Without hyperfine structure the single gray transition would be observed at a particular value of $\mathbf{B_o}$. With hyperfine splitting, the two cyan transitions will be observed.

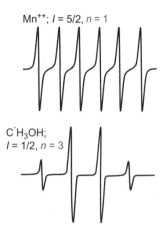

Mn^{++}; $I = 5/2$, $n = 1$

$C^{\cdot}H_3OH$;
$I = 1/2$, $n = 3$

Figure 6.2.5 Illustration of the observed isotropic hyperfine structure for two systems: the top spectrum is of manganese with six equal lines arising from an interaction with one nuclear spin of $I = 5/2$; bottom: a methanol radical split into four lines by interaction with three $I = 1/2$ protons (peak intensity ratio = 1:3:3:1).

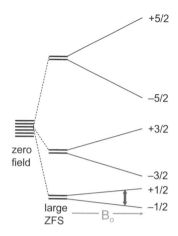

Figure 6.2.6 Energy levels for high-spin Fe(III) with a large ZFS. The cyan arrow indicates the size of an observable microwave transition. In this case, the ZFS is so large that this is the only observed transition for most available fields and frequencies.

There is, however, also a dependence on n, the *number* of nuclei to which the electron is coupled; this is related to the probability of finding the n neighboring nuclei in certain orientations; a doublet is observed when $n = 1$, a triplet when $n = 2$ (with intensity ratio 1:2:1), etc. (compare J multiplets in NMR); an example of the effect when $n = 3$ is given in Figure 6.2.5.

The dipolar or **anisotropic hyperfine interaction** term will be considered later in the **spectral anisotropy** section.

Zero-field splitting

While an applied magnetic field, $\mathbf{B_o}$, splits the electron-spin energy levels (removes the degeneracy), various local magnetic interactions in the molecule can also split energy levels even in the absence of $\mathbf{B_o}$. This effect is known as zero-field splitting (ZFS) (such effects are only observed in NMR for quadrupolar nuclei—Figure 6.1.36). The presence of ZFS can result in further fine structure in the EPR spectrum. The ZFS splitting of the energy levels can also be so large that it may be not be possible to observe

all the EPR transitions; this situation is observed, for example, in some high-spin Fe(III) complexes (see Figure 6.2.6).

Box 6.2 Spin labels

The term spin label was introduced by Harden McConnell to describe stable free radicals that are used as reporter groups or probes. The most commonly used spin labels contain the nitroxide radical, which is stable over a wide pH range and up to about 80°C. The EPR signal can be removed by exposure to mild reducing agents (e.g. ascorbate). These stable free radicals have a free electron in the vicinity of a nitrogen atom. The most abundant nitrogen isotope, ^{14}N, has $I = 1$, so three hyperfine lines of equal intensity (area) are observed (Figure A) for a rapidly tumbling spin label (see below). A reactive group $\mathbf{X}$ allows the label to be attached to a protein, e.g. via a free SH group.

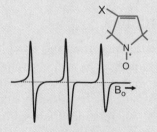

Figure A A nitroxide radical and its three-line spectrum (the lines here are not quite of equal intensity because of different linewidths).

Note In the methanol radical example in Figure 6.2.5, coupling is not seen to the ^1H in the OH group because of exchange of the labile proton on the OH with solvent (compare the spectrum of ethanol in Chapter 6.1).

Spectral anisotropy

EPR parameters often depend on the direction of $\mathbf{B_o}$ relative to the molecular axes. We will now discuss how anisotropy of the g-value, the dipolar A term, and the ZFS can make important contributions to the observed EPR spectra of transition metals and spin labels (Box 6.2).

Anisotropy of the g-value

The g-value can be characterized by three principal g-values, g_x, g_y, and g_z, defined as lying along the principal axes of the chemical group containing the unpaired electron; g_z is the observed value when the field is along the z-axis, etc. The assumption of axial symmetry can often be made where $g_x = g_y = g_\perp \neq g_z = g_\parallel$ (Figure 6.2.7).

For single crystals, the angular variation of the g-values can give information about the orientation of the principal axes. In a powder or a frozen sample, the molecules will be randomly oriented and a spread of g-values will be observed because the sum of the spectra from all molecules is observed, each with a different value of θ. This gives a "powder" spectrum (see also Chapter 6.1—solid state NMR). Figure 6.2.8 shows the expected absorption and derivative spectra for a rigid, randomly oriented, axially symmetric system.

Note that the derivative spectra highlight the edges of absorption spectra. These edges correspond to the principal g-values. More absorption is concentrated near the $g_\perp$ than the $g_\parallel$ direction because there are more molecules with $\mathbf{B_o}$ in the $g_\perp$ direction than $g_\parallel$, leading to a non-symmetric lineshape. If the molecule does not have axial symmetry then the powder spectrum will have three peaks rather than

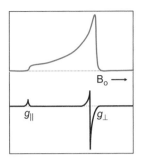

Figure 6.2.8 The absorption spectrum (cyan) obtained from a random orientation of molecules with axially symmetric g-values. The black derivative spectrum indicates the position of $g_\parallel$ and $g_\perp$.

two, with g-value edges corresponding to g_x, g_y, and g_z. This can be seen in Figure 6.2.9, which shows the EPR spectra of Fe(III) metmyoglobin and its azide complex. The Fe(III) metmyoglobin complex has only two g-value edges, at around 2 and 6, which suggests axial symmetry. The azide complex, however, has three g-value edges at 2.8, 2.25, and 1.75, suggesting non-axial symmetry around the metal ion.

Anisotropy of A

As illustrated in Figure 6.2.3, anisotropy of A can arise from an angular dependent dipolar interaction between μ_e and μ_n (the magnetic moments of the electron and nucleus). Similar to g, A-values are characterized by three principal A-values: A_x, A_y, and A_z. For axial symmetry, $A_x = A_y = A_\perp \neq A_z = A_\parallel$. An important example of A-value anisotropy is seen in EPR spectra of nitroxide spin labels (Box 6.2; Figure 6.2.10).

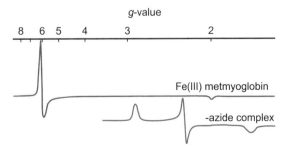

Figure 6.2.9 9.25-GHz spectra of Fe(III) metmyoglobin and its azide complex. The unliganded protein shows axial symmetry while the liganded protein does not.

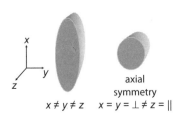

Figure 6.2.7 Representation of non-axial and axial symmetry where x and y axes ($\perp$) are the same.

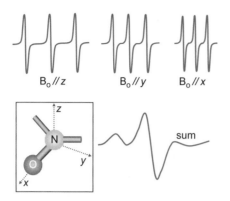

Figure 6.2.10 Spin label hyperfine anisotropy. The dipolar A interaction leads to different observed splitting of the three line spectrum for different directions of $\mathbf{B_o}$ relative to the axes of the nitroxide radical. The spectrum observed in a solid is the sum of the anisotropic spectra over all angles.

Significantly, different spin label spectra are observed at different angles with respect to $\mathbf{B_o}$; A_z is greater than A_x and A_y. The sum over all angles has very broad outer lines and a relatively narrow central line because the angular variation for that line is less than for the outer two. The high field line is particularly broad.

Anisotropy of the zero-field splitting

ZFS anisotropy also gives rise to angular dependent splitting and this effect can lead to the observation of broad lines from randomly oriented molecules. This effect contributes to inhomogeneous broadening. In some cases, such as in aqueous solutions of Mn(II), the molecular tumbling may be fast enough to average out the ZFS. On binding to a macromolecule, however, the tumbling rate falls, leading to line broadening (see below and example in Figure 6.2.14).

Applications of EPR

As unpaired electrons are required, EPR applications are mainly associated with systems containing free radicals, spin labels, or transition metal ions. EPR has made important contributions to our understanding of photosynthesis (see discussion of Feher in Journal Club) and the role of tyrosine radicals found in photosystem II and the enzyme ribonucleotide reductase. It has also been valuable in determin-

ing the geometric and electronic structure of iron sulfur proteins and metalloenzymes that contain molybdenum and other metals. Here, we illustrate some of the applications with a brief discussion of motional probes, some applications to transition metal ions, distance measurements using PELDOR and a brief introduction to the use of a technique called ENDOR. Direct investigation of free radicals in biological systems is usually difficult because the concentration of free radicals in cells is low; the trapping of radicals to make them more readily detectable is also discussed.

Time scales and molecular motion

The various spectral anisotropy effects discussed above can be significantly affected by motion in the sample. This can be used to give us information about the motion of the electron spin probe as the observed lineshape in a sample depends on the time scale of the motion and the spread in spectral parameter values of the spin probe.

Consider the case of a nitroxide spin label. When it tumbles rapidly in solution, the EPR spectrum consists of three narrow lines with almost equal heights. As the rate of motion decreases, the EPR spectrum alters because the averaging of anisotropy is incomplete. Figure 6.2.11 illustrates this effect by showing spectra of a spin label recorded at a wide range of temperatures. The estimated rotational correlation time for the motion is also shown (see Box 4.3 for definition of correlation time). Comparison of the observed spectra with theoretical spectra allows estimates of molecular motion to be made. This sensitivity of spin labels to motion makes them very useful probes.

An example of a motional probe is shown in Figure 6.2.12; this shows increasing spin label mobility towards the center of a lipid bilayer (compare Figure 6.1.23). The extent of motional averaging can be defined by an **order parameter**, $S = $ (observed anisotropy)/(maximum anisotropy); the maximum anisotropy values can be obtained when the label is rigidly oriented in a single crystal. It follows that $S = 1$ for a highly ordered system and $S = 0$ for rapid isotropic motion.

Spin label motional probes have been widely used to explore structure and function. One example is the

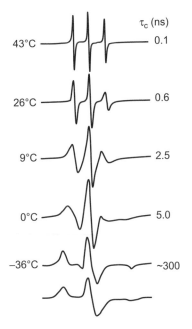

Figure 6.2.11 The effects of the rate of motion on the EPR spectra of a spin label that is rotating isotropically. The rotational tumbling time is altered by changing the temperature.

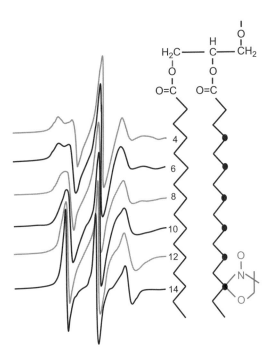

Figure 6.2.12 Spectra of a spin label attached at different positions to a lipid chain in a bilayer (original figure kindly provided by A. Watts). It can be seen that the probe mobility increases with increasing distance from the lipid headgroup (4 → 14).

systematic insertion of cysteine residues in membrane proteins by site-directed mutagenesis, followed by attachment of spin labels at these sites (see Problem 6.2.7). Analysis of the motion of the labels at specific sites gives information about the structure and motion at these sites (see also "Further reading").

Transition metal ion spectra

The transition metals that have been most frequently studied in biological systems are manganese, iron, cobalt, copper, molybdenum, and vanadium. The dependence of the EPR spectrum on the spin state and the arrangement ligands provides a "fingerprint" of the metal site. The identification of a metal-binding site in a biological compound is, in the first instance, usually made by comparing the observed spectra with those from well-characterized model complexes.

There are several characteristic features of transition metal ion EPR spectra:

- The observed spectrum depends on the number of unpaired electrons, which are primarily localized to d-orbitals, and on the arrangement and nature of the coordinated ligands.
- The g-values vary over a wide range (~1.4 -10).
- Very low temperatures (1 K) are often required to observe an EPR spectrum because of rapid T_1 relaxation.
- Crystal field theory (Tutorial 10) predicts that the ligand binding properties determine the number of unpaired electrons in a complex. This leads to the idea of high-spin and low- spin complexes (see Figure 6.2.13). For instance, Fe(III) has five d-electrons; these can either all be unpaired ($S = 5/2$) or have only one unpaired electron ($S = 1/2$) to give a low-spin complex, with a very different EPR spectrum.

Estimation of the distance between two paramagnetic centers

If two paramagnetic centers (e.g. a metal ion and a spin label) are close enough, their interaction can be can be detected by broadening of the EPR lines. In some cases, these interactions can be quantified to yield information about their separation. An early example (Figure 6.2.14) was a spin label attached to a

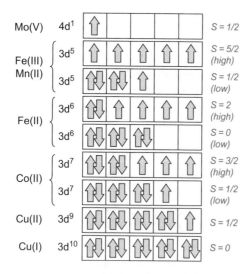

Figure 6.2.13 Arrangement of d-electrons in some transition metal complexes (Tutorial 10). The nature of the ligand determines whether the high- or low-spin arrangement is favored. The net spin is given by S.

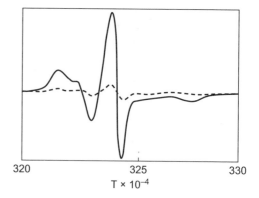

Figure 6.2.14 Spectrum of creatine kinase specifically labeled on an SH group and the spectrum obtained on addition of MnATP (dashed line) (adapted with permission from *Proc. Natl. Acad. Sci. U.S.A.* 64, 219 (1969)).

specific group in creatine kinase. The observed quenching of the EPR spectrum on addition of MnADP led to an estimate for the distance between the spin-label and the paramagnetic Mn^{2+} ion. A separation of 0.7 nm was obtained, assuming that the broadening arose from a dipolar interaction between the two paramagnetic centers (Tutorial 11).

The dipole–dipole interaction between two paramagnetic magnetic moments, μ_X and μ_Y, can be described by $\nu_{dd} = D_{dip}(1 - 3\cos^2\theta)/r_{XY}^3$, where r_{XY} is the separation between X and Y, and θ is the angle between the vector r_{XY} and the applied field direction (see Tutorial 11). ν_{dd} will vary between $-2D_{dip}/r_{XY}^3$ and $+D_{dip}/r_{XY}^3$ as θ goes from 0 to 90°. In an axially symmetric sample, θ is more likely to be 90° than 0° (see Figure 6.1.34), so the spectrum arising from a dipolar interaction between μ_X and μ_Y in a randomly oriented sample appears as shown in Figure 6.2.15. The separation between the two strong peaks, corresponding to $\theta = 90°$, is $2\nu_{dd}$. The interaction energy between the two magnetic moments is $D_{dip} = \mu_o\hbar\gamma_X\gamma_Y/8\pi^2$, where μ_o is the vacuum permeability, $\hbar$ is the reduced Planck constant, and γ_X and γ_Y are the magnetogyric ratios of the two spins X and Y. Entering the values for these constants gives $\nu_{dd} = 52.04(1 - 3\cos^2\theta)/r_{XY}^3$ MHz, where r is measured in nm. The value of r_{XY} can thus be readily calculated from the observed value of ν_{dd}.

In most liquid samples, the dipolar interaction illustrated in Figure 6.2.15 will be averaged out so frozen samples are usually used. Spectra of the sort in Figure 6.2.15 are, however, hard to observe with non-pulse methods because of strong inhomogeneous broadening effects that are often present in solid samples. As discussed in Chapter 6.1, however, spin echo experiments can be used to refocus many of the inhomogeneous broadening contributions (see Figure 6.1.18). Spin echo methodology is incorporated in a technique called **pulsed electron-double resonance (PELDOR)** [this method is also sometimes known as **double electron–electron resonance (DEER)**]. Quantitative estimates of distances between paramagnetic centers up to 8 nm apart can be obtained using this method. In a simple version of PELDOR (Figure 6.2.16), a two-pulse echo sequence with a fixed pulse separation time, τ, is used to monitor the echo intensity of the X spins

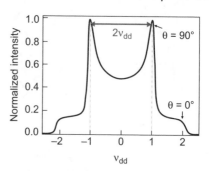

Figure 6.2.15 The spectrum observed from two interacting paramagnetic centers in a frozen or powder sample. The splitting between the two major peaks is $2\nu_{dd}$ MHz.

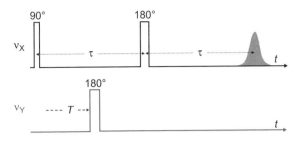

Figure 6.2.16 PELDOR pulse sequence. A simple 90°–τ–180°–τ spin echo pulse sequence is applied to X. A 180° inversion pulse is also applied to Y and the time T is varied. The Y pulse modulates the intensity of the X echo.

(at frequency v_X). An additional microwave pulse is used to flip the Y spins (at frequency v_Y) selectively at time T. This spin-flip induces a change in the X spins by v_{dd} which is related to the size of the dipolar coupling between X and Y. Varying the time position, T, of the Y pulse thus induces a periodic modulation of the X-spin echo intensity of the form $I(T) = I_0\cos(2\pi v_{dd}T)$, where I_0 is the echo intensity at $T = 0$. v_{dd} can thus be obtained from the modulation envelope by a Fourier transform and hence the distance r_{XY} can be deduced.

In an ideal situation, the spectra of X and Y would be resolved and could be irradiated separately, thus giving a maximum amount of echo modulation. In most biological applications, however, the paramagnetic species X and Y are identical nitroxide radicals, attached to specific cysteines so the X and Y signals are not well separated. It is, however, still pos-

sible to obtain echo modulation, albeit reduced, by selectively irradiating different regions of the spectrum in a frozen sample. As illustrated in Figure 6.2.10, the hyperfine structure is anisotropic; this means that a sub-population of X, corresponding to molecules at a particular angle to $\mathbf{B_0}$, can be irradiated at an edge of a spectrum. This irradiated subpopulation of X will interact with a non-irradiated subpopulation of Y to give the v_{dd} modulation required to measure a distance. Figure 6.2.17 shows two regions of a spin label spectrum (cyan and gray) typically used to represent the X and Y spins.

Example PELDOR of a histone octamer complex

DNA in eukaryotes is packaged in chromatin, a key component of which is the **nucleosome core particle**. This consists of about 150 base pairs of DNA wrapped twice around a protein octamer consisting of pairs of proteins called histones (H2A, H2B, H3, and H4); see Figure 6.2.18.

In a PELDOR experiment on the core particle, spin labels were inserted on specific cysteines, engineered into H3. The measured distances then correspond to the separation between specific points on the two copies of H3 in the histone octamer. When residue 49 on H3 was labeled (H3R49C) the experimentally measured distance was 6.3 nm while the distance expected from the X-ray structure was 6.4 nm; for H3Q76C, the observed and expected distances were both 7 nm (Figure 6.2.19). These experiments thus clearly demonstrate that PELDOR, like FRET (Chapter 5.5), offers a viable alternative to crystallography for obtaining structural information about the separation between pairs of spin labels.

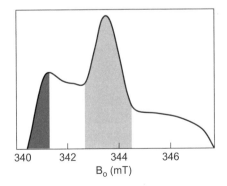

Figure 6.2.17 An absorption spectrum of an immobilized spin label (see Figure 6.2.11 for the derivative spectrum). The regions normally chosen to act as the X and Y spin systems in a PELDOR experiment are shown in cyan and gray, respectively.

Electron nuclear double resonance (ENDOR)

As we have seen, EPR hyperfine interactions are very informative. **Electron nuclear double resonance (ENDOR)** is a technique for determining hyperfine interactions between electrons and nuclear spins. It was introduced by George Feher in 1956 to detect interactions which are not readily accessible in normal EPR spectra. In ENDOR, the nuclear signals and the electron signals are both irradiated. In continuous-wave ENDOR, the intensity of an electron paramagnetic resonance signal, partially saturated with

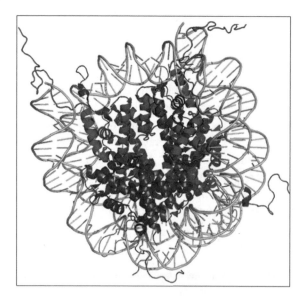

Figure 6.2.18 The crystal structure of the nucleosome core particle (PDB: 1KX5) showing the DNA wrapped around the eight core histones. The two H3 histones are shown in cyan.

microwave power, is monitored as a radiofrequency field is swept through nuclear resonance. In pulsed ENDOR, the radiofrequency is applied as pulses and the EPR signal is detected as a spin echo (see Figure 6.2.16).

Example   An ENDOR study of a bacterial nitrate reductase

The nitrate reductase from *Paracoccus pantotrophus* is a member of a group of enzymes that bind molybdopterin guanine dinucleotide (MGD), an enzyme cofactor that contains Mo(V). This enzyme has a 90 kDa catalytic subunit that binds an N-terminal [4Fe-4S] cluster and MGD. X-band proton-ENDOR associated with the Mo(V) EPR signal detected broad features that were assigned to weak interactions between the Mo(V) ion and four different classes of protons. Signals from two of the protons disappeared upon exchange into deuterated buffer (Figure 6.2.20) suggesting that these two arise from exchangeable OH groups. One of the other non-exchangeable protons gave a resolved feature at 22–24 MHz in the ENDOR spectrum (not shown). The anisotropy of this feature was interpreted as arising from an interaction corresponding to a molybdenum–proton distance of ~0.32 nm; this is consistent with this proton being a β-methylene on a Mo-Cys ligand.

EPR spin trapping

Many free radicals in biology have very short lifetimes, making them hard to detect. Spin traps are compounds that are used to stabilize the radical by

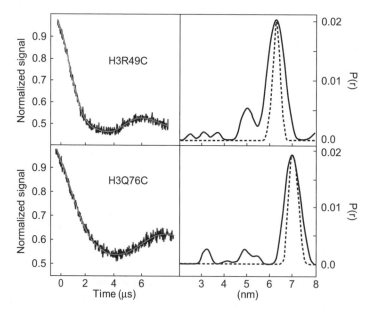

Figure 6.2.19 On the left are shown two examples of PELDOR echo modulations for spin labels attached to cysteines at positions 49 and 76 in H3. On the right are the distributions of distances that fit the PELDOR data. The dotted curves on the right indicate the expected variation in distance that would arise from motion of the spin label (adapted with permission from *J. Am. Chem. Soc.* **131**, 1348–49 (2009)).

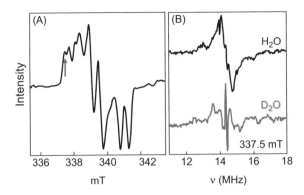

Figure 6.2.20 X-band ENDOR spectra of the Mo(V) signal from a nitrate reductase measured at 25 K: comparison of spectra before and after D_2O exchange. Note the loss of signal because of H–D exchange (adapted with permission from *Biochem. J.* 363, 817–23 (2002)).

carrying out a reaction of the form $R^* + ST \rightarrow RSA^*$, where R^* is a radical intermediate, ST is the spin trap and RSA^* is a radical spin trap adduct. Various spin traps, including 5,5-dimethyl-1-pyrroline *N*-oxide (DMPO), have been widely used to trap free radicals.

Example Spin trapping of NO

Nitric oxide (NO) is one of the smallest and simplest active molecules in nature but it controls multiple functions in the body. It has been implicated in numerous diseases, including hypertension, stroke, cardiac failure, and diabetes. There is a need to measure its distribution but this is difficult because of low concentrations and short half-life, especially in living tissues. NO is a free radical and is therefore, in principle, detectable by EPR, but it cannot be measured directly because the signal is too broad. The most reliable EPR-based approaches developed so far are based on trapping NO with various iron complexes, both intrinsic and extrinsic. Many of the standard spin traps such as DMPO do not form suitable NO adducts but NO does react with iron (especially Fe(II) ions) in solution, and it forms various types of paramagnetic nitrosyl–iron complexes with characteristic EPR signals. Hemoglobin can act as an intrinsic trap and EPR of NO bound to hemoglobin has provided useful information. *Extrinsic* iron dithiocarbamate has also been shown to be an effective NO spin trap (see Figure 6.2.21) that can be used *in vivo*.

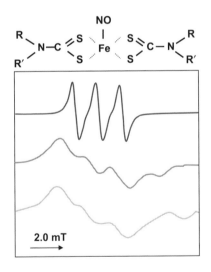

Figure 6.2.21 The general structure of mononitrosyl–iron complexes with RR′-dithiocarbamates is shown at the top. The top spectrum is of one such water-soluble compound, recorded at X-band and room temperature. The middle spectrum is the same but at liquid nitrogen temperature. The bottom spectrum is of a lipophilic dithiocarbamate–NO complex formed in a mouse aorta during incubation (37°C; 1 h). This spectrum gives a measure of the amount of NO in mouse blood (adapted with permission from *J. Chromatogr.* B 851, 12–20 (2007)).

Summary

1 Electron paramagnetic resonance (EPR) is a technique that detects unpaired electrons. Electron-spin reorientation causes absorption of incident microwave radiation when a sample is placed in a magnetic field.

2 In biological systems, unpaired electrons occur in free radicals and some transition metal ions. The use of synthetic, stable free radicals (spin labels) as probes extends the range of biological applications.

3 EPR spectra are usually displayed as the first derivative of the absorption spectrum. A spectrum is characterized by four main parameters: intensity, linewidth, *g*-value (which defines position), and multiplet structure.

5 The *intensity* of an EPR spectrum in solution can give information about concentration, the *linewidth* about dynamic processes, the *g-value* about the immediate environment of the unpaired electron, and the *multiplet structure* (characterized by a hyperfine splitting constant *A*) about interactions between electron spins and nuclear spins.

6 The *g*- or *A*-values are usually anisotropic, which means that the spectra depend on the angle between the molecular axes and the applied magnetic field. In a single crystal, the spectra are sharp, but in a randomly oriented (powder) sample, they are spread out because all possible angles exist. Information must then be obtained by interpreting the shape of the spectral envelope.

7 In molecules with more than one unpaired electron, the interaction between the spins leads to a splitting of the energy levels, even in zero field (zero-field splitting, ZFS).

8 EPR spectra of transition metal ions are characteristic of the number of unpaired electrons in the metal and the arrangement and nature of the coordinated ligands. The *g*-values range from about 1 to 10. Very low temperatures are often required to observe the spectra because relaxation times are very short.

9 Biological applications of EPR include exploration of the ligand environment around a metal site in a metalloprotein, the detection of free-radical intermediates (spin trapping), the measurement of distance between two paramagnetic centers, and measurement of both the rate and amplitude of molecular motion in a variety of systems.

Further reading

Books

Weil, J.A., and Bolton, J.R. *Electron Paramagnetic Resonance* (2nd edn) Wiley, 2007.

Schweiger, A., and Jeschke, G. *Principles of Pulse Electron Paramagnetic Resonance* Oxford University Press, 2001.

Reviews

Kleschyov, A.L., Wenzel, P., and Munzel, T. Electron paramagnetic resonance (EPR) spin trapping of biological nitric oxide. *J. Chromatogr. B* 851, 12–20 (2007).

Schiemann, O., and Prisner, T.F. Long-range distance determinations in biomacromolecules by EPR spectroscopy. *Q. Rev. Biophys.* 40, 1–53 (2007).

Columbus, L., Hubbell, W.L. A new spin on protein dynamics. *Trends Biochem. Sci.* 27, 288–95 (2002).

Problems

6.2.1 Predict the EPR spectrum of the benzene negative ion.

6.2.2 Three *g*-values were obtained for a spin label in an experiment carried out in a single crystal: $g_z = 2.0027$, $g_x = 2.0089$, and $g_y = 2.0061$. At what rate would the spin label have to tumble to average out the *g*-value anisotropy when it is in liquid at a magnetic field of 0.33 T? ($\beta = 0.927 \times 10^{-23}$ J T^{-1}; $h = 1.05 \times 10^{-34}$ J s.)

6.2.3 The structures of two basic types of 2- and 4-Fe sulfur centers are shown in the figure (8-Fe centers consist of two 4-Fe–S clusters). Some representative EPR spectra of Fe–S ferredoxins are also shown. Comment on the feasibility of using EPR as a diagnostic technique for the different types of protein centers. These spectra were obtained at very low temperatures—why?

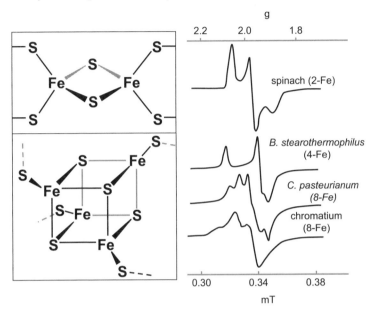

6.2.4 The figure shows the EPR spectra of three different spin-labeled Dnp ligands when bound to an anti-Dnp antibody. The approximate maximum hyperfine anisotropies that are observed are indicated by cyan arrows. What can you conclude about the rigidity and depth of the antibody binding site?

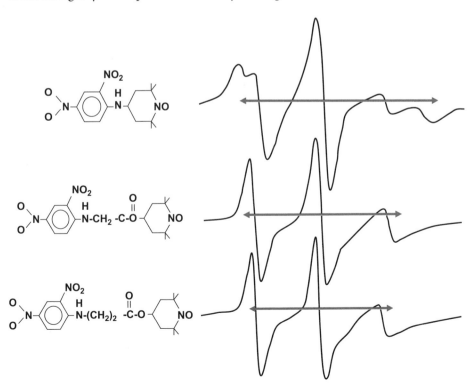

6.2.5 The various spin label spectra shown in this chapter arise from a hyperfine interaction with one ^{14}N nucleus. What would you expect the spectrum to look like if the interaction were with two ^{14}N nuclei?

6.2.6 Binding of the human immunodeficiency virus (HIV) trans-activation responsive (TAR) RNA to the Tat protein is essential for production of full length RNA transcripts during viral replication. Small molecules that disrupt this interaction have been shown to display antiviral activity *in vitro*. The figure shows the effect of Hoechst 33258 on the EPR spectra of spin-labeled TAR RNA at three different concentrations. Comment on the results.

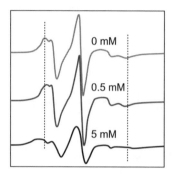

6.2.7 The group of Wayne Hubbell has developed techniques to explore protein dynamics using "nitroxide scanning" where spin labels are attached at a series of specific sites in proteins, changing residues, one at a time, to cysteine. A "scaled mobility factor" is derived, based on the observed width of the central resonance line of the attached spin label. The figure shows such a mobility plot for some residues in bacteriorhodopsin. The diagram on the right shows the corresponding residues to which the label was attached. Explain the results.

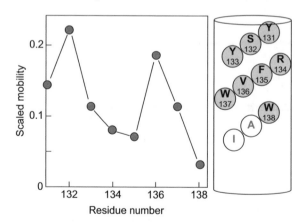

7 Microscopy and single molecule studies

"I ... saw, with great wonder, that ... there were many very little living animalcules, very prettily a-moving"

Antonie van Leeuwenhoek, on observing pond water with a home-made microscope in 1675

"Until 1930 or thereabout... the cell was as distant from us, as the stars and galaxies were from astronomers...... the microscope ... had ceased to be of any use, having reached, the theoretical limits of its resolving power."

Albert Claude, 1974

"Until not very long ago, it was widely accepted that lens-based optical microscopes cannot visualize details much finer than about half the wavelength of light. The advent of viable physical concepts for overcoming the limiting role of diffraction in the early 1990s set off a quest that has led to readily applicable and widely accessible fluorescence microscopes with nanoscale spatial resolution."

Stefan Hell, 2009

"Single-molecule techniques have emerged as powerful tools for deciphering mechanistic details of transcription and have yielded discoveries that would otherwise have been impossible to make through the use of more traditional biochemical and/or biophysical techniques."

Eric Greene, 2011

This section discusses microscopy and some classes of single molecule techniques.

A microscope (from the Greek: *mikrós*, "small" and *skopeîn*, "to see") is an instrument used to see objects too small for the naked eye. Optical and electron microscopes produce images with a magnification up to 1000- and 100 000-fold, respectively. Atomic force microscopes (AFM), developed in the mid 1980s, use a sharp probe to scan and magnify surface features of a specimen.

A new range of experiments has emerged in recent years that can investigate the properties of single molecules, an approach that gives new information compared to studies of a bulk collection of molecules, where only average characteristics can be measured.

7.1 Microscopy

Introduction

The main purpose of a microscope is to *magnify* an object and to *resolve* as much detail as possible in the resulting image. Magnification alone does not necessarily produce a more detailed picture as other factors are also important, such as the wavelength of the observing radiation and contrast variation between different parts of the specimen.

The first recorded user of a magnifying lens was Alhazan (Ibn al-Haitham; 965–1040) at a time when science flourished in the Arab world. Antonie van Leeuwenhoek (1623–1723) observed single living cells in pond water and a variety of other liquids with a single lens microscope that he had ground himself. Robert Hooke (1633–1703) was the first to use the word "cell" when he described the structure he observed in cork with a microscope. Microscopy improved significantly over the next 200 years and Ernst Abbé (1840–1905) developed a theory showing that the resolving power of microscopes was related to the wavelength of the observing light. In general, the shorter the wavelength the more detailed the image will be. Abbé's collaborators, notably Carl Zeiss (1816–1888) and Otto Schott (1851–1935), improved lenses to the point where they could just about achieve Abbé's calculated resolution limit (~200 nm). Pioneers also improved selective staining procedures that allowed them better to visualize different cellular components and, by the end of the nineteenth century, many important features of cell structure and neuronal networks had begun to emerge. Improvements continued to be made in optical microscopes; for example, Frits Zernicke developed the phase contrast microscope in the 1930s but, in the early 20th century, the ability to view the detailed structure of cells remained limited.

This situation was changed by the introduction of the electron microscope (EM). Electrons are particles, but they act like waves. Lenses for electrons can be made. Based on this principle, the first electron microscopes were constructed in the 1930s by Ernst Ruska and colleagues with lenses that "bent" the electron path with applied magnetic or electric fields. Electron microscopes have a theoretical resolving power much better than 1 nm, but technical problems (e.g. a requirement for an evacuated sample chamber) mean that this is hard to achieve for biological samples. In early EM experiments, however, Albert Claude, George Palade, and others used metal staining and chemicals, such as glutaraldehyde, to "fix" cells; this approach led to the discovery of many new cellular substructures and compartments. More recently, there have been other important technical advances in EM. As we will see, these are giving us valuable images of large macromolecules, macromolecular assemblies, and whole cells.

Light microscopy has also undergone a remarkable renaissance in the last 20 years. Beautiful images of fluorescent macromolecules in their cellular location are becoming familiar, as are "movies" of living cells, obtained using time lapse photography. These developments have been enabled by exploiting and developing fluorescent molecules (e.g. GFP, discussed in Chapter 5.5) and new microscope technology. Clever methods to select sub-sections of the sample for viewing and devices to get around wavelength limits mean that it is now possible to see detail in a sample well beyond that predicted by the Abbé "limit" with the fluorescence microscope. Relatively few microscopy images are reproduced here and the reader is directed to standard colored textbooks and excellent websites for image galleries (see list at end of this chapter and "Online resources").

While most microscopes use a series of lenses to produce the magnified images, scanning microscopes, developed in the 1980s, can now also produce high-quality images by scanning probes over the surface of a specimen.

Factors that influence resolution

In addition to technical considerations, such as the mechanical construction and quality of microscope lenses, the effective resolving power of a microscope is defined by diffraction effects. When radiation strikes an object, it is diffracted (scattered); for example, light passing through a small circular hole produces an **Airy disc** rather than a perfect circle of light (Figure 7.1.1). The diffraction patterns in Figure 7.1.1 illustrate the fact that any imaging system is imperfect, with the observed response being instrument dependent. The observed image of a point source is called the **point spread function** (this is related to the **contrast transfer function** used in electron microscopy—see Box 7.2). The simple treatment of diffraction at a slit in Box 7.1 shows that these effects arise from optical interference and they depend on both the wavelength (λ) and the slit width (a) or aperture.

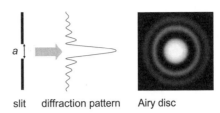

Figure 7.1.1 The diffraction pattern formed by a slit, width a, is shown in the center; the arrow represents a Fourier transform (Tutorial 6). An Airy disc, which is the diffraction pattern obtained from a circular aperture, is shown on the right (see also Box 7.2).

slit   diffraction pattern   Airy disc

the patterns are distinguishable when the first minimum of one diffraction pattern falls on the central maximum of the other (Figure 7.1.2).

The Rayleigh criterion and the relationship derived in Box 7.1 implies that the resolving power will be related to λ/a. In fact, Abbé showed that the resolving power of a circular aperture is $0.61\,\lambda/n\sin\theta$, where θ is the angle illustrated in Figure 7.1.3 and n is the refractive index of the medium between the specimen and the objective lens; $n\sin\theta$ is called the numerical aperture (NA). The Abbé limit can thus be written as $0.61\lambda/\mathrm{NA}$. For a light microscope, oil with high

---

Box 7.1 Diffraction at a slit

Some more insight into diffraction effects can be obtained by considering Figure A. This shows light rays scattered from the two edges of a slit, width a. The path difference between these two rays is p. For small θ, $p/a = d/D$, so that $p = ad/D$. The two scattered rays will add when $p = n\lambda$ ($n = 1, 2, 3$, etc.), thus $n\lambda = ad/D$. The position of the first maximum will thus occur when $d = \lambda D/a$ and the oscillation frequency at the edges of the diffraction pattern will depend on λ/a.

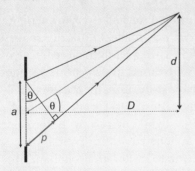

Figure A Diagram illustrating the path difference (p in cyan) between beams scattered from the edges of a slit of width a.

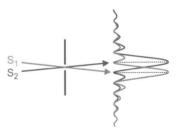

Figure 7.1.2 The Rayleigh criterion for resolving two objects S_1 and S_2.

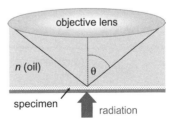

Figure 7.1.3 Definition of the numerical aperture (see also Figure 7.1.4). n is the refractive index (see Chapter 4.2) of the medium between lens and specimen.

The precise position at which it is possible to resolve two objects is somewhat uncertain, but the **Rayleigh criterion** is usually adopted; this states that

refractive index can be used, and θ can be large, with sin θ values of over 0.9 being achievable; this gives a maximum practical value of NA of ~1.5. Inserting this value of NA in Abbé's formula predicts a practical resolution limit of ~200 nm for light when λ ~ 500 nm.

The optical microscope

Many different lens arrangements are possible in microscopes. As well as magnification, a major consideration is to find an illumination scheme that maximizes the contrast generated in the sample. Light can be transmitted through the specimen or reflected by it. Fluorescence methods where molecules in the specimen emit light have especially good contrast properties (see below).

Light passes through the sample in a transmission microscope. There are two main lenses—an objective lens and a projector lens (see Figure 7.1.4). The overall magnification is the product of the magnification by each of these two lenses. Optimum illumination of the specimen with a condenser lens is also very important.

Lenses

A lens causes parallel light rays to focus at a point a distance f from the center of the lens (Box 4.4).

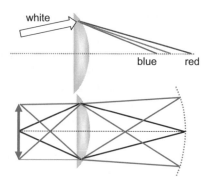

Figure 7.1.5 Illustration of chromatic (top) and spherical (bottom) aberrations in a lens.

Lenses are imperfect and usually have a variety of aberrations associated with them, additional to the diffraction effects discussed above.

Chromatic aberrations arise when different wavelengths are refracted through the lens differently, each wavelength having a different focal length (Figure 7.1.5). This problem can be greatly reduced by using a lens made up from layers of glass with different refractive indices (see Figure 4.2.5).

Spherical aberrations arise when the lens geometry is such that there are different focal lengths for rays striking different parts of the lens. These can be corrected by careful shaping of the lens or by using several separate weaker lenses. Another method of reducing aberrations is to reduce the aperture but, as we have seen, this reduces the resolving power.

Sample preparation for light microscopy

While we would generally wish to interfere as little as possible with a sample, in practice some degree of preparation is frequently required. Chemical fixatives are often used to prevent degradation and to maintain the structure of the cells, especially in studies of tissues in clinical work. The most common fixative for light microscopy is formaldehyde in phosphate-buffered saline. Typically, the sample is then dehydrated with an alcohol and the water is replaced with paraffin wax that solidifies, thus allowing thin sections to be cut. For light microscopy, a knife mounted in a microtome is used to cut ~5 μm thick tissue sections, which are then mounted on a glass slide.

Staining is employed to highlight particular features of interest. Hematoxylin and eosin are commonly

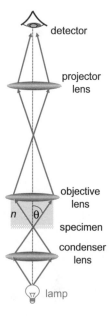

Figure 7.1.4 Simplified diagram of an optical microscope operating in transmission mode.

used stains; hematoxylin, a basic, positively-charged dye stains nuclei blue owing to an affinity for nucleic acids in the cell nucleus; eosin, an acidic dye, stains the cytoplasm pink. Methylene blue also binds nucleic acids. Rhodamine is often used to stain proteins and Nile red stains lipids.

The discipline of histology grew from the study of structures seen under the microscope. Histopathology, the study of diseased tissue with a microscope, is an essential tool in the diagnosis of cancer and other diseases. Histochemistry refers to the use of chemical reactions to distinguish components within tissue. A commonly performed histochemical technique is the Perls' Prussian blue reaction, used to demonstrate iron deposits in diseases like hemochromatosis. Histology samples can also be examined as an autoradiograph, an image on an X-ray film or on a nuclear emulsion, produced by a radioactive sample.

Contrast in the light microscope

Differences in the signal from different regions of a sample define the contrast in a microscope image. In light microscopy, contrast can arise from differential absorption from different regions of the sample (e.g. nucleus and cytoplasm); this contrast can be very high, especially with staining. Support systems and solvents that are transparent to light can readily be found and this helps produce good images.

A wide range of detection and illumination methods can also be used; the simplest is bright field where the sample is illuminated by white light from below and observed above. This method can detect absorbance differences, but factors other than absorption are often better for observing features of unstained samples. Phase-contrast and interference microscopes, for example, detect differences in refractive index (see below). Oblique illumination can also help improve contrast.

The phase-contrast microscope, developed by Zernicke in the 1930s, uses optics that highlight refractive index and scattering differences in a sample. As illustrated in Figure 7.1.6, the specimen scatters light (shown in gray) and there are direct transmitted rays (shown in cyan) that do not interact with the sample. The light scattered by the sample does not pass through the phase-plate annulus and it

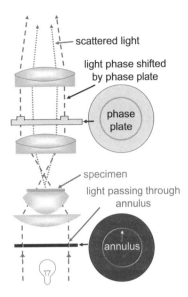

Figure 7.1.6 The phase-contrast microscope, showing direct rays passing through the annulus and the phase plate (cyan dashed lines) and light scattered by the specimen (light gray dotted lines).

is 90° out of phase with respect to the direct beam (see Tutorial 7; Chapters 4.1 and 4.2). The direct beam, which is a hollow cone of light produced by the annular mask, is deliberately retarded by 90° by the phase plate. There is, therefore, now a 180° phase shift between the direct and scattered rays. If the amplitude of the direct rays is adjusted, e.g. by a filter on the annulus, the net result is cancellation of the scattered and direct light waves and a resulting dark background. With a non-uniform specimen, variations in refractive index (scattering) result in some regions of the sample appearing bright in the image because the condition for destructive interference is removed.

Although historically and conceptually important, several improvements and variations have been introduced to phase-contrast microscopy. One of these is differential interference contrast (DIC) or Nomarski microscopy. In DIC, polarized light is passed through a prism below the condenser lens so that the beam is split into two. The two beams pass through the sample and their path lengths are altered by the variations in the specimen's refractive indices. The optics then recombines the two beams in a way that emphasizes these differences. DIC microscopy causes one side of an object to appear bright while the other side appears darker giving a pseudo three dimensional effect (Figure 7.1.7).

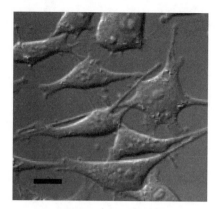

Figure 7.1.7 A DIC microscope image of HeLa cells (kindly provided by Mark Howarth). The scale bar is 10 μm.

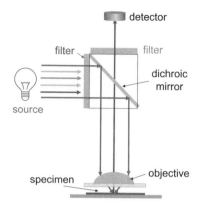

Figure 7.1.8 A schematic view of the optical system in a fluorescence microscope.

The fluorescence microscope

In recent years, application of fluorescence microscopy in the medical and biological sciences has expanded with astonishingly rapidity. One reason is that fluorescence microscopes have excellent contrast properties because light is detected from specific fluorophores at a wavelength that is different from the source illumination. The principles and many applications of fluorescence have already been discussed in Chapter 5.5. Information was also given about the wide range of currently available fluorophores and fluorescent proteins.

Some dyes bind to specific cellular compartments and fluorescently-labeled antibodies can be microinjected into cells. Antibodies are widely used to visualize specific components in a cell (see immunofluorescence discussed in Box 5.4). For fluorescence microscopy, a common DNA stain is DAPI (4′,6-diamidino-2-phenylindole) with a blue emission maximum. The ability to attach fluorescent proteins like GFP to other proteins by genetic manipulation has also been a major advance. Using these labels, fluorescence microscopy can define the precise location of intracellular components, as well as measure their diffusion properties and their interactions with other molecules. It can do this with extraordinary sensitivity—down to the single molecule level, as will be illustrated in Chapter 7.2.

A simplified schematic view of a fluorescence microscope is shown in Figure 7.1.8. This is often called an epi-fluorescence microscope ("epi" = above). As excitation and emission wavelengths are different, only the fluorescent light emitted by the sample reaches the detector, if suitable light filters are used. The fluorescence is produced using higher intensity light sources than in conventional light microscopes. A dichroic mirror which can both reflect and transmit light separates the excitation and emission light paths. These mirrors are usually incorporated in an exchangeable cube in the microscope along with associated excitation and emission filters for a particular wavelength range. The modern fluorescence microscope is under computer control, with digital images acquired using photomultipliers or charge coupled devices (CCD) (Chapter 4.3).

A significant feature of fluorescence microscopy is photobleaching, when the fluorophore is destroyed by the applied light. Loss of activity caused by photobleaching can be controlled (i) by reducing the intensity or time of light exposure, (ii) by increasing the concentration of fluorophores, or (iii) by employing fluorophores that are less prone to bleaching (e.g. Alexa Fluors, DyLight Fluors, or quantum dots). While photobleaching can be a problem, it can also be exploited to give information about diffusion and other properties of fluorophores, as illustrated below.

FRAP and FLIP

FRAP (fluorescence recovery after photobleaching) is the recovery of fluorescence in a defined region of a sample after photobleaching. The FRAP effect results from the movement of unbleached fluorophores from the surrounding area into the bleached region (Figure 7.1.9). The fluorophore in the FRAP experiment can be a lipid or a protein that has been specifically labeled with a fluorophore, e.g. by immunofluorescence (Box 5.4) or by attaching GFP. A selected area of a cell is briefly illuminated with high intensity laser light and the subsequent recovery of fluorescence is monitored as a function of time. The regeneration time is a measure of the fluorophore diffusion rate (see Problem 7.1.2). Sometimes the recovery of fluorescence is incomplete; this probably indicates the presence of an immobile population of proteins bound to structures within the photobleached region, e.g. the cytoskeleton.

Fluorescence loss in photobleaching (FLIP) is a complementary technique to FRAP that measures the decrease or disappearance of fluorescence in a defined region *adjacent* to a bleached region. The progress of fluorescence decay in the area adjacent to a bleached area is monitored as a function of time.

Both FRAP and FLIP can be used to measure molecular mobility, e.g. in diffusion, transport or any other kind of movement of fluorescence-labeled molecules in membranes or living cells.

The confocal microscope

A number of major developments and technical advances in fluorescence microscopy techniques have taken place in recent decades; these have greatly improved the quality of images that we can obtain. Marvin Minsky suggested a modification in 1957 which led to the development of the **confocal microscope**. Normally, when an object is imaged the signal produced is from the full thickness of the specimen with much of it out-of-focus. The confocal microscope eliminates this out-of-focus information with a "pinhole" in front of the image plane. Light from above and below the focal plane of interest does not appear in the final image (Figure 7.1.10). The ability of the instrument to exclude the out-of focus flare leads to remarkably clear images and there has been an explosion in confocal applications in recent years.

The confocal microscope selects a particular focal plane for viewing—often called an optical "section" (Figure 7.1.10). In a conventional wide-field microscope, the entire specimen is bathed in light, usually from a mercury or xenon source. In contrast, image formation in a confocal microscope is achieved by scanning a beam of laser light back and forward across the specimen (Figure 7.1.11).

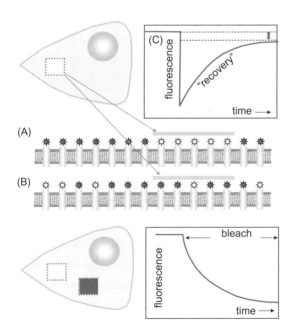

Figure 7.1.9 Illustration of FRAP (top). The white area surrounded by a dotted line is subjected to strong irradiation that causes photobleaching of surface membrane receptors in that region (A). The same area is used to monitor recovery (B and C). The double headed arrow in C corresponds to an immobile fraction in the FRAP experiment. In FLIP (bottom), the white area is irradiated and the cyan area is monitored.

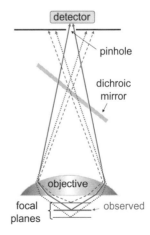

Figure 7.1.10 Confocal microscope. Fluorescent light that originates from outside the focal plane of the objective is rejected by the pinhole (dashed lines in diagram).

Figure 7.1.11 Illustration of the principle of a scanning confocal microscope. Conventional wide-field illumination is shown on the left and the point scanning confocal method on the right.

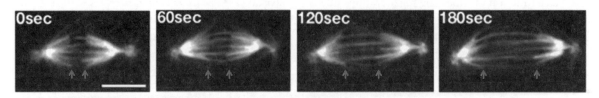

Figure 7.1.12 Time-lapse spinning disc confocal images of chromosome separation in a eukaryotic cell. Microtubules were visualized with GFP-labelled tubulin. The cyan arrows track the segregating chromatids, which appear as dark objects. The scale bar is 5 µm (adapted with permission from *Mol. Biol. Cell* 20, 4696–705 (2009)).

Scanning of the laser spot can be achieved in various ways but a common method involves the use of a spinning disc with holes in it to pass light. Examples of confocal images produced with a spinning disc are shown in Figure 7.1.12.

Total internal reflection fluorescence (TIRF)

TIRF microscopy exploits the properties of evanescent waves. As explained in Chapter 4.2, illumination that is totally internally reflected generates an evanescent wave which rapidly decays in the sample after only ~100 nm. The evanescent wave can be used to excite, selectively, a few fluorescent molecules near the surface of the coverslip (Figure 7.1.13). This is a powerful tool for visualizing events that take place near the surface of a cell, for example cell adhesion. TIRF is often used when imaging single molecules because the small excitation volume reduces the background fluorescence (Chapter 7.2).

Two-photon microscopy

It is possible to generate fluorescence by excitation that involves two or more photons rather than one (Figure 7.1.14). Multi-photon absorption was predicted to be possible in 1930, and was first demonstrated in the 1960s. Such a multi-photon absorption

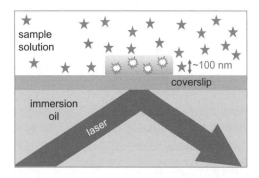

Figure 7.1.13 Illustration of the illumination method in the TIRF microscope. Only a few molecules exposed to the evanescent wave (shown above the coverslip) are excited.

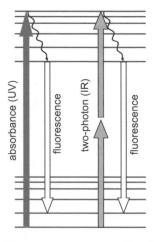

Figure 7.1.14 Illustration of one photon (cyan) and two-photon (gray) excitation of fluorescence. The radiationless transition from the excited to lower vibrational states is the same as that discussed in Chapter 5.5.

process has relatively low probability, so a very high density of photons is required to cause the event. A high-energy infrared laser, delivering bursts of short (~1 ps) pulses at high frequencies, is the usual source. As shown in the Figure 7.1.14, the fluorescent emission after two-photon IR excitation is the same as when UV excitation is used. The infrared radiation used for excitation in two-photon imaging penetrates most samples more efficiently than the shorter wavelengths used in conventional irradiation.

Images are built up by scanning the laser illumination across the specimen, as in the confocal laser-scanning microscope. The high intensity illumination necessary for two-photon excitation is, however, only achieved within a small volume. Thus, there is essentially no fluorescence from outside the focal plane and selectivity (optical sectioning) is obtained intrinsically by the two-photon effect, rather than by pinholes as in the confocal laser-scanning microscope. Two-photon imaging is thus particularly useful for imaging "thick" samples, such as found in tissues or tumors. Photon damage at the focal plane still occurs, but damage above and below the focal plane is avoided.

Fluorescence lifetime imaging microscopy (FLIM)

FLIM is a technique that maps the spatial distribution of fluorescent lifetimes within a microscopic image of fixed or living cells. If a population of fluorophores is excited with separated short laser pulses, an image can be produced that contains information about the different fluorophore decay rates. As lifetime is very sensitive to environment, FLIM images can resolve features not detectable using fluorescence intensity measurements alone. Differences detected in lifetimes can also be related to Förster resonance energy transfer (FRET) to give useful information about the separation between donor and an acceptor fluorophores in an image (see Chapter 5.5).

Super-resolution fluorescence microscopy

In confocal microscopy the specimen is illuminated by a laser-produced spot of light that is focused by the objective lens and scanned to produce the image. The spot size is controlled by the numerical aperture of the objective lens and the wavelength of the laser

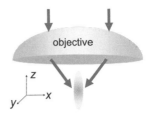

Figure 7.1.15 Generation by an objective lens of a diffraction limited spot, showing the x, y, and z directions.

used but diffraction limitations (Figure 7.1.1) result in finite spot dimensions both in the x–y plane (Figure 7.1.15) and the optical axis, the z direction. The dimensions along z are even less well defined than in the x–y plane.

The spot produced thus has a three-dimensional point spread function (PSF) with a finite diffraction-limited size. Such a spot produces, when scanned, images with a resolution only slightly better than the calculated limit for conventional optical microscopes. It is, however, possible to produce spots with better PSF profiles. Stefan Hell and others have developed various "tricks" to improve the spot characteristics in both the x–y and the z directions (see Journal Club). Here, we simply illustrate one approach that reduces the effective PSF dimensions in the x–y plane. This technique, called stimulated emission depletion (STED) microscopy, can be used to achieve ~30 nm resolution, which is significantly better than the calculated Abbé resolution limit.

In STED, the sample is not only irradiated with a normal excitation spot, as in Figure 7.1.15, but also with a ring of light at a wavelength that stimulates depopulation of some of the excited fluorophores (stimulated emission—see discussion in Box 5.2). The STED light ring, applied at an appropriate wavelength, effectively quenches the fluorescence from that area so that the combined irradiation from the ring and the objective lens gives a smaller spot (Figure 7.1.16).

Some of the "super-resolution" methods, like STED, result in low light levels reaching the detector. More time is then required to collect a digitized image with an acceptable signal to noise ratio. More sensitive detectors or increasing the intensity of the illuminating laser source can compensate for this effect, but increased intensity risks photobleaching the specimen. Rapid data collection is often a prerequisite, for example when monitoring dynamic

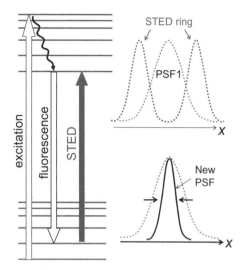

Figure 7.1.16 The basis of STED microscopy. A ring of light (cyan) is applied at a wavelength that stimulates depopulation of the fluorescent excited state. The sum of the depletion ring plus the normal point spread function gives a new, smaller PSF profile, thus reducing the effective size of the excitation spot.

molecular events in a cell—and some resolution may have to be sacrificed to achieve sufficiently fast data collection. In other words, a compromise has to be reached between resolution achieved and image collection time.

A brief discussion of another super-resolution technique, **photoactivated localization microscopy (PALM)**, is also given in Figure 7.2.6 in the next chapter and the topic appears in Journal Club.

Other forms of optical microscopy

Various other detection methods, apart from fluorescence, can be used in microscopes. Vibration spectroscopy (involving IR detection or Raman scattering) was introduced in Chapter 5.2. This and other kinds of spectroscopy can be incorporated in a microscope. Raman microscopy is perhaps the most widely applied. Light scattering is induced by illuminating a sample with a focused laser (e.g. 1064 nm Nd:YAG laser). Scanning the laser beam over the sample produces an image with information about molecular vibrations in defined small volumes (femtoliter). Raman microscopy needs no special sample preparation and it is possible to perform detailed analysis of cells in their natural state. Changes in cells, including bacteria and eukaryotes can be monitored as a function of time and comparisons between healthy and diseased tissue states made.

Electron microscopy

The electron microscope (EM) was developed as a result of the limitations of light microscopes. The transmission electron microscope (TEM) was the first type of EM to be constructed using a design that was based on light microscopes except that a focused beam of electrons was used instead of light to see through the specimen. The first designs were by Ruska and colleagues in the 1930s. As already mentioned, the EM has been very important for obtaining images of biological samples; images that have contributed greatly to our understanding of cellular structure. More and more information is also now being derived about large molecules and their complexes in a relatively new technology that we call "molecular electron microscopy" here. There are two main classes of electron microscope—transmission and scanning. Scanning electron microscopes were developed after TEMs, and we first deal with the transmission type here.

Transmission electron microscopy

Instrumentation

A simplified version of a transmission electron microscope (TEM) is illustrated in Figure 7.1.17. Electrons are produced by emission "guns" and are attracted to a positive anode by a very high voltage. Electrostatic and/or electromagnetic lenses are then used to focus the electron beam and form an image. In addition to the objective lens, there is usually an intermediate lens (not shown in Figure 7.1.17) and a final projector lens which projects the image on a detector.

The electrons for TEMs are produced by two main types of "gun". One is **thermionic**, where a filament, usually made from tungsten or lanthanum hexaboride (LaB_6), is heated to a high temperature. In contrast, **field-emission guns** (FEGs) emit electrons from a cold, sharp-pointed cathode, often made from tungsten. A FEG can produce a more intense,

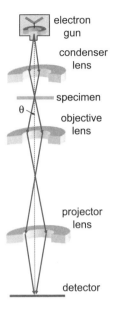

Figure 7.1.17 A simplified version of a transmission electron microscope. Compare the light microscope diagram in Figure 7.1.4 (note the light path is inverted). θ refers to the aperture angle (Figure 7.1.3).

more coherent, and smaller diameter beam than a thermionic gun. The EM must be evacuated because electrons interact strongly with all materials. This means that the samples have to be introduced via special airlock systems. FEGs have higher vacuum requirements than thermionic guns. Note that **radiation damage** in EMs is a serious problem, as exposure to an electron beam causes significant damage to delicate biological structures.

Electrons can be deflected by both electric and magnetic fields, although most **electron lenses** are magnetic and are made from a solenoid (Chapter 6.1; Box 6.1). These lenses deflect electron beams in a manner analogous to a glass lens "bending" light by refraction. As the electrons move through the lens, the magnetic forces cause them to spiral around and come to a focus. Note that the current supplied to the solenoid controls the electron lens properties in a smooth fashion. In contrast, a solid lens, e.g. glass, would be very coarse on a nm scale because of its atomic structure.

Detectors in the EMs can be charge coupled devices (CCD—see Chapter 4.3) which have good performance for lower resolution images but photographic film is still best for the highest resolution images.

Wavelength and resolution in the electron microscope

We saw in our discussion of resolution limits in the optical microscopy section that wavelength is an important parameter for defining the attainable resolution. What wavelengths can be achieved in an EM? When an electron, charge e, passes through a voltage difference, V, its energy is eV (Tutorial 3); this is equal to the kinetic energy, $1/2\,mv^2$, where m = mass and v = velocity of the electron. We also have the wave particle duality formula, derived by de Broglie: $\lambda = h/mv$, where h is Planck's constant (see Appendix A1.3 and Tutorial 9). Ignoring relativistic effects, although these can be important as the electron approaches the velocity of light, these equations lead to the approximate formula: $\lambda = 1.22/(V)^{1/2}$ nm (see Problem 7.1.1). For an accelerating voltage of 100 kV, this formula gives $\lambda = 0.0039$ nm—a value that can be substituted into Abbé's equation ($0.61\lambda/$NA) to estimate the achievable resolution.

NA values of ~1.5 are achievable for optical lenses. In contrast, a narrow beam with a very small value of numerical aperture (NA) has to be used in an EM, because of lens aberrations and other technical difficulties. As $n \sim 1$ for a vacuum and an angle θ of only around 0.5° is used in a transmission EM, this gives NA values as low as ~0.009. The predicted resolution for 100 kV is thus ~0.26 nm. In fact, the resolution attainable for biological samples is usually considerably worse than this because of various difficulties with sample preparation and signal to noise limitations (see below). Note that very small NA values mean that there is a relatively large range over which the image is approximately in focus.

Contrast in the EM

The main contrast in an EM arises from differential scattering between transmitted and scattered electrons from different regions of the specimen. There are two important kinds of electron scattering: (i) from nuclei, where there can be large deviations but little loss of energy (elastic scattering), and (ii) from electrons in the sample, where the deviations are small but the transmitted electrons do lose energy (inelastic scattering). Inelastically scattered electrons can give **phase contrast** in thin specimens as the lost energy results in electrons with a different wavelength with different phase compared to the

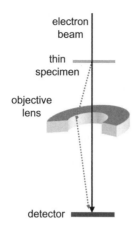

Figure 7.1.18 Illustration of contrast generation in TEM. Some electrons are scattered inelastically (cyan arrows) and are thus phase shifted compared to the unscattered beam (black arrow). This results in interference and contrast generation.

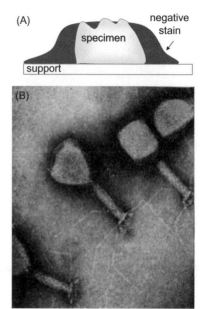

Figure 7.1.20 Negative staining. (A) Illustration of the distribution of stain around the sample. (B) T4 virus negatively stained with uranyl acetate (courtesy Michael Moody).

transmitted electrons in the main beam (Figure 7.1.18; Chapter 4.1).

Sample preparation for electron microscopy

Sample preparation can be quite elaborate in EM studies as harsh treatment (vacuum plus electron irradiation) requires good preservation. As with the optical microscope, chemical fixation is often used to stabilize the specimen by chemical crosslinking; for example, proteins are stabilized by glutaraldehyde and lipids by osmium tetroxide (Figure 7.1.19).

Fixation is followed by dehydration with organic solvents; the specimen is then freeze dried and/or *embedded* in a resin for sectioning. A resin rather than paraffin wax is used because it is tough enough to allow production of very thin sections with a microtome—60–90 nm thick. Because the intrinsic contrast in the EM is low, staining is routine. The most common stains are metals such as tungsten, lead, and uranium; osmium tetroxide acts both as a stain and a fixative. Specimens can either be stained before embedding in the resin or after they have been sectioned by the microtome. Uranyl salts sometimes bind preferentially to macromolecules, while lead

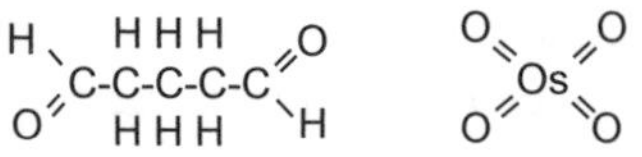

salts tend to bind to membranes. If the sample of interest binds the heavy metal, it is said to be positively stained. More commonly, if the heavy atoms bind to the background and not to the object of interest, the sample is then said to be negatively stained (Figure 7.1.20). Negative staining can result in a distortion of the molecule's shape and dehydration can cause the structure to collapse.

A preparation method that is particularly useful for examining lipid membranes and their incorporated proteins is freeze fracture or freeze etching. Fresh tissue or a cell suspension is frozen rapidly and then fractured with a microtome while still at liquid nitrogen temperature. The fracture occurs along the membrane plane because the forces holding the outer and inner halves of the bilayer together are relatively weak. One of the two cold-fractured inner surfaces is then coated with carbon, evaporated perpendicular to the average surface plane in a high vacuum evaporator (Figure 7.1.21C). This is followed by shadowing with evaporated platinum or gold at an angle of ~45° (Figure 7.1.21D). After shadowing, the underlying biological material is removed from the fragile metal replica by chemical digestion with acid or other reagents. The replica is thoroughly washed from residual chemicals, floated on water, and then

Figure 7.1.19 Glutaraldehyde and osmium tetroxide.

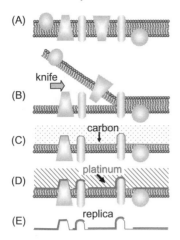

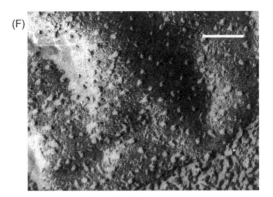

Figure 7.1.21 The technique of freeze fracture. (A) A membrane containing embedded proteins is rapidly frozen and (B) fractured with a knife; (C) one inner surface is exposed to a layer of evaporated carbon in a vacuum followed by; (D) platinum shadowing at an angle of 45°; (E) the biological material is digested away leaving a replica of the membrane. (F) An example, showing a freeze fracture replica of a membrane vesicle with proteins appearing like knobs; the bar length is equivalent to 0.5 μm.

lifted off by a fine copper grid, dried, and viewed in the transmission EM (Figure 7.1.21).

Cryo-fixation

So far, we have mainly described "traditional" electron microscopy, which visualizes the structure of stained, fixed cells or tissues with resolution of, at best, ~2 nm. To achieve higher resolution and better representation of structures it is important to avoid fixation and metal staining. A significant improvement over the chemical fixation methods discussed above is cryo-fixation. This is where a sample is rapidly cooled to liquid nitrogen (77 K) temperatures. The specimen is preserved in a snapshot of its solution state by cooling so rapidly that the water forms a vitreous (non-crystalline) solid. A solution containing the particles of interest is suspended as a thin layer across a hole in an EM grid. The grid is held in tweezers mounted on a plunger that is allowed to fall into a bath of liquid ethane at liquid-nitrogen temperature (Figure 7.1.22). The very rapid drop in temperature turns the water into a vitreous solid that has properties more like liquid water than a crystalline state. Provided the specimen is maintained at very low temperatures, this form of water is stable.

The discipline of cryo-electron microscopy has evolved from this technique, which was first demonstrated by Jacques Dubochet in 1982. It is now

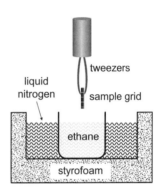

Figure 7.1.22 Illustration of a plunger for cooling cryo-EM grids. The sample is applied to the grid which is then suspended by fine tweezers and plunged rapidly into liquid ethane.

possible to observe samples from virtually any biological specimen *close to their native state* as specimens need not be stained or fixed in any way. Another advantage of operating at very low temperatures is that radiation damage from the electron beam is reduced.

Tomography

If electron micrographs are recorded with the specimen tilted at different angles with respect to the electron beam, then a series of different projections can be generated (Figure 7.1.23). Not all angles can be covered; the tilt angles are usually restricted, for

(A)

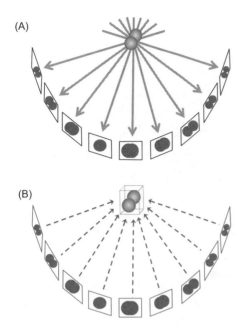

(B)

Figure 7.1.23 Tomography. A series of different projections of the object can be obtained by tilting the sample with respect to the transmission beam (top). The multiple images are then combined to give a three dimensional representation of the object (bottom).

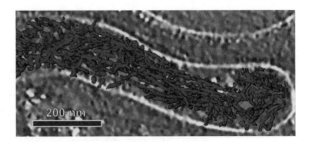

Figure 7.1.24 Image of a filopodium from a *Dictyostelium* cell that was subjected to cryo-electron tomography. The actin network is highlighted in cyan overlaid on a tomographic slice (adapted with permission from *Curr. Biol.* 17 (9) 79–84 (2007)).

(Figure 7.1.24). Tomography is also used in 2D electron crystallography, as described below and, relatively recently, X-ray tomography has become a viable method (Box 7.2).

Example Actin in filopodia

Filopodia are finger-like extensions of cells that form in a migrating cell. Once formed, these filopodia are stabilized by actin filaments. Figure 7.1.24 shows an image of a filopodium, illustrating the kind of resolution achievable using cryo-electron tomography.

practical reasons, to ~ ±60°. This means that, after reconstruction, the resolution in different directions is not the same because data are missing in some regions. The best resolution of the object is obtained for regions with many low angle projection images (e.g. in the plane of a flattened cell or a membrane). The images are digitized, followed by reconstruction. If the object has high symmetry, fewer tilt angles are required. Radiation damage limits the number of projections that can be collected, although, in some cases, it is possible to combine projections from different specimens.

Tomography is particularly effective in electron microscopy because of the high depth of focus (low numerical aperture). EM tomography has emerged as a very powerful technique for structure determination of cellular components which are too large and variable for the single-particle averaging approach. The group of Wolfgang Baumeister has produced remarkably detailed views of single cells, with single ribosomes and the structure of actin filaments resolved to a resolution of a few nanometers. This approach offers the exciting prospect of visualizing the detailed molecular architecture of cells

Box 7.2 X-ray tomography

Tomography using "soft" X-rays rather than electrons provides an alternative method for imaging sub-cellular architecture and organization, particularly in eukaryotic cells. "Soft" X-rays, with wavelengths of a few nanometers, penetrate biological materials more easily than electrons, allowing specimens up to 10 μm thick to be imaged. A spatial resolution of 50 nm or better can be achieved. High-contrast images of intact, fully hydrated cells can be obtained without added contrast-enhancing agents. Soft X-ray microscopes use synchrotron radiation sources (see Chapter 4.3) and zone plate lenses to focus the X-rays. Unlike a conventional light lens, which uses refraction, or an electron lens that uses electric and/or magnetic fields, zone plates rely on diffraction to focus the radiation; they consist of radially symmetric rings, alternating between opaque and transparent. The rings are sometimes known as Fresnel zones after Augustin-Jean Fresnel who invented a lens for lighthouses based on this principle in the early 19th century.

Image processing

Data processing of high-resolution EM data is an important part of image production, especially in tomography and molecular electron microscopy (see below). One use of processing is to improve noisy images. Noise arises from contributions to the image that are not directly from the specimen; it is generated by many factors but background noise levels in an EM are particularly high because very low electron doses have to be applied to avoid damage to the sample; this results in relatively weak signals from the sample compared with random background signals.

It is sometimes possible to improve the signal to noise ratio by averaging images in the electron micrograph when there are multiple copies of an object, or if the sample consists of a periodic or symmetrical array (see, e.g., Figure 7.1.25). As well as averaging, the data can also be filtered in various ways; for example with convolution functions (Tutorial 6). This kind of manipulation can either be done in real space or reciprocal space (Chapter 4.3). In EM studies, real space is the electron image itself while reciprocal space is obtained from the Fourier transformation (FT) of the image. Processing in reciprocal space is very powerful because it highlights periodic structures and allows the data to be manipulated readily.

Other important data processing procedures involve the production of a *corrected* image by taking account of the **contrast transfer function** (CTF). In all microscopes, the recorded image is not a direct representation of the real structure (Box 7.3). Consideration of the relationship between the real image and the one obtained is particularly important in high-resolution electron microscopy. The point spread function and its relative, the CTF, must be estimated and appropriate corrections applied to the images obtained.

Another important procedure in EM data processing is **reconstruction** of an image from multiple projections or views. This aspect of image processing will be discussed briefly below in the single particle studies section.

Molecular electron microscopy

We have discussed above how an EM can give high-resolution images of cell structures. In recent years, it has also been shown that an EM can be used to give detailed structural information about the three-dimensional structure of individual proteins or protein complexes. To achieve the required higher resolution and better representation of the object, staining is avoided; low electron doses and very low temperatures are used to reduce beam damage to the specimen, and computational averaging and extensive image processing are applied to generate the images. Application of this technology allows observations to be made at the molecular level. There are two main techniques in use for molecular electron microscopy: **single-particle studies** and **electron crystallography** of 2D crystals. Single-particle studies, largely pioneered by Joachim Frank, can currently achieve resolution of better than 1 nm while electron crystallography can give atomic resolution structures (0.2 nm). Single-particle electron microscopy is an exciting advance, with great potential for the study of large molecular assemblies. New insights into biological function have been obtained by trapping large molecular machines at different stages of their functioning cycle. The method is especially powerful when used in conjunction with functional

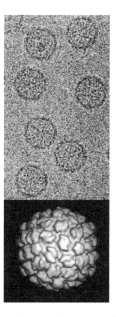

Figure 7.1.25 Cryo-electron microscopy of capsids of bacteriophage HK97. Original micrographs are shown (top). The outside surfaces of averaged reconstructions at ~2.5 nm resolution are shown below (adapted with permission from *J. Mol. Biol.* 253, 86–99 (1995)).

Box 7.3 The contrast transfer function (CTF)

No optical system is perfect—see Figure A. If we describe the image obtained after passing through an optical system as $g(r)$, we can write $g(r) = f(r)*h(r)$ where $*$ is a convolution operation (see Tutorial 6) and $h(r)$ is the point spread function which was mentioned earlier (see Figure 7.1.1). The contrast transfer function (CTF), $H(k)$, is the Fourier transform of $h(r)$, where r and k represent real and reciprocal space, respectively.

The shape of the CTF depends on the characteristics of a particular microscope and parameters that include lens aberrations; normally the only practical variable is the degree of defocus. Depending on the defocus setting, different features of the object appear enhanced or suppressed in the image because the CTF oscillates. A typical CTF is shown in Figure B. When it is negative, positive phase contrast occurs, with features appearing dark on a bright background. When it is positive, there is negative phase contrast, with features appearing bright on a dark background. The achievable resolution is related to the point where the CTF first crosses the k axis (see Figure B). For further details about the CTF see Frank's book listed in "Further reading".

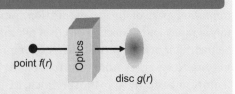

Figure A A point source, $f(r)$, appears as $g(r)$ after passing through an optical system.

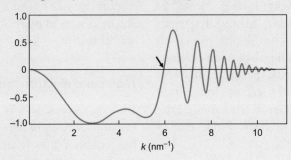

Figure B A CTF function showing the oscillatory behavior. The resolution limit is related to the point where the function crosses the axis as indicated by the arrow.

---

studies and higher resolution information derived from crystallography. Atomic resolution structures of individual components, determined by X-ray crystallography, can be fitted to the molecular shape of the complex determined by cryo-electron microscopy. An excellent example of this is the ribosome, where crystallography and single-particle electron microscopy have been combined to produce beautiful movies of the functioning ribosomal machinery (see websites provided in the online material).

Single-particle studies

Single-particle EM methods determine the structure of macromolecules from images of individual particles or single-particle projections. Unlike X-ray crystallography, crystals are not required and very small amounts of material are needed. The resolution that can be obtained depends on the sample and its symmetry but usually the highest resolution achievable is

of the order of 1 nm. The most informative results are obtained for large (>300 kDa), relatively rigid systems.

After cryo-fixation (see above) the frozen molecules have multiple possible orientations within the vitreous ice layer. The TEM then produces many different projection images of such a set of particles, as illustrated in Figure 7.1.26. The next task is to reconstruct a 3D structure from this set of noisy 2D projection images. This is done by processing the digitized images with a computer. Using two-dimensional alignment and classification techniques, single particles in the same orientation are clustered into classes. These clustered images are then averaged to reduce the noise and their relative 3D orientation is determined; one way to do this is to compare the experimental images with computer-simulated projections. Similar projections are identified, assigned to membership of a "class average" and their spatial

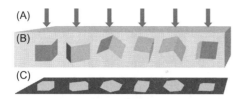

Figure 7.1.26 Schematic diagram of data collection from single particles in vitreous ice. (A) A collection of molecules, randomly distributed in vitreous ice, is exposed to an electron beam. (B) The images generated represent all possible projections of the particle. (C) Images of particles with similar orientations can be found and grouped together using computer-based procedures.

relationships defined. The procedure is illustrated in Figure 7.1.27 for the 80S rat ribosome. Provided the angular distribution is sufficiently uniform, a series of images, each formed from several hundred particles, gives the information necessary to reconstruct a model of the molecule (Figure 7.1.27).

Example   EF-G binding to the ribosome

As discussed in Chapter 4.3, the ribosome structure has been solved by X-ray crystallography, giving extraordinary detail about the cell's protein synthesizing machinery. The multistep details of how mRNA is translated into peptide by the ribosome are, however, still incomplete. We need to know how the ribosome interacts with its various ligands—mRNA, tRNAs, and various other factors. In *E. coli*, one of these other factors is called EF-G. Binding of this protein to the ribosome triggers breakdown of GTP and causes a rotation of the small subunit with respect to the large ribosomal subunit, thus driving the translation process forward. Figure 7.1.28 shows a cryo-EM reconstruction of EF-G binding to the 70S ribosome of *E. coli*. The study in which these images were derived helped develop a model of the molecular events that occur during protein synthesis.

2D electron crystallography

Nigel Unwin and Richard Henderson pioneered the use of 2D electron crystallography in their studies of

bacteriorhodopsin. This involved the collection of electron diffraction data, as well as images. Diffraction data can be collected in the TEM by adjusting the lens arrangement. If the sample is crystalline, Bragg's law applies and certain angles will produce enhanced electron scattering, resulting in a diffraction pattern (see Chapter 4.3). As the diffraction arises from an array of molecules, the signal to noise ratio obtained is much better than from single particles. As discussed in Chapter 4.3, diffraction data have lost phase information that needs to be recovered. In 2D EM crystallography studies, phase information can be extracted from high quality images produced using tomography and other processing methods.

As the first application to bacteriorhodopsin, substantial progress has been made in 2D crystallization methods; cryo-fixation, tomographic imaging technology, and data processing have all contributed to improvements. The main structural success has been with membrane proteins, some of which naturally form 2D crystals in lipid bilayers, e.g. bacteriorhodopsin and the light-harvesting complex, but other highly ordered molecular monolayers have been studied, e.g. Zn^{2+} ions promote monolayer crystal formation of tubulin. It is also possible to study molecules in helical arrays, as done, for example, with the acetylcholine receptor.

An alternative crystallization strategy, applicable to extrinsic membrane proteins, involves the use of lipid monolayers. These provide a suitable surface for the assembly of ordered protein arrays if the proteins have an affinity for the monolayer surface. This affinity may either be natural or induced by appropriate chemical modification of the protein (e.g. a histidine tag, binding to a membrane via a Ni^{2+} ion or binding to a 2D DNA lattice). This "crystallization" technique has already been applied successfully in a number of systems including annexins and α-actinin. As illustrated by a recent structure determination of the membrane water channel protein, aquaporin, it is possible to achieve very high resolution structural information (<0.2 nm) using 2D electron crystallography.

The scanning electron microscope

In the scanning electron microscope (SEM), a fine beam of electrons is scanned back and forth across

---

Note Multiple views/projections in electron microscopy can be obtained either from random orientations of identical particles (Figure 7.1.26) or from tomography (Figure 7.1.24).

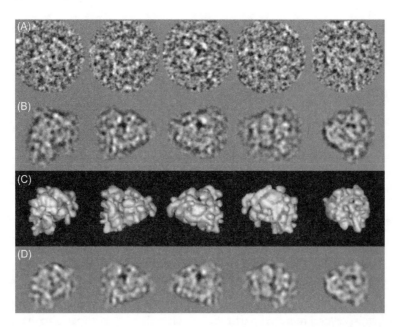

Figure 7.1.27 Processing of images of the 80S rat liver ribosome embedded in vitreous ice. Around 6000 raw EM images (five are shown in (A)) were aligned and grouped into noise reduced class averages (B). These class averages were then used to reconstruct the 3D structure of the ribosome (C). The 3D reconstruction was based on 180 different orientations of the kind shown in (D) (adapted with permission from *Structure* 6, 389–99 (1998)).

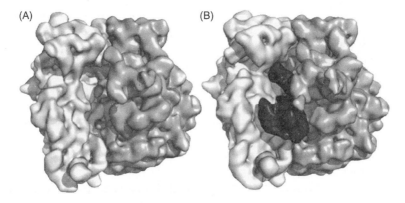

Figure 7.1.28 (A) Cryo-EM reconstruction of the ribosome. The large ribosomal subunit is shown in gray and the small subunit in white. (B) Cryo-EM reconstruction of the EF-G 70S complex, showing EF-G in black (adapted with permission from *PNAS* 104, 19671–8 (2007)).

the sample (Figure 7.1.29). This technique gives striking images with 3D effects (Figure 7.1.30). Unlike transmission microscopes, image magnification in the SEM is not dominated by the lenses; the function of the condenser and objective lenses is to focus the beam to a spot, rather than image the specimen. Magnification depends on the scanning system and the diameter of the electron beam focused on the sample. Living cells and tissues usually require

chemical fixation to preserve and stabilize their structure. The dry specimen is frequently mounted on a specimen stub using an adhesive and coated with gold by "sputtering" in a vacuum before examination in the microscope.

The SEM has a large depth of field, which allows more of a specimen to be in focus at one time. A comparatively large specimen area can be viewed. Another advantage is that analytical techniques can

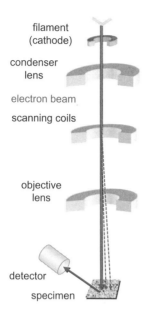

Figure 7.1.29 A schematic view of an SEM. The scanning coils sweep the electron beam back and forward across the specimen, and the resulting secondary radiation is collected and analyzed.

be applied. For example, characteristic X-rays are emitted when the electron beam removes an inner shell electron from the sample, causing a higher energy electron to fill the shell and release energy (see Chapter 5.6). These characteristic X-rays can be used to identify and measure the elemental composition of

the sample. Scanning electron microscopy can achieve ~3 nm resolution of cellular structures. It has also been used to visualize large macromolecular structures, such as nuclear pore complexes.

Scanning probe microscopy

We have now seen two kinds of microscope that scan an irradiation spot over the sample – the fluorescence confocal microscope and the scanning electron microscope. Gerd Binnig and Heinrich Rohrer developed alternative scanning microscopes in the 1980s. Rather than using a beam of light or electrons, these scan a fine tip over a surface. Atomic resolution is possible for "hard" specimens but this is rarely achievable with biological samples.

Scanning probe microscopy is a general term used to describe a variety of techniques that employ a sharp scanning probe to measure some property of a surface. The first developed was **scanning tunneling microscope** which depends on a quantum mechanical effect, but this requires the specimen to have a conducting surface. The **atomic force microscope** (AFM) which was devised to overcome this conductivity problem is the method described here.

The operational principles of an AFM are illustrated in Figure 7.1.31. In most AFMs, a laser is reflected from the back of a reflective cantilever (Figure 7.1.31)

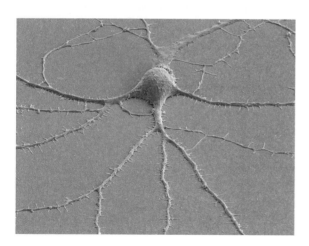

Figure 7.1.30 SEM image of a lone neuron with dendrites radiating from it—compare Figure 3.6.14. This cell was chemically fixed, dried with CO_2, lightly sputter coated with gold/palladium, and imaged at 2 keV (image kindly provided by Thomas Deerinck, Varda Lev-Ram, and Mark Ellisman, NCMIR).

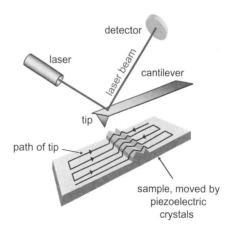

Figure 7.1.31 Schematic view of an atomic force microscope. The sample is moved by a piezoelectric device, so that the tip goes back and forth across the sample. The tip senses the surface properties and cantilever deflections are detected by laser light.

Figure 7.1.32 High-resolution AFM imaging of outer membrane protein OmpF adsorbed on mica. The protein was reconstituted in the presence of lipid as an orthogonal lattice with alternating rows of trimers extracellular side up. The inset shows the pore openings of an OmpF trimer from its extracellular side. The scale bar is 50 nm long (adapted with permission from *Biophys. J.* 96, 329–38 (2009)).

to a position-sensitive detector. AFM tips and cantilevers are fabricated from Si or Si_3N_4. A typical tip radius is ~ 10 nm. A key component in these scanning devices is a piezoelectric material that is used to scan the sample. A voltage applied across a piezoelectric crystal elongates or compresses the crystal at the atomic level. This means that the specimen can be moved with atomic resolution by applying a voltage. A scanner is constructed from three independent piezoelectric crystals, each being responsible for movements along the x, y, or z axes. A typical piezoelectric material used in the AFM is lead zirconium titanate.

Different AFM scanning modes are possible; a particularly useful one for delicate biological samples is the tapping mode, where the cantilever is made to oscillate rapidly. When the scanned oscillating tip comes into contact with the surface a change in the oscillation can be detected. A big advantage of the AFM is ease of sample preparation and the ability to use buffers and near physiological conditions. AFM has been applied to a wide range of problems (see further reading and Figure 7.1.32). It is particularly useful for studying ordered bacterial membranes including bacteriorhodopsin, porins, and aquaporins.

 ## Summary

1 Microscopes magnify an object with several lenses. The usual illumination sources are light or electrons.

2 The resolution of conventional light microscopes is limited by diffraction phenomena to around 200 nm.

3 Contrast in the object can be enhanced in the light microscope by staining with dyes and/or detecting refractive index differences (phase contrast).

4 The fluorescence microscope gives excellent contrast as the sample itself is effectively the light source; a wide range of fluorescent probes are now available. The fluorescence microscope can detect specifically labeled regions of a cell and measure molecular interactions using Förster resonance energy transfer (FRET). Fluorophore movements can be analyzed using photobleaching and particle tracking.

5 A new generation of light microscope is giving improved resolution. Microscope images of a "section" of the specimen can be obtained using confocal, total internal reflectance (TIRF) and two-photon methods. "Super-resolution" techniques that reduce the 200 nm resolution limit are also available.

6 In the electron microscope, the sample is often prepared on some sort of grid, and is fixed, stained, and/or shadowed with metals.

7 The resolution of an electron microscope is usually limited by sample preparation

procedures and radiation damage. New methods have been introduced to obtain much higher resolution images of unstained samples; important technical improvements include rapid cooling, tomography, and extensive image processing procedures.

8 The scanning electron microscope gives images of metal-coated objects with 3D effects.

9 The atomic force microscope scans a probe over the surface of a specimen to give an image.

 ## Further reading

Useful websites

http://www.micro.magnet.fsu.edu/primer/

http://www.itg.uiuc.edu/technology/atlas/

http://www.mpibpc.mpg.de/groups/hell/

http://jcb.rupress.org/site/misc/annotatedvideo.xhtml

http://ncmi.bcm.tmc.edu/homes/wen/ctf/ctfapplet.html;
  contrast transfer function calculation

http://www.mrc-lmb.cam.ac.uk/ribo/homepage/mov_and_overview.html;
  ribosome movie

http://pubs.acs.org/cen/multimedia/85/ribosome/elongationcycle.html;
  ribosome movie

http://www.matter.org.uk/tem/

Optical/fluorescence/super-resolution

Wilt, B.A., Burns, L.D., Wei Ho E.T., Ghosh, K.K., Mukamel, E.A., Schnitzer, M.J. Advances in light microscopy for neuroscience. *Annu. Rev. Neurosci.* 32, 435–506 (2009).

Paddock, S. Over the rainbow: 25 years of confocal imaging. *Biotechniques* 44, 643–4, 646, 648 (2008).

Lichtman, J.W., and Conchello, J.A. Fluorescence microscopy. *Nat. Methods* 2, 910–19 (2005).

Schneckenburger, H. Total internal reflection fluorescence microscopy: technical innovations and novel applications. *Curr. Opin. Biotechnol.* 16, 13–18 (2005).

Goldman, R., Swedlow, J., and Spector, D. *Live Cell Imaging: A Laboratory Manual* (2nd edn) Cold Spring Harbor, 2009.

Electron

Frank, J. *Three-dimensional Electron Microscopy of Macromolecular Assemblies* Oxford University Press, 2006.

Flegler, S.L., Heckman, J.W., and Klomparens, K.L. *Scanning and Transmission Electron Microscopy* Oxford University Press, 1993.

Leis, A., Rockel, B., Andrees, L., and Baumeister, W. Visualizing cells at the nanoscale. *Trends Biochem. Sci.* 34, 60–70 (2009).

Hite, R.K., Raunser, S., and Walz, T. Revival of electron crystallography. *Curr. Opin. Struct. Biol.* 17, 389–95 (2007).

X-ray

McDermott, G., Le Gros, M.A., Knoeche,l C.G., Uchida, M., and Larabell, C.A. Soft X-ray tomography and cryogenic light microscopy: the cool combination in cellular imaging. *Trends Cell Biol.* 19, 587–95 (2009).

AFM

Frederix, P.L., Bosshart, P.D., and Engel, A. Atomic force microscopy of biological membranes. *Biophys. J.* 96, 329–38 (2009).

 ## Problems

7.1.1 The surface of a single fibroblast cell was labeled with fluorescein isothiocyanate (this attaches covalently to proteins). A small area of the surface ($\sim 10^{-7}$ cm^2) was briefly exposed to an intense laser pulse to bleach the fluorophores in that region. The fluorescence emission from the irradiated region was then monitored as a function of time (the same laser was used to provide the monitoring excitation source but at much lower intensity). The fluorescence was observed to increase approximately exponentially with time with a half-life for recovery of 1 min. Estimate the diffusion coefficient for the fluorophores diffusing back into the irradiated region from this FRAP experiment. (The mean free path length $\bar{a}$ of a molecule is $(2Dt)^{1/2}$—see Chapter 3.1; a useful approximation is that half the intensity will reappear when $\bar{a} = r/\sqrt{2}$.)

7.1.2 How does two-photon microscopy differ from confocal laser scanning microscopy?

7.1.3 Derive the approximate relationship $\lambda = 1.22/(V)^{1/2}$ nm for the wavelength of an electron accelerated by a voltage V. What is the electron wavelength in a 400 kV microscope?

7.1.4 The images shown in the figure are three of the projections obtained in EM micrographs of a large complex made from similar domains. Can you guess what structure the assembly has?

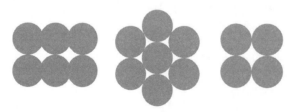

7.1.5 A sample of microtubules consists of tubes, 25 nm in diameter, made up from 13 tubulin monomers around the perimeter of the tube. What sort of EM micrographs do you expect to observe from microtubules?

7.1.6 A cantilever in an AFM has a 0.2 N m^{-1} spring constant (and a tip radius of $\sim$10 nm). How much force does it take to deflect such a cantilever by 1 nm? If an area of 10 µm × 10 µm is scanned with 512 lines and 512 points per line, what is the resolution of the image (in nm)?

7.2 Manipulation and observation of single molecules

Introduction

The last two decades have seen the development of a number of ultrasensitive biophysical methods that allow the manipulation and study of individual molecules. Most techniques measure the properties of an ensemble of many molecules. The ability to watch *one molecule at a time* can give unique information about heterogeneous samples; there is no need to synchronize a population in order to study the molecular kinetics and it is possible to study some of the stochastic (random) aspects of life at the molecular level. Bulk methods tell us about the mean movements of a crowd, which may or may not be useful for finding your brother at a football match; obviously it is better to have a tracking device for him alone.

Most measurement techniques are not sensitive enough to observe single molecules but a few do have this ability. These include mass spectrometry and the patch clamp technique, which were discussed in Section 3, and surface-enhanced versions of Raman spectroscopy—see Figure 5.2.17.

In this chapter, we focus on methods to manipulate single molecules and ways of observing them, such as fluorescence detection. Single molecules can be trapped and manipulated by a beam of laser light (optical "tweezers") or small magnets. Atomic force microscopy can mechanically stretch single molecules and record their elastic responses. Fluorescence is sensitive enough to detect light emission from single molecules in their near physiological state.

Biophysical techniques that allow the manipulation and study of individual biomolecules are relatively novel. The ability they give us to monitor biological processes at this fundamental level is an exciting development and likely to greatly increase our understanding of molecules and cells in the next decade. These methods are particularly valuable for studies of motility, molecular motors, DNA transcription, and protein folding. We begin with a description of the ways in which single molecules can be manipulated.

Manipulation by force

Single molecule forces are very small; the force generated by the action of a single myosin head in muscle is around 2 pN (2×10^{-12} N); for comparison, the gravitational force exerted by the Earth on a golf ball is ~0.5 N. The forces required to trap and move individual molecules are also in the pN range. A number of techniques are now available that can do this, including optical traps, magnetic traps, and atomic force microscopy (AFM). The small forces generated by single molecule motors can be measured using the same methods.

Typically, one end of the molecule studied is attached to a surface, and the free end is attached to a probe (e.g. an optically trapped bead or an AFM tip) through which the force is applied. Distances can be calibrated by moving the probe through a known distance while recording the position. Optical traps can generate forces in the range ~0.1–100 pN, while the AFM range is 10–10^4 pN.

Optical traps (tweezers)

In the 1970s, Arthur Ashkin noticed that small latex beads could be trapped with a laser. The reason is that a change in the direction of light, caused by reflection or refraction, generates a small force that can interact with the bead (see below). Optical trapping methodology has since been developed into a very versatile tool. It can exert forces on a wide range of particles and measure the displacement of the trapped particle with high accuracy (<nm) and good time resolution (~ms).

An optical trap is created by focusing a laser to a diffraction-limited spot with a high numerical aperture (NA) microscope objective (Chapter 7.1). This gives a light beam with a very high gradient at the focal point (the intensity changes rapidly with radial distance). If the particle is located symmetrically at the beam center, light passes through the particle with no net lateral force. If the particle is displaced radially from the center of the beam, however, it will experience differential forces that drive it back towards the center of the beam (Figure 7.2.1). Light-refracting particles are thus attracted along the gradient to the region of strongest light intensity at the center of the beam. Axial trapping (along the light direction) arises from a balance between the scattering force (which pushes the bead in the direction of the light beam) and gradient forces (there is a light gradient in the axial as well as the radial direction). The strength of a trap is related to the light gradient and thus the NA of the objective lens. The force generated by the trap on a bead can be calibrated; it is then treated as a linear spring with force constant k ($F = -kx$; Tutorials 3 and 7). High values of k give strong or "stiff" traps.

Traps are usually built from a modified commercial optical microscope. High power infrared laser beams are frequently used to achieve high trapping stiffness with minimal photo-damage to biological samples. Precise steering of the optical trap is accomplished with lenses, mirrors, and electro-optical devices that can be controlled via computer; traps usually operate in the direction shown in the diagram with gravity acting towards the objective lens.

Optical traps have been applied to manipulate a wide range of systems, including bacteria, organelles, proteins, and DNA. They have also been used to measure the displacement of and force generated by various molecular motors, e.g. kinesin-coated beads moving along microtubules.

Example RNA polymerase

RNA polymerase, which transcribes DNA into RNA, was studied in action with a pair of traps, as illustrated in Figure 7.2.2. A single, transcriptionally active molecule of RNA polymerase was attached to a bead held in relatively weak trap (right) and tethered via DNA (cyan) to a larger bead held in a strong trap (left). Displacements of the weakly trapped bead were measured while RNA transcription occurred. Remarkably, individual base-pair steps of RNA polymerase translocation could be observed. The observed step size was ~0.35 nm, a distance equivalent to the mean rise per base found in B-DNA.

There are some limitations to optical trapping. Trap stiffness depends on the light gradients, so any optical perturbations that affect the light intensity will degrade performance. In the example shown in Figure 7.2.2, a very stable system was obtained by

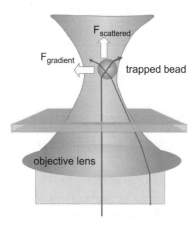

Figure 7.2.1 Schematic diagram of bead trapping by a laser beam brought to a focus by a lens (see text). The radial gradient force the axial scattering force and two beams (cyan) refracted by the bead are shown.

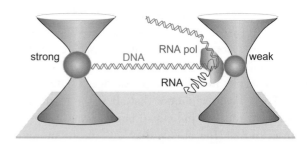

Figure 7.2.2 The arrangement used to measure the activity of a single RNA polymerase activity directly using two optical traps. Movement of the weakly trapped bead could be observed as the DNA was transcribed to RNA (see *Nature* 438, 460–65 (2005) for further details).

using a sealed optical enclosure with helium gas replacing air. The high intensity at the focus of the trapping laser can also result in local heating and optical damage.

Magnetic traps (tweezers)

Magnetic traps (see Figure 7.2.3) can be constructed from a pair of permanent magnets placed above the sample holder of an inverted microscope with a camera with a CCD detector (charge coupled device—see Chapter 4.3). The magnetic particles can be manipulated with a force range of 10^{-3}–10^2 pN. The paramagnetic beads are typically composed of magnetic particles, ~1 µm diameter, embedded in a protective polymer shell. They are designed to be non-magnetic until exposed to a strong magnetic field. Unlike optical traps, magnetic traps do not create a stable three-dimensional trapping position. Instead, the position has to be controlled with electronics and a feedback loop which also determines the effective "stiffness" of the probe. Magnetic traps are well suited to the study of rotary mechanisms that are found, for example, in DNA topoisomerases and the rotary motor F_0F_1 ATP synthase, as rotation forces can be applied (see Figure 7.2.4).

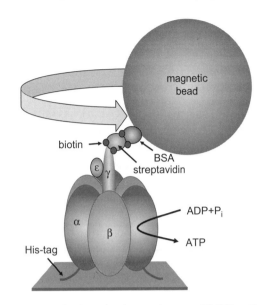

Figure 7.2.4 Single molecule experiment on F1 ATP synthase. ATP was synthesized by rotating the magnetic bead attached to the γ subunit with electromagnets (see *Nature* 427, 407–8 (2004) for more details); BSA, bovine serum albumin.

Example ATP synthase

ATP, the main biological energy currency, is synthesized from ADP and inorganic phosphate by the membrane-spanning enzyme, ATP synthase. Energy is normally supplied by a proton gradient across the membrane. The enzyme structure is known to be constructed from an F_0 part in the membrane and a globular F_1 part outside the membrane, with three subunits, α, β, and γ, and stoichiometry $\alpha_3\beta_3\gamma$ (see also Chapter 4.3). It had been shown that, *in vitro*, F_1 can be made to act in reverse, hydrolyzing ATP and behaving like a rotary molecular motor with the $\alpha_3\beta_3$ subunits rotating around the γ subunit. This was done at the single molecule level by direct observation of a large fluorescent actin fragment attached to F_1 ATP synthase on a coverslip. A later experiment, using a magnetic trap, demonstrated that ATP could be synthesized directly by mechanical energy. A magnetic bead was attached to the γ subunit of isolated F_1 on a glass surface, using biotin and streptavidin tags (see Box 4.5), and the bead was rotated using magnets (Figures 7.2.3 and 7.2.4). Driven rotation in the appropriate direction resulted in the appearance of ATP in the medium.

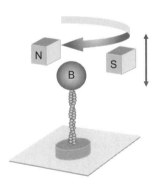

Figure 7.2.3 Magnetic traps based on permanent magnets. A paramagnetic bead (B) is attached to a surface by a single molecule of interest (outlined in cyan). A pair of small permanent magnets (N and S) produces a magnetic field which results in a force on the bead directed up toward the magnets. The height can be controlled by moving the magnets (bi-directional arrow). Rotation of the magnets produces rotation of the magnetic bead and the molecule. A microscope with a CCD camera is used to track the position of the bead with time.

Magnetic traps are not as versatile as optical tweezers or AFM, and the relatively large size of the magnets makes manipulation and the detection of very fast or very small displacements more difficult.

Atomic force microscopy

The atomic force microscope (AFM) was introduced in Chapter 7.1. A specialized version of an AFM can be used for force measurements, when it is it is sometimes called a molecular force probe. The sample holder here only moves in the vertical direction rather than being scanned. This device allows measurement and generation of inter- and intramolecular interaction forces with pN resolution. It can detect force-induced conformational changes in macromolecules and measure their mechanical stability.

In the force mode of the AFM, single molecules, or pairs of interacting molecules, are stretched between the tip of a microscopic cantilever and a flat, gold-covered substrate whose position is controlled by a high-precision piezoelectric positioning device (Figure 7.2.5).

Force generation is essential in biology; cell motility and cell division are key processes that require force. Mechanical stretching could also be an important control mechanism. It has been shown, for example, that application of physiologically relevant forces can cause stretching of cell migration proteins

and the creation of new binding sites. Protein stretching with resulting exposure of new binding sites may be a general mechanism in biology (see further reading). The availability of new experiments that can induce force changes and measure them and their consequences is thus an important advance.

Fluorescence methods

An early demonstration that single molecules in aqueous solution could be observed at room temperature by fluorescence was achieved by Funatsu et al. in Japan in 1995 when they used a TIRF microscope (Chapter 7.1) to observe single, fluorescently labeled myosin molecules, and to detect individual ATP turnover reactions. Since then, there has been a huge growth in studies of single molecules by fluorescence methods. Such experiments are clearly very challenging technically, with inherently low signal to noise ratio. Success has been possible because of: (i) the availability of brighter and more photostable fluorophores; (ii) technology advances that include more sensitive CCD detectors, better lenses, and better laser sources; and (iii) microscopes (TIRF, two-photon) that can focus on a very small regions of the specimen.

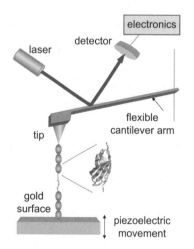 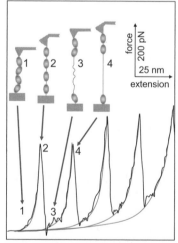

Figure 7.2.5 Force measuring version of an AFM (left). A single molecule is attached between the tip and a gold surface. Controlled movement of that surface can cause deflections in the cantilever arm. The view on the right shows a series of force extension curves obtained from an engineered protein consisting of a number of identical domains. As the protein is stretched, the force on the cantilever increases until one of the domains unravels causing the force on the cantilever to drop (position 2 to 3). The force begins to increase again when the unfolded domain is fully elongated (position 4) and then the next domain unfolds and so on (adapted with permission from *Nat. Struct. Biol.* 7, 719–24 (2000)).

Single-molecule super-resolution imaging and tracking

Specifically labeled fluorescent proteins and digitized images make it possible to gather large quantities of data on the position of particles in a living cell as a function of time. Nanometer-scale particle tracking is very informative, for example in measuring diffusion constants of proteins in cell membranes or the step displacements of beads attached to molecular motors. One method often used to track small spherical fluorescent particles is to fit the x–y dimensions to a two-dimensional Gaussian distribution (Figure 7.2.6; Appendix 3). This method, called **centroid fitting**, can identify the position of the centers of single particles in a series of time resolved images. Calculation of the centroid of the images of individual fluorescent particles allows localization and tracking with a precision that is about an order of magnitude better than the resolution of most microscopes.

Centroid fitting, which can give nanometer accuracy, is also used in **photoactivated localization microscopy (PALM)**, where very high resolution is achieved by illuminating only a few fluorescent molecules at a time. Remember that resolution is diffraction limited (Chapter 7.1 and Box 7.2) and any lens produces a point spread function (PSF) rather than a perfect image. If several fluorophores close enough in space to have overlapping PSFs, they cannot be resolved or identified individually. If, however, proteins are tagged with fluorescent molecules that can be activated selectively and at low enough density to avoid PSF overlap, individual molecules can be localized relatively precisely by centroid fitting. After the location of the activated fluorophores is noted, they are switched off by photobleaching, and a new set is activated. A summed map of the sequentially localized molecules can then give a very high resolution image.

Fluorescence correlation spectroscopy

Fluorescence correlation spectroscopy (FCS) was introduced in the 1970s. It measures light fluctuations in a sample caused by the random Brownian diffusion of a fluorophores through a small volume. Analysis of these light intensity fluctuations as a function of time gives a correlation function $g(\tau)$ with characteristic correlation time τ_D which characterizes the diffusional motion (Figure 7.2.7; see Box 4.3 and Chapters 4.1 and 6.1 for discussions of correlation times). In the example in Figure 7.2.7 the motion is dominated by diffusion alone. Small molecules have high diffusion coefficients and short τ_D values. If they bind to a larger molecule, their τ_D values increase. Measurement of τ_D thus offers a method for detecting binding because it is sensitive to changes in size (see

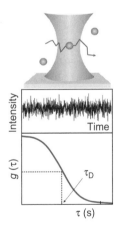

Figure. 7.2.7 A small volume element is illuminated by a high numerical aperture lens, as shown in Figure 7.2.8. Any fluorescent molecule diffusing through this volume will emit light. If the solution is very dilute, it can be arranged that only ~1 molecule diffuses through this small volume at a time. The fluorescent intensity/time plot will give information about the residence time of the molecule in the illuminated volume. The autocorrelation function $g(\tau)$ can be evaluated by analyzing the fluctuations, thus giving information about the correlation time τ_D.

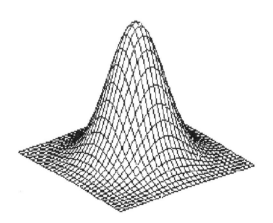

Figure 7.2.6 A 2D Gaussian distribution can be fitted (centroid fitting) to fluorescence emission points in microscopy images.

Problem 7.2.5). Molecules also often have conformational flexibility, as well as diffusion properties. $g(\tau)$ is then more complex than shown in Figure 7.2.7, but the different kinds of motion can often be resolved.

FCS studies have been widely applied to investigations of molecular motion and binding interactions *in vitro*. Examples include the DNA binding properties and conformational flexibility of the tumor suppressor protein, p53. FCS has also been applied to studies of intact cells. Examples include binding to cell surface receptors; intracellular diffusion rates of proteins are then often found to be significantly less than expected, probably because of binding to intracellular networks.

Single molecule studies of Förster resonance energy transfer (FRET)

Some of the most powerful applications of single molecule fluorescence methods entail FRET. As explained in Chapter 5.5, FRET involves energy transfer from a donor to an acceptor fluorophore. FRET can be used to determine the proximity of fluorescence probes attached to two proteins. It is also the basis of many sensors, e.g. for Ca^{2+}—see Problem 5.5.8. FRET efficiency depends on various factors including the 6th power of the distance (r) between the donor and acceptor.

A microscope suitable for detecting single molecule FRET events is illustrated in Figure 7.2.8. There are two main modes of operation. One is similar to that described for FCS above, where the fluorophore diffuses through the focal point of the beam. The other is where the molecule is attached to the surface of the coverslip. Freely diffusing molecules give short (ms) light bursts while immobilized molecules give longer bursts, characteristic of the kind of event being observed (Figure 7.2.9). The light irradiation that is focused on the immobilized molecule is

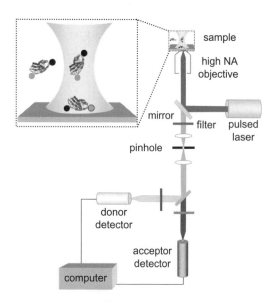

Figure 7.2.8 An illustration of the main components of a confocal single-molecule instrument that can be used to measure FRET. Two detectors, tuned to the emission maxima of the donor and acceptor are shown. Sometimes these detectors are duplicated to detect both ∥ and ⊥ polarized light. Only one excitation laser is shown but there could be two, each tuned to either donor or acceptor excitation frequencies. As in Figure 7.2.7, a high numerical aperture (NA) objective focuses the light to a diffraction limited spot (Chapter 7.1) with a very small volume (Problem 7.2.2).

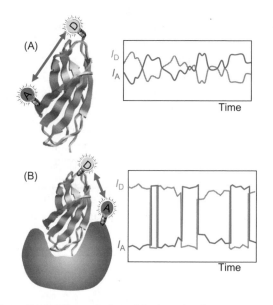

Figure 7.2.9 Schematic view of single-molecule FRET measurements obtained using an instrument of the sort shown in Figure 7.2.8. (Top) The molecule is undergoing conformational changes that modulate the FRET signals between donor (D) and acceptor (A) as a function of time. This results in changes in I_D and I_A—the donor and acceptor emission intensities. (Bottom) D binds to A so large and discrete intensity changes are seen because of larger distance changes on association/dissociation. Note that donor and acceptor intensities are anticorrelated in both cases; this is because emission from A takes away emission from D when their separation is small but not when distances are large (adapted with permission from *Nat. Struct. Biol.* 7, 724–9 (2000)).

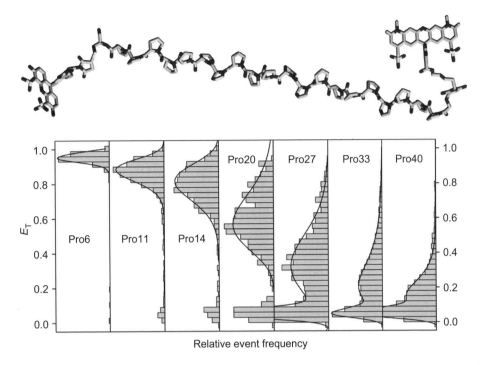

Figure 7.2.10 Single-molecule FRET efficiency measurements on polyproline peptides. A model of a 20 residue polyproline peptide with the donor and acceptor at each end is shown above. FRET efficiency (E_T) histograms, obtained from confocal single-molecule measurements on polyproline peptides of various lengths are shown below. It can be seen that the variation is small for short lengths and large for intermediate lengths (e.g. Pro20). As the separation length gets very long (e.g. Pro40), most E_T values become very small (adapted with permission from *PNAS* 102, 2754–9 (2005)).

relatively high, resulting in photobleaching after a few seconds.

As shown in Figure 7.2.9, the donor and acceptor fluorescence are usually anticorrelated (one increases when the other decreases) in these experiments; when the FRET efficiency is high, the emission from the acceptor (A) increases while the donor (D) emission decreases (see also Figure 5.5.15 which shows how fluorescence is relayed from donor to acceptor).

Single molecule FRET experiments have included studies of the conformational dynamics of DNA and RNA. These molecules can be immobilized on a surface coated with biotinylated bovine serum albumin (BSA) via biotin-streptavidin linkage (Box 4.5). This approach has been used to study the dynamics of DNA recombination and the folding transitions of RNA ribozymes. Polyethylene glycol (PEG) surfaces have also been used for protein immobilization; one example is the observation of repetitive translocation of a helicase on single-stranded DNA using single molecule FRET.

Example The proline spectroscopic ruler

We discussed in Chapter 5.5 (see Figure 5.5.16), a "classic" FRET experiment where a donor and an acceptor molecule were separated by a proline "rod". That 40-year-old experiment, which was used to relate FRET efficiency (E_T) in bulk solution to distance between donor and acceptor, has been revisited using single molecule methods. As shown in Figure 7.2.10, donor and acceptor dyes were attached to the ends of synthesized polyproline molecules of various lengths. In a single molecule experiment, carried out in the way illustrated in Figures 7.2.8 and 7.2.9(A), energy transfer was observed between a green-fluorescing donor dye (Alexa Fluor 488), attached at the carboxy terminus, and a red-fluorescing acceptor dye (Alexa Fluor 594) attached at the amino terminus. In the older, bulk phase experiment, only one value for E_T was obtained; it was also assumed that the "rod" was rigid and that a specific distance could be ascribed to a particular E_T value. The single molecule experiments, however, clearly show that there is a distribution of E_T values (Figure 7.2.10) rather than one; the polypeptides diffusing through the illuminated volume exhibit a range of different distances between donor and acceptor. The observed mean values of E_T were also found to differ from the values expected from Förster theory if polyproline is treated as a rigid rod (see also Problem 7.2.4). At donor–acceptor distances much less than the Förster radius, R_0, the observed efficiencies were lower than predicted, whereas at distances comparable to, and greater than R_0, they were higher. Molecular dynamics simulations and NMR experiments later showed that these results can be well explained by a "rod" that is somewhat flexible.

A similar approach to that shown in Figure 7.2.10 is being increasingly applied to measure protein folding pathways. As the polyproline example shows, single molecule studies give more information about the trajectory of a molecule moving between the folded and unfolded states than bulk average measurements. Single-molecule experiments are currently mainly performed *in vitro*. The ultimate challenge is to describe the conformational dynamics and bio-molecular interactions of individual molecules in the living cell. New methods are being developed for this difficult task, and the prospects are good; astonishing progress has been made in the last decade and new labels, such as quantum dots, and new microscopes are still emerging.

Summary

1 Single molecules can be manipulated and observed under near physiological conditions.

2 The most common manipulation techniques are optical and magnetic traps and a version of the atomic force microscope—a force probe. These tools can measure the forces generated by single molecules. Force probes can cause single protein domains to unfold.

3 Fluorescence is the most powerful tool to observe single molecules directly because of high sensitivity and powerful microscopes (e.g. TIRF) that have an ability to focus on very small volumes.

4 Monitoring donor and acceptor emission simultaneously with a two channel microscope allows distance fluctuations caused by FRET to be measured. This can give information about a variety of events, including binding/dissociation, conformational changes, and protein folding.

5 Single molecule experiments give information on the distribution of random events rather than an ensemble average; this provides new information about mechanisms.

Further reading

Reviews/articles

Michalet, X., Weiss, S., and Jäger, M. Single-molecule fluorescence studies of protein folding and conformational dynamics. *Chem. Rev.* 106, 1785–813 (2006).

Chang, Y.P., Pinaud, F., Antelman, J., and Weiss, S. Tracking bio-molecules in live cells using quantum dots. *J. Biophotonics* 1, 287–98 (2008).

Neuman, K.C., and Nagy, A. Single-molecule force spectroscopy: optical tweezers, magnetic tweezers, and atomic force microscopy. *Nat. Methods* 5, 491–505 (2008).

Chen, Q., Groote, R., Schönherr, H., and Vancso, G.J. Probing single enzyme kinetics in real-time. *Chem. Soc. Rev.* 38, 2671–83 (2009).

Joo, C., Balci, H., Ishitsuka, Y., Buranachai, C., and Ha, T. Advances in single-molecule fluorescence methods for molecular biology. *Annu. Rev. Biochem.* 77, 51–76 (2008).

Patterson, G., Davidson, M., Manley, S., and Lippincott-Schwartz, J. Super-resolution imaging using single-molecule localization. *Annu. Rev. Phys. Chem.* 61, 345–67 (2010).

Selvin, P.R., and Ha, T. *Single Molecule Techniques: a Lab Manual* Cold Spring Harbor, 2007.

Schuler, B., and Eaton, W.A. Protein folding studied by single-molecule FRET. *Curr. Opin. Struct. Biol.* 18, 16–26 (2008).

Joo, C., Balci, H., Ishitsuka, Y., Buranachai, C., and Ha, T. Advances in single-molecule fluorescence methods for molecular biology. *Annu. Rev. Biochem.* 77, 51–76 (2008).

Blanchard, S.C. Single-molecule observations of ribosome function. *Curr. Opin. Struct. Biol.* 19, 103–9 (2009).

del Rio, A., Perez-Jimenez, R., Liu, R., Roca-Cusachs, P., Fernandez, J.M., and Sheetz, M.P. Stretching single talin rod molecules activates vinculin binding. *Science* 323, 638–41 (2009).

 ## Problems

7.2.1 FRET experiments were carried out to measure the interactions between a protein receptor and its hormone ligand. A fluorescent donor was attached to the hormone and an acceptor to the receptor. Initially, a bulk experiment was carried out and a FRET efficiency, $E_T = 0.6$, was obtained. When repeated under single molecule conditions two unequal distributions of E_T were observed, one with a mean value of 0.8, the other with a mean value of 0.2. Can you suggest an explanation for these results?

7.2.2 For many single molecule studies using fluorescence, it is important to illuminate as small a volume as possible (Figure 7.2.7). Calculate the size of a diffraction limited volume for NA= 1.5 and $\lambda = 500$ nm. Assume the radius (x) of the Airy disc is given by the Abbé formula and that the axial resolution is threefold worse than this. How many molecules will be in this volume if the concentration of the solution is 10nM? (The volume of an ellipsoid is $4\pi x^2 z/3$, where z defines the axial dimension.)

7.2.3 Single molecule detection of a fluorescent ATP analogue (Cy3-EDA-ATP) using a TIRF microscope showed binding to isolated myosin S1 molecules and ATP turnover. The lifetime of individual bound fluorophores could be measured. In an analysis of about 100 events the following observations were made:

Lifetime range (s)	0–4	4–8	8–12	12–16	16–20	20–24	24–28	28–32	32–36	36–40
Number of events	20	15	12	9	8	6	5	4	4	3

From this distribution, estimate the catalytic turnover rate (k_{cat}) that might be expected to be seen in an enzyme kinetic experiment in solution.

7.2.4 In the example illustrated in Figure 7.2.9, the end to end chromophore distances were calculated assuming a polyproline type II trans helix with a pitch of 0.31 nm per residue in aqueous solution. Using the expected separations for Pro20 and Pro27 (if it is rigid this can be obtained by simply multiplying the number of prolines by 0.31 nm), calculate the expected E_T values (remember $E_T = R_0^6/(R^6 + R_0^6)$), given that $R_0 = 5.4$ nm. Compare these values to the observed means in the distribution in Figure 7.2.9.

7.2.5 The results shown in the figure are autocorrelation functions obtained from FCS experiments. The cyan curve was derived from a GFP tagged protein alone. The gray curve was obtained from the same protein in the presence of a receptor. Explain the curves and suggest a way to estimate the affinity between protein and receptor.

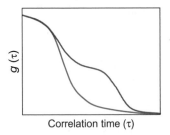

Correlation time (τ)

8 Computational biology

"Computers are useless. They can only give you answers."
Pablo Picasso, ~1970

"As far as the laws of mathematics refer to reality, they are not certain; and as far as they are certain, they do not refer to reality."
Albert Einstein, 1956

"Systems biology ... is about putting together rather than taking apart, integration rather than reduction. It requires that we develop ways of thinking about integration that are as rigorous as our reductionist programmes, but different .... It means changing our philosophy, in the full sense of the term"
Denis Noble, 2006

"Everything that living things do can be understood in terms of the jiggling and wiggling of atoms"
Richard Feynman lectures, 1962–63

This single chapter section briefly introduces some of the methods of computational biology. The computer is an essential component of all molecular biology and biophysics research, one that has an increasingly important impact on a huge range of topics—from instrument control to making sense of huge quantities of biological data (*bioinformatics*). The coverage in this chapter is uneven. The *mathematical modeling of systems* and *bioinformatics* sections are dealt with relatively superficially; the *systems* section is very a brief introduction to an important and growing topic that can be supplemented by the "Further reading" material provided. The *bioinformatics* section indicates how useful online sources provide information about genes, macromolecules, and structures that can be used as a starting point for the application of some of the tools described in other chapters. The *molecular modeling* section is dealt with in more detail than the other two topics because there is a direct link between techniques like *molecular dynamics simulations* and many of the experimental tools discussed in other chapters in this book, including NMR and X-ray crystallography.

8.1 Computational biology

Introduction

Computational biology is the application of computational methods to all levels of exploration; molecules to ecosystems. This is huge area of research that can be subdivided in various ways. The computer is, of course, also an indispensible tool in every branch of biophysics discussed in this book, both for controlling the instrumentation and for analyzing and visualizing the results. This chapter aims to introduce, very briefly, some of the main features of three major areas of computational biology, namely modeling of systems, bioinformatics, and molecular modeling.

The computer

Computing has a long history, from the abacus (~2400 BC) and Charles Babbage's mechanical engine (1833), to human "computers", people employed to compute in the early 20th century. Modern digital computing dates from the 1940s, when mechanical devices began to be replaced by electronic ones (vacuum tubes) and when John von Neumann designed new computer architecture, incorporating Alan Turing's ideas about a "universal computing machine". The first commercial computers were introduced in the 1950s; these were enormous heat-generating machines with tiny memories. The introduction of the transistor in the mid 1950s began to reduce the size and cost, a process greatly speeded up by the introduction, in the 1960s, of the "integrated circuit" with many components to one "integrated circuit" or "chip". This led to the prediction by Gordon Moore, a co-founder of Intel, that the number of transistors on a chip would double every two years. This prediction has been remarkably accurate and is expected to apply until at least 2015. As computer power (processing speed and memory) is linked to the number of transistors and the cost is linked to size, the power of computers has dramatically increased since 1960, while their cost has decreased. This has led to the computer becoming a major driving force for technological and social change in the last 50 years.

The internet and the World Wide Web

The internet is a global data communications system, an infrastructure of hardware and software that connects computers, a system that is now in most homes and whose use is considered routine, even by young children. It consists of millions of linked networks, carrying a vast array of information and services including the World Wide Web and E-mail. The Web arose from a proposal in 1989 by Tim Berners-Lee working at CERN in Geneva, to marry "hypertext" (computer text with inbuilt "hyperlinks") to the internet. The Web has since become a primary means of communicating scientific information, not least in the biological sciences where there has been an extraordinary explosion of data in the last 20 years. Collections of useful relevant websites are listed in the online resources associated with this book and can be found at **http://mcb.harvard.edu/BioLinks.html** and **http://www.pasteur.fr/~tekaia/biohypertext. html**.

Mathematical modeling of systems

Mathematical models can give insight into a wide range of biological processes. The system studied is considered as a network, with network components that can be proteins, cells, tissues, or individuals. The goal is to produce a model that predicts, numerically, the behavior of the network in a way that gives new insight into the system. The thinking is that a collection of interacting components can have **emergent properties** that do not fully manifest themselves unless a system is studied in its entirety. In other words, the aim is to form an integrated (holistic) view of how component parts make up the whole system. This is part of a general change in the philosophy of molecular and cellular biology in the last 50 years. To use a car analogy, we first knew what a car did—get us from A to B. We then isolated and characterized components parts—e.g. the battery. We are now figuring out how the various components are assembled and interact to make something that can be driven from A to B.

This field, where a predictive model is constructed using quantitative data, has come to be known as **systems biology**. The data can come from a wide range of sources. We live in an age where huge amounts of data are available (see bioinformatics later in this chapter) so the ambitious goals of systems biology are becoming less of a dream and closer to realization.

It is clearly unrealistic to model a system at all levels simultaneously. For example, modeling an ecosystem is more likely to be useful if it focuses on species predation rather than the molecular networks in a single cell. Models of different levels in a system thus use different assumptions and simplifications.

Modeling at the **molecular network** level is often done by describing the system with a set of **differential equations** (Appendix A2.5); these incorporate knowledge of some basic properties of the system, for example concentrations, rate constants, and equilibrium constants. Solutions to the equations then predict how the system might respond to different initial conditions. An example would be a model for how the cell cycle is regulated; current models of this kind can reproduce the physiological properties of normal cell division and the aberrant properties of mutant cells; see Further reading. (It should be noted, however, that for most of the intricate molecular networks in the cell, many of the required rate constants and local concentrations of molecular network components are not yet known.) At a different, **organ** level, models of the heart can provide quantitative predictions of normal function and dysfunction, including arrhythmia. The input parameters now include factors like flow rates, volumes, and electrical signals, rather than molecular information.

Some models are termed **deterministic**—they always give the same result for the same set of initial conditions. Others are **stochastic** (probabilistic), where a random number generator produces a variety of possible outcomes with different probability. **Sensitivity** and **robustness** are features used to describe aspects of a particular model. If a system is sensitive to a certain parameter, small changes in its value will have large effects on the system. In contrast, robustness defines whether or not the system settles to a solution over a broad range of input conditions. A robust system has a low sensitivity towards the input, like a buffered solution is insensitive to the addition of acid. Many system models also depend on **feedback**, where responses to various stimuli can lead to **amplification** or **adaptation** of the system.

Model hypotheses for how a system works can be constructed and tested. Model testing with a computer (*in silico*) can be considered as working in a "virtual laboratory" where a wide range of "experiments" can be conducted. Modeling is an **iterative process**. Output from a model can be compared with existing data or new experimental data can be generated to compare predictions with reality. The model can then be refined, by adjusting parameters, or by changing the model, thus starting another iteration cycle (see Figure 8.1.1).

Computer-based models have contributed, and are likely to contribute more in the future, to our understanding of living systems at all levels. This subject can be explored using the "Further reading" section at the end of this chapter. A range of programs and software is also available on the Web for carrying out mathematical modeling of systems (e.g. **http://www.wolfram.com/products/mathematica/index.html**; **http://www.cellml.org/getting-started/**

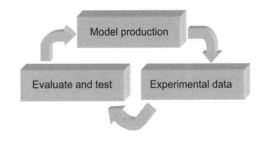

Figure 8.1.1 Illustration of the iterations between model development, experimental design and data collection, model testing, model refinement, and so on round the cycle.

tutorials/cellml-1.1-tutorial; **http://www.copasi.org/tiki-view_articles.php**; **http://vcell.org/vcell_software/login.html**; and **http://www.e-cell.org/ecell/**).

Bioinformatics

Bioinformatics, a term first used in 1979, involves the creation and curation of databases and the development of computer methods and statistical techniques to analyze the data held in those databases. Its primary use has been in "omics", mainly **genomics** (the DNA content of cells), but also **transcriptomics** (the messenger RNA produced by active genes), **proteomics** (the proteins in the cell), and **metabolomics** (the small compounds in cells).

Major research efforts in bioinformatics include **genome annotation**, in which genes are identified and likely biological function ascribed to them. Gene sequences can be translated into a protein sequence. One way to classify a new gene is thus to scan a database to see if there is a protein of known function with a similar sequence. Such sequence analysis searches can thus help to identify genes and distinguish **orthologs** (genes with the same function in different species) from **paralogs** (genes with different but related functions in one organism).

Other bioinformatics activities include **protein structure prediction**, prediction of **protein–protein interaction networks**, **drug discovery**, and **phylogenetic relationships** among organisms. A sense of the dramatic increase in available biomolecular data can be obtained by viewing some of the many available bioinformatics tools and databases (**http://www.ebi.ac.uk/Tools/**; **http://cmr.jcvi.org/tigr-scripts/CMR/CmrHomePage.cgi**; **www.ensembl.org**; and **www.rcsb.org/pdb**). More information about other aspects of the world of bioinformatics can be found in numerous websites and the literature listed in the "Further reading" section.

Sequence analysis

Part of genome annotation activity involves the application of **sequence alignment** tools. FASTA, for example, is a computational tool that takes a given nucleotide or amino-acid sequence and searches for a similarity with another sequence, represented as rows within a matrix. Gaps are inserted between the residues so that identical or similar residues are aligned in successive columns. Numerous different alignment programs, including FASTA, BLAST, FASTA, CLUSTAL-W, and T-Coffee, are available on the Web and are straightforward to use; see, for example, the list at **http://expasy.org/tools/**. An example of sequence alignment output from the Uniprot website (**http://www.uniprot.org/**) is given in Figure 8.1.2. This was obtained using the program CLUSTAL-W, which aligns multiple sequences of DNA or proteins. Note that multiple alignments of three or more sequences are generally more informative than pairwise alignments.

The **significance** of a sequence alignment can be assessed with a **scoring** scheme. Similar sequences have high scores. For nucleic acids it is common to use a simple scheme with +1 for a match and –1

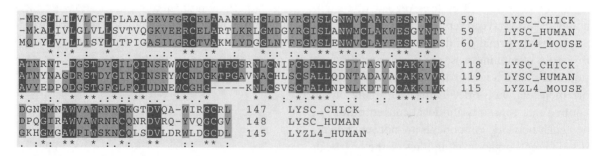

Figure 8.1.2 Alignment of three lysozyme sequences. You can obtain these for yourself as follows: start at the Uniprot website; type "lysozyme" in the query box; a large number of entries "reviewed" appear; tick the boxes next to the mouse (Q9D925), human (P61626), and chick (P00698) entries; click on "retrieve" at the bottom of the page, and then on "align"; the aligned sequences shown here then appear. Dark gray residues are completely conserved between the three sequences and light gray are conservative substitutions; – indicates a gap and the *., :, ., notation gives another measure of conservation.

for a mismatch. Proteins require a more complex scoring system that involves classification of amino-acid types.

It is important to realize that there is no "correct" solution for a sequence alignment. Different algorithms (computer procedures) give different answers. Consensus-based methods are thus often used where a number of approaches are applied to a problem and the results compared.

Phylogeny determines relationships among species or genes by comparing sequences. The results of a phylogenetic analysis of sequences are usually presented as a "tree" showing evolutionary relationships, as illustrated in the following simple example for four proteins q, r, s, and t.

The first step is to construct a matrix with some measure, d_{qr}, etc. of how far any two sequences have diverged. These evolutionary distances may, for example, simply be the number of amino-acid differences between the two compared sequences. If q and r are most similar, i.e. d_{qr} is smallest, followed by their distances to s and then to t, we would obtain a tree of the sort shown in Figure 8.1.3A. Note that this is a very simplistic approach; quite sophisticated algorithms can be used to assess distances and relationships between sequences (see, for example http://phylomedb.org/).

Structure prediction

Proteins

Since Christian Anfinsen showed that proteins can fold to a unique 3D structure in an aqueous environment, a large effort has been expended in finding ways to compute protein structure from a known sequence. These efforts go on at several levels and there are online servers available to help and to educate (e.g. http://www.expasy.org/ and http://www.click2drug.org/).

Secondary structure prediction

Estimates of the local secondary structures of proteins involves assigning regions of the amino-acid sequence to α-helices, β-strands, or turns. The algorithms used to do this operate with different degrees of sophistication. One of the first approaches to be applied was the Chou–Fasman method; this relies on probability parameters, determined from the relative frequencies that amino acids are found in known secondary structures, e.g. alanine and glutamate are commonly found in α-helices. More complex approaches use neural networks (constructions that attempt to mimic the properties of biological neurons) to make the predictions. Modern secondary structure prediction methods can give over 70% accuracy; the predictions are more reliable if multiple sequence alignments are used. These relatively accurate secondary structure predictions can then be used in predictions of overall (tertiary) folds of proteins although tertiary structure prediction is generally much less reliable (see below).

To illustrate the use of secondary structure prediction tools that are available on the Web, let us return to the Uniprot website and the lysozyme example. Click on the human lysozyme sequence (P61626); the page that appears contains a secondary structure prediction, as shown in Figure 8.1.4. [Note that the lysozyme page also contains a large amount of other useful information about human lysozyme, including the sequence (18 residues in the "signal peptide" plus 130 residues in the "chain", etc.), as well as the molecular weight and the isoelectric point.]

As well as secondary structure, bioinformatics tools have been developed to detect other patterns in sequences that can provide useful information. A straightforward example is a hydropathy plot. Kyte and Doolittle gave each amino acid a score related to its hydrophobicity, for example isoleucine, with a very hydrophobic sidechain, was assigned a score of +4.5, while arginine, a very hydrophilic residue, has a score of −4.5. These scores can be averaged over a "window" of n amino acids to analyze any given sequence for hydrophobic character. Regions with a

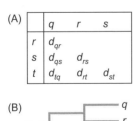

Figure 8.1.3 (A) A matrix, constructed for four proteins with the evolutionary distances that separate them. (B) A tree constructed on the basis that q and r are most similar, followed by s and then t.

HELIX	23–32
HELIX	43–54
STRAND	61–63
TURN	65–67
STRAND	70–72
TURN	73–76
TURN	79–81
STRAND	82–84
HELIX	99–103
STRAND	104–106
HELIX	108–119
HELIX	123–126
HELIX	128–133
TURN	134–136
HELIX	140–143

Figure.8.1.4 Secondary structure prediction for human lysozyme using Uniprot.

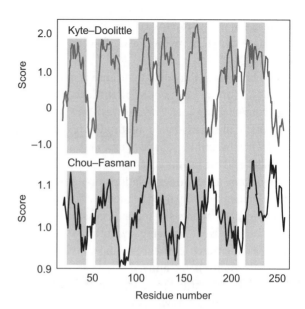

high hydrophobicity are likely to be embedded in a membrane or in the core of a large protein. Returning again to the Uniprot site we can find accession code (P02945) for a well-known membrane protein, bacteriorhodopsin (see Chapters 5.2 and 5.3). If we now go to the ExPASy proteomics server (**http://expasy.org/cgi-bin/protscale.pl**), we can extract various plots for bacteriorhodopsin (P02945). For example, the Kyte–Doolittle plot (obtained by clicking on Hphob/Kyte & Doolittle) gives a measure of hydrophobicity, while alpha helix/Chou–Fasman predicts the likelihood of a helix being found. These outputs are shown in Figure 8.1.5. It can be seen that combining these complementary kinds of prediction method (hydrophobicity and helix) gives more confidence in the result than one type of prediction on its own.

Note that a number of different options are available in ExPASy, with several versions of hydropathy and secondary structure prediction programs. (Compare, for example, the lysozyme prediction result in Figure 8.1.4 with the predictions made by other Web-based secondary structure tools listed in **http://expasy.org/cgi-bin/protscale.pl**.)

Tertiary structure prediction

Tertiary protein structure is formed when different secondary structural elements (sheet, helix, turn) pack together to give a folded protein **domain**. There are two major classes of method for predicting protein tertiary structure, namely **comparative modeling** and *ab initio* prediction.

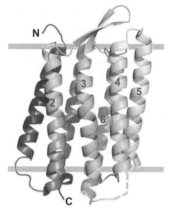

Figure 8.1.5 Output from the ExPASy server for bacteriorhodopsin (Uniprot entry P02945), a protein with seven membrane-embedded α-helices. Kyte–Doolittle and Chou–Fasman predictions for hydrophobicity and helix propensity are shown. A window of nine residues was applied to residues 14–262. The true helix positions, derived from an X-ray structure are also indicated by gray shading. The structure itself is also shown below: PDB: 2NTU. The approximate position of the membrane is indicated on the structure by gray lines.

Comparative modeling, also known as **homology modeling**, identifies evolutionarily related proteins with similar sequences and assumes that they have similar structures. As a large number of experimentally determined structures are now available in the Protein Data Bank (PDB; **http://www.rcsb.org/**), it is becoming increasingly likely that a homolog with a known structure will be found by searching the sequence database; this known structure is sometimes called the "parent". Using a sequence alignment of parent and the unknown target protein, an initial model can be constructed, using main-chain and sidechain coordinates of aligned residues in the core of the parent. Sidechains and backbone must then be modeled for residues that are different in the target and parent, or where there are gaps or insertions in the alignment. Some of the tools described later in the "molecular modeling" section, e.g. molecular dynamics, are often employed at this stage.

An example of output from a Web-based protein structure prediction server is shown in Figure 8.1.6. This arose from submission of the amino-acid sequence of *Drosophila* lysozyme (P37161) to SWISS-MODEL (**http://swissmodel.expasy.org/**). The server carried out a sequence alignment and decided that the house fly had the closest sequence with a known 3D structure; it then built a structure of *Drosophila* lysozyme, using the house fly coordinates (PDB: 2fbd) as a template.

A variation on comparative modeling is called **threading** or **fold recognition**. This is used for proteins that do not have obvious homologous protein structures available in the PDB. Protein threading works because the number of different protein folds found in nature is finite; indeed, most new structures now submitted to the PDB have similar structural folds to ones that are already there. Known structures are used as templates and ones predicted to be related to the sequence of interest are identified. Each

amino acid in a target sequence is "threaded" to a position in a template structure. How well the target fits the templates is evaluated with a list of comparative "scores". A variety of non-trivial scoring methods are used; here we will just say that the structural fits are ranked and the fold with the best score is assumed to be the one adopted by the sequence. After the best-fit template is selected, the structural model of the sequence is built, based on the alignment with the chosen template and modeling as described above for the lysozyme example.

An alternative to comparative modeling approaches is to use an *ab initio*, or *de novo*, approach where physical principles (e.g. Lennard-Jones potentials) and empirical force fields (see next section) are alone used to deduce the way a protein folds to a three dimensional structure. A search procedure is required to explore possible conformations and a scoring system to assess whether the structures found are "correct". *Ab initio* methods tend to require huge computational resources, and they have thus only been carried out for very small proteins. Current knowledge and computational power are not sufficient to solve the problem reliably. The success of *ab initio* methods for reasonably sized proteins therefore remains low compared with comparative modeling methods.

Critical assessment of structure prediction (CASP)

Every two years since 1994, it has been possible for research groups to test their structure prediction methods and performance in a worldwide experiment (**http://predictioncenter.org**). To indicate the scale of this exercise, CASP8 in 2009 had 112 human expert groups registered and 121 automatic prediction servers. 128 targets were released for prediction, generating a total of 80 560 submitted models. This large exercise has helped drive progress towards better structure prediction protocols and online server analysis.

Comparison of protein structures

When a new structure is being assessed it is useful to compare its structure with structures that are already in the database to see how similar they are. One useful program that does this is called DALI (distance-matrix alignment; **http://ekhidna.biocenter.helsinki.fi/dali_lite/start**). Pairwise comparisons of the coordinates of structures in PDB format can be processed

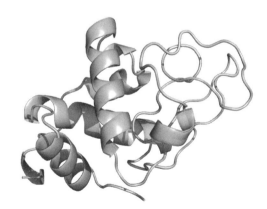

Figure 8.1.6 A structure of *Drosophila* lysozyme, predicted by the SWISS-MODEL server. This was achieved using comparative modeling methods with the house fly structure as the parent template.

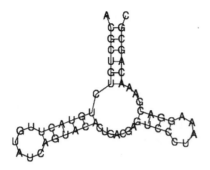

Figure 8.1.7 Predicted secondary structure for a hammerhead ribozyme using the online RNAfold server.

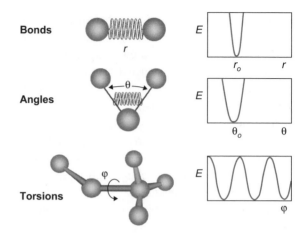

Figure 8.1.8 An illustration of the bonded terms and their contributions to an empirical potential energy function. The bonded energy (E) and the E dependence (in cyan) are shown as a function of the parameters r, θ, and φ. It can be seen that in the "bonds" term, E rapidly increases when r is less than or greater than the equilibrium value r_o. Similarly, there is an equilibrium energy for the "angles" term at θ_o. The "torsions" energy increases and decreases sinusoidally as the torsion angle, φ, is rotated.

interactively; the result is usually returned within a minute.

RNA structure

As with proteins, the structure of RNA is important for its function. RNA molecules are associated with enzyme function, gene transcriptional regulation, and protein biosynthesis. RNA secondary structure, which is primarily defined by stems (base-paired regions) and loops, can be predicted with ~75% accuracy by computer methods and a number of Web-based servers with RNA structure prediction software are available (see: **http://en.wikipedia.org/wiki/List**).

Hammerhead ribozymes are small RNAs that have a conserved "hammer"-shaped structural motif. They are able to cleave themselves enzymatically. Figure 8.1.8 shows the structure prediction output, produced by the following sequence of a hammerhead RNA: ACGCUGUCUGUACUUGUAUCAGUACACUGA-CGAGUCCCUA AAGGACGAA ACAGCGC by the server at **http://rna.tbi.univie.ac.at/cgi-bin/RNAfold.cgi**.

Molecular modeling

Introduction

Molecular modeling is where molecules are represented numerically and their structure and behavior is simulated using physical laws. The energy of any arrangement of atoms can be calculated using a model that applies molecular (classical) mechanics (e.g. Newton's laws), quantum mechanics, or a combination of the two. Molecular modeling is a powerful tool for predicting molecular behavior in biological systems and for analyzing and interfacing with experimental results.

In molecular mechanics (MM), an empirical potential energy function is constructed for the molecule and the rules of classical physics are applied to determine energetically favorable molecular conformations. More rigorous molecular models can be produced by minimizing the total electronic energy of a molecule, essentially using Schrödinger's wave equation (Tutorial 9). Such quantum mechanical (QM) methods are, however, still limited to a relatively small number of atoms because of the computational demands and the difficulty they face in searching conformational space to find the lowest energy configuration. Hybrid QM/MM methods are, however, being used increasingly to get around these problems (see below).

Molecular dynamics adds motion to molecular mechanics and computes a dynamic trajectory based on Newton's laws of motion (Tutorial 3). Molecular dynamics (MD) is very powerful tool and is the dominant method used for searching, finding global minima, and incorporating experimental data. Alder and Wainwright first introduced MD in the late

1950s. The first molecular dynamics simulation of a realistic system, liquid water, was done by Rahman and Stillinger in 1974, while Karplus and McCammon carried out the first atom level simulation of a protein in 1977. These studies built on earlier developments of potential energy functions (Morse curves, Lennard-Jones potentials—Chapter 2.1) and conformational analysis of organic molecules in the 1950s.

MD simulations allow conformational changes and binding to be assessed in systems of known structure, thus aiding our understanding of a range of important processes such as protein folding, enzyme catalysis, ion transport, and drug design. MD simulations are also widely used to incorporate known experimental restraints and to predict and validate experimental results (see below). Simulation techniques have advanced remarkably in recent years; they can be very demanding on computer time but computers are getting faster and cheaper. Depending on the system size, simulations of solvated proteins that represent molecular time courses up to μs can be achieved, although this may take months of computing time. Simulations in a more experimentally accessible (ms) range can also be achieved using further approximations such as **coarse grain modeling** (see below). Some of the important features of molecular modeling will now be described in a little more detail.

Potential energy function (force fields) and molecular mechanics

A key step in molecular modeling is to represent the molecular system by an energy function that is a good approximation to reality. A **force field** refers to a mathematical function used to describe the potential energy of a system. As we will see, force fields and their parameters are derived both experimentally and by calculation.

The energy of any atom in a molecular system will depend on the properties of both its *non-bonded* and *bonded* atom neighbors: $E = E_{\text{bonded}} + E_{\text{non-bonded}}$. Let us now consider these two energy terms in more detail.

Non-bonded terms

As described in Chapter 2.1, interactions with non-bonded neighbors (non-covalent interactions) are dominated by van der Waals and electrostatic interactions:

$$E_{\text{non-bonded}} = E_{\text{electrostatic}} + E_{\text{van der Waals}}.$$

The **van der Waals potential** between a pair of atoms can be approximated by the Lennard-Jones potential function; this can be extended to all non-bonded pairs of atoms labeled i and k:

$$E_{\text{Lennard-Jones}} = \sum_{\substack{\text{non-bonded} \\ \text{pairs}}} \left(A_{ik} / r_{ik}^{12} - B_{ik} / r_{ik}^{6} \right),$$

where A and B are constants that depend on atom type.

We also saw in Chapter 2.1 that the **electrostatic interaction** between charges q_i and q_k, separated by a distance r_{ik}, is $E_{\text{electrostatic}} = \sum_{\substack{\text{non-bonded} \\ \text{pairs}}} q_i q_k / \varepsilon \varepsilon_o r_{ik}$, where ε is the dielectric constant.

The basic non-bonded force field is thus given by the sum of the $E_{\text{Lennard-Jones}}$ and $E_{\text{electrostatic}}$ energy terms (see also below).

Bonded terms

To describe the *bonded* terms, we use a representation that assumes that bond stretching and bending can be treated in terms of springs. Equations of the following sort are used:

$$E_{\text{bonded}} = \sum_{\text{bonds}} k_r (r - r_o)^2 + \sum_{\text{angles}} k_\theta (\theta - \theta_o)^2 + \sum_{\text{torsions}} A_n (1 + \cos(n\varphi - \varphi_o))$$

The first term is a bond stretching term and the k_r values are related to spring constants (see Tutorials 3 and 7); the k values depend on the atoms involved. This term is summed over all the bonds of length r. The second term refers to bond angles, defined by θ and three atoms. The third term is summed over torsion angles, φ, defined by four atoms; it models barriers between rotations around the middle bond and can be described by a periodic cosine function. These various terms can be understood by referring to Chapter 2.1 (Figure 2.1.4 and 2.1.5) and to Figure 8.1.8.

Force fields

The potential energy function ($E_{\text{bonded}} + E_{\text{non-bonded}}$) discussed above is the basis for **empirical force fields** where the various constants (k_r, A_{ik}, B_{ik}, etc.) have been determined by experiment, together with calculations and exhaustive testing. Well-tested force fields are now available for proteins, nucleic acids, lipids, and carbohydrates. These are essential features of the good predictive power of current MD simulations. There are numerous MD simulation programs available, each with their own force field parameters (e.g. CHARMM, AMBER, and GROMOS).

Effects other than the basic $E_{bonded} + E_{non\text{-}bonded}$ terms can be accounted for by adjusting the force field parameters. An "improper" dihedral term is often added to the bonded energy term to maintain the chirality and planarity of peptide bonds. Hydrogen bonds are often incorporated by adjusting the constants in the $E_{non\text{-}bonded}$ terms and the non-bonded terms can also be adjusted empirically to take account of some of the properties of solvent water (see solvation section below).

An important consideration in choice of force field is the time taken for calculations. Sometimes the force fields are *truncated* and smoothed to speed these up. It is especially important to find good approximations for electrostatic interactions, as the computing cost for them is high. One approximation, widely used for electrostatics, is known as the particle-mesh Ewald (PME) summation (**http://www.gromacs.org/Documentation/Terminology/Particle_Mesh_Ewald**). A major advantage of PME is that it scales as $N\ln N$ where N is the number of atoms, as opposed to N^2 for other types of approximation (an N^2 scale simply means that if there are N atoms the calculation time is proportional to N^2).

Energy minimization

We have considered the energy force fields for the various atoms. What about the energy of the whole macromolecular system? It is unlikely that initial models or experimental structure determinations will have an appropriate global energy minimum. Numerous local energy minima usually exist and these can trap conformations and obscure the global minimum that is sought (see Figure 8.1.9). To minimize the energy we employ a starting set of coordinates and the empirical force fields. Box 8.1 shows the relationship between force and potential energy. An energy minimum corresponds to a point where $F = -\nabla E$ is zero.

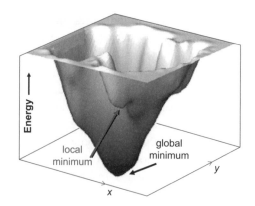

Figure 8.1.9 Illustration of global and local minima in a macromolecule's energy landscape.

Box 8.1 Force and potential energy

Force is related to the differential of the potential energy. Consider the gravitational potential energy, PE = mgh, of a mass m at height h (Tutorial 3). Differentiating with respect to h gives $mg(dh/dh) = mg$; this is mass × acceleration which equals force. It is conventional to use a – sign to indicate that if PE increases with increasing h, the force will tend to move it toward a smaller value of h, decreasing the PE.

This simple treatment can be extended to three-dimensional potential energy functions (force fields) using the equation $F = -\nabla$(PE) where ∇ is the three-dimensional differential operator (Appendix A2.5). We can thus write the force on any atom in a force field, E, as $F = -\nabla E$.

Evaluation of $F = -\nabla E$ can be used to determine the direction and magnitude of a step towards an energy minimum. Small changes in the position of every atom are made and the new energy is calculated after every move. The move is kept if the energy is lowered, otherwise the atom is returned to its original position. This process is repeated many times until an overall energy minimum is reached. Typically, thousands of iterations are required for a large macromolecule to convergence to a minimum. There are three major protocols for energy minimization: steepest descent, conjugate gradient and Newton–Raphson. In the steepest descent method (not the most efficient) the first derivative is calculated and its direction determined. A line is then drawn in that direction across the potential energy contour map. This gives a one dimensional curve whose minimum can readily be found. Moving to that new minimum, another derivative is calculated and the process is repeated until an overall minimum is found (see Problem 8.1.5).

Molecular dynamics simulations

As discussed in the introduction to this section, molecular dynamics simulations provide a very powerful tool for assessing experimental results and for predicting many properties of molecules and complexes. This is done by using Newton's equations of motion to calculate the time evolution of molecules, the **trajectory**, in the empirical force fields discussed above. (It would be helpful to look at Tutorial 3 and Appendix A2.5 before reading this section.)

Remember that Newton's second law is $F_i = m_i a_i$, where F_i is the force exerted on a particle i, m_i is the particle mass, and a_i is the acceleration. To see how a trajectory might develop as a function of time, consider the case when a_i is constant. We can write this as $dv_i/dt = a_i$ which integrates to $v_i = a_i t + v_{i0}$, where v_{i0} is the velocity when $t = 0$. As $v_i = dr_i/dt$, we have $\int dr_i = \int(a_i t + v_{i0})dt$ or $\boldsymbol{r_i = a_i t^2/2 + v_{i0}t + r_{i0}}$ where r_{i0} is the position at $t = 0$. This is a parabolic function that can be used to compute the value of r_i at time t as a function of the acceleration, a_i, r_{i0}, and v_{i0} (compare this to the somewhat different derivation of the trajectory equation for gravitational acceleration in Tutorial 3).

To calculate a molecular trajectory using this equation we need to know the a_i, r_{i0}, and v_{i0} terms. The **acceleration** a_i is the gradient of the potential energy function, $F_i = -dE/dr_i$ or $-\nabla_i E$ in three dimensions (Box 8.1). The initial **coordinates** r_{i0} can be derived from an existing, experimentally determined, set of coordinates in the Protein Data bank (PDB) or from preliminary experimental data from NMR data (see below).

The initial velocities v_{i0} can be related to the **temperature** of the system. Simple kinetic gas theory (Box 2.2) gives the average internal energy of a system as $U = 3/2 N_A kT$ where N_A is Avogadro's number, k is the Boltzmann constant, and T is the temperature. U is also the sum of the kinetic energies of the atoms, $\sum_i m_i v_i^2/2$. Initial velocities with a Maxwell–Boltzmann distribution (Box 2.2) are randomly assigned to the atoms. This distribution is usually set to correspond to a low temperature at first but

this can be scaled upwards by increasing the velocities during a number of calculated iterations; the temperature is usually kept constant by scaling the velocities during the main simulation period (Figure 8.10).

How do we compute the trajectory efficiently? We saw above, that the equation $r_i = a_i t^2/2 + v_{i0}t + r_{i0}$ could be used to describe the system. In fact, the acceleration terms, a_i, depend on the differential of a complex force field and they are not constant, as was assumed in deriving this equation. A numerical integration is therefore carried out with a function of the form $r(t + \delta t) = r(t) + v(t)\delta t + 1/2 a(t)\delta t^2$, where r is the position, v is the velocity (dr/dt), and a is the acceleration (higher order terms have been neglected). A typical time step, δt, in these calculations is 1 fs. There are a number of tricks that can be used to simplify these calculations. One is called the "leap frog" method; another is the "Verlet algorithm".

MD simulations can thus be summarized as shown in the flow chart in Figure 8.1.10. The temperature adjustment and equilibration stages are typically carried out for a time period that is equivalent to ~10 ps

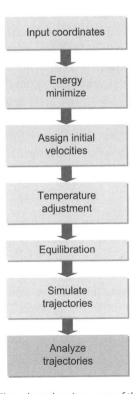

Figure 8.1.10 Flow chart showing some of the important steps in a MD simulation.

(10 000 steps of 1 fs). The trajectory simulations take the longest time with ~100 ns being a typical simulation time scale. Note that this time scale refers to a large number of fs steps; they do not necessarily translate to real experimental times. A calculation corresponding to 100 ns may also take many weeks of computing time.

Analysis of trajectories

Trajectories are valuable because they tell us about possible molecular movements as a function of time and about the stability of a system; if the coordinates change greatly during a trajectory, the initial structure is unstable for some reason; possible explanations are a poorly determined experimental structure or a structure that has other, relatively low energy, conformational states. To analyze the trajectories, the coordinates and velocities of the system are saved during an MD simulation. Time-dependent properties, such as energy, position, radius of gyration, etc. can be displayed graphically. Average structures can be calculated and compared with experimental structures. Molecular dynamics simulations can thus help us visualize and understand conformational changes when combined with a molecular visualization program (see below). Examples of simulation trajectories can be viewed on the Web (e.g. **http://www.ks.uiuc. edu/Research/aquaporins/** and **http://www.mpibpc. mpg.de/groups/de_groot/gallery.html**). Figure 8.1.11 shows a simple example of a MD simulation of a protein domain (Src SH2). The root mean square deviation of the domain coordinates from the original coordinates is shown as a function of trajectory time, for a liganded and an unliganded domain.

Solvation

Water is a very important solvent, one that has a huge influence on the structure and dynamic properties of macromolecules. We saw, for example, in Chapter 2.1 that hydration decreases electrostatic interactions by increasing the effective dielectric constant. One way of dealing with the solvent in an MD simulation is to treat the water *implicitly;* water molecules are not included directly in the simulation but are accounted for by empirical adjustment of the force fields, for example a distance-dependent dielectric constant can be used to mimic solvent effects.

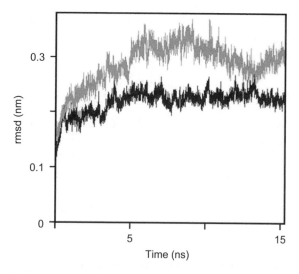

Figure 8.1.11 The time course of molecular dynamics simulations of a ligand-bound (black) and ligand-free (gray) protein domain (SH2). The starting structures were PDB: 1AYA (bound form) and 1AYD (free). The root mean square deviation (rmsd) from these starting structures is plotted as a function of simulation time. It can be seen that the ligand-bound structure deviates less from the starting structure than the unbound form, suggesting that it is more stable (adapted with permission from *In Silico Biol. 2*, 0028 (2002)).

It is clearly better to include water molecules *explicitly* in the simulation, although this is more computationally demanding. The dielectric constant is then taken to be constant with the water molecules providing the electrostatic shielding, as they do in reality. To prevent the water molecules from diffusing away during the simulation and to limit the number of solvent molecules in the calculations, periodic boundary conditions are imposed. These conditions enable a simulation to be performed using a relatively small number of particles, although they still experience forces as if they are in a bulk solution. As shown in the two-dimensional representation in Figure 8.1.12, the coordinates of the particles in the central box are related to those in the surrounding boxes by simple translations. If a particle leaves the box during a simulation it is replaced by a particle that enters from the opposite side, thus keeping the number of particles in the box constant. The shape and size of the box can be adjusted to fit the system being studied.

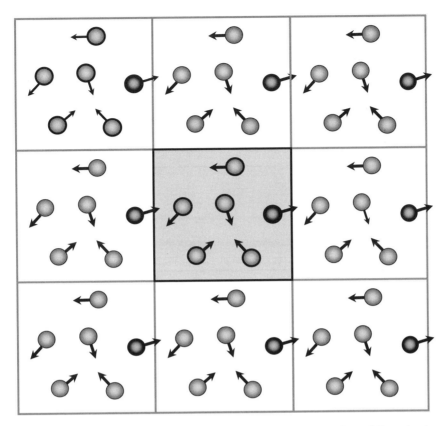

Figure 8.1.12 Explicit solvent molecules can be represented by "periodic boundary conditions". Six molecules are shown, moving randomly in the central box which is surrounded by eight image boxes. A leaving particle is replaced by an entering one.

Coarse-grained methods

Computational time is a major limitation in MD simulations and various reasonable approximations are sought to speed up the calculations so that longer time windows (more simulation steps) and larger systems can be explored. In coarse-grained (CG) methods, "pseudo-atoms" or "beads" are used to represent groups of atoms rather than every atom of the system. For example, the aliphatic tails of lipids can be represented by gathering 2–4 methylene groups into one bead. The parameters used in these coarse-grained models are adjusted empirically, by matching the behavior of the model to appropriate experimental data or all-atom (atomistic) simulations. In addition to just requiring fewer calculations, CG simulations take less time because the energy landscape is smoother, because of fewer

beads. CG models are being increasingly applied to a wide range of questions in structural biology. An example illustrating the dimerization of two α-helices embedded in a membrane is shown in Figure 8.1.13. It can be seen in the lower diagram, which plots the helix separation as a function of time, that the two helices initially fluctuate around their initial position but they form a stable dimer with a small, constant separation (d) after about 1 μs in the simulation.

Simulated annealing

The word annealing was originally applied to the heating and cooling of a metal to change its properties. In computed simulated annealing, a starting structure is "heated" to a high temperature by giving the atoms a high kinetic energy. The system is then

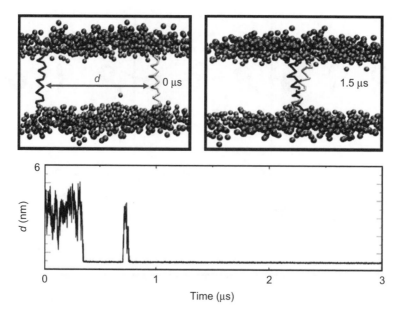

Figure 8.1.13 Coarse-grained simulations of glycophorin trans-membrane helices. The initial system configuration (0 μs) consists of two helices (light and dark gray) inserted in a lipid bilayer in a parallel orientation with a separation of $d \approx 5$ nm. The choline, phosphate, and glycerol backbone particles of the lipid are shown. The snapshot at 1.5 μs illustrates the stable trans-membrane helix dimer that is formed after ~1 μs. The interhelix distance (d) as a function of time during the simulation is shown below (adapted with permission from *Acc. Chem. Res.* 43, 388–96 (2010)).

allowed to cool slowly with the starting structure evolving towards a lower energy final structure. The potential energy landscape of a protein is complex, with many local minima. Simulated annealing protocols can overcome these local minima so that a global minimum (Figure 8.1.9) is found. This process, which can be repeated several times, is also useful when MD simulations are being used to calculate NMR structures (see below). Note this approach is generally much more likely to find the global minimum that the steepest descent and other methods introduced above.

Experimental data and molecular dynamics simulations

The results of MD simulations can be compared with a wide range of experimental techniques. For example, ion conductance experiments in electrophysiology (Chapter 3.6) can be related to calculated ion paths. Large conformational changes (as observed, for example, in the enzyme adenylate kinase) can be computed and compared to experimentally observed distance changes measured by FRET (Chapter 5.5) or PELDOR (Chapter 6.2).

MD simulations have also become an integral part of structure determination protocols in both NMR and X-ray crystallography. It is relatively easy to add **experimental restraints** to MD simulations, as energy minimization and exploration of available conformational space are intrinsic to MD protocols. Versions of MD simulation packages, specifically designed to incorporate experimental restraints, are available (e.g. "Crystallographic and NMR System (CNS)" from Axel Brünger's laboratory (**http:// cns-online.org/v1.3/**)).

NMR protein structure calculations begin by generating a coordinate set with the correct amino acid sequence. This will initially bear no relation to the actual structure. A new experimental energy term, E_{NMR}, is added to the empirical force fields so that they have the form $E = E_{bonded} + E_{non\text{-}bonded} + E_{NMR}$. MD is then applied, with various experimental restraints (NOE, J, RDC; see Chapter 6.1) included in the E_{NMR} term. The restraints are not entered as precise values; instead, they are allowed to take up a range of values, depending on the accuracy of the experimental information. This is done with a "pseudo-potential" of the sort shown in Figure 8.1.14. This particular example is

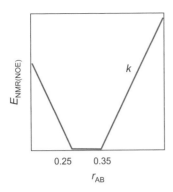

Figure 8.1.14 An example of a pseudo-potential representing an NOE experimental restraint.

for an experimental NOE observed between A and B; an energy penalty is introduced if A and B are further apart than 0.35 nm or closer than 0.25 nm. The slope of the penalty function, k, can also be adjusted.

In NMR structure determination procedures, an ensemble of structures is created by repeating the calculation several times. This is done to explore therange of conformations that are found to be consistent with experimental restraints. To create an ensemble (see Figure 6.1.47), we either begin with different starting structures or with the same starting structure but different initial conditions (e.g. velocities), the variation is generated by a random number generator.

X-ray **crystallography refinement** is also possible using MD. In this case, the experimental information comes from the difference between the observed and the calculated diffraction pattern. The energy term added to simulated annealing protocols is thus of the form $E_{crystal} = \Sigma(|F_o| - |F_{calc}|)^2$ (see Chapter 4.3).

Other aspects of molecular modeling

Monte Carlo methods

An alternative approach to MD simulations for exploring conformational space is the Monte Carlo (MC) method which uses statistical searching procedures. The name, derived from the famous Monaco casino, emphasizes the importance of chance in the methods. **Monte Carlo** methodology is applied in many fields, from economics to traffic control. MC treatments of molecules aim to explore the conformational space available to a system, so that some average property can be determined. MC methods are good at finding the global energy minimum in a system (Figure 8.1.9); they do this by exploring energy landscapes (Figure 8.1.9) with random probing of the molecular geometry.

One MC sampling scheme is called the Metropolis **algorithm**, named after a Los Alamos physicist; it generates a new configuration at each step of the minimization procedure. A random change is made to the coordinates of an atom. The size of the change is controlled by a parameter δr_{max}, so that $-\delta r_{max} < \delta r < +\delta r_{max}$. After the change in r, the nergy of the new configuration, E_v, is calculated using the empirical force fields described above. E_v is then compared to the old energy, E_σ. If $E_v < E_\sigma$ the move is accepted and the new energies are used as the starting point for the next step. If $E_v > E_\sigma$ then the new state is also accepted, but with a probability that depends on the Boltzmann distribution (Box 2.2 Chapter 5.1), $\exp[(E_\sigma - E_v)/kT]$. This probability will be very small if $E_v \gg E_\sigma$. A new move δr is then made and so on. The freedom to choose any move, δr, makes the method very flexible.

Statistical mechanics

Statistical mechanics gives a view of energy (heat) in terms of the structure and dynamics of matter. Statistical mechanics can be used to relate the output of MD or MC simulations to the observed macroscopic properties of a system, for example the energetics of conformational changes, the free energy of a drug binding to a receptor (see below). It also provides a way of evaluating thermal properties (Chapter 3.7) in terms of molecular motion (Box 2.2).

Statistical mechanics can be used to characterize the thermodynamic state of a system using the large number of geometries and their potential energies generated by MD or MC methods. Experimental observables, such as heat capacity, are **ensemble averages**, while the output from an MD simulation is a **time average**. If we measure the position of N atoms as a function of time, the ensemble average is the average position of all atoms at one specific time; the time average is the average position of one molecule over the time period. The **ergodic hypothesis** in statistical mechanics states that the time average equals the ensemble average as long as the system explores all states (i.e. if the simulation is long enough).

Free energy perturbation/binding

There are many instances where information about the differences in free energy between two states would be useful. Examples include the energy differences between two protein conformations, or a ligand-bound and a ligand-free state. We might also want to calculate the effect of protein environment on the pK_a of a histidine sidechain or the effect of changing one amino-acid for another in a protein. Energy differences between states can be computed in several ways. One popular method is **free energy perturbation** (FEP). FEP uses statistical mechanics to compute free energy differences from MD or MC simulation data. The free energy difference on going from state A to state B is defined as $\Delta G_{AB} = G_B - G_A = kT \ln K$, where K is related to an average observed energy difference in a simulation. FEP calculations converge more readily when the difference between the two states is relatively small, for example exchange of one ligand for a similar one, as shown in Figure 8.1.15, where the relative binding affinity of L1 and L2 is $\Delta G_2 - \Delta G_1$, often written as $\Delta\Delta G$. Because Figure 8.1.15 represents a thermodynamic cycle, we also have $\Delta G_2 - \Delta G_1 = \Delta G_4 - \Delta G_3$.

In the example in Figure 8.1.15, the system is biased to move along a non-physical "alchemical" reaction coordinate which changes A into B (alchemy tried to change lead into gold!). It is, however, also possible to calculate free energy profiles using "real"

space, as well as "alchemical" space. For example, an ion can be forced to sample the entire pore of an ion channel (the reaction coordinate); this allows the potential of mean force to be computed as the ion moves along this reaction coordinate with a 1D energy profile.

Docking

An ability to predict the binding of ligands to macromolecules would be of great practical value in characterizing interacting partners and in predicting which drugs might bind to a receptor. Computational docking involves the prediction of the three-dimensional structure of a complex, starting from the structures of the individual molecules in their unbound forms. The interactions between receptor and ligand can mainly be described by non-covalent interactions and force fields of the kind already described above and in Chapter 2.1. Computer-based docking protocols combine a "search algorithm" and a "scoring" function (see, e.g., **http://www.pyrosetta.org/tutorials**). The search algorithm explores possible forms of the complex and the scoring function tests the energy of different forms and ranks them.

Different levels of approximation are applied; the level chosen depends on the application and the computational demands. In conformational searching, the rigid-body approximation, where both ligand and receptor are assumed to be rigid, is one of the simplest methods as there are many fewer degrees of freedom. A more realistic approach than the rigid-body one is to allow a flexible ligand to dock on a rigid receptor. Search algorithms then sample conformational space by performing random changes to the ligand geometry. The alteration performed at each step is accepted or rejected based on a probability function similar to that used in Monte Carlo methods. Ideally, receptor flexibility would be taken into account as well as ligand flexibility, but efficient handling of protein receptor flexibility is a major computational challenge. One approach to the receptor flexibility problem is to use an ensemble of conformations as a target for docking instead of a single structure.

Scoring the docked complex is based on calculations of binding energy ($\Delta G = \Delta H - T\Delta S = -RT \ln K$). This is best done by a rigorous approach, e.g. by FEP, but computational demands mean

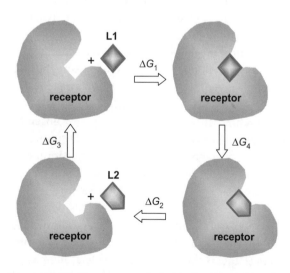

Figure 8.1.15 Free energy cycle for binding ligands L1 and L2 to a receptor.

that simplified empirical scoring schemes are used for most systems, e.g. when screening libraries of compounds to identify potential drugs. Solvent is also often only treated implicitly with the hydrophobic effect, the major driving force in complex formation, treated by calculating the hydrophobic surface area that becomes buried upon ligand binding.

Example NMR and computer based docking

Experimental data about ligand/receptor interfaces can be included in docking programs. This information can come from several sources, including mutagenesis coupled with binding data, or from NMR experiments. NMR shifts are induced at a receptor ligand interface when a complex is formed (see, for example, Figure 6.1.52). An example of an NMR docking experiment is shown in Figure 8.1.16. The structure of the complex was obtained here using computer-based docking procedures of the sort described above, combined with NMR shift perturbations, observed when the complex formed. One widely used data-driven biomolecular docking software package is HADDOCK (**http://www.nmr.chem.uu.nl/haddock/**).

Hybrid methods

Classical molecular mechanics has been very successful in giving information about molecular conformation and conformational changes, but some features, such as bond making or breaking cannot be modeled by classical methods alone. Aspects of quantum mechanics (Tutorial 9) are required. In principle, the wave functions of all the atomic particles in a system could be solved, but this is not a practical proposition for systems larger than a few hundred atoms. To address this limitation, mixed quantum and molecular mechanical approaches are being developed. In these hybrid methods, a small part of the system is treated quantum-mechanically, frequently an enzyme active-site, while the remainder of the system is treated classically. An important advantage of hybrid methods is the reduction in computation time. Simple *ab-initio* quantum mechanical calculations typically scale as N^3, where N is the number of atoms involved, while molecular mechanics calculations scale roughly as N^2.

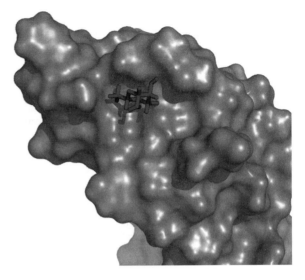

Figure 8.1.16 A complex of a domain from latrophilin and a carbohydrate rhamnose (cyan), obtained using NMR and docking methods (PDB: 2JXA; see *Structure* 16, 944–53 (2008) for further details).

Visualization of molecules

Visualization of molecules has become an increasingly ubiquitous tool in the life sciences and numerous free and commercial visualization programs are available. "Seeing" models of atomic resolution molecular structures helps us to understand a wide range of properties, including catalysis, binding, and dynamics. Developments in visualization tools mirror developments in computer technology. Kendrew and coworkers (1958) built brass and balsa wood models to represent their structure of myoglobin. Fred Richards (1968) introduced an optical comparator that helped later crystallographers to build brass models into electron density maps (see Chapter 4.3). In parallel, Cyrus Levinthal and colleagues at MIT (**http://www.umass.edu/molvis/francoeur/levinthal/lev-index.html**) developed a system that could display macromolecular structures on an oscilloscope in the 1960s. During the 1980s, the most popular computer visualization system was manufactured by Evans & Sutherland (costing about $250 000 in 1985!). The main users then were crystallographers, and specialized programs, such as FRODO and "O", written by Alwyn Jones, were used. Visualization tools were made available to a wider audience in 1997 by Roger Sayle with the freeware program Rasmol.

Although computer screens are two-dimensional, the computer can give a good impression of three dimensions as the operator has the ability to manipulate and rotate the object, and "depth cueing" reduces the intensity of more distant parts. Numerous excellent freeware programs, including PyMOL, COOT, VMD, and Swiss PDB viewer, are now readily available and can be used on any personal computer. Lists of available visualization programs can be found at **http://molvis.sdsc.edu/visres/** and **http://www.umass.edu/microbio/chime/**. Any set of coordinates can be downloaded from the PDB and viewed. Various examples of Pymol-generated images have also been given throughout this book.

Data analysis and statistics

It seems appropriate in this computational chapter to re-emphasize the importance of proper analysis of data in all aspects of biophysics and molecular biology. Regression methods were mentioned in Chapter 2.1 and some very basic statistics formulae are given in Appendix 3. A number of data analysis software packages are available, such as Origin, Matlab, and Microsoft Excel; these have powerful data fitting, graphing, and statistical analysis capabilities. Some familiarity with such programs is highly recommended.

 ## Summary

1 The computer has become an indispensable tool in biological sciences.
2 Mathematical modeling of molecular networks and systems, systems biology, is an increasingly powerful way of exploring properties of a system that emerge when it is viewed as a whole.
3 The World Wide Web gives ready access to a wide range of databases and analysis software that form the basis of bioinformatics. Numerous computational tools are available online that provide ready access to curated information and perform functions like sequence alignment and macromolecule structure prediction.
4 Molecular modeling can be carried out using quantum or classical mechanics. Classical mechanics is the more tractable and versions of this method, especially molecular dynamics (MD) simulations, have become a major tool in studies of biological molecules.
5 MD simulations are often incorporated in structure determination procedures for NMR and X-ray crystallography. The output from MD simulations can be compared to a wide range of data derived from a wide range of experimental data, thus giving insight into molecular mechanisms.
6 Other important simulation methods include Monte Carlo, free energy perturbation, and docking procedures.
7 Visualization of molecules with software that operates on personal computers is a powerful aid to understanding molecules.

 ## Further reading

Useful websites
http://vcell.org/vcell_software/login.html
www.ebi.ac.uk/embl/index.html
http://cmr.jcvi.org/tigr-scripts/CMR/CmrHomePage.cgi

www.ensembl.org
http://www.ebi.ac.uk/uniprot/
www.rcsb.org/pdb/home/home.do
http://vit-embnet.unil.ch/MD_tutorial/
http://molvis.sdsc.edu/visres/

Mathematical modeling of systems

Phillips, R., Kondev, J., and Theriot, J. *Physical Biology of the Cell* Garland, 2008.

Allman, E.S., and Rhodes, J.A. *Mathematical Models in Biology: An Introduction* Cambridge University Press, 2004.

Novák, B., Tyson, J.J., Győrffy, B., and Csikász-Nagy, A. Irreversible cell cycle transitions due to systems-level feedback. *Nat. Cell Biol.* 9, 724–8 (2007).

Noble, D. Systems biology and the heart. *Biosystems* 83, 75–80 (2006).

Bioinformatics

Lesk, A.M. *Introduction to Bioinformatics* (2nd edn) Oxford University Press, 2005.

Kelley, L.A., and Sternberg, M.J. Protein structure prediction on the Web: a case study using the Phyre server. *Nat. Protocols* 4, 363–71 (2009).

Molecular modeling

Leach, A.R. *Molecular Modeling: Principles and Applications* (2nd edn) Pearson, 2001.

Karplus, M., and McCammon, J.A. Molecular dynamics simulations of biomolecules. *Nat. Struct. Biol.* 9, 646–52 (2002).

Sherwood, P., Brooks, B.R., and Sansom, M.S. Multiscale methods for macromolecular simulations. *Curr. Opin. Struct. Biol.* 18, 630–40 (2008).

Gilson, M.K., and Zhou, H.X. Calculation of protein-ligand binding affinities. *Annu. Rev. Biophys. Biomol. Struct.* 36, 21–42 (2007).

Kitchen, D.B., Decornez, H., Furr, J.R., and Bajorath, J. Docking and scoring in virtual screening for drug discovery: methods and applications. *Nat. Rev. Drug Discovery* 3, 935–49 (2004).

Visualization

O'Donoghue, S.I., Gavin, A.C., Gehlenborg, N., Goodsell, D.S., Hériché, J.K., Nielsen, C.B., North, C., Olson, A.J., Procter, J.B., Shattuck, D.W., Walter, T., and Wong, B. Visualizing biological data: now and in the future. *Nat. Methods* 7(3 Suppl.), S2–4 (2010).

 ## Problems

8.1.1 A good way to learn about much of the content in this chapter is to explore the databases, tools, and tutorials that are available on the Web. Some bioinformatics examples, concerning sequences and structure prediction, were already given in Figures 8.1.1–5. Choose a protein or gene of interest and analyze it using some of the other available tools—for example look for domain structure with Pfam (**http://pfam.sanger.ac.uk/search**), find homologues (**http://expasy.org/tools/scanprosite/**), carry out a phylogenetic analysis (**http://www.phylogeny.fr/**), predict glycosylation sites (**http://www.cbs.dtu.dk/services/NetOGlyc/**), etc.

8.1.2 There are a number of useful tutorials on the Web about how to use various molecular dynamics and docking packages. These include one on the formation of a drug enzyme complex (**http://davapc1.bioch.dundee.ac.uk/prodrg/gmx.pdf**), one on NMR structure calculation (**http://tesla.ccrc.uga.edu/courses/BioNMR2006/labs/apr15_cns-mac.pdf**), and one on using AMBER (**http://ambermd.org/tutorials/**). Explore some of these tutorials.

8.1.3 Use a freeware molecular visualization tool (e.g. Pymol) to observe and manipulate a protein of interest; see **http://www.ks.uiuc.edu/Training/Tutorials/vmd/tutorial-html/**.

8.1.4 It is important to have some measure of assessing sequence similarity. One method is to use a Hamming distance, which is the number of positions with mismatching characters in two strings of equal length. Another measure is the Levenshtein distance, which is the minimum number of "edit operations" required to change one string into another. The strings need not be equal length for the Levenshtein distance method and the edit operations include deletion, insertion, or alteration of a single character. What is the Hamming distance between the strings in the words *animal* and *mammal?* What is the Levenshtein distance between *mammal* and *marsupial?*

8.1.5 A two-dimensional harmonic potential energy surface can be represented by $E(x,y) = 4x^2 + y^2$, as shown in the figure. The plots of this function are shown as contours with energy values between 0.4 and 4 in steps of 0.4 units (black, lowest energy, outer ellipse; light gray, highest energy, central ellipse). Carry out a simple "steepest decent" minimization. Start at the cyan point on the $E = 4$ contour, where $y = 1$ and $x = \sqrt{((4 - 1)/4)} = 0.87$ and draw a line (e.g. on a photocopy) that represents the direction of the derivative of $E(x,y)$. Estimate the position of the minimum and the direction of the next line in the iteration.

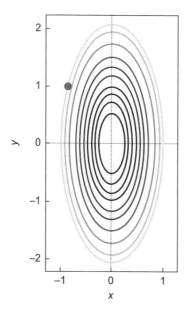

9 | Tutorials

"For the things we have to learn before we can do, we learn by doing"

Aristotle, ~300 BC

"The integrals which we have obtained are not only general expressions which satisfy the differential equation, they represent in the most distinct manner the natural effect which is the object of the phenomenon... when this condition is fulfilled, the integral is, properly speaking, the equation of the phenomenon"

Jean-Baptiste-Joseph Fourier, 1822

"All science is either physics or stamp collecting"

Attributed to Lord Rutherford, 1871–1937

"Those who are not shocked when they first come across quantum theory cannot possibly have understood it."

Neils Bohr, 1958

This section presents 11 tutorials on various topics that are useful for understanding other parts of the book. They range from an introduction to the major molecules found in biology to aspects of classical physics and quantum mechanics. The concepts are reinforced with some worked examples.

Tutorial 1 Biological molecules

The aim of this tutorial is to remind readers about some of the molecules of life, their nomenclature and some of their properties.

Nucleic acids are made from linear chains of nucleotides (Figure T1A). Phosphate groups bridge the 3′ and 5′ positions of successive ribose rings in ribonucleic acid (RNA); four different bases, adenine (A), cytidine (C), guanine (G), and uracil (U), are attached to the ribose. In deoxyribonucleic acid (DNA), the ribose rings have one oxygen atom less (deoxyribose) and the uracil base is replaced by thymine; see also Figure 2.1.7.

Proteins are made from linear chains of amino acids (Figure T1B). Twenty amino acids are naturally found in proteins; amino acids are defined by their sidechains (R_1, R_2, etc.); see also Table T1A and Figure 2.1.5.

A polysaccharide is usually a branched chain of sugar residues (Figure T1C). Around eight different sugar groups can combine in various ways; a relatively common branched structure is shown on the right.

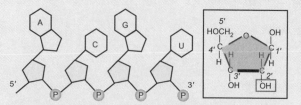

Figure T1A RNA with four bases. Ribose groups are shaded in gray; a more detailed view of a ribose ring is also shown on the right; in DNA, one oxygen is removed from the OH marked with a cyan rectangle.

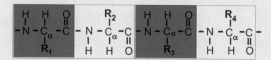

Figure T1B A protein with four amino acids and sidechains R_1–R_4.

Table T1A The 20 amino acids, their sidechains, and properties

Name	"Aaa" abbreviation	"A" abbreviation	Mass*	pK_a sidechain**	R (sidechains) (bold = joined in rings)
Alanine	Ala	A	71.08	–	CH_3-
Arginine	Arg	R	156.19	12.5	$H_2N^+ = CNH_2-NH-(CH_2)_3-$
Asparagine	Asn	N	114.1	–	$H_2N-CO-CH_2-$
Aspartic acid	Asp	D	115.09	3.9	$^-OOC-CH_2-$
Cysteine	Cys	C	103.14	8.4	$HS-CH_2-$
Glutamine	Gln	Q	128.13	–	$H_2N-CO-(CH_2)_2-$
Glutamic acid	Glu	E	129.12	4.1	$^-OOC-(CH_2)_2-$
Glycine	Gly	G	57.05	–	$H-$
Histidine	His	H	137.14	6	$NH-CH = H^+N-CH = \mathbf{C}-CH_2-$
Isoleucine	Ile	I	113.16	–	$CH_3-CH_2-CH(CH_3)-$
Leucine	Leu	L	113.16	–	$(CH_3)_2-CH-CH_2-$
Lysine	Lys	K	128.17	10.5	$H_3N^+-(CH_2)_4-$
Methionine	Met	M	131.2	–	$CH_3-S-(CH_2)_2-$
Phenylalanine	Phe	F	147.18	–	$Ph-CH_2-$
Proline***	Pro	P	97.12	–	$\mathbf{N}_2H-(CH_2)_3-$
Serine	Ser	S	87.08	–	$HO-CH_2-$
Threonine	Thr	T	101.11	–	$CH_3-CH(OH)-$
Tryptophan	Trp	W	186.21	–	$\mathbf{Ph}-NH-CH = \mathbf{C}-CH_2-$
Tyrosine	Tyr	Y	163.18	10.5	$HO-Ph-CH_2-$
Valine	Val	V	99.13	–	$(CH_3)_2-CH-$

* The masses given here refer to the species NHCH**R**CO; this means that the mass of a protein of any known sequence can be determined by adding together these values together plus the mass of one H_2O molecule.

** These pK_a values refer to free amino acids. Actual pKa values can be very different when incorporated in the environment of a folded protein.

*** Proline is a special amino acid with a cyclic connection between the C_α and NH groups.

Non-covalent assemblies of lipids are the major constituent of cell membranes. A major form is a bilayer structure where the hydrophilic head groups of the lipids are in the aqueous phase while their hydrophobic fatty acid tails are in the bilayer core (Figure T1D). Biological membranes are dynamic and have many embedded proteins that carry out a range of functions (Chapter 3.6).

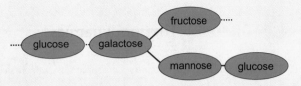

Figure T1C A branched polysaccharide consisting of four different sugar residues.

Worked example

Using the amino-acid table and the amino-acid composition of a small trypsin inhibitor ($A_6C_6D_2E_2F_4G_6I_2K_4L_2MN_3P_4QR_6$ ST_3VY_4), what is the molecular weight and the pI of the protein (pI is defined in Chapter 2.1)? Note that the six cysteines form three cystine disulfide bridges.

Answer: Mol. wt = $6 \times 71.08 + 6 \times 103.14 + 2 \times 115.09 + 2 \times 129.12 + 4 \times 147.18 + 6 \times 57.05 + 2 \times 113.16 + 4 \times 128.17 + 2 \times 113.16 + 131.2 + 3 \times 114.1 + 4 \times 97.12 + 128.13 + 6 \times 156.19 + 87.08 + 3 \times 101.11 + 99.13 + 4 \times 163.18 = 6499.59$. We need to add one water molecule (18) and subtract six protons (6) because of SS bridge formation. Thus, the molecular weight is 6511.6 Da. The pI = $(6 \times 12.5 + 4 \times 10.5 + 2 \times 4.1 + 2 \times 3.9)/14 = 9.5$.

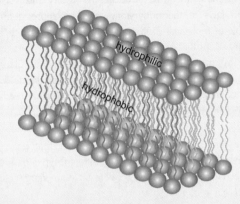
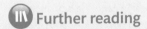

Figure T1D Schematic illustration of a lipid bilayer with hydrophilic head groups outside and hydrophobic fatty acid chains pointing into the interior.

Further reading

Berg, J.M., Tymoczko, J.L., and Stryer, L. *Biochemistry* (6th edn) WH Freeman, 2006.

Voet, D. and Voet, J.G. *Biochemistry* (4th edn) Wiley, 2010.

Tutorial 2 **Thermodynamics**

The aim of this tutorial is to introduce some key concepts about thermodynamics, which is the study of the energetics of a system.

Life depends on the redistribution of energy. The study of thermal energy (heat) has a fascinating history; the first and second laws of thermodynamics and an appreciation of the role of molecular motion (Box 2.2) emerged in the 19th century. Contributions to developments in the field were made by many famous names, including Rankine, Clausius, Lord Kelvin, Maxwell, and Boltzmann. Some key points about thermodynamics are listed below (see also "Further reading" material and worked examples).

- The first law of thermodynamics states that energy is conserved. Conversions between different types of energy (mechanical, chemical, electrical) result in no energy change ($\Delta U = 0$).

- There are two main types of energy: work, which is associated with orderly movement (e.g. climbing a stair), and heat, which is associated with disorderly movement (the hotter a liquid the faster the molecules move). From the first law, $\Delta U = \Delta q + \Delta w$ where Δw is the work done on a system and Δq is the change in heat when it goes from an initial to a final state.

- Enthalpy is a measure of the total energy of a thermodynamic system. Enthalpy change (ΔH) at constant pressure is $\Delta H = \Delta U + P\Delta V$, where $P\Delta V$ is the energy required for expansion (work). For most biological reactions, the pressure is constant and the expansion is small so $\Delta H \approx \Delta U \approx \Delta q$. This heat change can be measured directly in a calorimeter (Chapter 3.7). ΔH is measured in units of kJ mol^{-1} (see Appendix A1.2).

- A reaction is endothermic if heat flows into the sample (ΔH is positive); it is exothermic if heat flows out of the sample (ΔH is negative).

- Entropy is a measure of the disorder of a system (the hotter a liquid, the more disordered it is and the higher its entropy). Entropy change (ΔS) multiplied by temperature is a measure of heat energy change. The units of S are J K^{-1} mol^{-1}.

- The Gibb's free energy, G, is defined as $G = H - TS$. For a reaction at constant temperature and pressure, the free energy change is $\Delta G = \Delta H - T\Delta S$. ΔG must be negative for a reaction to proceed spontaneously. $\Delta G = 0$ at equilibrium.

- For an isolated system at constant temperature and pressure, $\Delta H = 0$ (the first law); thus the $\Delta G < 0$ condition for a reaction to proceed means that $\Delta G < T\Delta S$ or $\Delta S > 0$.

This implies that the entropy, or disorder, of a system always increases. This statement is a version of the second law of thermodynamics.

- For the equilibrium reaction $A + B \rightleftharpoons C + D$, the equilibrium constant is defined as $K_{eq} = [C][D]/[A][B]$ where [ ] represents the concentrations of A, B, C, and D at equilibrium.

- The free energy change for a reaction at any concentration is $\Delta G = \Delta G° + RT\ln K_{eq}$, where $\Delta G°$ is the value when A, B, C, and D are in their standard states; R is the gas constant, T is the temperature in Kelvin, and ln is the natural logarithm. As $\Delta G = 0$ at equilibrium, $\Delta G° = -RT \ln K_{eq}$.

- The standard state of a solute is 1M (1 mole per liter of solution) at pH = 1. For convenience, the biochemical standard state of H$^+$ is defined as 10^{-7}M (pH 7). Standard states are denoted by the superscript ° and the biochemical standard state is denoted by °'.

- Instead of free energy, ΔG, the chemical potential, μ, is often used to describe mixtures of components where μ_x is the free energy of the component x; $\mu_x = \mu_x° + RT\ln[X]$, where [X] is the concentration of the solute component, X, and $\mu_x°$ is the chemical potential of its standard state.

- Redox reactions: Some reactions involve the transfer of electrons between reactants. Electron loss corresponds to oxidation, electron gain to reduction. These reactions generate an electrical potential E that is related to ΔG by the equation $\Delta G = -nFE$ where n refers to the number of transferred electrons and F, the Faraday constant, is a conversion factor equal to 96.5 kJ V^{-1} mol^{-1}.

- The Nernst equation follows from the relationship between ΔG for a reaction and the equation $\Delta G = -nFE$, namely $nFE = nFE° - RT\ln([C][D]/[A][B])$. This equation describes the relationship between the electrical potential and the concentrations of the components in a redox reaction (see Problem 2.1.3).

- Relation of K_{eq} to rate constants: For the equilibrium reaction $A + B \underset{k_{-1}}{\overset{k_{+1}}{\rightleftharpoons}} C + D$, the forward rate ($k_{+1}[A][B]$) must equal the backward rate ($k_{-1}[C][D]$); thus $[C][D]/[A][B] = K_{eq} = k_{+1}/k_{-1}$.

The thermodynamic concepts, summarized above, allow us to predict whether a chemical reaction will proceed or not and to quantify heat related interactions (see Chapter 3.7). The following worked examples illustrate how we can use thermodynamics to give useful information about equilibrium reactions.

Worked example 1

For the hydrolysis of ATP to ADP, $\Delta G^{\circ\prime} = -33$ kJ mol^{-1}; what is the [ATP]/[ADP] ratio at equilibrium if the inorganic phosphate (P$_i$) concentration is 1 mM, pH = 7, temperature = 25°C? Why is the [ATP]/[ADP] reaction not at equilibrium in the cell?

Answer 1:   We have $\Delta G = \Delta G^{\circ} + RT\ln([C][D]/[A][B])$ for the reaction $A + B \rightleftharpoons C + D$, where ΔG° is the free energy change when A, B C, and D are in their standard states (1 M; 1 atmosphere pressure). $\Delta G^{\circ\prime}$ refers to the biochemical standard state where the H$^+$ concentration is 10^{-7} M (pH = 7) rather than 1 M (when pH = 1).

In this example, the reaction is ATP → ADP + P$_i$ so that $\Delta G = \Delta G^{\circ} + RT\ln([ADP][P_i]/[ATP])$.

At equilibrium and pH = 7, $\Delta G = 0$.

$\therefore \Delta G^{\circ\prime} = -RT\ln([ADP][P_i]/[ATP]) = -8.31 \times 298 \times \ln([ADP][10^{-3}]/[ATP])$

since $\Delta G^{\circ\prime} = -33\,000$ we have $\exp(33\,000/8.31 \times 298) = 6.13 \times 10^5 = [ADP][10^{-3}]/[ATP]$

$\therefore [ADP]/[ATP] = 6.13 \times 10^8$ (or [ATP]/[ADP] = 1.6 × 10^{-9}).

No work could be done in the cell by ATP if ATP and ADP were at equilibrium.

Worked example 2

Calculate ΔS° and K_{eq} for a reaction at 17°C if $\Delta G^{\circ} = -20$ kJ mol^{-1} and $\Delta H^{\circ} = -40$ kJ mol^{-1}. What is the value of K_{eq} at 40°C?

Answer 2: $\Delta G^{\circ} = -RT\ln K \therefore \ln K_1 = \Delta G^{\circ}/RT_1 = \Delta H^{\circ}/RT_1 + \Delta S^{\circ}/R$

$\therefore$ if ΔH° and ΔS° are independent of temperature, $\ln K_1 = -\Delta H^{\circ}/RT_1 + \Delta S^{\circ}/R$ and $\ln K_2 = -\Delta H^{\circ}/RT_2 + \Delta S^{\circ}/R$;

$\therefore \ln K_1/K_2 = -\Delta H^{\circ}/R(1/T_1 - 1/T_2)$

since $\Delta G^{\circ} = \Delta H^{\circ} - T\Delta S^{\circ}$

$-20\,000 = -40\,000 - 290\Delta S^{\circ}$

$\therefore \Delta S^{\circ} = -20\,000/290 = -68.97$ J K^{-1} mol^{-1}

$K_{290} = \exp(-\Delta G^{\circ}/R \times 290) = \exp(20\,000/8.31 \times 290) = 4020$

$K_{313} = \exp(-\Delta G^{\circ}/R \times 313)$ where $\Delta G^{\circ} = -40\,000 + 68.97 \times 313 = -18\,412$

$\therefore K_{313} = \exp(18\,412/8.31 \times 313) = 1187$

 Further reading

Atkins, P., and de Paula, J. *Physical Chemistry for the Life Sciences* (2nd edn) Oxford University Press, 2010.

Price, N.C., Dwek, R.A., Ratcliffe, R.G., and Wormald, M.R. *Physical Chemistry for Biochemists* (3rd edn) Oxford University Press, 2001.

Tutorial 3 Motion and energy of particles in different force fields

The aim of this tutorial is to introduce some basic concepts about how various applied forces (gravity, electric and magnetic fields, and a spring) affect the motion and energy of a particle. These ideas are important for understanding many topics, including the transport phenomena discussed in Section 3.

Newton's laws

First law

If there is no resultant force, a stationary object will stay at rest and a moving object will maintain a constant velocity.

Second law

The resultant force = mass × acceleration (unit: newton).

Third law

If A exerts a force on B then B exerts an equal and opposite force on A.

Other aspects of motion

Velocity is a vector, v = displacement (x)/time = $dx/dt = \dot{x}$ (units: m s^{-1}).
Acceleration = $dv/dt = d^2x/dt^2 = \ddot{x}$ (units: m s^{-2}).

Drag and terminal velocity

When an object moves through a gas or a liquid there is resistance in the form of drag (friction); this frictional force, F_f, is proportional to the velocity (v), i.e. $F_f = fv$ where f is the frictional or drag coefficient. As discussed in Chapter 3.1 (Box 3.1), f is related to the diffusion coefficient, D, by $f = kT/D$ or RT/N_AD, where T is the absolute temperature, k is the Boltzmann constant, R is the gas constant, and N_A is Avogadro's number. The object will reach a terminal (equilibrium) velocity when the applied force, F, is equal to the drag force. It follows that the terminal velocity is $v = F/f$. (Note that a parachute increases f, thus reducing the terminal velocity!)

Circular motion

The rate of rotation, ω, of a particle moving in a circle (Figure T3A), radius r, is measured in radians per second (rad s^{-1}). As 2π radians = 360°, the frequency of rotation is $v = 2\pi/\omega$. The circumference of the circle is $2\pi r$ so the velocity $v = 2\pi r/v = \omega r$. This velocity is not in a straight line and if the velocity at P is v it will be v' at some time later. Figure T3B shows that the centripetal acceleration is $\mathbf{a} = \delta v/\delta t$ and $\delta\theta = \delta v/v$. Thus, $\mathbf{a} = v\delta\theta/\delta t$. We also have $\delta\theta = \omega\delta t$ (Figure T3A), so $\mathbf{a} = v\omega = r\omega^2$. The particle thus undergoes constant centripetal acceleration, caused by a centripetal, or centrifugal, force $F = m\omega^2r = mv^2/r$ that is directed towards the center of the circle.

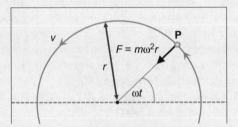

Figure T3A Representation of circular motion. ωt is the angle turned in time t. F is the centripetal force.

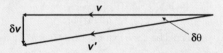

Figure T3B The change in velocity caused by motion in a circle (see text).

Linear momentum

The momentum of an object is p = mass × velocity = mv. Momentum is conserved (units: kg m s^{-1}).

A torque or moment is the tendency of a force to rotate an object about an axis; (units: newton m); see Figure T3C.

Figure T3C The torque or moment of P about A is $F \times r$ (N m) where F is an applied force.

Angular momentum

Angular momentum (L) is a rotational analog of linear momentum. Like p, it is conserved in the absence of friction and it has inertia; an example of the conservation of L is the increase in rotational speed of a spinning skater as their arms are contracted (units: kg m^2 s^{-1}).

Kinetic and potential energy (KE) is the energy something has because it is moving; for an object, mass m, velocity v, KE = $1/2mv^2$ (unit: joule). Potential energy (PE) is the energy something has because of its position or state; PE exists when a force acts upon an object tending to restore it to a lower energy configuration. For a particle mass m in a gravitational field (see below), at height h, the PE $= mgh$ (unit: joule).

The effect of a gravitational force

Newton's law of gravitation states that two masses, m_1 and m_2, a distance r apart have an attractive force between them, defined by $F = Gm_1m_2/r^2$, where G is a universal constant (6.67428×10^{-11} N m^2 kg^{-2}). For a mass m on the surface of the Earth (mass M, radius R) we can write this as $GmM/R^2 = mg$. $GM/R^2 = g$ is the acceleration owing to gravity at the Earth's surface; $g \sim 9.8$ m^2 s^{-1}.

The potential energy of an object, mass m, height h in gravity g, is mgh (unit: joules).

The trajectory of a particle in gravity

A trajectory is the position of an object as a function of time. An example is shown in Figure T3D, where a ball, thrown at angle θ with initial velocity v_o, is subject to gravitational acceleration, g. The acceleration in the y direction is $-g$, while it is 0 in the x direction if air friction is negligible. The initial velocities are v_{oy} and v_{ox} in the x and y directions. In the x direction, we also have $v_x = v_{ox}$ and in the y direction $v_y = v_{oy} - gt$. We can also work out the distance moved (velocity × time) in the x and y directions, i.e. $x = v_{ox}t$ and $y = \int(v_{oy} - gt)$ d$t = v_{oy}t - gt^2/2$. We can combine these equations and eliminate t by setting $t = x/v_{ox}$ to obtain $y = v_{oy}x/v_{ox} - gx^2/2v_{ox}^2$. As $v_{oy}/v_{ox} = \tan \theta$ and $v_{ox} = v_o\cos \theta$, we obtain $y = x\tan \theta - gx^2/(2v_o^2\cos^2\theta)$. This function, which depends on x and x^2, is a parabola that describes the trajectory of the ball.

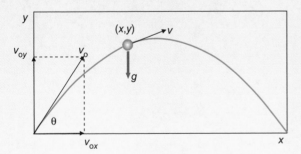

Figure T3D The trajectory of a ball in the Earth's gravity.

The effect of an electric field

A voltage V applied between two plates separated by a distance d generates an electric field $E = V/d$ (Figure T3E). The units of E are V m^{-1} or N C^{-1} (see Appendix A1.2 and Box 2.3).

The potential energy of q brought from infinity to a position where the voltage is V is Vq.

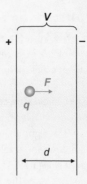

Figure T3E The force on a charge q in a voltage V.

The trajectory of a particle in an electric field

Beams of electrons and other charged particles are deflected by an electric field. A particle, charge q and mass m in an electric field E, experiences a force, $F = Eq$. From Newton's second law we also have $F = ma$, so $E = ma/q$; the application of E thus causes a deflection that depends on m/q. The resultant path of particles experiencing a constant electrical field is a parabola, just like the ball thrown in a gravitational field.

The effect of a magnetic field

A magnetic field (**B**) generates a force (**F**) on a current (**I**). (Properties of magnetic fields are also discussed in Box 6.1; **B**, **F**, and **I** are also written in bold here to emphasize their vector properties). **F** can be described by a vector cross product (Appendix A2.2): $\mathbf{F} = \mathbf{B}l \times \mathbf{I}$ where l is the length of the wire. The direction of the force is given by Fleming's left hand rule (Figure T3F).

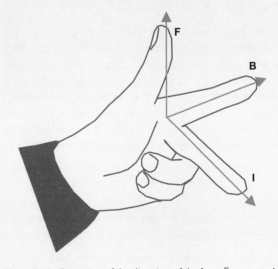

Figure T3F Illustration of the direction of the force **F** generated by a current **I** in a field **B**.

The trajectory of a particle in a magnetic field

A particle of charge q moving at a steady velocity, v, is equivalent to a current of q/t in a wire of length vt. The magnitude of the magnetic field-induced force in a uniform magnetic field is $BIl = B(q/t)(vt) = Bqv$, where B is the magnitude of the magnetic field. The current, and the force generated, thus *depend on the velocity* of the particle. Note also that the force on q is at right angles to B (Figure T3G); such a force generates a constant acceleration of q towards the centre of a circle (see Figure T3A above), which is a centripetal force $F = mv^2/r = Bqv$. The radius of q's circular path is $r = mv/qB$, which depends on the particle momentum (velocity × mass).

The equation $r = mv/qB$ describes the path of an ion that is trapped in a circle by a magnetic field; this is what happens in a cyclotron. The ion velocity is $v = Bqr/m$ and as the path of the trapped ion is circular we also have $v = 2\pi vr$, where v is the angular frequency (see above); the circular frequency of q is thus $v = Bq/2\pi m$.

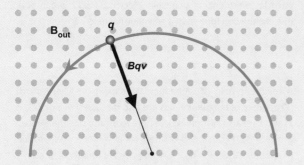

Figure T3G A charged particle q, travelling at speed v, in a magnetic field B that is directed out of the paper, moves in a circle.

The effect of a spring force

Springs obey Hooke's law which means that the extension, x, is proportional to the applied force (F). To find the potential energy stored in a stretched (compressed) spring, we calculate the work done in stretching the spring. The force to stretch a spring varies from 0 at $x = 0$, to $-kx$ at x. The average force applied is thus $\bar{F} = 1/2(0 - F_x) = -1/2kx$. The work done by this force is thus $-1/2kx \times -x = 1/2kx^2$. This work is stored in the spring as potential energy, so the potential energy of a spring is $1/2kx^2$.

Worked example 1

Give two similarities, and one difference, between electric and gravitational fields.

Answer: Both obey an inverse r^2 law for force and an inverse r law for potential. Electric fields can be attractive or repulsive. Gravity is only attractive.

Worked example 2

Why is the path of an electron in a magnetic field circular while it is parabolic in an electric field?

Answer: The electron moves at right angles to **B** and the force is of fixed magnitude and perpendicular to the direction of motion (conditions for circular motion). In an electric field there is constant acceleration at right angles to E but constant velocity perpendicular to E (conditions for a parabolic path).

Worked example 3

At what distance of separation must two 1.00 μC charges be positioned in order for the repulsive force between them to be equivalent to the weight (on Earth) of a 1 kg mass? (see also Box 2.3 and assume g = 9.8 m^2 s^{-1}).

Answer: We have $q_1 = q_2 = 1.0 \times 10^{-6}$ C. From Coulomb's law (Box 2.3), the electrical force is $F_e = q_1q_2/4\pi\varepsilon_o r^2 = (10^{-6})^2/(4\pi \times 8.85 \times 10^{-12}r^2) = 0.009/r^2$. The gravitational force is $F_g = mg = 9.8$ N. Since F_e and F_g are equivalent, we have $r^2 = 0.009/9.8$ or $r = 0.03$ m.

Further reading

Grant, I.S., and Phillips, W.R. *The Elements of Physics* Oxford University Press, 2001.

Tutorial 4 **Electrical circuits**

The aim of this tutorial is to introduce some basic ideas and nomenclature for electrical circuits. These concepts are helpful for understanding aspects of Chapters 3.4 and 3.6.

Some of the nomenclature used in this field is illustrated in Figure T4A; this shows an electrical circuit, consisting of a battery with positive and negative poles, a resistor, and a capacitor.

The current, I, is the flow of charge per unit time. The SI unit of charge is the coulomb (C) and the unit for current is the ampere (A). Current conventionally flows in the opposite direction to electrons.

The unit used to measure electric potential difference is the volt. Meters that measure the current (A) and the potential differences or voltages are shown in Figure T4A; the voltmeters are placed to measure the potential difference across the battery (V_1) and the resistor (V_2).

The resistance of a conductor is represented as a resistor, with resistance R in units of ohms (Ω). The reciprocal of R is the conductance G (unit: Ω^{-1} or siemens).

A capacitor stores energy by separating charges. Capacitance, C, is defined as charge/potential difference = Q/V in units of C V^{-1} (unit: farads).

Ohm's law states that the current, I, passing through a resistor, R, is directly proportional to the voltage, V, across it, i.e. $R = V/I$ or $V = RI$.

The power (unit: watt) dissipated by the resistor is $P = V^2/I = I^2R$ (V obtained by V_2 in Figure T4A).

Circuit analysis

Circuits can be analyzed using some simple rules. One is that for resistors placed in series the total resistance is $R = R_1 + R_2 + $ etc.;

for resistors in parallel, R is given by $1/R = 1/R_3 + 1/R_4 + $ etc. (see Figure T4B). Note that for conductance, $G = G_3 + G_4$.

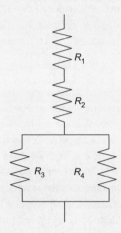

Figure T4B Two resistors in series (R_1 and R_2) and two in parallel (R_3 and R_4).

Worked example

What is the total resistance and current in the circuit shown? What is the power generated: (i) in the 4 Ω resistor; (ii) in the 6 Ω resistor?

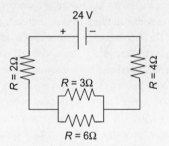

Answer: The resistance of the parallel pair of resistors is given by $1/R = 1/3 + 1/6$, i.e. $R = 18/9 = 2$ Ω. The total resistance of the circuit is 8 Ω so the current is 3 amps. The power dissipated in the 4 Ω resistor is thus 36 watts. The 3 amps will pass through the parallel resistors in the ratio 1:2, i.e. 1 amp will go through the 6 Ω resistor. It will thus dissipate 6 watts.

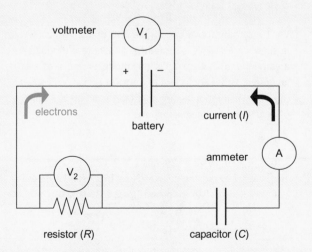

Figure T4A An electrical circuit with a battery, resistor, capacitor, monitoring meters, and a current flow (I).

 Further reading

Grant, I.S., and Phillips, W.R. *The Elements of Physics* Oxford University Press, 2001.

Tutorial 5 Mathematical representation of waves

The aim of this tutorial is to show how waves can be represented and summed. These ideas are important for understanding electromagnetic radiation, scattering, and spectroscopy experiments (Tutorial 8, Sections 4 and 5).

As shown in Figure T5A, a wave can be characterized by three parameters, the amplitude a_o, wavelength λ, and phase ϕ. The wave propagates with velocity c; $\nu = c/\lambda$, where ν is the frequency (Tutorial 8). The equation describing such a wave is $x = a_o\cos(2\pi\nu t + \phi)$, where $2\pi\nu = \omega$ is the angular frequency (Tutorial 3).

Wave summation

When two waves come together, the net observed disturbance is the sum of the individual waves. This superposition gives rise to interference phenomena. Consider first, waves that differ only in phase. We can describe them by $x_1 = a_o\cos(\omega t)$ and $x_2 = a_o\cos(\omega t + \phi)$. Their sum, $x_r = x_1 + x_2$, can be found using the cosine sum formula given in Appendix A2.4, namely, $x_r = 2a_o\cos(\omega t + \phi/2)\cos(\phi/2)$. This resultant wave has angular frequency ω, amplitude $2a_o\cos(\phi/2)$, and phase $\phi/2$; points to note are that x_r has the same frequency as x_1 and x_2 (ω) and that the amplitude is $2a_o$ when $\phi = 0°$ and zero when $\phi = 180°$ (see also Figure 4.1.5).

In general, the summation of waves is tedious using this cosine summation method. It is simpler to represent a wave by a vector with amplitude $|a_o|$ and a direction, defined by a phase angle ϕ. Such a vector has components $a_o\sin\phi$ along the x ($\phi = 0°$) and $a_o\cos\phi$ along the y ($\phi = 90°$) axes, as shown in Figure T5B. We can write this as $z = a_o\cos\phi + ia_o\sin\phi = a_oe^{i\phi}$ using the formulae that define the exponential forms of sines and cosines given in Appendix A2.4. This representation, where the 90° phase-angle axis (y) is labeled as the "imaginary" axis, and the 0° axis the "real" axis (x) is sometimes called an Argand diagram.

Let us now look at wave summation again but, this time, using vectors. We showed above, mathematically, that $x_2 + x_1$ was $2a_o\cos(\omega t + \phi/2)\cos(\phi/2)$. Using vector notation, we first draw a vector $\mathbf{a_o}$ at an angle ϕ, followed by a vector parallel to the x axis ($\phi = 0$). The amplitude of the resultant vector is $2\mathbf{a_o}\cos(\phi/2)$ and the phase angle is $\phi/2$, as shown in Figure T5C; this is the same result as before. This methodology can be

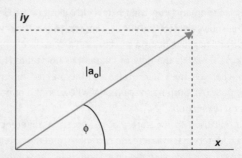

Figure T5B A vector representation of a wave with amplitude $|a_o|$ and phase ϕ.

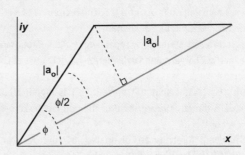

Figure T5C Illustration of the sum of two waves using vectors.

generalized and any number of waves can be summed relatively easily using end to end vectors.

Worked example

We have from above and Figure T5B that $z = a_o\cos\phi + ia_o\sin\phi = a_oe^{i\phi}$. This is sometimes known as a modulus–argument form where $\mathbf{a_o}$ is the modulus and ϕ is the argument. This is equivalent to the form $z = x + iy$. Express (i) $z = 2 + 5i$ in modulus–argument form and (ii) $|z| = 3$ and $\phi = -\pi/10$ in real–imaginary form.

Answer:

(i) $a_o = (2^2 + 5^2)^{1/2} = \sqrt{29}$; $\cos\phi = 2/\sqrt{29}$; $\sin\phi = 5/\sqrt{29}$; $\therefore \phi = 1.19$ radians; $\therefore z = \sqrt{29}(\cos 1.19 + i\sin 1.19)$.

(ii) $z = 3\{\cos(-\pi/10) + i\sin(-\pi/10)\} = 2.853 - 0.927i$.

Further reading

Jordan, D., and Smith, P. *Mathematical Techniques: an Introduction for the Engineering, Physical, and Mathematical Sciences* (4th edn) Oxford University Press, 2008.

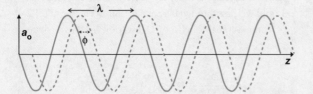

Figure T5A Representation of two waves, amplitude a_o, wavelength λ, and phase difference f, travelling in direction z, with velocity c.

Tutorial 6 **Fourier series, Fourier transforms, and convolution**

The aim of this tutorial is to introduce some mathematical concepts that are essential for understanding many parts of this book. The Fourier transform is very important for understanding material in Sections 4, 5, 6, and 7.

In 1807, Joseph Fourier introduced the idea that a periodic (repeating) function $F(t)$, may be represented by an infinite series of sine and cosine functions: the Fourier series. A square wave, for example, can be represented by the series:

$$f(x) = \frac{4}{\pi}\left(\sin\omega x + \frac{1}{3}\sin 3\omega x + \frac{1}{5}\sin 5\omega x + \ldots\right)$$

This function is plotted in Figure T6A with different terms included in the summation. Note that the representation of the square wave becomes better and better as we add more high frequency (3ω, 5ω, etc.) terms. If we add another term (7ω), the cyan sum will be an even better representation of a square wave. To illustrate the importance of phase, however, the black dashed line in the lower panel in Figure T6A is a plot of the function:

$$f(x) = \frac{4}{\pi}\left(\sin\omega x - \frac{1}{3}\sin 3\omega x + \frac{1}{5}\sin 5\omega x + \frac{1}{7}\sin 7\omega x\right)$$

This illustration, showing the large effect of phase reversal of the 3ω term, emphasizes the importance of phase when representing an object by a Fourier series. This aspect is further discussed in Chapter 4.3.

An extension of the Fourier series discussed above is the Fourier transform (FT). For a function $f(x)$ this is defined as $F(r) = \int_{-\infty}^{\infty} f(x)e^{-2\pi i x s}\,dx$. This has the reversible property $f(x) = \int_{-\infty}^{\infty} F(r)e^{2\pi i x s}\,dr$.

Two examples of FT pairs are shown in Figure T6B. The square wave, the decaying oscillation, and their transforms are very commonly encountered in biophysical experiments.

(i) A square wave transforms to a $\sin r/r$ function. Physical examples of this are diffraction at a slit (Chapter 7.1) and the response to a pulse (Chapter 6.1).

(ii) An oscillating wave, exponentially damped, with decay rate a transforms as:
$e^{-xa}\cos \omega x \Leftrightarrow (a - i(\omega - \omega_o))/(a^2 + (\omega - \omega_o)^2)$ where $\Leftrightarrow$ indicates a FT in either direction. This FT of a decaying oscillation is known as a Lorentzian function; it has both

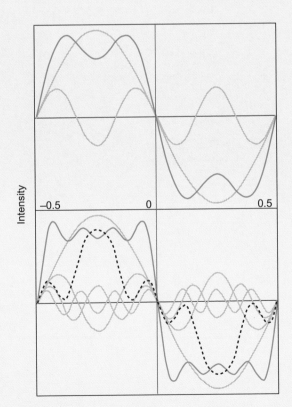

Figure T6A The top panel shows the sum (cyan) of the first two terms in this series (w and 3w shown in gray). The bottom panel shows the sum of the first three terms (w, 3w, and 5w in cyan). The black dashed line illustrates the effect of inverting the phase of one term in the summation (see text).

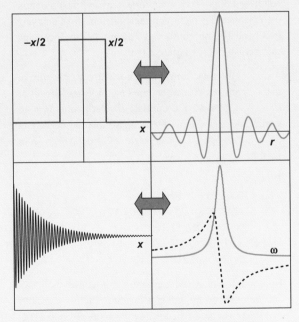

Figure T6B Two Fourier transform pairs: top—a step function of width x, transforms to a $\sin r/r$ function, which oscillates at a frequency related to $1/x$; bottom—a transient oscillation beginning at $x = 0$, with angular frequency ω and decay rate a transforms to a Lorentzian line shape whose real (cyan) and imaginary (dashed line) components are shown. The dashed line is often called the dispersion spectrum and the solid cyan line is the absorption spectrum.

real $[a/(a^2 + (\omega - \omega_o)^2)]$ and imaginary $[(\omega - \omega_o)/(a^2 + (\omega - \omega_o)^2)]$ components. A physical example of this FT pair is a decaying transient, which transforms to an absorption spectrum (e.g. in NMR).

The convolution of two functions $F(r)$ and $G(r)$ is a function . with the property $F(r) . G(r) = f(x) \times g(x)$ where $\times$ is simple multiplication, and $f(x)$ and $g(x)$ are the Fourier transforms of $F(r)$ and $G(r)$, respectively.

Let us use the Lorentzian function above as an example: if $f(x)$ is the function $e^{-xa}\cos r_o x$ and $g(x)$ is the function e^{-xb}, the exponent of their product is related to the sum $(a + b)$. As the linewidth of a Lorentzian line is proportional to a, the convolution will now have a line that is broader or narrower depending on the sign of b. This kind of convolution is often used to change line shapes in spectroscopy. Similarly, a crystal can be considered as the convolution of a lattice and a unit cell. Thus, the observed diffraction pattern is the product of the diffraction pattern of the lattice and the diffraction pattern of the unit cell (see Chapter 4.3).

Worked example

An exponentially decaying cosine wave has a decay constant of 20 ms (this corresponds to the second example of an FT pair illustrated in Figure T6B). What are the linewidths at half height after Fourier transformation and multiplication with the following decay constants: (a) none; (b) a decaying exponential, time constant $\tau = 30$ ms; (c) an increasing exponential, decay constant –30 ms.

Answer: With no convolution applied, the linewidth at half height will occur when $a^2 = (\omega - \omega_o)^2$ (see above) or $a = \omega - \omega_o$. We need to translate angular frequency ω to ν (Hz): $(\omega - \omega_o) = 2\pi\Delta\nu$; we also have that the decay rate $a = 1/\tau$. The linewidth at half height will also run from $-\omega$ to $+\omega$ through ω_o, so we have: $\Delta\nu_{1/2} = 1/\pi\tau$.

Thus, for
  (a) $= 1000/20\pi = 15.9$ Hz;
  (b) $\tau' = 1/(1/20 + 1/30) = 600/50 = 12$ ms
    $\therefore \Delta\nu_{1/2} = 1000/\pi12 = 26.5$ Hz.
  (c) $\tau'' = 1/(1/20 - 1/30) = 60$
    $\therefore \Delta\nu_{1/2} = 1000/\pi60 = 5.3$ Hz.

Note that the observed linewidths, after convolution, can also be obtained by adding and subtracting the linewidth of the convolution function ($\Delta\nu_{1/2} = 1000/\pi30 = 10.6$ Hz). This example shows how spectral lines can be manipulated by simple multiplication in one domain before Fourier transformation to another domain.

 ## Further reading

Jordan, D., and Smith, P. *Mathematical Techniques: an introduction for the Engineering, Physical, and Mathematical Sciences* (4th edn) Oxford University Press, 2008.

Tutorial 7 Oscillators and simple harmonic motion

The aim of this tutorial is to introduce concepts about simple harmonic motion and driven oscillating systems. These concepts are important for an understanding of a number of topics, including scattering and vibrational spectroscopy.

Oscillating systems are ones that exhibit periodic or cyclic motion. One common type of oscillation is simple harmonic motion (SHM). Familiar examples are the pendulum and a mass on a helical spring. A simple harmonic oscillator is one in which the frequency (ν) of oscillation is independent of the oscillation amplitude (a). Consider a mass (m) on a spiral spring. The strength of the spring is defined by a force constant (k). The force exerted on the mass, after it is displaced a distance (x) from its equilibrium position is $F = -kx$ (see also Tutorial 3). The negative sign indicates that the force is to the left in Figure T7A.

A mass subjected to a force undergoes acceleration according to Newton's second law; acceleration is defined as d^2x/dt^2 which is more conveniently written as $\ddot{x}$. Thus, the motion of the mass is described by the equation $m\ddot{x} = -kx$; this is a differential equation that can be solved by a variety of methods; it is, however, instructive to solve it here by considering the circular motion in Figure T7B, as this gives us insight into both the frequency and energy of the system.

Consider a particle (P) rotating at a constant angular speed $\omega = 2\pi\nu$ in a circular path, and particle Q which follows the projection of P, oscillating about O along the horizontal axis (Figure T7B). The velocity of Q is given by differentiation of $x = a\cos\omega t$, i.e. $dx/dt = \dot{x} = -a\omega\sin\omega t$. When $x = a$, the motion will be zero and it will be fastest when $x = 0$. A second differentiation gives the acceleration of Q: $\ddot{x} = -a\omega^2\cos\omega t = \omega^2 x$. Since the time period for a revolution of P is $2\pi/\omega$, the frequency of oscillation is $\nu = \omega/2\pi$. This equation for motion in a circle ($\ddot{x} = \omega^2 x$) has the same form as the equation derived for the motion of the spring ($\ddot{x} = -kx/m$). Applying this solution for the circle case ($\nu = \omega/2\pi$) to the case of the spring, gives the oscillation frequency of the spring as $(k/m)^{1/2}/2\pi$. The kinetic energy of this system is then $KE = 1/2m\,\dot{x}^2 = 1/2ma^2\omega^2\sin^2\omega t$ and the potential energy is $PE = 1/2kx^2$ (Tutorial 3).

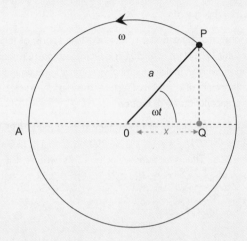

Figure T7B Simple harmonic motion (SHM). P undergoes a circular path. Q, at the projection of P on the horizontal axis through the center, undergoes SHM, back and forth through O and A.

Damped and forced oscillations

The energy of most simple harmonic oscillators decreases with time (it is damped). An external driving force is often also imposed on the oscillating system, e.g. in induced light scattering. A differential equation describing a damped, forced oscillator can be obtained by adding terms to the equation for a spring that we discussed above, i.e. $m\ddot{x} = -kx - f\dot{x} + F_o\cos(\omega t)$ where, as before, $-kx$ is the restoring force from the spring, $\dot{x} = -fdx/dt$ is the damping force opposing the motion, and $F_o\cos(\omega t)$ is the applied driving force.

The graphical solution of this differential equation, a plot of amplitude as a function of the driving frequency ω, is shown in Figure T7C. This response "resonates" (has maximum

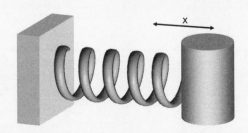

Figure T7A A mass on a spiral spring displaced a distance x with respect to its equilibrium position.

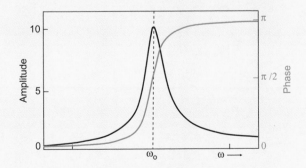

Figure T7C Amplitude (gray) and phase (cyan) of a driven damped oscillator as a function of driving frequency.

amplitude) when $\omega = \omega_0$. The phase difference between the driving oscillation and the output wave is also shown in Figure T7C. Note that at low ω, the amplitude ($x(t)$) and driver ($F_o \cos \omega t$) are in phase; at $\omega = \omega_0$ they are out of phase by 90°; at high ω, they are out of phase by 180°.

Worked example

An object of mass 6.8 kg is attached to a spring of force constant 1780 N m^{-1}. The object is set into simple harmonic motion, with an initial velocity of $v_o = 2$ m s^{-1} and an initial displacement of $x_o = 0.1$ m. Calculate the maximum velocity the object has during its motion.

Answer: The energy of the system at $t = 0$ is KE + PE = $1/2mv_o^2 + 1/2kx_o^2 = 3.4 \times 2 \times 2 + 890 \times 0.1 \times 0.1 = 13.6 + 8.9 = 22.5$. The maximum velocity v_m is when PE = 0. $\therefore a1/2mv_m^2 = 22.5$ or $v_m = (2 \times 22.5/6.8)^{1/2} = 2.57$ m s^{-1}.

 ## Further reading

Grant, I.S., and Phillips, W.R. *The Elements of Physics* Oxford University Press, 2001.

Tutorial 8 Electromagnetic radiation

The aim of this tutorial is to convey the basic properties of electromagnetic radiation: these concepts are essential for understanding Sections 4, 5, 6, and 7 in this book.

Electromagnetic waves

James Clerk Maxwell (1831–1879) combined the laws of electricity and magnetism with those of light to develop the theory of electromagnetic radiation. He postulated that electromagnetic radiation behaves as two wave motions at right angles. One of these waves is magnetic (M) and the other is electric (E) (see Figure T8A). Electromagnetic waves are generated by oscillating electric or magnetic dipoles and are propagated through a vacuum at the velocity of light (*c*). The energies associated with E and M are equal but interactions with the electric wave are usually dominant. As discussed in Tutorial 5, a wave can be characterized by an equation of the form $x = a_o\cos(2\pi vt + \phi)$.

Polarization

As the **E**- and the **M**-components are always perpendicular to each other, it is often sufficient to consider only the **E**-vector when describing an electromagnetic wave. Unpolarized light contains oscillations of the **E**-vectors in all directions perpendicular to the direction of propagation. The degree of polarization can be defined by $P = (I_\parallel - I_\perp)/(I_\parallel + I_\perp)$, where $I_\parallel$ and $I_\perp$ are the intensities of the parallel (∥) and perpendicular (⊥) E waves. Plane-polarized light can be considered to arise from a source that only oscillates parallel to one axis (e.g. $I_\perp = 0$) (Figure T8B).

Waves and their summation were discussed in Tutorial 5. An example involving the summation of similar plane-polarized *y*-axis and *x*-axis waves with the *same phase* is shown in Figure T8C. This produces a plane-polarized wave oriented at 45° to the *x*-axis.

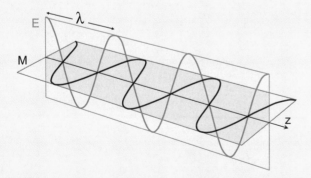

Figure T8A Electromagnetic radiation is made up of magnetic (M) wave and electric (E) waves at right angles to each other. The waves, of wavelength λ, are propagated along the *z*-direction with velocity *c*.

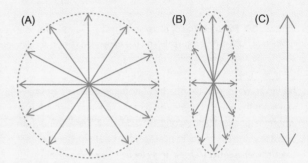

Figure T8B Illustration of polarized light. In (A) the light is unpolarized, i.e. equal in all directions. When the E vectors only oscillate in one direction, as shown in (C), the radiation is said to be plane polarized. In (B), the light is partly polarized.

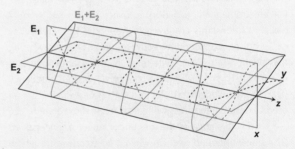

Figure T8C The addition of two E waves perpendicular to each other. The resultant (cyan wave) is more intense and is at 45° to the direction of the component vectors.

When two E waves have *different* phases, their superposition leads to interesting effects. For example, if the phase difference is π/2, then the path of the **E**-vector is helical and the resultant wave is said to be circularly polarized. If the two components of oscillation along *x* and *y* have different amplitudes the resultant wave is said to be elliptically polarized (these effects are discussed and illustrated in Chapter 5.4).

Types of electromagnetic radiation

Electromagnetic radiation covers a wide range of energies and wavelengths (Figure T8D). The names of the different types—X-rays, visible, radiofrequency, etc. have largely arisen from the origin of the radiation and there is no sharp transition between these names or between the properties of the radiation.

The inclusion of neutrons and electrons in Figure T8D is not strictly correct, as these are particles with finite rest mass (*m*) while electromagnetic radiation has zero rest mass. There is,

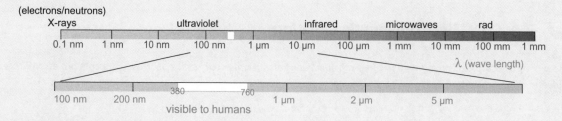

Figure T8D The spectrum of electromagnetic radiation. The relatively small region visible to humans (~380–760 nm) is shown as a white box both on the top scale and the expanded region shown below. Electrons and neutrons can be considered as waves but are not strictly electromagnetic radiation.

however, wave–particle duality (Tutorial 9) and particles have wavelengths given by the de Broglie relationship $\lambda = h/p$, where h is the Planck constant and p is the momentum (mass × velocity). In the context of this book, it is convenient to consider neutron waves and electron waves together with electromagnetic radiation.

Interconversion between scales

The frequency ν and wavelength (λ) of a wave are related by the equation $\nu = c/\lambda$ where c is the velocity of propagation. For electromagnetic radiation in a vacuum, $c \approx 3 \times 10^8$ ms^{-1} (values of constants are given in Appendix 1.3). Frequency can also be converted directly to units of energy using the relationship $E = h\nu$ where h is the Planck constant ($h = 6.63 \times 10^{-34}$ J s). Units of energy (J mol^{-1}), frequency (Hz), and wavelength (m), electron volts (eV), and cm^{-1} are all used to describe the energy of electromagnetic radiation. The electron volt is equal to the amount of kinetic energy gained by a single unbound electron when it accelerates through an electric potential difference of one volt, i.e. 1 volt (1 joule per coulomb) × electron charge (1.602×10^{-19} C) $= 1.602 \times 10^{-19}$ J. These scales can all be interconverted as shown by the following example:

For λ (wavelength) = 600 nm:

ν (frequency) $= c/\lambda = 3 \times 10^8/6 \times 10^{-7} = 5 \times 10^{14}$ Hz (s^{-1} or Hertz)

E (energy/photon) $= h\nu$/photon $=$
$6.63 \times 10^{-34} \times 5 \times 10^{14} = 3.32 \times 10^{-19}$ J

E (energy/mol) $= 3.3 \times 10^{-19} \times 6 \times 10^{23}$ (Avogadro's number) $= 200$ kJ mol^{-1} for photons

$\bar{\nu}$ (wavenumber) $= 1/\lambda = 1/600 \times 10^{-9} =$
1.67×10^6 m$^{-1} = 1.67 \times 10^4$ cm^{-1} (used in IR)

electron volts $= 3.32 \times 10^{-19}/1.602 \times 10^{-19} = 2.069$ eV.

Worked example

Calculate the frequency and wavenumber that correspond to light of wavelength 400 nm. What is the energy of this light?

Answer: frequency: $\nu = c/\lambda$; thus, $\nu = 3 \times 10^8/400 \times 10^{-9} = 7.5 \times 10^{14}$ Hz (or s^{-1}). Wavenumber ($1/\lambda$; cm^{-1}): 400 nm $= 400 \times 10^{-9}$ m $= 400 \times 10^{-7}$ cm. Thus, $\nu = 2.5 \times 10^4$ cm^{-1}

Energy: $E = h\nu = 6.63 \times 10^{-34} \times 7.5 \times 10^{14}$ (J s. s^{-1}) $= 4.97 \times 10^{-19}$ J.

This is the energy associated with one atom. To express it as energy per mol, we must multiply this number by Avogadro's number (6.03×10^{23}). Thus, $E = 1.99 \times 10^5$ J mol$^{-1} = 313$ kJ mol. This substantial amount of energy can be utilized, for example, in photosynthesis.

 ## Further reading

Grant, I.S., and Phillips, W.R. *The Elements of Physics* Oxford University Press, 2001.

Atkins, P., and de Paula, J. *Physical Chemistry* (9th edn) Oxford University Press, 2010.

Tutorial 9 Quantum theory and the Schrödinger wave equation

The aim of this tutorial is to introduce some ideas about quantum mechanics. These are important for understanding spectroscopy (Sections 5 and 6) and aspects of molecular modeling (Section 8).

We can adequately describe the behavior of most objects around us with observable quantities, such as position, mass, and velocity, but it was realized around 100 years ago that this view could not explain all the observations. Some key developments and ideas in this area are summarized below.

Quanta

To explain observations made on the distribution of heat radiation, Max Planck hypothesized in 1900 that light was emitted or absorbed in packets called photons, or quanta, with energy $E = h\nu$, where ν is the frequency of a light wave and h is the Planck constant, 6.62×10^{-34} J s.

The photoelectric effect

Electromagnetic radiation causes some substances to emit electrons (Chapter 5.6). Using the classical theory of light, the energy of the ejected electrons should be proportional to the intensity of the incident light. Instead, it was found that the energies of the emitted electrons are independent of the intensity of the incident radiation. Albert Einstein (1905) resolved this paradox by proposing that incident photons, energy $h\nu$, would release an electron provided $h\nu > h\nu_o$, where $h\nu_o$ is a photon threshold.

The Bohr model

To explain spectroscopic observations, Niels Bohr proposed, in 1913, that electrons could only travel in certain orbits around the nucleus at a certain discrete distances from the nucleus with specific energies.

Wave–particle duality

In 1922 de Broglie pointed out that a particle with momentum p also has the property of a wave, with wavelength λ where $\lambda = h/p$ (see also Tutorial 8).

Uncertainty principle

Heisenberg postulated in 1927 that we cannot measure precisely both the position and momentum of an electron simultaneously (the uncertainty principle). If the momentum of the electron is p, the uncertainty in position (Δx) and momentum (Δp) are related by $\Delta x \Delta p > h/2\pi$ where h is Planck's constant.

Wave mechanics

One of the most powerful ways of reconciling many of the observations and ideas described above is to use wave mechanics; the Schrödinger wave equation, in particular, can be used to describe particles in terms of waves. Consider a particle of mass m with potential energy V and kinetic energy $mv^2/2 = p^2/2m$, where v is the velocity and p the momentum (see Tutorial 3). The total energy E is the sum of the kinetic and potential energies, $E = p^2/2m + V$.

Let us now consider p to be a momentum operator (an operator is a symbol that tells us to do something, e.g. log or d/dx; see Appendix A2.5.) and write p^2 as $-\hbar^2(\frac{d^2}{dx^2} + \frac{d^2}{dy^2} + \frac{d^2}{dz^2})$ (the term in parentheses here is written as ∇^2 and is known as the Laplacian operator—see Appendix A 2.5); $p^2/2m + V = E$ can thus be rewritten as $\hbar^2\nabla^2/2m + V = E$. The left hand side of this equation ($\hbar^2\nabla^2/2m + V$) is called the Hamiltonian of the system ($\mathcal{H}$); it is a wave mechanical description of the kinetic energy plus the potential energy.

To complete our transcription of this energy equation, we regard E as another operator ($i\hbar = d/dt$), where $i = \sqrt{-1}$ and $\hbar = h/2\pi$.

The operators E and H must have something to operate on, so we introduce a wave function ψ on each side of the transcribed equation. This leads to a version of the Schrödinger wave equation, $H\psi = E\psi$. The wave function ψ describes how the system behaves as waves; $\psi^2\hbar V$ is a measure of the probability that a particle is in a certain volume of space, $\hbar V$. Solutions of the wave equation for electrons in atoms give insight into their possible distributions and leads to the concept of orbitals (see below).

For a particular wave function, E can be found by imposing certain boundary conditions on the system; this simply means that the solution is made to satisfy some condition at some chosen point in space. As illustrated in the following simple example, the imposition of boundary conditions allows the Schrödinger equation to be solved. It also shows that solutions are only possible for discrete (quantized) energy values.

Example Particle in a box

Some of the predictions of the Schrödinger wave equation can be illustrated by considering a particle confined in a one-dimensional box, with walls at positions $x = 0$ and $x = L$ (Figure T9A). The potential energy of the particle is assumed to be zero between the walls and infinity at the walls; i.e. the particle has to lie somewhere between $x = 0$ and $x = L$.

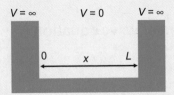

Figure T9A Representation of a one-dimensional box, dimension x, containing a particle with potential energy $V = 0$ inside the box.

The behavior of the particle is specified by the Schrödinger wave equation, $\mathcal{H}\psi = E\psi$ where $H = \mathcal{H}^2\Delta^2/2m + V(x)$. As $V(x) = 0$ in the box, $\mathcal{H} = -\dfrac{\hbar^2}{2m}\dfrac{d^2\psi}{dx^2} = E\psi$. This is a differential equation (Appendix A2.4) with a solution of the form $\psi(x) = A\sin(kx) + B\cos(kx)$. We have the boundary condition that $y = 0$ when $x = 0$; this means that the cos term in the solution is not possible (as $\cos 0 = 1$) and the solution must have the from $\psi = A\sin(kx)$. To obtain $\psi = 0$ at L, k must be an integral multiple of π, thus, the allowed wave functions must be of the form $\psi_n = A\sin(n\pi x/L)$ where n is an integer, greater than zero; we call n a quantum number. Using this wave function to solve the Schrödinger equation gives the following discrete values for the energies: $E_n = n^2h^2/8mL^2$ for $n = 1, 2, 3, \ldots$. This result means that the energy is *quantized*, characterized by a series of energy levels, each associated with a wave function, ψ_n (Figure T9B).

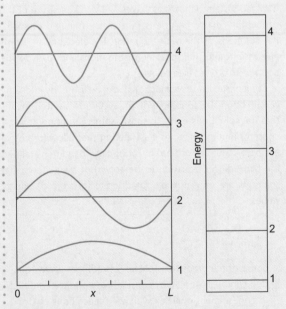

Figure T9B Representation of some possible waves in a box (note they are all zero at the walls); the energy levels for quantum numbers 1–4 are also shown.

Some things to note about this result are: (i) the larger the particle mass (m) the closer the energy levels are together; (ii) the larger the box (L) the closer the energy levels. This means that quantization is less important for large mass and large size. Another concept to emerge from this simple treatment is that n cannot be zero so there is a minimum energy when $n = 1$ ($h^2/8mL$). This zero point energy is energy that cannot be given up. Yet another revealing feature of this simple example is that the quantized energies arose from the imposition of boundary conditions ($V = \infty$ when $x = L$, etc.). The above arguments can be extended from one to three dimensions with the energy states then defined by three quantum numbers rather than one.

Worked example

From the information about the energy levels for an electron in a one-dimensional box, calculate the energy separation (in joules): (i) between levels 2 and 3 in a box of length 0.3 nm; (ii) between levels 10 and 11 in a box length 3 nm. Convert the 10 to 11 energy to wavelength using the relationships $h\nu = \Delta E$ and $c = \nu\lambda$ (Tutorial 8). Why are the energy levels of large conjugated ring systems like porphyrin closer together than for a single aromatic ring such as tyrosine?

Answer For $L = 0.3$ nm, we have $E_n = n^2h^2/8mL^2 = n^2(6.626 \times 10^{-34}\,\text{J s})^2/\{(8 \times 9.11 \times 10^{-31}\,\text{kg}) \times (0.3 \times 10^{-9}\,\text{m})^2\} = n^2 6.69 \times 10^{-19}\,\text{J}$.

For the $2 \rightarrow 3$ transition we have $\Delta E = (3^2 - 2^2) \times 6.69 \times 10^{-19} = 3.35 \times 10^{-18}\,\text{J}$
For the $10 \rightarrow 11$ transition we have $\Delta E = (11^2 - 10^2) \times 6.69 \times 10^{-19} = 1.405 \times 10^{-17}\,\text{J}$
$\therefore \nu = \Delta E/h = 1.405 \times 10^{-17}/6.626 \times 10^{-34} = 2.12 \times 10^{16}\,\text{s}^{-1}$
$\therefore \lambda = c/\nu = 3 \times 10^8/2.12 \times 10^{16} = 1.43 \times 10^{-8}\,\text{m} = 14.3\,\text{nm}$
For $L = 1.2$ nm, $\lambda = (1.2/0.3)^2 \times 14.3 = 357.5$ nm

This kind of calculation is relevant for spectroscopic transitions in molecules with dimensions around 1 nm (see Chapter 5.3). The energies of a particle restricted to a box of width L are proportional to $1/L^2$. In a conjugated ring system, the electrons are extensively delocalized, i.e. L is larger; hence, the energy levels are closer together than for smaller rings with spectra in the visible range. Note also that for larger L more electrons can fit in the box so transitions between higher energy levels (e.g. between 10 and 11) are appropriate as L increases.

 ### Further reading

Atkins, P., and de Paula, J. *Physical Chemistry for the Life Sciences* (2nd edn) Oxford University Press, 2010.

Tutorial 10 Atomic and molecular orbitals, their energy states, and transitions

The aim of this tutorial is to introduce ideas about the shapes and energies of the wave functions that describe the distribution of electrons in atoms and molecules. This is important for understanding many of the properties described elsewhere, especially Section 5, which is concerned with spectroscopic transitions.

Atomic orbitals

The spatial distribution and energy of electrons in atoms can be characterized by a three-dimensional wave function, or orbital. Orbitals can be described by the three quantum numbers n, l, and m, which come from solution of the Schrödinger equation (Tutorial 9). The principal quantum number, n, determines the energy of the orbital. The orbital angular momentum quantum number, l, has values 0, 1, 2, …. $n - 1$. The magnetic quantum number, m, has values l, $l - 1$, $l - 2$…, $-l$. Successive electron shells, defined by n, are denoted K, L, M…. Orbitals with the same value of n but different value of l are denoted s, p, d, f…. The shapes of s, p-, and d atomic orbitals are shown in Figure T10A.

Molecular orbitals

In molecules, the orbitals describing the energy levels of the electrons can be assumed to arise from a *combination of atomic orbitals*. Two simple examples are shown in Figure T10B; one where σ and σ* molecular orbitals (MOs) are formed from a combination of two s-type atomic orbitals (AOs) and one where π and π* MOs are formed from two p-type AOs.

Electronic transitions

We now consider how ideas about orbitals can be used to help us understand some spectroscopic observations. Two

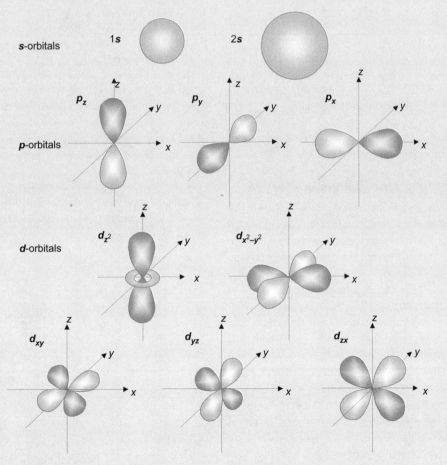

Figure T10A Representations of s, p, and d atomic orbitals.

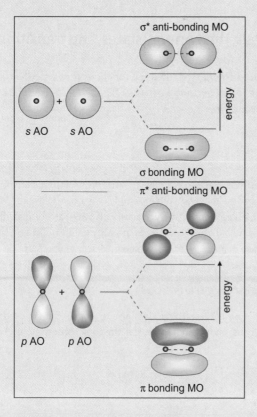

Figure T10B Combination of atomic orbitals (AO) to form molecular orbitals (MOs): *s*-type AOs form σ and σ* bonding and antibonding orbitals; σ has lower energy than σ*; *p*-type AOs form π and π* MOs. The black dots represent nuclei.

illustrative examples are given, one for a simple organic molecule the other for *d*-block metal ions.

Electronic transitions in organic molecules

Consider a simple organic molecule: formaldehyde ($H_2C=O$). The configuration of the 6 electrons in carbon can be written as $1s^2 2s^2 2p_y^1 2p_x^1 2p_z^0$. This means that there are 2 electrons in the $1s$ orbital, 2 in the $2s$ orbital and 2 in the three p orbitals. Similarly, the configuration of the 8 electrons in oxygen can be written as $1s^2 2s^2 2p_y^2 2p_x^1 2p_z^1$. Mixing (hybridization) of the s and p orbitals occurs to give sp, sp^2 or sp^3 hybrids which have different arrangements of bonds around the carbon. For formaldehyde, with three carbon bonds, $3sp^2$ hybrids best describe the situation and a better electronic representation for carbon in formaldehyde is $1s^2 2(sp^2)^3 2p_z^1$. Recall that s and p type AOs form σ and π type bonding and antibonding MOs. The $3sp^2$ hybrid orbitals of carbon thus form 3 σ-bonds with oxygen and the two hydrogens; the remaining $2p_z$ orbital forms a π bond with the oxygen. This leaves oxygen with a $2p_y$ orbital that is not involved in bonding, i.e. it contains a **non-bonding pair** of electrons.

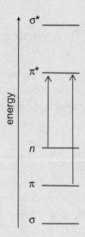

Figure T10C Schematic view of possible energy states in organic molecules, showing the observable $\pi - \pi^*$ and $n-\pi^*$ transitions.

From this discussion of the expected orbitals in formaldehyde we can see that a number of energy states are possible, involving the orbitals σ, σ*, π, π*, as well as n, a non-bonding orbital. There are six possible transitions between σ, σ*, π, π*, and n levels (Figure T10 C) but, for most organic molecules, only the $\pi - \pi^*$ and $n-\pi^*$ transitions lie within the 200–800 nm range of the UV/visible spectrum.

The $\pi - \pi^*$ transition involves a charge displacement so it is "allowed" (Chapter 5.1). The $n-\pi^*$ transition is "forbidden", although it is often observed because of the breakdown of symmetry rules; it is, however, relatively weak, with only ~1% of the intensity of $\pi - \pi^*$ transitions. Figure T10D illustrates the shape of orbitals involved in certain $\pi - \pi^*$ and $n-\pi^*$ transitions.

Note that energetically favored electron promotion will generally be from the **highest occupied molecular orbital** (HOMO, e.g. π and n) to the **lowest unoccupied molecular orbital** (LUMO; e.g. π*).

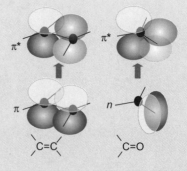

Figure T10D Illustration of a π–π* transition in a molecule with a carbon–carbon double bond and an $n-\pi^*$ transition in a C=O group.

Electronic transitions in d-block metals

The metals Mn, Fe, Co, Ni, Cu, and Mo are important in biology; their electronic spectra generally arise from transitions involving the d-electrons. A simple but informative treatment of d-orbitals that gives insight into observed spectra is **crystal field theory**. The theory supposes that the ligands impose an electric field on the central metal ion. In the free metal ion, the five d-orbitals have the same energies (they are degenerate); the ligand field removes this degeneracy by changing the energies of some of the d orbitals relative to the others. Consider an octahedral field in which six negative charges are placed symmetrically on each of the x, y, and z-axes. The d-electrons in the d_{xy}, d_{yz}, and d_{xz} orbitals will experience less repulsion from the negative charge than will the d_{z^2} and the $d_{x^2-y^2}$ orbitals, whose lobes lie along the axes. The consequence is that these two orbitals are more destabilized than the other three, as illustrated by the energy level diagram in Figure T10E. Similar arguments lead to the energy level diagram for a tetrahedral field. The splitting between the levels in an octahedral field is given the symbol Δ; the value of Δ is less in a tetrahedral than an octahedral field. The energy of the Δ transition often corresponds to wavelengths in the UV/visible range.

The observed value of Δ depends on the ligand; complexes with CN^- ligands, for example, have a larger Δ than those with halide ion ligands because CN^- binds more strongly that halide ions. Each d-orbital can contain up to two electrons with opposed spins. The d orbitals initially fill up on the principle that electrons tend to remain unpaired as long as possible, thus minimizing electron-electron repulsion; see also Chapter 6.2.

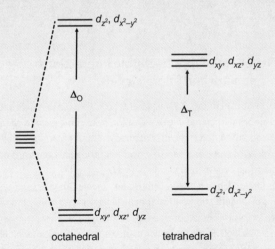

octahedral   tetrahedral

Figure T10E Illustration of the crystal field splitting of the d-orbital energies with octahedral and tetrahedral ligand geometries.

Worked example 1

Zn^{2+} is classed as a d-block element but it does not show any d–d electronic spectra. Why?

Answer 1: The electronic configuration of zinc is an argon-like core plus $3d^{10}4s^2$. It thus has a filled $3d$ level with no d–d transitions possible.

Worked example 2

Hemoglobin contains Fe(II) which has six electrons in d-orbitals with approximate octahedral symmetry. Explain why the spin state goes from $S = 2$ in deoxyhemoglobin to $S = 0$ in oxyhemoglobin.

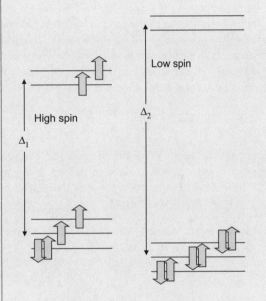

Answer 2: In an octahedral field the energy levels split with three down and two up. The six electrons in the d orbitals distribute according to the size of Δ. When O_2 is bound, Δ is large, the spin-paired arrangement is favored and the net spin = 0 (low spin). (This also leads to a relatively small Fe radius.) In the absence of O_2, Δ is lower and it is more favorable for the electrons to occupy the higher levels rather than pair up. This produces a high-spin configuration and leads to deoxyhemoglobin being paramagnetic.

Further reading

Atkins, P. and de Paula, J. *Physical Chemistry* (9th edn) Oxford University Press, 2010.

Atkins, P., Overton, T., Rourke, J., Weller, M., and Armstrong, F. *Inorganic Chemistry* Oxford University Press, 2009.

Tutorial 11 Dipoles, dipole–dipole interactions, and spectral effects

Dipoles, dipole moments, and their interactions are very important for explaining many properties of molecules that are considered in this book, including non-covalent interactions and spectra. This tutorial sets out some of the key points about dipoles and their interactions.

An electric dipole consists of two charges $+q$ and $-q$ separated by a distance a (Figure T11A). The electric dipole moment of the pair is defined as $\boldsymbol{\mu_e}$ (written in bold because it is a vector with a direction, defined here as going from $-q$ to $+q$). The magnitude of the dipole moment $\boldsymbol{\mu_e}$ is qa which in SI units is coulomb meter (C m) but the Debye ($D = 3.335 \times 10^{-30}$ C m) is often used, as it has a more convenient size.

A single point charge, q, generates a field, E, that falls of as $1/r^2$ (see also Box 2.3). For a dipole, $\boldsymbol{\mu_e}$, the field E at P is the sum of the fields from $+q$ and $-q$ and this falls off as $1/r^3$ (the positive and negative charges of the dipole tend to cancel at large distances so the field falls more rapidly). The value of E at P in Figure T11B also depends on the relative direction of P with respect to the dipole. For the case in Figure T11B this is

$$\mathbf{E} = (\frac{\mu_e}{4\pi\varepsilon\varepsilon_0})\frac{(1 - 3\cos^2\theta)}{r^3}$$

Let us now consider a more general interaction between two dipoles, with respective angles as defined in Figure T11C. The energy of interaction between the dipoles is $\mathbf{E} = \frac{1}{4\pi\varepsilon\varepsilon_0}\left\{\frac{(\boldsymbol{\mu_1}\cdot\boldsymbol{\mu_2})}{r^3} - 3\frac{(\boldsymbol{\mu_1}\cdot\mathbf{r})(\boldsymbol{\mu_2}\cdot\mathbf{r})}{r^5}\right\}$ (for a derivation of this equation, see, for example, Atkins and de Paula, *Physical Chemistry* Oxford University Press, 2010).

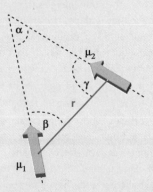

Figure T11C Angles α, β, and γ define the relative orientation of two dipoles $\boldsymbol{\mu_1}$ and $\boldsymbol{\mu_2}$. α is the angle between two dipoles $\boldsymbol{\mu_1}$ and $\boldsymbol{\mu_2}$, β is the angle between $\boldsymbol{\mu_1}$ and $\mathbf{r}$, the distance vector between them, and γ the angle between $\boldsymbol{\mu_2}$ and $\mathbf{r}$.

Using these definition for the angles, the interaction energy is $\mathbf{E} = (\frac{\mu_1\mu_2}{4\pi\varepsilon_0})\frac{(\cos\alpha - 3\cos\beta\cos\gamma)}{r^3}$. If $\alpha = 0$ and $\beta = \gamma$, (parallel dipoles) this reduces to the same form as described as described for Figure T11B.

We can also have a magnetic dipole moment ($\boldsymbol{\mu_m}$) generated by a magnetic field (Box 6.1). The field produced by $\boldsymbol{\mu_m}$ is $\mathbf{B} = (\frac{\mu_m\mu_o}{4\pi})\frac{(1 - 3\cos^2\theta)}{r^3}$, where μ_o is the permeability of free space.

The effect of dipole–dipole interactions on spectra and energy levels

As discussed in Box 5.1 and Chapter 5.1, the transition dipole moment associated with a transition between a ground state, g, and an excited state, e; is defined as μ_{ge}; this can have both μ_e and μ_m components. The transition probability depends on the square of the magnitude of μ_{ge}: $D_{ge} = |\mu_{ge}|^2$.

Interactions between transition dipole moments have important effects in optical spectra (Section 5; exciton splitting, hyperchromism, fluorescence resonance energy transfer, etc.), as well as magnetic resonance spectra (Section 6; NOE, PELDOR). These effects can be computed from equations for dipolar interactions. Consider two identical interacting transition dipoles $\boldsymbol{\mu_1}$ and $\boldsymbol{\mu_2}$ with directions and separation, as defined in Figure T11C.

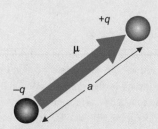

Figure T11A Two charges q, separated by a generate a dipole $\boldsymbol{\mu}$.

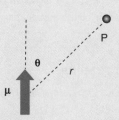

Figure T11B $\boldsymbol{\mu}$ generates a field at P that depends on r and θ.

Let the frequency of the chromophore-containing **monomer** spectrum be ν_o. When two monomers are brought near each other, the absorption bands of the monomers will be split into two new bands, denoted plus and minus, by the dipole–dipole interaction between $\mathbf{\mu}_1$ and $\mathbf{\mu}_2$ (Figures T11D and T11E). The frequencies of the new absorption bands are $\nu_\pm = \nu_o \pm \nu_{12}$, where ν_{12} is a measure of the dipole–dipole interaction between $\mathbf{\mu}_1$ and $\mathbf{\mu}_2$, as indicated by the formulae for $\mathbf{E}$ given above. The value of ν_{12} and the transition probabilities, $D_\pm$, can be computed for any angle and distance between the dipoles. The angle α in Figure T11C turns out to be especially important with $D_\pm = D_o(1 \pm \cos \alpha)$.

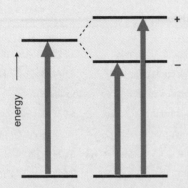

Figure T11D When two monomers are near enough for their transition dipoles to interact, the absorption spectrum of a monomer is split into two bands corresponding to transitions to the + and − state.

Figure T11E Illustration of the orientation of transition dipoles in two chromophores that correspond to the minus state and plus states. Note the inversion of the direction of the dipole in the upper chromophore in the minus state.

The **optical activity** of a molecule involves both electric ($\mathbf{\mu_e}$) and magnetic ($\mathbf{\mu_m}$) transition dipole moments. Certain relative geometries of a dipole pair give rise to both kinds of moment and hence optical activity.

Worked example 1

Proteins have partial charges, especially on the C=O and NH groups; these generate a dipole moment μ that can contribute to a number of structural and functional properties of proteins. The approximate value of μ is 2.7 D with direction as shown in the figure. Calculate the potential of the dipolar interaction between two peptides separated by 0.5 nm at an angle 180° to each other. (Assume $\varepsilon = 1$)

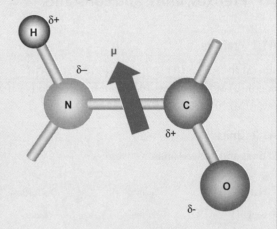

Answer 1 $U = (-2) \times (2.7 \times 3.335 \times 10^{-30}\ \text{C m})^2 / \{4\pi \times 8.854 \times 10^{-12}\ \text{J}^{-1}\ \text{C}^2\ \text{m}^{-1}) \times (0.5 \times 10^{-9}\ \text{m})^3\} = -1.032 \times 10^{-21}\ \text{J}$

Worked example 2

Find the "magic" angle where $1 - 3\cos^2\theta = 0$ (see Chapter 6.1).

Answer 2: $\cos^2\theta = 1/3 \therefore \cos\theta = 0.57735 \therefore \theta = 54.736°$

📖 Further reading

Atkins, P., and de Paula, J. *Physical Chemistry* (9th edn) Oxford University Press, 2010.

APPENDICES

A1 Prefixes, units, and constants

A1.1 Prefixes

atto (a) 10^{-18}; femto (f) 10^{-15}; pico (p) 10^{-12}; nano (n) 10^{-9}; micro (μ) 10^{-6}; milli (m) 10^{-3}; centi (c)10^{-1}; kilo (k) 10^{3}; mega (M) 10^{6}; giga (G) 10^{9}; tera (T) 10^{12}

A1.2 Units

SI units and derived units

Name	Symbol	Definition
Length	1 m	Meter
Mass	1 kg	Kilogram
Time	1 s	Second
Current	1 A	Ampere
Temperature	1 K	Kelvin
Amount	1 mol	Mole
Newton	1 N	$kg\,m\,s^{-2}$
Pascal	1 Pa	$kg\,m^{-1}\,s^{-2}$
Volt	1 V	$J\,C^{-1}$
Tesla	1 T	$kg\,s^{-2}\,A^{-1}$
Ohm	1 Ω	$V\,A^{-1}$
Coulomb	1 C	$s\,A$
Joule	1 J	$kg\,m^{2}\,s^{-2}$
Watt	1 W	$J\,s^{-1}$
Ampere	1 A	$C\,s^{-1}$
Frequency	1 Hz	s^{-1}

"Operational" units (in common use but not SI):

1 liter (L) = 10^{-3} m^3
1 Angstrom (Å) = 10^{-10} m
Molar (M) = mol dm^{-3} = mol L^{-1}
1 calorie = 4.184 J
1 atm = 101.325 kPa
1 Debye = 3.335×10^{-30} C m

1 Dalton (universal mass unit) = 1/12 the mass of nucleus of the ^{12}C isotope.
1 Poise = 1 g cm^{-1} s^{-1}

A1.3 Constants

Speed of light in vacuum	c	2.99792458×10^{8} m s^{-1}
Gravitational constant	G	6.67428×10^{-11} Nm^2 kg^{-2}
Acceleration of free fall	g	9.80665 m s^{-2}
Avogadro's constant	N_A	6.02214×10^{23} mol^{-1}
Boltzmann's constant	k	1.38066×10^{-23} J K^{-1}
Gas constant	$R = N_A k$	8.31451 J K^{-1}mol^{-1}
Faraday constant	$F = N_A e$	9.64853×10^{4} C mol^{-1}
Planck constant	h	6.62608×10^{-34} J s
Reduced Planck constant	$\hbar = h/2\pi$	1.05457×10^{-34} J s
Vacuum permeability	μ_o	$4\pi \times 10^{-7}$ N A^{-2}
Vacuum permittivity	ε_o	8.85419×10^{-12} J^{-1} C^2 m^{-1}
Elementary charge	e	1.602177×10^{-19} C
Bohr magneton	$\beta_{\varepsilon\,(\mu B)}$	9.27402×10^{-24} J T^{-1}
Nuclear magneton	μ_N	5.05079×10^{-27} J T^{-1}
g value	g_e	2.00232
Mass electron	m_e	9.10939×10^{-31} kg
Mass proton	m_p	1.67262×10^{-27} kg
Mass neutron	m_n	1.67493×10^{-27} kg
Atomic mass constant	m_u	1.66054×10^{-27} kg

A2 Some mathematical functions

A2.1 Trigonometry

In a right angled triangle, as shown in Figure A2.1, we have the following relationships: $\sin \theta = o/h$; $\cos \theta = a/h$; $\tan \theta = o/a$.

A2.2 Vectors

A quantity that is completely determined by its magnitude is called a **scalar** quantity. A quantity that is determined by

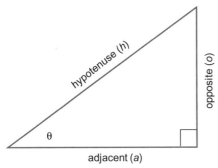

Figure A2.1 The names of the sides of a right angled triangle and their abbreviations *a*, *o*, and *h* are shown.

its magnitude and direction is called a **vector**, usually represented by a boldfaced letter, such as **V**.

A vector **V** is often written as $a\mathbf{i} + b\mathbf{j} + c\mathbf{k}$, where **i**, **j**, and **k** are unit vectors along the *x*-, *y*-, and *z*-axes, respectively. The magnitude is $(a^2 + b^2 + c^2)^{1/2} = |\mathbf{V}|$. The sum of two vectors $\mathbf{V}_1$ and $\mathbf{V}_2$, $\mathbf{V}_r$, can be represented graphically, as shown in Figure A2.2.

A **scalar product** of two vectors $\mathbf{V}_1$ and $\mathbf{V}_2$ is $|\mathbf{V}_1||\mathbf{V}_2| \cos \theta$, where θ is the angle between $\mathbf{V}_1$ and $\mathbf{V}_2$.

A **vector (or cross) product.** of two vectors $\mathbf{V}_1$ and $\mathbf{V}_2$ is written as $\mathbf{V}_1 \times \mathbf{V}_2$ and is defined to be $|\mathbf{V}_1||\mathbf{V}_2|\mathbf{1} \sin \theta$, where **1** is a unit vector perpendicular to the plane of $\mathbf{V}_1$ and $\mathbf{V}_2$; it is directed such that a right-handed screw axis in the direction of **1** would carry $\mathbf{V}_1$ to $\mathbf{V}_2$.

A2.3 Complex variables

A complex number *c* is of the form $a + ib$ where *a* and *b* are real numbers and $i = (-1)^{1/2}$ ($i^2 = -1$) is the imaginary unit.

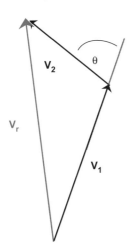

Figure A2.2 The sum of the two vectors V_1 and V_2 is V_r.

The magnitude of *c* ($|c|$) is defined as cc^* where c^*, the complex conjugate of *c*, is $a - ib$:

$$|c| = ((a + ib)(a - ib))^{1/2} = (a^2 + b^2)^{1/2}.$$

A2.4 Logarithms, exponentials, and trigonometric functions

Logarithms

The natural logarithm of *x* is written as $\ln x$.

$\ln x + \ln y = \ln xy$; $\ln x - \ln y = \ln x/y$; $a\ln x = \ln x^a$

$\ln x = 2.303\log x$ where log is the "common" logarithm base 10, now largely obsolete except in dealing with acid and base chemistry (Chapter 2.1).

Exponentials

The exponential function of *x* is written as e^x or $\exp(x)$; $e^x + e^y = e^{x+y}$; $e^x/e^y = e^{x-y}$.

e^x and $\ln x$ are **inverse functions**, which means that if $e^x = y$ then $\ln y = x$. e^x is defined such that it is equal to its own derivative, $e = 2.71828$.

Logarithm, exponential, sine, and cosine functions can be defined by infinite power series:

$$\ln x = (x - 1) - \frac{1}{2}(x - 1)^2 + \frac{1}{3}(x - 1)^3 - \frac{1}{4}(x - 1)^4 + \dots$$

$$e^x = 1 + x + \frac{x^2}{2!} + \frac{x^3}{3!} + \dots$$

$$\sin x = x - \frac{x^3}{3!} + \frac{x^5}{5!} - \frac{x^7}{7!} \dots$$

$$\cos x = 1 - \frac{x^2}{2!} + \frac{x^4}{4!} - \frac{x^6}{6!} \dots$$

where $n! = n \times (n - 1) \times (n - 2) \dots 2 \times 1$

The following relationships arise from these definitions:

$$e^{ix} = \cos x + i\sin x; \quad e^{-ix} = \cos x - i\sin x$$
$$\cos^2 x + \sin^2 x = 1; \quad \cos(x_1 + x_2) = \cos x_1 \cos x_2 - \sin x_1 \sin x_2$$
$$\text{and } \cos x_1 + \cos x_2 = 2\cos(\frac{x_1 + x_2}{2})\cos(\frac{x_1 - x_2}{2})$$

A2.5 Calculus

Differentiation of a function $f(x)$ with respect to *x* is written as $df(x)/dx$. It defines the slope of $f(x)$ at the point *x*. When $df(x)/dx = 0$ the gradient is zero, i.e. the function has a maximum, minimum, or turning point.

Useful examples of differentiation are:

$dx^n/dx = nx^{n-1}$; $d(\ln ax)/dx = a/x$; $d(\sin ax)/dx = a\cos ax$; $d(\cos ax)/dx = -a\sin ax$.

The **second derivative** of $f(x)$, written as $d^2f(x)/dx^2$, is the derivative of the derivative of $f(x)$. Calculus can also be applied to vectors and three-dimensional systems. The symbol ∇ (nabla) is used for the first derivative: $\nabla = \dfrac{d}{dx}i + \dfrac{d}{dy}j + \dfrac{d}{dz}k$, where $\mathbf{i}, \mathbf{j}, \mathbf{k}$ are unit vectors in the x, y, z directions; in this context ∇ is called an **operator**; ∇^2 (the second derivative) is called the **Laplace operator**.

Integration is a way of finding the area of a function $f(x)$ between the points a and b, written as $\int_a^b f(x)\,dx$. Useful examples are:

$$\int x^n dx = x^{n+1}/(n+1); \int e^{ax}dx = e^{ax}/a; \int adx/x = a\ln x;$$
$$\int \sin ax dx = -\cos ax/a.$$

Differential equations

A differential equation is an equation involving an unknown function and its derivatives. The *order* of the differential equation is the order of the highest derivative of the unknown function involved in the equation. For example, a **linear differential equation** of order 3 has the form $a_3(x)d^3y/dx^3 + a_2(x)d^2y/dx^2 + ady/dx = f(x)$. This type of differential equation is called an ordinary differential equations (ODE).

A simple example of a differential equation is a description of a concentration decaying with time: $d[A]/dt = -k[A]$, where k is a rate constant, $[A]$ is the concentration, and t is time. We can rearrange this and integrate to give $\int (d[A]/[A]) = -\int kdt$. Using the values for integrals given above, we have $\ln[A] = -kt + C$. We can determine the unknown constant, C, from the initial boundary conditions, namely that $[A] = [A_o]$ when $t = 0$, i.e. $\ln([A]/[A_o]) = -kt$. As ln and exp are inverse functions, we can also write this as $[A] = [A_o]e^{-kt}$.

A3 Basic statistics

Statistical methods are very valuable for analyzing data collected using biophysical tools. There are many books on the subject and most of us have ready access to programs such as Microsoft Excel with powerful inbuilt statistical analysis functions, including linear regression (see below and Chapter 2.1). Several good online statistics tutorials are also available. Some basic definitions of statistical terms used are given below.

Given a list with n elements, x_i, the **mean** is defined as $\mu(x) = \bar{x} = \Sigma x_i/n$ where Σ means the sum over all x_i.

The **standard deviation** is $\sigma(x) = \{\Sigma(x - \dot{x})^2\}/(n-1)$.

If the elements in the list are thought of as being selected at random according to some probability distribution, then the mean gives an estimate of where the center of the distribution is located, while the standard deviation gives an estimate of how wide the dispersion of the distribution is.

The **median** of a range of data lies in the middle of the spread of values; it effectively gives the value at the halfway point in a sorted version of a list. It is often considered a more robust measure of the center of a distribution than the mean, as it depends less on outlying values.

Simple linear regression plots a line through a cloud of points. Given a set of values, $x_0 \ldots x_{n-1}$ and $y_0 \ldots y_{n-1}$, linear regression calculates the constant coefficients a and b for the line $y' = a + bx$. The coefficient b (the slope of the line) is $b = \{\Sigma(x - \bar{x})(y - \bar{y})\}/\Sigma(x - \bar{x})^2$, where $\bar{x}$ is the mean of the x values and $\bar{y}$ is the mean of the y values. Once the coefficient b is known, the coefficient a, the y-intercept, can be calculated using the equation $a = \bar{y} - b\bar{x}$.

The **correlation coefficient** r is usually used to assess the scatter in a set of data. It assesses whether a correlation exists between values in a dataset. The value of r can be calculated from $r = \{\Sigma(x - \bar{x})(y - \bar{y})\} / [\{\Sigma(x - \bar{x})^2\}\{\Sigma(y - \bar{y})^2\}]^{1/2}$. r lies in the range $-1 \leq r \leq 1$. An r value of zero means that there is no correlation between respective values of x and y. Positive and negative values of r indicate positive and negative correlations. r^2 rather than r is often used. Data are found to be distributed in a number of key ways. These include:

Uniform distribution: $P(x) = $ constant $(a \leq x \leq b)$; 0 otherwise.

Normal (Gaussian) distribution: $P_G(\mu,\sigma;x) = \exp[-(x-\mu)^2/(2\sigma^2)]/(\sigma\sqrt{2\pi})$. A Gaussian distribution of data is shown in Figure A3.1; 68.3% of the values will lie within $\mu \pm \sigma$ and 99.7% within $\mu \pm 3\sigma$, where σ is the standard deviation.

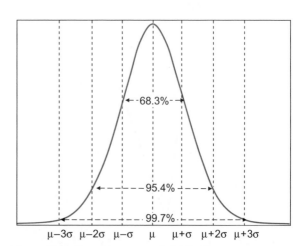

Figure A3.1 Gaussian or normal distribution.

SOLUTIONS TO PROBLEMS

Chapter 2.1 Interactions

2.1.1 From $(nM\hat{u}^2)/3 = nRT$ we obtain $\hat{u} = (3RT/M)^{1/2}$ so the velocity varies as $\sqrt{T}$. The kinetic energy is $1/2M\hat{u}^2 = 3RT/2 = 3N_AkT/2$ mol^{-1} or $3kT/2$ per molecule. The molar mass of N_2 is 28×10^{-3} kg.

$$\hat{u} = (3 \times 8.315 \times 298/28 \times 10^{-3})^{1/2} \times 10 = 515.3 \text{ m s}^{-1}.$$

2.1.2 In a vacuum, the potential energy $U = q_1q_2/4\pi\varepsilon_0 r = (1.6 \times 10^{-19})^2/(4\pi \times 8.85 \times 10^{-12} \times 0.28 \times 10^{-9}) = 8.22 \times 10^{-19}$J. To convert to J mol^{-1}, we multiply by $N_A = 6.02 \times 10^{23}$ $\therefore U = 494.8$ kJ mol^{-1}. On the surface, $\varepsilon \sim 80$ so $U = 6.2$ kJ mol^{-1}. In the interior, $\varepsilon \sim 4$ so $U = 123.7$ kJ mol^{-1}. Thus, on the surface, the interaction is weaker than an H-bond but it is much stronger than an H-bond in the protein interior.

2.1.3 The amino acids are chiral: they lack an internal plane of symmetry. Of the standard α-amino acids, all but glycine, with an H sidechain, can exist in either of two optical isomers, called l or d. l-amino acids represent all of the amino acids found in proteins during translation in the ribosome, although d-amino acids are found in some proteins produced by enzyme post-translational modifications. Because of the geometry of l-amino acids, the formation of a left-handed helix is much less favorable than a right-handed helix (see also Figures 2.1.4 and 2.1.5).

2.1.4 We can perform the $\bar{n}$ vs [L], $\bar{n}/$[L] vs $\bar{n}$, and $\ln(\bar{n}/(n - \bar{n}))$ vs. $\ln$[L] plots with the following data. Direct, Scatchard, and Hill plots of the data are shown below. The four subunit protein shows positive cooperativity with a Hill coefficient of ~3.0. Open circles correspond to the monomer data.

2.1.5 We have $HA \rightleftharpoons H^+ + A^-$; $K_a = [H^+][A^-]/[HA]$; and $-\log_{10}K_a = -\log[H^+] - \log([A^-]/[HA])$

$\therefore pK_a = pH - \log([A^-]/[HA])$ or $pH = pK_a + [\log[A^-]/[HA]$.

Assuming that the buffering in the reaction $ATP^{4-} + 2H_2O \rightarrow ADP^{3-} + HPO_4^{2-} + H_4O^+$ is dominated by tris, we have, from the Henderson–Hasselbalch equation, that $8.0 = 8.1 + \log[tris]/[tris^+]$

$\therefore$ [tris]/[tris$^+$] = 0.7943; but we also know that

[tris] + [tris$^+$] = 0.1 mol dm^{-3}. We can solve these two equations to obtain [tris] and [tris$^+$] before the reaction: [tris] = 0.04427 and [tris$^+$] = 0.05573 mol dm^{-3}.

After hydrolysis of ATP, which produces 0.001 mol dm^{-3} of H$^+$, we have: [tris] = 0.04427 – 10^{-3} and [tris$^+$] = 0.05573 + 10^{-3} $\therefore$ pH = 8.1 + log(0.04327/0.05673) = 7.98

With only 0.01 mol dm^{-3} buffer and still assuming tris buffering dominates the process, we have, initially, [tris] = 0.004427 mol dm^{-3} and [tris$^+$] = 0.005573 mol dm^{-3}

After addition of 10^{-3} H$^+$, we have [tris] = 0.003427 and [tris$^+$] = 0.006573

$\therefore$ pH = 8.1 + log(0.003427/0.006573) = 7.82

Chapter 3.1 Introduction to transport

3.1.1 From the formula $\bar{a} = (2Dt)^{1/2}$ we have:

(a) $t = \bar{a}^2/2D = 10^{-4}/(2 \times 2.26 \times 10^{-9}) = 0.221 \times 10^5$ s ~ 6.1 hours;

(b) $t = 10^{-12}/(2 \times 2.26 \times 10^{-9}) = 2.21 \times 10^{-4}$ s.

3.1.2 $f = kT/D$; $k = 1.38 \times 10^{-23}$ J K^{-1}; $T = 298$; $D = 12 \times 10^{-11}$ m^2 s^{-1}

$\therefore f = 1.38 \times 10^{-23} \times 298/12 \times 10^{-11} = 34.3 \times 10^{-12}$ J m^{-2} s $= 3.43 \times 10^{-11}$ kg s^{-1} (see Appendix 1.2 for definition of units).

The mass of one protein molecule $= 15/6.023 \times 10^{23} = 2.49 \times 10^{-23}$kg. The volume, V, is the mass $\times$ the partial specific volume, $V = 2.49 \times 10^{-23} \times 0.71 = 1.77 \times 10^{-23}$ L $= 1.77 \times 10^{-26}$ m^3.

As $V = 4\pi r^3/3$, $r = (1.77 \times 3 \times 10^{-26}/4\pi)^{1/3} = (4.23 \times 10^{-27})^{1/3} = 1.62 \times 10^{-9}$ m.

The expected f value for a sphere of this radius is $f_o = 6\pi r\eta = 6\pi(10^{-3}$ kg m^{-1} s$^{-1})(1.62 \times 10^{-9}m) = 3.05 \times 10^{-11}$ kg s^{-1}. Differences between f and f_o can arise from shape (Box 3.1) or hydration (bound water) effects.

Table of data for solution to 2.1.4

[L] (μM)	0.25	1.0	2.5	4	7	10	13	16	19	25
$\bar{n}$ (m)	0.18	0.46	0.70	0.79	0.86	0.90	0.92	0.935	0.95	0.96
$\bar{n}$ (m)/[L]	0.72	0.46	0.28	0.198	0.123	0.09	0.071	0.0584	.05	0.0384
$1 - \bar{n}$	0.82	0.54	0.3	0.21	0.14	0.1	0.08	0.065	0.05	0.04
$\ln(\bar{n}/(1 - \bar{n}))$	-1.52	-0.16	0.85	1.32	1.81	2.197	2.442	2.666	2.944	3.178
$\ln$[L]	-1.39	0	0.916	1.386	1.946	2.303	2.565	2.773	2.944	3.219
$\bar{n}(t)$	0.0	0.05	0.18	0.40	1.08	2.0	2.72	3.12	3.42	3.69
$\bar{n}(t)/$[L]	0	0.05	0.072	0.1	0.154	0.2	0.209	0.195	0.18	0.148
$4 - \bar{n}$	4	3.95	3.82	3.6	2.92	2	1.28	0.88	0.58	0.31
$\ln(\bar{n}/(4 - \bar{n}))$	-	-4.37	-3.06	-2.20	-0.995	0	0.775	1.266	1.774	2.476

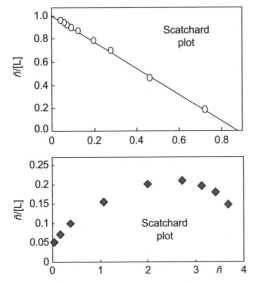

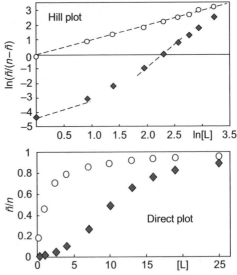

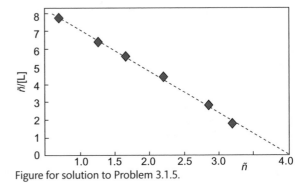

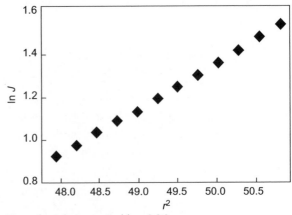

Figure for solution to Problem 2.1.4.

The f/f_o ratio here is $3.43/3.05 = 1.12$.

3.1.3 From Fick's first law we have that $J = -Ddc/dx = 5.2 \times 10^{-10}$ $(m^2 s^{-1}) \times (0.2 \text{ mol L}^{-1} \text{ cm}^{-1})$

{$1 \text{ L}^{-1} = 10^3 \text{ m}^{-3}$ and $1 \text{cm}^{-1} = 10^2 \text{ m}^{-1}$ so $\text{L}^{-1} \text{ cm}^{-1} = 10^5 \text{ m}^{-4}$}
$\therefore J = 1.04 \times 10^{-5} \text{ mol m}^{-2} \text{ s}^{-1}$

The amount passing through 1 cm² per minute is thus $JA\Delta t = 1.04 \times 10^{-5} \times (10^{-2})^2 \times 60 = 6.24 \times 10^{-8}$ mol.

3.1.4 Glucose has the formula $C_6H_{12}O_6$, thus the molecular weight is $6 \times 12 + 12 + 6 \times 16 = 180$. The concentration is thus $18/180 = 0.1$M. The osmotic pressure is $\Pi = RTc$ (J K⁻¹ mol⁻¹ K mol L⁻¹) $= 8.314 \times 293 \times 0.1 = 243.6 \text{ J L}^{-1} = 243,600 \text{ J m}^{-3} = 243.6 \text{ kPa}$.

0.1M $MgCl_2$ produces 0.3M ions therefore the new osmotic pressure would be $243.6 \times 4 = 974$ kPa.

3.1.5 We can obtain the amount of drug bound from [total] – [free]

[Total] (mM)	0.125	0.26	0.38	0.61	1.14	2.15
[Free] (mM)	0.09	0.197	0.297	0.50	0.98	1.75
[Bound] (mM)	0.035	0.063	0.083	0.11	0.142	0.40
ň		0.7	1.26	1.66	2.2	2.84	3.2
ň/[free]	7.78	6.39	5.59	4.4	2.90	1.78

and the fraction bound to enzyme, ň, from [bound]/[enzyme].

A Scatchard plot is then obtained from a plot of ň/[free] against ň. The slope is $-1/K_d$ and the intercept on the ň axis is the stoichiometry. This gives a straight line (indicating non-cooperative binding) with $K_d \sim 0.4$mM and an intercept on the ň axis of around 4. The enzyme thus has four binding sites and is probably a tetramer.

3.1.6 We have $R_e = vL\rho/\eta$; $\eta = 10^{-3} \text{ kg m}^{-1} \text{ s}^{-1}$ and $\rho/\eta = 10^6 \text{ m}^{-2}\text{s}$.

$\therefore$ (i) for *E. coli* $R_e = (10^{-5} \text{ m s}^{-1})(10^{-6}\text{m})(10^6 \text{ m}^{-2}\text{s}^{-1}) = 10^{-5}$;

(ii) for the human, $R_e = 10^6$.

This means that the human motion is dominated by inertial effects while the bacterium motion is dominated by viscosity and diffusion.

Figure for solution to Problem 3.1.5.

Chapter 3.2 Centrifugation

3.2.1 Extrapolation of the $s^{0.x\%}_{20,w}$ values to zero concentration gives $s^0_{20,w} \sim 7.95$ S. From the Svedberg equation we then have $M = RTs/(D(1 - \bar{v}\rho)) = 8.314 \times 293 \times 7.95 \times 10^{-13}/4.75 \times 10^{-11}(1 - 0.73) = 151 \text{ kg mol}^{-1}$; the molecular weight is thus 151 kDa.

3.2.2 (i) In the sedimentation equilibrium experiment, $M(1 - \bar{v}\rho)\omega^2 = 2RTd(\ln c)/dr^2$; so M can be derived from a plot of $\ln c$ (or $\ln J$) against r^2.

Figure for solution to Problem 3.2.2.

Data for solution to Problem 3.2.2											
J	2.514	2.651	2.812	2.974	3.108	3.307	3.497	3.699	3.912	4.150	4.437	4.688
$\ln J$	0.922	0.975	1.034	1.090	1.134	1.196	1.252	1.308	1.364	1.423	1.490	1.545
r (cm)	6.925	6.943	6.962	6.981	6.999	7.018	7.036	7.054	7.072	7.091	7.110	7.129
r^2	47.95	48.21	48.47	48.73	48.99	49.25	49.50	49.76	50.02	50.28	50.55	50.82

Note on units: The slope is in units of mol cm^{-2}; $(1 - \bar{v}\rho)\,\omega^2/2RT$ has units = s^{-2}/(J K^{-1} mol^{-1} K) = mol s^{-2} kg^{-1} m^{-2} s^2 = mol kg^{-1} m^{-2}. We must therefore convert the slope to units of m^2 by multiplying by 10^4. The molecular mass will then be in kg.

(i) Plotting the above data ($\ln J$ vs r^2) gives a slope of ~0.219 (mol cm^{-2}). This is equal to $M \times (1 - 0.74 \times 1.014) \times (2\pi \times 4908/60)^2/(2 \times 8.314 \times 293) = M \times (0.2494) \times (264077)/(4872) = 13.51M$, i.e. mol. wt = $0.219 \times 10^4/13.51$ kg mol^{-1} = 162 kDa.

(ii) We are given that $s^\circ_{20,w} = 8.0$ S for $M = 162$ kDa at pH 7.0; $s^\circ_{20,w} = 3.5$ S for $M = 81$ kDa at pH 3.0 and M = 40 kDa when denatured. This suggests that the protein is a tetramer, with monomer subunits ~40 kDa. Note the s value at pH 3 is less than might be expected; this could arise if the pH 3 dimer is not as compact as the tetramer and it has a larger frictional coefficient.

3.2.3 The sedimentation equilibrium equation is $\omega^2 M(1 - \bar{v}\rho)/2RT = d(\ln c)/dr^2$. Integration from $r = r_m$ to $r = r_b$ gives $\ln(c_b/c_m) = \omega^2\{(M(1 - \bar{v}\rho)(r_b^2 - r_m^2))\}/2RT$

$\therefore \ln(100) = 4.605 = \omega^2 \times 5 \times 10^4 \times 0.27 \times (7.0^2 - 6.7^2) \times 10^{-4}/2 \times 8.314 \times 298 = \omega^2 \times 0.0011197$

$\therefore \omega^2 = 4.605/1.1197 \times 10^{-4} = 4.1129 \times 10^4 \therefore \omega = 64.1316 \therefore$ rpm $= 2\pi \times 60 \times \omega = 24\,177$ rpm

3.2.4 The sedimentation coefficients are given by the Svedberg equation, $s/D = M(1 - \bar{v}\rho)/RT$. As M is constant, we have $s_1/s_2 = (D_1/D_2)((1 - \bar{v}\rho_1)/(1 - \bar{v}\rho_2)) = (D_1/D_2)((0.2422)/(0.1944))$, where position 1 is 10% sucrose and position 2 is 25%. D is related to $1/\eta$ from Stokes equation, thus $D_1/D_2 = \eta_2/\eta_1 = 2.201/1.167$ so that $s_1/s_2 = 2.35$ (sediments slower in the higher viscosity and density further down the tube). The velocities are given by $v = s\omega^2 r$ so the ratio of velocities at the 10% and 25% points will depend on r (8/15) = 0.533 (faster when r is larger) so the net velocity ratio change is $2.35 \times 0.533 = 1.25$ (a decrease in velocity at position 2 compared to position 1).

3.2.5 The crab DNA is an example where there is "satellite" DNA, consisting of repetitive DNA sequences that can be separated on a density gradient because of their unusual GC or AT content.

3.2.6 The increase in M_{app} at lower concentrations is the kind behavior expected for a self-associating system, e.g. a monomer/dimer equilibrium. The fact that three different concentrations give a similar pattern of results suggests that there is no significant other sample contamination. The decrease in M_{app} at high concentrations is a reflection of "non-ideality" in the solution which arises from the finite size of macromolecules and the charge they carry, leading to intermolecular interactions; non-ideality is expected to reduce the apparent molecular weight with increasing concentration.

Chapter 3.3 Chromatography

3.3.1 Proteins have (i) a charge that varies with pH; (ii) a size characterized by a Stokes radius; (iii) specific binding properties; and (iv) hydrophobic patches on their surface. These can be exploited in the following chromatography separations: (i) ion exchange, (ii) size exclusion, (iii) affinity, and (iv) reverse phase HIC.

3.3.2 An anion exchanger interacts with negatively charged proteins and a cation exchanger interacts with positive ones. Low pH adds H$^+$ groups to ionizable groups such as COO$^-$, increasing the + charge while high pH adds negative OH$^-$. Thus a low to high pH gradient will elute from a cation exchanger and a high to low one will elute from an anion exchanger. Note that local effects (e.g. resulting from the Donnan effect, Chapter 3.1) mean that the pH of the buffer near the ionizable groups on the column is not equal to the bulk pH.

3.3.3 (i) Use size exclusion chromatography, (ii) use ion exchange chromatography or isoelectric focusing. Assuming D is the only DNA binding protein, affinity chromatography could be used, e.g. by covalently attaching a sequence-specific DNA-binding fragment to a solid support (this can be done with cyanogen bromide (CNBr) activation of an agarose support).

3.3.4 As mentioned in the ion exchange section, DEAE is an anionic exchanger with a pKa ~9.5. It is thus positively charged up to pH 9.5 but at higher pH values the charge will reduce until it will be ~90% neutral at pH 10.5. DNA has negatively charged phosphate groups that will bind to DEAE until the DEAE is neutral.

3.3.5 When $V_e = 180$ mL, $\log_{10}M \sim 4.4$ thus $M \sim 25$ kDa.

3.3.6 We have $\tau_r \sim 12.5$ min and $W \sim 2$ min $\therefore N = 16 \times (12.5/2)^2 = 625$; $H = L/N = 0.016$ cm. One way to improve the resolution would be to increase the number of plates by decreasing the bead size. Optimizing the flow rate to give a minimum H value using the van Deemter plot might also help.

3.3.7 The A term of the van Deemter equation will be smaller with smaller particles because packing and path differences will be less. The curve is thus shifted to lower H. The C term will also be less because the transfers between mobile and stationary phases will be more efficient. This results in a shallower slope for the curve at higher u_m values (see figure).

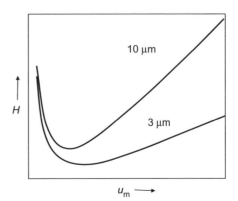

Figure for solution to Problem 3.3.7.

Chapter 3.4 Electrophoresis

3.4.1 The approximate distances migrated by the marker proteins are given in the table:

M(kDa)	14.2	20	24	29	36	45	66
cm	8.6	7.1	6.0	5.4	4.3	3.6	1.9
ln(M)	2.65	2.99	3.18	3.37	3.58	3.81	4.19

As shown, a plot of distance against ln(M) gives a straight line. From the graph, the molecular weight of a protein that migrates to 3.3 cm will be around 48 kDa (cyan spot). The expressed protein will therefore be around 22 kDa when GST is removed.

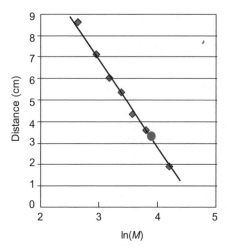

Figure for solution to Problem 3.4.1.

3.4.2 The protein migrates towards the cathode if it is negative and towards the anode if it is positive. When there is no migration it is neutral (pI = pH). From the table, this occurs when pH ~4.75.

3.4.3 In standard electrophoresis methods, DNA molecules larger than 20 kb show essentially the same mobility in normal electrophoresis, so the 30 and 50 kb fragments are unlikely to be resolved. Pulse field gel methods would be required to separate these. The 1- and 2-kb fragments should be resolved by 1% agarose (Figure 3.4.2) but 2% would be better.

3.4.4 EMSA (Figure 3.4.7) could be used to measure the fraction of protein/DNA complex at different concentrations. More accurate methods for measuring K_D (e.g. ITC, SPR) in the solution state rather than in a matrix are described in other chapters.

3.4.5 (i) The most likely explanation is uneven temperature across the gel because of local heating effects. Lower running voltage or better cooling procedures are likely to help. (ii) The most likely explanation is that the protein has not been fully denatured and is not evenly coated with SDS. Higher SDS and 2-mercaptoethanol may help, as may boiling the sample in the presence of these reagents. (iii) The protein may be degrading because of protease activity in the sample, it may be heterogeneously glycosylated or the gel may have been overloaded with sample.

3.4.6 Electro-osmotic flow (EOF) is observed when an electric field is applied to a solution in a capillary that has fixed charges on its interior wall, e.g. in a fused-silica capillary, silanol (Si–OH) groups attached to the interior wall of the capillary are ionized to negatively charged silanoate (Si–O⁻) when pH > 3. The EOF of the buffer solution is generally greater than that of the electrophoretic flow of the analytes; all analytes are carried along with the buffer solution toward the cathode in the order: cations, neutrals, anions. Tricks to help separate the neutrals include addition of charged ligands to the buffer to form complexes with the neutral molecules e.g. adding borate to sugars to form anionic complexes. Another is to use an anionic detergent such as sodium dodecyl sulfate (SDS). Above a concentration of about 5–10mM SDS self-assembles into micelles. Neutral molecules that partition between these micelles and the surrounding buffer are slowed down compared with neutral molecules that do not partition into the micelles.

Chapter 3.5 Mass spectrometry

3.5.1 Gly to Ser corresponds to a mass change of +30 Da and His to Gln to one of –9 Da. The net expected change is thus +21 Da. Acetylation plus phosphorylation will give an increased mass of +123 Da.

3.5.2 A plausible solution is that there are two proteins with molecular weights of 20 350.8 and 35 317.3 Da. The peaks at 10 176.5 and 17 659.7 Da, which are approximately half these values, probably correspond to the proteins with a charge of +2. The small peak at 40 702.5 is likely to be a dimer of the 20 350.8 Da molecule.

3.5.3 We need to calculate the charge on each peak and m the molecular weight. Assuming that the mass of H⁺ = 1 and that n is the charge on the 1773.396 peak, we have $(m + n)/n$ = 1773.396 or $m/n + 1$ = 1773.396, giving $m = 1772.396n$. The adjacent peak will have a charge of $n + 1$; thus $(m + (n + 1))/(n + 1)$ = 1576.292 or $m = 1575.292n + 1575.292$; subtracting equation (ii) from (i) gives 197.104n = 1575.292

∴ n = 7.992 = 8. This allows us to work out m for each peak as follows:

$(m + 8)/8$ = 14 179.2; $(m + 9)/9$ = 14 177.6; $(m + 10)/10$ = 14 177.0; $(m + 11)/11$ = 14 180.0; $(m + 12)/12$ = 14 180.1, which gives an average value and standard deviation of 14 178.8 ± 1.41 Da for the molecular weight.

3.5.4 The mass of the b_2 ions is the sum of the two mass values given in Tutorial 1 (remember H_2O has been eliminated) + 1 (NH is now NH_2). Gly (57) + Ala (71) + 1 =129 and Asp (115) + Asn (114) + 1 = 230. The two dipeptides thus consist of Gly–Ala and Asp–Asn.

3.5.5 From the Aldente output, we can see that the most likely protein to have been fragmented is myoglobin.

3.5.6 When samples are pooled, the same peptide from different samples will appear at the same mass-to-charge ratio in a mass spectrum because the tags are isobaric. However, on fragmentation of labeled peptides by collision-induced dissociation, peptide fragments generate amino-acid sequence information (giving peptide identity), while the iTRAQ tags break up to release tag-specific (and sample-specific) reporter ions. The ratios of these reporter ions are representative of the proportions of each peptide in the individual samples. Up to eight samples can be analyzed in tandem using this methodology. The primary advantage over an *in vitro* labeling strategy such as SILAC (Figure 3.5.19) is that iTRAQ strategies are applicable to directly extracted samples, e.g. human biofluids or biopsy material. In iTRAQ, each sample is processed in an identical manner before labeling with different iTRAQ reagents; errors can thus be quantified. SILAC, in contrast, is best for quantitatively comparing two related but different cells, or tissue cell preparations.

Chapter 3.6 Electrophysiology

3.6.1 The movements of ions across the membrane are mainly influenced by their concentrations on both sides of the membrane and the voltage across the membrane. $[K^+]_{in} > [K^+]_{out}$ and K^+ tends to flow out of the cell, while $[Na^+]_{in} < [Na^+]_{out}$ and Na^+ tends to flow into the cell. A negative voltage of ~ -92 mV is needed to stop the K^+ outward flow, while a positive voltage of ~ $+70$ mV is needed to stop Na^+ from entering the cell.

3.6.2 $E_{Ca} = 26.7 \times \ln(1.25 \times 10^{-3}/0.1 \times 10^{-6}) = 251$ mV. $E_{HCO3} = -26.7\ln 2 = -18.5$ mV (note this is a negative ion). The electrochemical gradient is $E_M - E_{Ca} = -70 - 251 = -321$ mV.

3.6.3 The charge on the membrane at -100 mV is $Q = E_M \times C_M = 0.1 \times 10^{-6}$ C cm^{-2} (coulomb cm^{-2}). Dividing by the Faraday constant gives moles of charge: $Q/F = 10^{-7}/96\,485 \sim 10^{-12}$ mol cm^{-2}.

3.6.4 Activation refers to the opening of the activation gate, allowing ions to flow through the channels. Deactivation is when this activation gate closes again. The inward Na^+ current is kinetically faster than the outward K^+ current (see Figures 3.6.8 and 3.6.11). As the Na^+ current is faster, and as it activates at more hyperpolarized levels than the K^+ channels, the membrane potential is able to depolarize rapidly before exit of K^+. Inactivation of the Na^+ current allows the action potential to repolarize via K^+ flow.

3.6.5 Using $V_C = Q/C_M$ and $V = IR_M$, where I is the current, the voltage across the circuit is $IR_M + Q/C_M$. But $dQ/dt = I$, so we have the equation $V_C = R_M dQ/dt + Q/C_M$. This is a differential equation with solution $Q = V_C C_M(1 - e^{-t/R^M C^M})$ (check by differentiating and see Appendix A2.5). This function decays from $V_C C_M$ at $t = 0$ to 0 at $t = \infty$. When $t = R_M C_M$, $Q = V_C C_M(1 - e^{-1})$ and Q is 0.63 times its original value. $R_M C_M$ is defined as the time constant of the decay.

3.6.6 Taking the relative permeability for K^+ as 1 and p for Na^+. Then the Goldman equation for the $E_M = -70$ mV case is $26.7 \times \ln((4.5 + p140)/(p15 + 140))$; $\therefore -2.622 = \ln((4.5 + p140)/(p15 + 140))$ or $(4.5 + p140)/(p15 + 140) = 0.07268$; or $p = (10.1752 - 4.5)/(138.91) = 0.04086$.

For the $E_M = +50$ mV case we can calculate p again from $50 = 26.7 \times \ln((4.5 + p140)/(p15 + 140))$

$\therefore 1.8727 = \ln((4.5 + p140)/(p15 + 140))$;

$\therefore (4.5 + p140)/(p15 + 140) = 6.5056$; or $p = (4.5 - 910.78)/(97.584 - 140) = 21.336$.

The permeability ratio for K^+/Na^+ thus changes from 0.04 to 21, a factor of 525 during the action potential!

Chapter 3.7 Calorimetry

3.7.1 The heat change is 4.18 J g^{-1} °C$^{-1} \times 100$ g $\times (31.0$°C $- 25.0$°C$) = +2.508 \times 10^3$ J. In other words, 2508 J of heat was evolved in the reaction with a resulting increase in water temperature. The enthalpy change, ΔH, for the reaction is equal in magnitude but opposite in sign to the heat change:

$\Delta H_{reaction} = -(q_{water})$. This is an exothermic reaction.

3.7.2 0.5 mg of a 25-kDa protein corresponds to $5 \times 10^{-4}/25\,000 = 2 \times 10^{-8}$ mol protein. The heat generated by the unfolding transition is thus $2 \times 10^{-8} \times 50 \times 10^3 = 1 \times 10^{-4}$ J. This will cause a temperature increase of $10^{-4}/4.18 = 2.4 \times 10^{-5}$ °C!

3.7.3 We have $c = nK_A P_t = 1 \times P_t /(10^{-4})$. To obtain a c-value of 10, $P_t = 10^{-3} = 1$mM; this is equivalent to $30 \times 1.4 = 42$ mg (a large quantity!).

3.7.4 We have $N + L \rightleftharpoons NL$, with dissociation constant $K_D = [N][L]/[NL]$ $\therefore [NL] = [N][L]/K_D$.

We also have $K_f = [D]/[N]$ $\therefore [D] = K_f[N]$.

The apparent folding equilibrium constant in the presence of ligand will be $K_{af} = [D]/([N] + [NL]) = K_f[N]/([N] + [N][L]/K_D) = K_f/(1 + [L]/K_D) \sim K_f K_D/[L]$ when $[L] \gg K_D$.

$\Delta G = -RT\ln K = -RT\ln(K_f) - RT\ln K_D/[L]$. The approximate change in ΔG is thus $RT\ln K_D/[L]$; i.e. higher $[L]$ and smaller K_D increase ligand stabilization.

3.7.5 The data are consistent with two domains unfolding relatively independently, one with a $T_m \sim 300$ K the other with $T_m \sim 325$ K.

Chapter 4.1 Scattering

4.2.1 If there is a distribution of molecular weights rather than a homogeneous solution then the Rayleigh ratio $R_\theta = KcM_w$ where $M_w = \sum_i N_i M_i^2 / \sum_i N_i M_i$ is the "weight-average" molecular weight (Box 4.1).

$\therefore M_w = [0.001 \times 10^6 + (40)^2]/ [0.001 \times 10^3 + 40] = [10^3 + 1600]/41 = 63.4$ kDa.

4.1.2 For the circle, the circumference is $2\pi r = 500$ nm; therefore, $R_G = 500/2\pi = 79.58$ nm. For the rod, $R_G = 250/\sqrt{12} = 72.17$ nm. An approximate form for the scattering is $\exp(-Q^2 R_G^2/3)$ (strictly this only applies for small θ):

(i) when $\theta = 10°$, $Q = 4\pi\sin(\theta/2)/\lambda = 2.738 \times 10^{-3}$ nm^{-1}; $\therefore Q^2/3 = 2.50 \times 10^{-6}$ nm^{-2}

$\therefore$ scattering ratio $I(\theta)_{circle}/ I(\theta)_{rod} = \exp\{-2.5 \times 10^{-6} \times \{R_{Gc}^2 - R_{Gr}^2\}\} = \exp\{-2.5 \times 10^{-6} \times (6333 - 5208.5)\} = \exp\{-2.5 \times 10^{-6} \times 1124.5\} = 0.9972$

(ii) when $\theta = 30°$, $Q = 8.131 \times 10^{-3}$ nm^{-1}; $\therefore Q^2/3 = 2.2038 \times 10^{-5}$.

$\therefore$ the ratio of scattering $I(\theta)_{circle}/ I(\theta)_{rod} = \exp\{-2.2038^{-5} \times (1124.5)\} = 0.9755$

Note the increased sensitivity at larger angles.

4.1.3 At 70% D$_2$O, the observed scattering is essentially all from protein, while at 41% D$_2$O it is from RNA. If the exchange properties are the same throughout we can assume the volume of the scatterer is proportional to the match point. Thus, the ratio of protein volume to RNA volume is given by $70x + 41(1 - x) = 44.1$, where x is fraction of RNA and $1 - x$ is fraction of protein. This gives $x = 3.1/29$ so the ratio of RNA/protein is $3.1/25.9 = 12\%$.

4.1.4 The freshly prepared sample is mono-disperse and non-aggregated and gives a straight line. The older sample has a higher apparent value of R_G (steeper slope) and it is curved. This suggests that the sample has formed a mixture of higher molecular weight aggregates during irradiation (see, e.g. Svergun, D.J., and Koch, M.H. *Rep. Prog. Phys.* 66, 1735–82 (2003)).

4.1.5 R_G is reflecting the average size of the molecule in solution and the changes observed indicate the folded and unfolded states in absence and presence of GuHCl (see Plaxco et al. *Nat. Struct. Biol.* 6, 554–6 (1999) for more detail).

4.1.6 Using the crude formula given, we get a rotational diffusion coefficient of $10^{13}/(5 \times 15 \times 10^6) = 10^6/7.5$. The relaxation time constant will thus be 7.5 μs. Therefore, the 10 μs time constant probably corresponds to rotational diffusion of the DNA. The slower relaxation times could correspond to overall deformations of the DNA coil in solution.

Chapter 4.2 Refraction/plasmons

4.2.1 The apparent light velocity decreases on entering the new medium ($c_{medium} = c_{vac}/n_{medium}$); the wavelength of the light in medium 2, λ_2, is thus less than the wavelength in medium 1, λ_1. From the triangle ABC in the figure, $AC = BC\sin\theta_1 = 2\lambda_1$. From the triangle BDC, $BD = BC\sin\theta_2 = 2\lambda_2$. Dividing these equations gives $\lambda_1/\lambda_2 = \sin\theta_1/\sin\theta_2$. Since $\lambda_1 = c_1 T$ and $\lambda_2 = c_2 T$, where T is a fixed time period, it follows that $(c_{vac}/n_1 T)/(c_{vac}/n_2 T) = \sin\theta_1/\sin\theta_2$, or $n_1\sin\theta_1 = n_2\sin\theta_2$.

4.2.2 $\sin\theta_c = n_2/n_1 \therefore \theta_c = 61.045°$.

As $4\pi n_2[\{\sin\theta_1/\sin\theta_c\}^2 - 1]^{1/2} = 16.71 \times 0.26989 = 4.511$; $\therefore \delta = 488/4.511 = 108.2$ nm.

4.2.3 We have, $\nu = 1.381 \times 10^{-23} \times 298/(8\pi^2 \times 10^{-3} \times R_H^3) = 5.212 \times 10^{-20}/R_H^3$ or $R_H = (5.212 \times 10^{-20}/\nu)^{1/3}$.

At pH 5, $R_H = (5.212 \times 10^{-20}/8.9 \times 10^6)^{1/3} = (5.86)^{1/3} \times 10^{-9}$ m = 1.8 nm; at pH 9, $R_H = 2.5$ nm. The high pH seems to change the molecule to a more open conformation or cause aggregation.

4.2.4 $R_{max} = (M_{analyte}/M_{ligand}) \times R_L \times S$; $\therefore 100 = (11/150)R_L \times 2$; $\therefore R_L = 100 \times 150/(2 \times 11) = 682$.

$\therefore$ An RU of 100 is obtained with 682 pg/mm^2 of β2-microglobulin.

4.2.5 For acetyl pepstatin, the K_D vales are 7280nM and 806nM at pH 7.4 and 5.1 (calculated the ratios of k_{off}/k_{on}). For indinavir, the K_D values are 1.18nM and 1.44nM. Note the large pH dependence for acetyl pepstatin but much smaller pH changes for indinavir, which is also a better inhibitor (for further details see Gossas, T., and Danielson, U.H. *J. Mol. Recognit.* 16, 203–12 (2003)).

Chapter 4.3 Diffraction

4.3.1 The volume V of the unit cell is $a \times b \times c$. Thus, $V = 7.02 \times 4.23 \times 8.54 \times 10^{-27}$ m^3 = 2.54×10^{-25} m^3. The mass of the unit cell is therefore $1.29 \times 2.54 \times 10^{-25} \times 10^6 = 3.28 \times 10^{-19}$ g. As 36% of the crystal is water, the mass of the protein in the unit cell is $0.64 \times 3.28 \times 10^{-19} = 2.1 \times 10^{-19}$g. If there is one protein per unit cell, then the molecular weight = $2.1 \times 10^{-19} \times N_A$, where N_A is Avogadro's number (6.02×10^{23}). Thus, the apparent molecular weight is 126 000 Da. (Note that the real molecular weight could be less or more than this depending on the number of molecules in the unit cell.)

4.3.2 The volume of the unit cell is $6.65 \times 8.75 \times 4.82 = 280.5 \times 10^{-27}$ m^3. The mass of the unit cell is $\times$ volume = $1.26 \times 2.805 \times 10^{-25} \times 10^6 = 3.53 \times 10^{-19}$ g. The mass of protein per unit cell is 50% of this value, or 1.77×10^{-19}g. As the molecular weight is 55 000, the mass of one protein molecule is 55 000 divided by Avogadro's number (6.023×10^{23}) or 0.913×10^{-19} g. Thus, there are 1.77/0.913 or about 2 molecules per unit cell. If there are eight asymmetric units per unit cell, the protein probably has four subunits per molecule.

4.3.3 The volume of the reciprocal lattice that must be sampled will increase by $(0.6/0.2)^3 = 27$. Thus, about $27 \times 400 = 10\,800$ reflections will need to be analyzed to obtain a resolution of 0.2 nm.

4.3.4 Bragg's law relates the d-spacing ($2d\sin\theta = n\lambda$) to the wavelength so if more than one wavelength is incident the d-spacing determination will be smeared out. If the beam is divergent or convergent, the angle θ will be ill-defined and the d-spacing inaccurate. Bragg's law relies on interference between waves created at molecular sites; incoherent waves thus do not give interpretable interference effects.

4.3.5 The de Broglie relation is $\lambda = h/mv = 6.62 \times 10^{-34}/(1.67 \times 10^{-27} \times 2.8 \times 10^3) \therefore \lambda = 0.142$ nm.

4.3.6 If a group takes up several different positions in different molecules in the crystal, or if it moves while the diffraction data are being collected, then the scattering contribution will be averaged out over the crystal. This means that conformational flexibility or crystal packing, which allows different conformations, will lead to undetected regions of the polypeptide chain.

4.3.7 As the enzyme is still active in the crystal, the real substrate is cleaved. It is thus normally only possible to study inhibitor binding and extrapolation must be made to the situation with substrate. Cryo-cooling, rapid collection and mutations to slow the reaction can be used in attempts to observe real ES complexes.

4.3.8 The G-6-P presumably causes conformational changes in the molecule that cannot be accommodated by the crystal lattice.

4.3.9 The $\lambda = 0.229$ nm line is nearest to resonance ($\lambda_o = 0.52$ nm) and would thus give the best anomalous scattering data.

4.3.10 $a = 1/p; b = 1/h, c = \theta$.

4.3.11 With 65% D$_2$O, the scattering from DNA is "matched" to the solvent and scattering (diffraction) essentially comes from the protein alone. At 39% D$_2$O, the observed scattering is predominantly from the DNA, which is clearly wound around the histone core (further information about this experiment can be obtained in *J. Mol. Biol.* 145, 771 (1981)).

Chapter 5.1 Spectroscopy introduction

5.1.1 The probability depends on ν^3. The ratio will thus be $(30 \times 10^{-2}/300 \times 10^{-9})^3 = 10^{18}$, i.e. spontaneous emission is very much more likely to occur for the electronic transition.

5.1.2 The populations are given by the Boltzmann distribution: $n_u/n_g = \exp(-\Delta E/RT)$ where

$R = 8.314$ J mol^{-1}, $T = 298$ ($\therefore RT = 2477.6$ J mol^{-2} K):

(a) $n_u/n_g = 0.999995$; (b) 7.88×10^{-3}; (c) 9.23×10^{-22}

5.1.3 From $\delta E \sim h/2\pi\tau$ and $h\nu = E$, we obtain $\delta\nu \sim 1/2\pi\tau$. When $\tau = 10^{-13}$, $\delta\nu = 1.6 \times 10^{12}$ Hz;

$\bar{\nu} = 1/\lambda = \nu/c = 1.6 \times 10^{12}$ (s^{-1})/2.998×10^8(m s^{-1}) = 5.34×10^3 m^{-1} = 53.4 cm^{-1}.

When $\tau = 10^{-1}$, $\delta\nu = 1.6$ Hz = 53.4×10^{-12}cm^{-1}. The difference between NMR and rotational energy levels is their lifetime. In NMR, the relaxation processes are fluctuating magnetic fields, and these are very inefficient compared with the collisional relaxation processes for rotational levels. Typical NMR level lifetimes are about 1 s, whereas rotational levels have a lifetime of 10^{-13} s.

5.1.4 One 800 nm photon has energy $E = hc/\lambda = 6.626 \times 10^{-34} \times 3 \times 10^8/800 \times 10^{-9} = 2.483 \times 10^{-19}$ J. One watt = 1 J s^{-1}, so the energy in one pulse is $1/(8 \times 10^7) = 0.125 \times 10^{-7}$ J. This means that there are $0.125 \times 10^{-7}/2.483 \times 10^{-19} = 5.03 \times 10^{10}$ photons per pulse. The energy per pulse is 0.125×10^{-7} J. If this is delivered in 10 fs (10^{-14} s), then the wattage is 1.25×10^6 watts! A short pulse will give a wide spread of frequencies due to Fourier transform/Heisenberg uncertainty principles (Tutorial 9 and Problem 5.1.4).

5.1.5 In the absence of weed killer, most of the energy in the excited state is used to make ATP. The weed killer inhibits the ATP generation pathway thus leading to greater use of the fluorescent pathway.

Chapter 5.2 Infrared

5.2.1 Allowed vibrational transitions require a change in dipole moment. Symmetric stretching of CO_2 leaves the dipole moment unchanged, so this mode is infrared inactive. The two bending modes have the same frequency because of the symmetry of the molecule. Two unresolved transitions are said to be degenerate. Note that stretching frequencies are higher than the bending frequencies because it is generally easier to distort a molecule by bending than by stretching.

5.2.2 The asymmetric stretch is expected to be very sensitive to binding, e.g. coordination to a metal ion is expected to give large changes in frequency. The fact that no change is observed implies that the CO_2 is not binding to the enzyme's zinc ion, but rather is in a symmetric hydrophobic pocket that does not perturb the vibration frequency.

5.2.3 This is an example of the isotope effect on the reduced mass (μ). Treating the C–H vibration as independent of the rest of the molecule, we have $1/\mu = 1/M_C + 1/M_H = 1/12 + 1/1 = 13/12$. On substitution with deuterium, $1/\mu = 1/12 + 1/2 = 7/12$. The ratio of reduced masses is thus 7/13 and the ratio of frequencies will be $\sqrt{7/13} = 0.734$. Thus, deuteration alters the fundamental vibrational frequency from 1308 to 960 cm^{-1}.

5.2.4 The reduced mass, μ, of $^{12}C^{16}O$ is $6.857 \times 1.66 \times 10^{-27}$ kg $= 11.383 \times 10^{-27}$ kg

$\therefore v_{vib} = (k/\mu)^{1/2}/2\pi = (1850/1.1383 \times 10^{-26})^{1/2}/2\pi = 64.16 \times 10^{12}$Hz or $\lambda = 3 \times 10^8/64.15 \times 10^{12} = 4.676 \times 10^{-6}$ m or 4676 nm or $\bar{v} = 2138 cm^{-1}$.

5.2.5 (i) The reduced masses of $^{12}C^{16}O$ and $^{13}C^{16}O$ are proportional to 192/28 and 208/29, respectively. The isotope effect on the stretching frequency of $^{13}C^{16}O$ is thus $\sqrt{(192 \times 29)/(28 \times 208)} = 0.9778$. This gives the stretching frequency of $^{13}C^{16}O$ as $0.9778 \times 2140 = 2092$ cm^{-1} when free and 1902 cm^{-1} when bound to myoglobin (assuming no other isotope effects).

(ii) The data can be explained by CO binding in two opposite orientations (C≡O and O≡C) in the docking site (for further details, see *Nat. Struct. Biol.* 4, 209–14 (1997)).

5.2.6 Figure 5.2.8, which shows the position of the amide I band for bacteriorhodopsin, indicates that the protein is predominantly α-helical. The dichroism data suggest that these helices are perpendicular to the plane of the membrane.

5.2.7 Resonance Raman involves excitation of an electronic transition of a chromophore. However, absorption into the excited energy level is also probable and emission, i.e. fluorescence, can then occur (see Chapter 5.5). The high intensity of the emitted fluorescence interferes with the relatively weak resonance Raman spectrum. Various techniques have been used to try to reduce this problem. They include quenching of the fluorescence by addition of I^- or making use of the much shorter lifetime ($\sim 10^{-13}$ s) of scattering compared with fluorescence ($\sim 10^{-9}$ s). If very short laser pulses (ps) are used, Raman photons can be observed before most fluorescence photons reach the detector.

5.2.8 The spectra show that as the rat ages from 28 days to 7 months, there is a large drop in the number of free S–H groups in the eye lens. Monitoring of other peaks, e.g. at 508 cm^{-1} (not shown), suggests that this is due to S–S bond formation (for further details see *J. Biol. Chem.* 253, 1436–41 (1978)).

Chapter 5.3 UV/visible

5.3.1 $I_t = (20.8/100)I_o$. Thus $A = \log_{10}(I_o/I_t) = 0.682$.

As $A = \varepsilon cl$, $c = 0.682/(6.22 \times 10^3 \times 1) = 0.11 \times 10^{-3}$ mol L^{-1}.

5.3.2 (i) $A = \varepsilon cl = (6220)(1 \times 10^{-3})(1) = 6.22 = \log_{10}(I_o/I_t)$

$\therefore I_t = 6.01 \times 10^{-5}\% I_o$.

(ii) $A = \varepsilon cl = (6220)(1 \times 10^{-6})(1) = 0.00622 = \log_{10}(I_o/I_t)$

$\therefore I_t = 98.6\% I_o$.

These values of I_t show that it will be difficult to measure the concentration of both solution. In (i) I_t is negligibly small, and in (ii) the amount of radiation absorbed is very small. In practice, I_t should be between 10% and 90% of I_o (i.e. A should be between 1.0 and 0.05). The most accurate measurements will be when $I_t \sim 50\%$ I_o. These conditions could be met by altering c, the path length, or changing the wavelength of the radiation (and hence ε).

Note: The units of ε depend on those chosen for c and l. If c is given in mol L^{-1} and l is given in cm, then ε has units of mol^{-1} L cm^{-1}. Molar (M) is often used instead of mol L^{-1}.

5.3.3 There must be some specific interaction in the antibody binding site, as a blue shift is normally observed on going to a less polar environment. (On the basis of model studies with tryptophan, the red shift was deduced to arise from an interaction of Dnp with tryptophan residues in the antibody binding site.)

5.3.4 The peak at 292 nm in the solvent difference spectrum for lysozyme arises predominantly from Trp. The ratio of the absorbances (0.034/0.042) suggests that the relative exposure of Trp residues in lysozyme is ~0.8 (that is, about five Trp residues). Note that this method measures only the average exposure. All six Trp may in fact be exposed some of the time. Note also that one could choose other standards, such as the denatured protein, for the 100% exposed standard.

5.3.5 We can determine the concentration of nitrated tyrosine in the protein from the equation $A = \varepsilon cl$. As $\varepsilon = 4100$ L mol^{-1} cm^{-1}, $l = 1$ cm, and $A = 0.154$, $c = A/\varepsilon l = 0.154/(4100 \times 1) = 3.75 \times 10^{-5}$ mol L^{-1}. As the protein concentration is 4.0×10^{-5} mol L^{-1}, this means that 3.7/4.0, or 0.94, tyrosine residues are nitrated. From the plot of absorbance versus pH, the pK_a of this tyrosine is about 7.3.

5.3.6 Viral DNA is often single stranded while the replicative DNA is double stranded and has an increased hypochromism in the native structure arising from a higher degree of base stacking.

5.3.7 The plane of the bases must be approximately perpendicular to the long axis.

5.3.8 The product is produced during the reaction with rate constant of ~40 s^{-1}. There appears to be no pre-steady-state kinetic burst and the progress curves can be fit to a kinetic mechanism that includes the formation of a transient intermediate whose breakdown is rate limiting (see *Biochemistry* 38, 1547–53 (1999) for further details). This example illustrates the ability of UV/visible stop flow to give detailed information about rapid reactions in a straightforward way.

Chapter 5.4 Optical activity

5.4.1 The most likely explanation is that the inactive sample is a balanced mixture of l- and d-amino acids; while the other samples contain either mainly l- or mainly d-amino acids.

5.4.2 The membrane spanning hydrophobic peptide 60–89 is predominantly α-helix. The water-soluble peptides seem to be mainly random coil, although the 1–40 peptide may have some β-sheet character (further details can be found in *Biochemistry* 18, 275–9 (1979)).

5.4.3 Prion propagation involves the conversion of cellular prion protein (PrP^C) into a disease-specific isomer, PrP^{Sc}, shifting from a predominantly α-helical to β-sheet structure. These CD spectra showed that recombinant human PrP can switch between the native α conformation and a conformation rich in β structure (further details can be found in *Science* 283, 1935–7 (1999)).

5.4.4 The data show that folding in the presence of Ca^{2+} is similar to that in its absence, although the rate is faster with Ca^{2+}. The far UV spectra (gray) indicate that a significant fraction of the protein backbone folds before observation can take place. The refolding process involving sidechains (observed using near UV—blue) seems to have a similar time constant to the slow far UV backbone component. The sidechains (the near UV) do not seem to be involved in the early invisible backbone folding stage (further details can be found in *J. Mol. Biol.* 288, 673–88 (1999)).

5.4.5 The MCD spectra are sensitive to coordination and the synthetic compound appears to be a good model for the coordination in the enzymes. Both enzymes shown appear to have both 5 and 6 coordination complexes although enzyme c has more of the 5 coordination site (further details can be found in *Inorg. Chem.* 48, 8822–9 (2009)).

Chapter 5.5 Fluorescence

5.5.1 An energy level diagram is shown below for three excited vibrational levels in the ^0S and ^1S states. We get excitation and emission to/from different vibration levels, depending on relative positions of electrons and nuclei. The vibrational states are similar in ground and excited states; most excitation and emission processes are from lowest vibrational level in ground and excited states.

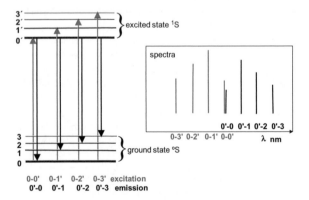

5.5.2 When the fluorophores are free in solution, their rapid rotational motion depolarizes the emitted light. On binding to antibody, the rotational motion is much reduced and the polarization increases.

5.5.3 From the spectra in the figure, we can measure the Tb(III) fluorescence quenched in the presence and absence of Co(II). The ratio of intensities for the Co to Zn spectra is ~0.25.

Thus, $E_T = 1 - Fl_d/Fl_d^\circ = 0.75$. We also have $E_T = R_o^6/(R^6 + R_o^6)$, so $0.75R^6 = R_o^6 - 0.75R_o^6$ or $R^6 = 0.25R_o^6/0.75 \therefore R = (1/3)^{1/6}R_o$. When $R_o = 1.63$ we have $R = 1.36$ nm. A value of 1.39 nm was reported from a crystal structure (for further details, see *Biochem. Biophys. Res. Commun.* 66, 763–8 (1975)).

5.5.4 In the photosynthetic unit, energy transfer occurs to other chlorophyll units, thus depopulating the excited state and shortening the lifetime. The fluorescence yield in solution is given by $\Phi_F = \tau/\tau_F = 7 \times 10^{-9}/25 \times 10^{-9} = 0.28$. In the photosynthetic unit, $\Phi_F = \tau/\tau_F = 10^{-10}/25 \times 10^{-9} = 0.004$.

5.5.5 In the fluorescence cuvette, Dnp is present in two environments—free in solution and bound to the protein. The fluorescence from the bound state is relatively low so that the fluorescence intensity is inversely correlated with the fraction bound. A significant correction has to be made, especially for low affinity binding to the protein, because the fluorescence of free Dnp is quenched with increasing concentration.

5.5.6 (i) The tryptophan emission spectrum overlaps with the absorption spectrum of the DNS, so there is energy transfer. The enzyme tryptophan fluorescence is therefore quenched when the enzyme substrate complex (ES) is formed; the fluorescence returns to its normal value as this complex breaks down. The dansyl fluorescence is enhanced initially because of the energy transfer in the ES complex but this then decays as the complex breaks down.

(ii) The fluorescence emission of the dansyl (DNS) group overlaps the Co^{2+} absorption spectra. Energy transfer from DNS to Co^{2+} occurs and results in a 100% quenching of the DNS fluorescence. This quenching can be used to calculate the distance between DNS and Co^{2+} (for more details see *Biochemistry* 11, 3015–22 (1972)).

5.5.7 The most likely explanation is efficient quenching of Trp fluorescence by the Tyr residues in the folded protein. This quenching is removed when the protein is denatured and the Tyr and Trp fluorescence (emission maxima ~303 and ~348 nm, respectively—Table 5.5.1) can be seen separately. Usually Trp efficiently quenches Tyr fluorescence (for details see *Biochim. Biophys. Acta* 427, 20–7 (1976)).

5.5.8 In the presence of Ca^{2+}, calmodulin binds the M13 peptide thus causing the BFP and GFP proteins to move closer to each other, increasing the FRET peak ~540 nm while decreasing the peak ~470 nm (for further details see *Nature* 388, 882–7 (1997)).

Chapter 5.6 XAS

5.6.1 The Fe(II) edge occurs about 2 eV lower in energy than the Fe(III) edge. Higher oxidation states have higher positive charge, making it slightly more difficult to eject a $1s$ electron thus shifting the K edge to higher energy.

5.6.2 Three protein domains binding to 1 DNA molecule would give a P:S ratio of 4.67 which is near the experimental value.

Chapter 6.1 NMR

6.1.1 From Table 6.1.1, $\gamma_H = 2\pi \times 500 \times 10^6/11.74 = 2.675 \times 10^8$ rad T^{-1}. Thus, $B_o = \omega_o/\gamma_H = 2\pi \times 400 \times 10^6/2.675 \times 10^8 = 9.394$ T.

Precession frequency of M_z around B_1 is $\omega_1 = \gamma_H B_1$; for π/2 rotation $\pi/(2t) = \gamma_H B_1$;

$\therefore B_1 = \pi/(2\gamma_H t) = \pi/(2 \times 2.675 \times 10^8 \times 10^{-5}) = 0.587 \times 10^{-3}$ T.

6.1.2 Assuming there is no current induced when B_o is parallel to the ring, $\mu = 0$ and

$B_{ring} = (\mu_o\mu_\perp/4\pi)(3\cos^2\theta - 1)/3r^3$.

There are six free electrons in the ring, so $\mu_\perp = 6B_oe^2a^2/4m_e$

$\therefore B_{ring} = 6\,\mu_oa^2B_oe^2(3\cos^2\theta - 1)/(4 \times 4\pi m_e \times 3r^3)$

$\therefore B_{ring}/B_o = 1.4 \times 10^{-15}(3\cos^2\theta - 1)a^2/r^3$.

In (i) $\theta = 0°$ $\therefore 3\cos^2\theta - 1 = 2$, $a = 0.14$ nm, and $r = 0.3$ nm,

$\therefore B_{ring}/B_o = 2.8 \times 10^{-15} \times (0.14 \times 10^{-9})^2/0.3 \times 10^{-9})^3 = +2.03$ ppm.

In (ii) $\theta = 90°$ $\therefore 3\cos^2\theta - 1 = -1$, $a = 0.14$ nm, and $r = 0.249$nm,

$\therefore B_{ring}/B_o = -1.78$ ppm.

Note the change in sign and the magnitude of the shifts.

6.1.3 From Table 6.1.1, the ratio of γ for ^1H:^{13}C:^{31}P is 500:125.72: 202.4. Taking ^1H as 1 thus gives us the relative sensitivity of ^{13}C and ^{31}P as 0.016 and 0.066, respectively. Taking the natural abundance of ^{13}C (1.1%) into account, the relative sensitivity is 1.76×10^{-4}. If the ^1H spectrum takes 1 min, the ^{31}P spectrum will take $(1/0.066)^2 = 229$ min, while the ^{13}C spectrum will take $(10^4/1.76)^2 = 3.2 \times 10^7$min! In practice, the differences are not as great as implied, especially for ^{13}C, where the spectra are usually sharp singlets and an amplifying NOE effect is produced by ^1H irradiation. Modern methods also often detect weak nuclei like ^{13}C *indirectly* though coupled ^1H nuclei giving greatly enhanced sensitivity (see Problem 6.1.10). Isotope substitution, giving nearly 100% ^{13}C, rather than 1.1%, is also routine.

6.1.4 ^1H COSY detects J coupling effects (through bond) up to three bonds away. This predicts the cross-peak pattern shown in the diagram.

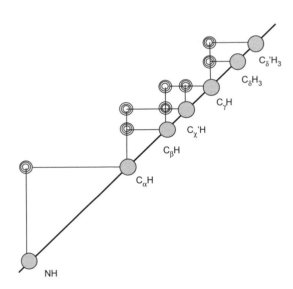

6.1.5 The shift difference is $\delta_A - \delta_B = 80$ Hz; $\therefore \Delta\omega = 2\pi \times 80 = 502.6$ rad s^{-1}.

Shifts: As k_a, $k_b > \Delta\omega$, fast exchange is observed and $\delta_{obs} = f_A\delta_A + f_B\delta_B$ where $f_A = [A]/([A] + [B])$ and $k_a[A] = k_b[B]$ at equilibrium. This gives $f_A = k_b/(k_a + k_b) = 1/3$ and $f_B = k_a/(k_a + k_b) = 2/3$.

$\therefore \delta_{obs} = 1/(3 \times 820) + 2/(3 \times 740) = 788$ Hz

Linewidth: As k_a, $k_b < \Delta\omega$, slow exchange is observed (separate lines).

For A: $1/T_{2obs} = 1/T_{2A} + i_a = 3 + 3 = 6$ s^{-1}. For B: $1/T_{2obs} = 1/T_{2B} + k_b = 2 + 1.5 = 3.5$ s^{-1}.

The linewidths are given by $1/\pi T_2$ $\therefore \Delta\nu_{1/2}(A) = 6/\pi = 1.91$ Hz, $\Delta\nu_{1/2}(B) = 3.5/\pi = 1.11$ Hz.

6.1.6 We have for the P-creatine → ATP direction $I'_{ATP}/I_{ATP} = (1/1.5)/(1/1.5 + k_b) = 0.7$; therefore, $k_b = 0.67/0.7 - 0.67 = 0.29$ s. For the ATP → P-creatine direction, $I'_{PC}/I_{PC} = 1/3.2/(1/3.2 + k_a) = 0.82$; therefore, $k_a = 0.313/0.82 - 0.313 = 0.07$s. The relative flux is therefore 0.07×6:$0.29 = 1.45$:1 for the forward and backward reactions.

6.1.7 In the fast tumbling limit, $W_{23} = C^2\tau_c/20$, $W_{14} = 3C^2\tau_c/10$, and $W_B = 3C^2\tau_c/40$.

$\therefore \eta = \gamma_A/\gamma_B\{(3/10 - 1/20)/(3/10 + 1/20 + 6/40)\} = \gamma_A/2\gamma_B$. For the ^1H pair, $\gamma_A = \gamma_B$, thus $\eta = 1/2$. Using the information in Table 6.1.1 about γ, η will be 1.98 for the ^{13}C-^1H pair. Note that sign of γ for ^{15}N is negative which inverts the heteronuclear NOE.

6.1.8 As seen in Figures 6.1.25 and 6.1.49 and Problem 6.1.7, the ^1H-^{15}N NOE is a measure of flexibility. In the prion protein there is a stretch of residues ~23–125 where these values are small or negative. This strongly suggests that this region of the protein is unfolded and very flexible, whereas the C-terminal 130–220 region is relatively inflexible and probably forms a globular domain.

6.1.9 Strong d_{NN} ^1H NOEs are characteristic of an α-helix, while strong $d_{\alpha N}$ NOEs are characteristic of a β-sheet (see Figure 6.1.43). Inspection shows that residues 6–15 and 25–37 are α-helical while 50–55 is probably a β-strand. The solvent exchange data are consistent with these secondary structure predictions as regions with well-defined secondary structure tend to have slow NH exchange (e.g. 6–15, 25–37 and 50–55) because of well-defined H-bonding.

6.1.10 This is an example where selective labeling by ^{19}F can give useful information about kinetics and dynamics of an enzyme. It can be seen that at 25°C the ligand, and presumably the loop containing ^{19}F-Trp 168 that closes down on the substrate, is in fast exchange while it is in slow exchange at 5°C. To have fast exchange we need $k \gg 2\pi \times 340$Hz $= 2136$ s^{-1} (see Figure 6.1.26) (see *J. Mol. Biol.* 310, 271–80 (2001) for further details). An obvious interpretation is that the loop is opening and closing at these rates.

6.1.11 If we assume the 90°$_x$ pulse on ^1H is applied in the middle of the NH doublet, the two doublet components will diverge, precessing at a rate 1/2J with respect to each other. After time 1/4J, the two components will be at right angles to each other (see also Figure 6.1.17). The ^1H 180°$_y$ pulse will then advance flip the slow component to the position of the fast component and the fast to the slow position. By applying a ^{15}N 180° pulse in the INEPT sequence at the same time as the ^1H 180° pulse, what is essentially a double flip keeps the magnetization of the two doublet components evolving in the same direction. The field inhomogeneity is removed, however, and at time 2τ we obtain an echo with the two components of the doublet at 180° with respect to each other. A 90° pulse now creates a maximum population disturbance with one component along z and the other along $-z$. As the ^1H signals are about 10× more intense than the ^{15}N lines we get a 10-fold enhancement in the ^{15}N signal (see also **http://www.chem.queensu.ca/facilities/nmr/nmr/webcourse/inept.htm**).

Chapter 6.2 EPR

6.2.1 The electron is delocalized over all six carbon atoms, It will therefore interact with the six protons giving a seven-line spectrum with relative intensities 1:6:15:20:15:6:1.

6.2.2 The maximum g-value anisotropy Δg is given by $2.0089 - 2.0026 = 0.0062$. To average this out, the rate $1/\tau$ must be $\gg \Delta g \beta B_o/\hbar$. That is, $1/\tau \gg 0.0062 \times 0.927 \times 10^{-23} \times 0.33/(1.05 \times 10^{-34})$.

$$\therefore 1/\tau \gg 1.88 \times 10^6 \text{ s}^{-1}.$$

6.2.3 In spite of the differences in the number of Fe atoms at the active site, these EPR spectra are quite similar but historically such spectra were very useful in indicating structural classes of Fe–S proteins before many crystal structures were known. The structural information was obtained by careful comparison with spectra from model compounds. The linewidths are dominated by a short value of T_1 at ordinary temperatures. At low temperatures, T_1 lengthens and its contribution $(1/T_1)$ to the linewidth decreases.

6.2.4 Comparison of the spectra shows that as the distance from the nitroxide to the Dnp ring increases, both the linewidths and anisotropy of the spectra decrease. This corresponds to an increased amplitude and rate of motion of the nitroxide group, which partially averages out the anisotropy in the EPR spectrum. The shortest label–Dnp distance gives the largest value of anisotropy with a spectrum that is similar to that expected when rigid. This, together with the broad lines, indicates that the label is held tightly in the binding site.

6.2.5 A five-line spectrum with intensities 1:2:3:2:1 is expected; see figure.

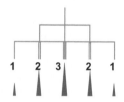

1   2   3   2   1

6.2.6 The vertical lines in the figure illustrate that the spectral width of the spin label *decreases* at low concentrations of the drug but *increases* at higher concentrations. These data with their "dynamic signature" were obtained for a range of ligands like the drug Hoechst 33258, thus giving information both about internal motion in the RNA and reduced flexibility at higher ligand concentration (for further details see *Biochemistry* 41, 14843–7 (2002)).

6.2.7 The "scaled mobility" factor is a measure of the flexibility of the nitroxide attached to the equivalent position in the bacteriorhodopsin helix. When Cys is at the Trp138 position it has the least mobility while Ser132 and Val136 have the highest mobility. One explanation is that 132 and 136 are on one side of an exposed helix thus allowing greater flexibility in these sidechains (further details can be found in *Trends Biochem. Sci.* 27, 288–95 (2002)).

Chapter 7.1 Microscopy

7.1.1 The area of the irradiated spot is $\pi r^2 = 10^{-7}$ cm^2. We have half the intensity when $\bar{a}^2 = 2Dt \sim r^2/2 = (10^{-7})/2\pi$ cm^2 $\therefore D = 10^{-7}/(2\pi \times 2 \times 60)$ cm^2 s^{-1} $\sim 1.3 \times 10^{-10}$ cm^2 s^{-1}

7.1.2 Confocal laser scanning microscopy (CLSM) was developed to overcome the problems associated with out-of-focus light; a single diffraction limited spot from a laser is scanned across the specimen. Two photon microscopy depends on the two-photon effect, by which a chromophore is excited not by a single photon of visible light, but by two lower-energy (infrared) photons that are absorbed at the same time (within 1 femtosecond). Fluorescently-labeled specimens are illuminated by short (less than 200 fs) pulses of infrared light with large peak amplitude at application high rates (>50 MHz). CLSM lasers cause photobleaching of the fluorophore. Because the pinhole aperture blocks most of the light emitted by the tissue, including some light coming from the plane of focus, the exciting laser must be very bright to allow an adequate signal-to-noise ratio. Photobleaching causes the fluorescence signal to weaken with time. Phototoxicity is also a problem—excited fluorescent dye molecules generate toxic free-radicals. Short wavelength CLSM photons are diffracted by cellular components and cannot travel deep into tissues; with two-photon excitation it is reduced, special UV optical components are not necessary and longer wavelengths penetrate tissue better. The large disparity between excitation and emitted florescence wavelengths also means that there can be very efficient suppression of laser light by filters, increasing signal/background ratios. Resolution is achieved solely by focal excitation with the two-photon method with no need for the pinhole in confocal systems.

7.1.3 We have the de Broglie relationship $(\lambda = h/mv)$ and the conservation of energy $(eV = 1/2 mv^2)$. This gives $\lambda = h/(2emV)^{1/2} = 6.626 \times 10^{-34}/(2 \times 1.602 \times 10^{-19} \times 9.109 \times 10^{-31} V)^{1/2} = 1.22 \times 10^{-9}/V^{1/2}$ m.

For $V = 400\,000$ V, $\lambda = 0.019$ nm

7.1.4 Eight spheres at the corners of a cube.

7.1.5 One would expect to see structures similar to those shown in the figure although the probability of seeing the ring structure is much lower than the elongated structure.

7.1.6 The force in an AFM is given by $F = -kx$, where x is the displacement and k is the force constant. Thus, 1 nm movement = 200 pN. In fact a tip is first attracted to a surface by van der Waals forces (non-contact mode) and is then repelled (contact mode). In a 10 μm × 10 μm scan with a resolution of 512 × 512 points, the width of each image pixel = $10 \times 10^{-6}/512$ m ~ 20 nm.

Chapter 7.2 Single molecule

7.2.1 This is an example of the difference between an ensemble measurement, where an average over all species is obtained, compared to a distribution of single molecule events. The single molecule experiments suggest that there are two main types of FRET-inducing event that can be observed and these have different probabilities. In this situation a number of species could exist: unlabeled hormone or receptor, inactive hormone, aggregated receptor, etc. Many of these will not contribute to FRET and a probable explanation of the results is that there are two forms of the complex.

7.2.2 Airy disc size = $0.61 \times 500 \times 10^{-9}/1.5 = 203$ nm.

Volume of ellipsoid = $4\pi x^2 z/3 = 4\pi (0.203 \times 10^{-6})^2 \times (0.609 \times 10^{-6})/3 = 0.105 \times 10^{-18}$ m$^3 = 1.05 \times 10^{-16}$ L.

Number of molecules = $6.02 \times 10^{23} \times 10^{-8} \times 1.05 \times 10^{-16}$ ~0.6 molecules. In practice, the diffraction limited volume is usually more like 10^{-15} L than 10^{-16} L.

7.2.3 A histogram plot of events/lifetime is approximately an exponential decay with a half-life of around 10 s. This predicts that k_{cat} would be ~0.1 s^{-1} in a bulk experiment.

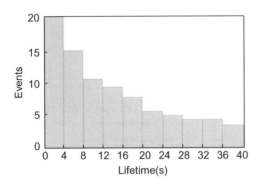

7.2.4 Expected separation for Pro20 = 6.2 nm. ∴ $E_T = 5.4^6/(6.2^6 + 5.4^6) = 0.304$. The mean is ~0.6. Expected separation for Pro27 = 8.37 nm. ∴ $E_T = 5.4^6/(8.37^6 + 5.4^6) = 0.067$. The mean is ~0.35. These large discrepancies can be explained if the polyproline "rod" is in fact flexible. The width of the distribution also suggests flexibility (see *PNAS* 102, 2754–9 (2005)).

7.2.5 When the tagged protein binds to its receptor, the correlation function changes and is characterized by a longer value of τ_D (slower diffusion). At the low concentrations used, free and bound forms coexist with a significant fraction in both forms. An estimate of the affinity could be obtained by changing the relative concentrations of tagged protein to receptor (see Chapter 2.1).

Chapter 8 Computing

8.1.4 Hamming distance = 3; Levenshtein distance = 5.

8.1.5 We can find the slope from the derivative; $\nabla E(x,y) = 8x + 2y$; so the slope in an (x,y) plot will be –1/4; the line will thus have the form $y = -0.25x + c$. Taking the point $y = 1$, $x = -0.866$ on the $E = 4$ curve allows us to obtain $c = 1 - 0.25 \times 0.866 = 0.783$. The line $y = -0.25x + 0.783$ passes through $(1, -0.866)$ and $(0, 0.783)$. The intersection with the contours will look something like the curve shown below with the cyan square being the minimum, somewhere near $(0.2, 0.77)$, where $E \sim 0.75$. If the process is repeated from this point (cyan arrow) the next point will be nearer the local minimum (for this particular function, the local minimum equals the global minimum).

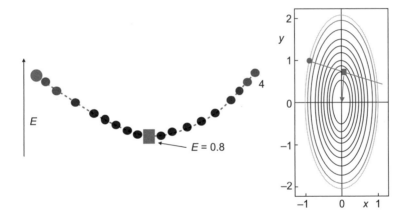

INDEX

A

A, hyperfine splitting in EPR 245–7
ab initio modeling methods 295, 305
Abbé formula for resolution 259–60
Abbé, Ernst 258
absorption
   concentration dependence 143
   dependence on direction of
      transition dipole 144
   spectra 27, 140–143, 203, 320, 331
acceleration 298, 314
actin 270
action potential 75–6
   depolarization 75–6
   propagation 79–81
   stimulus threshold 75–6
Airy disc 259
aldolase 190
algorithm 233, 293, 334
Alhazan 258
alpha-lactalbumin 67
amino-acids, properties 310
analyte 36–8, 56, 112–5
analytical ultracentrifuge; *see also*
    AUC 26–35
analytical uses of XRF 206–208
Andrew, Raymond 228
Anfinsen, Christian 22, 293
angular momentum 211, 314
anomalous scattering, multiple
    wavelength (MAD) 128
antibody
   binding site dimensions by spin
     labeling 255
   coupling 114
   molecular motion 192–3
   shape 100
aquaporin 273, 276, 300
Ashkin, Arthur 279
assessment of cancer drug toxicity by
    NMR 226
assignment of ¹H protein spectra 232
Aston, Francis 55
asymmetric unit 120
atomic force
   manipulation 282

microscopy 275–6
atomic orbitals 327
atomic resolution 3, 119–130, 230, 203,
    271, 275, 305
atoms 6
attenuated total reflectance (ATR) 111
AUC instrumentation 26–8
   optical systems 26–7
   Rayleigh interference optics 27
autoradiograph 261

B

B factors 130–1
Babbage, Charles 290
bacteriochlorophyll special pair 170
bacteriophage 271
bacteriorhodopsin 133, 172, 273
   photocycle 153
Baumeister, Wolfgang 270
Beer–Lambert law 143, 162, 204
Berners–Lee, Tim 290
beta–lactamase kinetics by UV/
    visible 175
biarsenical fluorescent reagents 188
binding data analysis 11–14
bioinformatics 290, 292–5
biological molecules,
    nomenclature 310
bioluminescence resonance energy
    transfer (BRET) 195–6
Black, Joseph 84
Bloch, Felix 210
body fluids by NMR 226
Bohr model 325
Bohr, Niels 140
boundary conditions 325
Bragg equation 120
Bragg, Laurence 94, 122
Brownian motion 18
buffers 12
Bunsen, Robert 140

C

calculus 333–4
calmodulin 41, 194–5, 202
calorimetry 84–91

capillary electrophoresis 51–2
carbohydrates 311
carbon dioxide 160
carbonic anhydrase 160, 168–9
carboxypeptidase 201
catabolite gene activator protein 193
centroid fitting 283
Chadwick, James 94
charges on macromolecules 44
   isoelectric point 13, 20, 39, 44
   phosphate backbone of DNA 44
charge coupled device (CCD) 124
chemical potential 19–20, 312
chemical shifts; *see also* NMR 214–6
chlorophyll 200
chromatography 36–43
   mobile phase 36
   stationary phase 36
chromatography techniques 38–40
   affinity 39–40
   analyte detectors 38
   chromatofocusing 38–9
   fast protein liquid (FPLC) 39
   guard column 38
   high performance liquid (HPLC) 39
   hydrophobic interaction (HIC) 39
   immobilized metal ion affinity
     (IMAC) 40
   ion exchange 38
   reverse phase 39
   size exclusion (SEC) 39
   support and columns 38
   tandem affinity purification
     (TAP) 40
chromatography, quantitative
    uses 40–2
   association with ligand, Hummel–
     Dryer method 42
   molecular weights 41
   self–association 41–2
chromophore 28, 155–7, 162–7, 331
chromosomes 138, 196, 264
circuit analysis 317
circular dichroism (CD) 144, 177
circular motion 314
circular polarization 323

circularly polarized light 176–77
class average in single particle EM 273
Claude, Albert 258
coarse grain methods in
        simulations 297, 301
column resolution in
        chromatography 37
complex variables 333
computational biology 289–308
computers 290
conductance 317
confocal microscopy 263–4
contrast and illumination in light
        microscopy 261–2
    bright field 261
    differential interference contrast
        (DIC) 261–2
    phase contrast 261
contrast transfer function 259, 272
convolution 320
cooperativity 12
correlation functions 103
COSY spectra 231, 240
Coulomb's law 8
coupling between electronic
        transitions 170–1, 330
    exciton splitting 170, 330
crambin 137
creatine kinase 240, 250
Crick, Francis 125
critical assessment of structure
        prediction (CASP) 295–6
cryo-fixation 269
cryo-electron microscopy 269
crystal lattice 119–20
crystal planes 122
crystallization; see also diffraction
        sample preparation 122–3
current, electrical 317
cuvette 163

D

damped and forced oscillations 321–2
data analysis 13, 306
de Broglie, Louis 325
denaturing gels 47–8
density gradient centrifugation 31–3
    CsCl gradient 32–3
    equilibrium or isopycnic
        method 33–3
    rate-zonal method 32
detectors for X-ray diffraction
        experiments; see also neutron
        and electron detection 124
    area 124
    charge coupled device (CCD) 124
    image plate 124

detectors for mass spectrometry 62
determination of molecular model
        from diffraction data 125–30
dialysis 21
diamagnetic 210
dichroic mirror 149, 262
di-cobalt enzyme 184
dielectric constant 8, 15, 108, 297–9
dielectric spectroscopy 110
differential equations 334
differential scanning calorimetry
        (DSC) 87–90
    complex formation 89–90
    folding/unfolding transitions 89
    instrumentation 88
differentiation 333
diffraction 119–38
diffraction at a slit 259
diffraction, data collection 124
diffraction, sample preparation 122–3
    crystals 123
    fibers 122
    growth 122–3
    hanging drop 123
    nucleation 123
    sitting drop 123
    twinning of crystals 123
diffusion 18–19
    concentration gradient 18
    Fick's laws of diffusion 19
    root mean square displacement 18
dimensions of life 1
dipole 330
dipole moment 8–9, 142–4, 330–1
dipole–dipole interactions 330–1
Dirac, Paul 209
discontinuous gel
        electrophoresis 46–7
disorder in crystals 130
dispersion 109
dissociation constant 11
distance between two paramagnetic
        centers by EPR 249–51
DNA 310
    CD spectra 179–80
    fiber diffraction 125, 137
    hybridization 11
    hypochromism 170–1
    linear dichroism 170
    microarrays 11
    non-covalent forces 10–11
    nucleosome 252
    satellite 35
docking of molecules 304
    computer based 304
    HADDOCK 305
    use of experimental restraints 304

Donnan effect 20
double beam spectrophotometer 162
drag 314
Dubochet, Jacques 269
dynamic light scattering (DLS) 103–4
    diffusion coefficient 103
    polydispersity 104
dynamic processes in biology 1

E

effective mass in solution 21
EGF signaling 65–6
electric dipole moment 330
electric transition dipole moment 177
electrical circuits 317–8
electrical circuit model of a
        membrane 74–5
electrical properties of gels 45
electromagnetic radiation 95, 323–4
electromagnetic spectrum 323–4
    interconversion between
        scales 324
electromagnetic waves 323
electron crystallography 273
electron density map 130
electron detection 124
electron diffraction from 2D
        crystals 124, 132–3
electron lens 267
electron magnetic moment 243
electron microscopy (EM) 266–75
electron nuclear double resonance
        (ENDOR) 251–3
electron paramagnetic resonance
        (EPR); see also EPR 243–56
electronic energy levels and
        transitions 163–4
    in organic molecules (n–π,
        π–π*) 163–4, 327–8
    in transition metals (d–d) 166,
        328–9
electrospray ionization (ESI) 56
electro-osmotic flow 51–2
electrophoresis apparatus 45–6
    anionic and cationic systems 45
    slab gels 45
electrophoresis 44–54
electrophysiology 72–83
electrostatics 8
elements 6
emergent properties 291
encephalopathy diseases 183, 241
ENDOR 251–2
energy classification 1, 140
energy minimization 298–9
energy state populations 141
    Boltzmann distribution 141

energy states
    degenerate 140, 160, 181, 329
    ground and excited states 141–6,
        156, 158, 164, 185, 188, 193, 266
    quantization 6, 140, 327
enthalpy 312
entropy 312
environmental effects on
        fluorescence 188–9
    ANS binding to protein 189
enzyme kinetics 87–8
eosin 261
EPR applications 248–53
EPR, measurement 243–44
    cavity 243
    continuous wave 244
    gyrotron 243
    klystron 243
    microwaves, waveguides and
        bands 243
    pulsed 244
EPR spectra of transition metal
        ions 249–50
EPR spectral parameters 244–7
    g-value 244–5
    hyperfine splitting (A) 245–6
    intensity 244
    inhomogeneous broadening in
        EPR 245
    linewidth and relaxation in EPR 245
    multiplet structure 245–6
    zero-field splitting (ZFS) 246
EPR spin trapping 252–3
equilibrium constant 312
Ernst, Richard 229
erythrosine 198
evanescent wave 110–12
Ewald, Paul 122
exciton splitting 170, 330
exponential functions 333
extended X-ray absorption fine
        structure (EXAFS) 203–6

F

Feher, George 251
Fenn, John 55, 56
Fermat's principle of least time 117
ferredoxin 255
ferromagnetic 210
fiber diffraction 125
fibronectin domain 237
Fick's laws of diffusion 19
field emission gun (FEG) 266
filopodia 270
filtration 19
fluorescence 185–202

fluorescence activated cell sorting 197
fluorescence anisotropy 191–3
    time resolved 192
fluorescence correlation spectroscopy
        (FCS) 283–4
    autocorrelation function 283
fluorescence detection of single
        particles 282
fluorescence in situ hybridization
        (FISH) 196
fluorescence lifetime 185–6
fluorescence lifetime imaging
        microscopy (FLIM) 265
fluorescence loss in photobleaching
        (FLIP) 262–3
fluorescence measurement 186–7
    emission spectrum 186
    excitation spectrum 186
    inner filter effect 186
    photobleaching 186
fluorescence microscope 262–4
fluorescence polarization 191–2
fluorescence recovery after
        photobleaching (FRAP) 262–3
fluorescent biosensor 202
fluorophores 187–8
    extrinsic/synthetic 187
    fluorescent proteins (GFP etc.)
        187–8
    intrinsic 187
    quantum dots 188
    terbium 187
Förster resonance energy transfer
        (FRET) 145, 193–6
    donor and acceptor spectra 194
FoF1 ATP synthase 134, 281
force and potential energy 298
force fields 297–8
    bonded terms 297–8
    empirical 298
    H-bonds 298
    improper dihedral term 298
    non-bonded terms 297
    truncated 298
forces on a molecule in solution 21
formate dehydrogenase 206
Fourier series 319
Fourier synthesis 126
Fourier transform relationships 120–1,
        126, 213–4, 319
fraction bound 11
Franck–Condon rule 164, 185
Frank, Joachim 271
Franklin, Rosalind 125
free energy perturbation/binding 304
freeze fracture 268–9

frequency/wavelength dependence of
        CD 178
frictional coefficient 18, 23
    shape dependence 23
Friedel's law 128
FTIR difference spectra 153–4
functional MRI 223–4
    blood oxygenation 224

G

g factor in EPR 244–5
Galvani, Luigi 72
Gaussian distribution 283, 334
GC base pairing 171
gel media 46
    acrylamide 46
    agarose 46
genomics 292
GFP 139, 158, 187–8
Gibbs free energy 312
glutaraldehyde 268
glutathione–S transferase (GST) 53
glycogen phosphorylase 137
glycophorin 183, 302
Goldman equation 74
gold sputtering 275
goniometer 124
gravitational field 315
green fluorescence protein (GFP) 158,
        187–8, 195
Guinier plot of scattering data 99
gyroscope 211

H

Hamiltonian 325
Hamming distance 308
Harker construction 127–8
heat capacity 89
heavy metal, X-ray scattering 127
Heisenberg uncertainty principle 144,
        325
Heisenberg, Werner 140
helical displacement of charge and
        optical activity 177
hemoglobin 157
Henderson, Richard 273
Henderson–Hasselbalch equation 12
highest occupied molecular orbital
        (HOMO) 328
Hill plot 13–14
histochemistry 261
histology 261
histone octamer 252
histopathology 261
Hodgkin, Alan 72
Hodgkin–Huxley model 75

Hooke, Robert 258
HSQC spectrum 236–7
Hubbell, Wayne 256
Huxley, Andrew 72
hybrid methods 296, 305
hydrogen bonds 8
hydrogen/deuterium exchange 63,
        152, 234
hydropathy plot (Kyte and
        Doolittle) 293–4
hydrophobic effect 9–11, 39, 304

I

image processing in EM 271
immunofluorescence 196
INEPT (insensitive nuclei
        enhanced by polarization
        transfer) 242
infrared spectra; see also
        IR 148–155
inner electron shells 204
integration 334
integrin domains, binding
        studies 115–6
inter- and intra-molecular de-
        excitation of excited
        states 144–5
interactions 6–16
    ionic 7–8
    molecular 6–16
    non-covalent 7–11
    van der Waals 9
internet 290
internuclear bond length 148
ion channels and pumps 72–3
    electrogenic Na^+/K^+ pump 72–3
    gated channel 72–3
    ion asymmetry 72–3
    passive, active, and coupled
        transport 72–3
ion mobility in mass
        spectrometry 67
IR spectra
    amide I and amide II bands
        151–2
    amyloid fibrils 154–5
    characteristic frequencies 151
    group frequencies 151
    H–bonding effects 151
    oriented samples 153–4
    polyatomic molecules 150–1
    protein secondary
        structure 151–2
    solvent effects 151
IR spectra, measurement 149–50
    attenuated total reflection 150

femtosecond IR laser pulses 150
Fourier transform (FTIR) 149–50
    interferogram 150
    two dimensional methods 150
isoelectric point 12, 13, 20, 39, 44
isomorphous replacement, multiple
        (MIR) 126–7
isosbestic points 169
isotachophoresis 47
isothermal titration calorimetry
        (ITC) 84–91
    enzyme kinetics 87–8
    experimental design 86–7
    instrumentation 84–5
    ligand binding 85–6
isotope labeling in NMR (^2H, ^{15}N,
        and ^{13}C) 232
isotopes 6
isotopic substitution effects on IR
        spectra 148

J

Jones, Alwyn 305
Jun/Fos structure by CD 181
J-splitting in NMR 216–7

K

Karplus, Martin 216, 297
Kendrew, John 127
kinetic energy 314
kinetic model of gases 7
Kirchoff, Gustav 140
Kratky scattering plot 101
Kretschmann configuration in
        SPR 111–2

L

lactate dehydrogenase 172
Laemmli, Ulrich 47
Lamm equation 29
Langmuir plot 12
Laplacian operator 325
laser 145
    gain medium 145
    population inversion 145
Lauterbur, Paul 223
Lavoisier, Antoine 84
Lennard-Jones potential 9, 297
lens aberrations 260
    chromatic 260
    spherical 260
lens protein 161
Levenshtein distance 308
Levinthal, Cyrus 305
ligand binding
    by AUC 31–2

by CD 180
by computer docking 303–4
by EPR 249, 255
by FEP 303
by fluorescence 190, 193
by IR 160
by ITC 85–86
by NMR 236
by SPR 113–116
by UV/visible 174
light scattering 99
light-driven reactions 172–3
linear dichroism 144, 169–70
linear momentum 314
linear regression 13
linewidth of absorption
        spectrum 143–4
lipids 220, 227, 228, 249, 311
logarithms 333
lowest unoccupied molecular orbital
        (LUMO) 328
lysozyme 89, 137, 175, 241, 295

M

magic angle spinning 228
magnetic CD 181
magnetic dipole moment 210, 330
magnetic resonance imaging
        (MRI) 223–4
    contrast, T_1 and T_2 223–4
magnetic resonance of cells and tissues
        (MRS) 224–5
    ^1H studies 225
    ^{31}P studies 224–5
    localization of signals 225
magnetic susceptibility 210
magnetic transition dipole
        moment 177
magnetic traps 281–2
magnetism 210
magnetization 210–2
Mansfield, Peter 223
mass spectrometry applications 62–9
    calculation of molecular weight 63
    hydrogen exchange 63
    isobaric tagging for relative and
        absolute quantification
        (iTRAQ) 71
    mass analysis 62–3
    post-translational modifications 62
    protein identification/
        sequencing 64
    quantification 64–6
    stable isotopic labeling in cells
        (SILAC) 65–6
mass spectrometry data analysis 62

mass spectrometry, ion sorting 57–60
  by electric and magnetic fields 58
  by ion cyclotron resonance 59–60
  by ion mobility 61
  by orbitrap 60
  by quadrupole analyzers 58–9
  by quadrupole traps 59
  by time of flight 58
mass spectrometry methodology 55–71
  collision-induced dissociation (CID) 61
  electrospray (ESI) 56
  fragmentation 60–1
  matrix-assisted laser desorption ionization (MALDI) 55–6
  parent, daughter, and neutral ions 61
mathematical functions 332–3
mathematical modeling of systems 291–2
  deterministic 291
  feedback 291
  input parameters 291
  molecular networks 291
  organs 291
  robustness 291
  sensitivity 291
  stochastic 291
matrix-assisted laser desorption ionization (MALDI) 55–6
mathematical representation of a wave 318–9
matter 6
maximum likelihood 130
Maxwell, James Clerk 93
Maxwell–Boltzmann distribution 7
mechanical stretching of protein domains by AFM 282
membrane potential 72–4
Meselson–Stahl experiment 33
metabolomics 292
microscopy 258–78
microtome 260
migration in an electric field 44–5
Miller indices 122
Minsky, Marvin 263
molecular biology tools 11
molecular dynamics simulations 130, 234, 297, 299–303
  fitting to experimental data 302–3
  NMR structure calculations 302–3
  trajectory 299, 300
  Verlet algorithm 300
  X-ray crystallography refinement 303
molecular EM 271–3

molecular mechanics 296
molecular modeling 296–303
molecular motion and ^{1}H–^{15}N NOEs 235
molecular orbitals (MO) 327–8
molecular replacement 128
molybdenum edge 206
momentum operator 325
monochromator 163
Monte Carlo methods 303
  Metropolis algorithm 303
Moore, Gordon 290
Morse curve 148–9, 164
  anharmonic vibrations 148
  parabolic potential 148–9
motion and energy of particles in different force fields 314–5
myoglobin 131, 160
myosin 101, 282, 287

N

NADH absorbance 174
native gels 47–9
  ampholytes 49
  electrophoretic mobility shift (EMSA) 48–9
  Ferguson plots 48–9
  isoelectric focusing 49
  native mass spectrometry 66–7
  sensitivity to conformation (ESI) 66–7
negative staining in EM 268
Neher, Erwin 72
Nernst equation 73, 312
neuron 79
  axon 79
  dendrites 79, 275
  myelin sheath 79
  node of Ranvier 79
neutron detection 124
neutron diffraction 124, 132
neutron scattering 101–2
Newton, Isaac 107
Newton's laws of motion 314
Newton–Raphson minimization 299
nile red 261
nitrate reductase 252–3
nitric oxide 253
NMR applications 223–37
NMR, chemical exchange effects 221–2
  fast, slow, intermediate regimes 221–2
NMR, chemical shifts 214–6
  induced 214–6
  intrinsic 214–5
  parts per million 214
  ring current shift 216

NMR measurement 212–3
  B_1 magnetic field 212
  B_0 magnetic field 212
  rotating reference frame 213
NMR, multidimensional spectra 229–31
  COSY 231
  NOESY 231
  off-diagonal cross-peaks 230–1
NMR parameters 214–9
NMR relaxation in xy plane (T_2) 216–7
  B_0 inhomogeneity 216–8
  Lorentzian lineshape 216–7
  spin echo 218
NMR relaxation in z direction (T_1) 218–9
NMR size limit 231
  linewidth and molecular weight 231
NMR structure calculation
  experimental restraints 233–5
  restrained molecular dynamics 234, 302
NMR technical problems 229
Nobel Prizes in crystallography 133
NOE patterns and protein secondary structure 232
NOESY spectrum of tyrosine 231
non-covalent forces in proteins 10
non-linear regression 13
nuclear magnetic resonance; *see also* NMR and magnetic resonance 209–42
nuclear Overhauser effect (NOE) 221, 240–1
nucleic acids 310
nucleosome 137–8
nucleotide bases 310

O

Ohm's law 317
omp F 276
opsins 168
optical activity 176–84
  induced 177, 179
  intrinsic 177
optical microscope 260
optical rotatory dispersion (ORD) 176
optical sections 263, 265
optical traps (tweezers) 279–81
optics, classical geometric 108
  dispersion by a prism 108
  lens focal length 108
  reflection 108
  refractive index 107–9
  Snell's law 107–8
  total internal reflection and critical angle 108

optogenetics 81
orbitrap 60
ornithine carbamoyl transferase 68
orthologs 292
oscillators 321
osmium tetroxide 268
osmometer 20
osmosis 19–20
osmotic pressure 19

P

Palade, George 258
paralogs 292
paramagnetic 210
paramagnetic center effects in
          NMR 222
    dipolar 222
    gadolinium, europium,
          manganese 222
particle in a box 325–6
particle-induced X-ray emission
          (PIXIE) 207
partition coefficient 37
patch clamp experiments 77–9
Patterson difference map 127
Patterson function 126
Paul, Wolfgang 55
Pauli, Wolfgang 140, 209
PELDOR 248, 250–2
Perutz, Max 127, 129
phase coherence in NMR 212
phase contrast in EM 267–8
phase problem 126, 133
phosphoglycerate kinase 202
phosphorescence 197–8
photoactivated localization
          microscopy (PALM) 266, 283
photobleaching 262
photodiode 163
photoelectric effect 325
photomultiplier 163
photoselection 192
photosynthesis 147
photosynthetic reaction center
          (PCR) 134, 172–3
phylogenetic tree 293
phylogeny 292, 293
physical constants 332
piezoelectric crystal in AFM 275
pKa 12, 236, 310
plasmid DNA conformation 104
point spread function (PSF) 259
polarity 8
polarizability 155
polarization 323
population differences between energy
          levels 143

potential energy 314–5
potential energy of a spring 316
powder spectrum 226–7
power dissipated by a resistor 317
precession 211
prion protein 183, 241
PROCHECK 131
proline spectroscopic ruler 284–5
propagation of an action potential
          79–81
    cable theory 80–1
    firing rate 80
    measurement of propagation rate 81
    orthodromic 80
    refractory period 79–80
    saltatory 80
protease
    cleavage 40
    HIV 118
    serine 132, 236
protein CD spectra 178–9
    multicomponent analysis 178–9
protein data bank (PDB) 295, 299
protein-DNA stoichiometry by
          PIXIE 208
protein, non-covalent forces 10–11
protein structure determination by
          NMR 228–36
protein structure prediction 292–4
proteins 310
proteomics 64, 292
Prussian blue 261
pulsed electron-double resonance
          (PELDOR) 250–1
pulsed field gel (PFG) 50–1
Purcell, Edward 210

Q

quadrupole analyzer 58–9
quanta 325
quantitative real time PCR 190–1
quantum dots 188
quantum mechanics 296
quantum number 326
quantum theory 325–6
quantum yield 185–6
quenching of fluorescence 189–90
    Stern–Volmer plot 189–90

R

Rabi, Isador 209
radiation damage 267
radiationless processes 186
radiofrequency (rf) pulses 213
radius of gyration (R_G) 41, 98, 299
Ramachandran plot 10–11, 131
Raman microscopy 266

Raman scattering 155
    anti-Stokes lines 156
    fluorescence effects 156
    Stokes lines 156
Raman spectroscopy applications
          157–158
rapid reaction monitoring 171–3, 184
Rayleigh, Lord 94, 96
redox reactions 312
reduced mass in a spring 148
refinement and R factors 130
refraction 107–18
relaxation mechanisms in NMR
    Brownian motion and correlation
          times 219
    chemical shift anisotropy 219
    dipolar interactions 219–20
    fluctuating magnetic fields 219
relaxation of populations among
          energy levels 142–4, 185–6,
          216, 219–20
resistance in a circuit 45, 74, 317
resolution achievable by a
          microscope 259
    electron 267
    Rayleigh criterion 259
resolution of diffraction
          experiments 129
resonance condition of oscillator 142
resonance frequency for different
          nuclei 211
resonance Raman spectra 155–6
retention time in chromatography 37
Reynolds number 23, 25
    inertia and viscosity 23
rhodamine 261
ribosome 28, 102, 134, 273–4
Richards, Fred 305
ring current shifts 239
RNA 310
    duplex formation 86
    structure determination by
          NMR 236
    structure prediction 296
RNA polymerase 280
Röntgen, Wilhelm 94
root mean square deviation
          (rmsd) 235
rubredoxin 208
Ruska, Ernst 258

S

Sachman, Bert 72
sample preparation for EM 268–9
sample preparation for light
          microscopy 260–1
Sanger, Fred 36

saturation transfer 240
Sayle, Roger 306
scanning electron microscopy
    (SEM) 273–5
scanning probe microscopy 275–6
Scatchard plot 13
scattering 95–107
    angular dependence 98
    from larger particles 98
    multi-angle light scattering 99
scattering theory 96
    isolated particle 96
    molecular array 96
    molecular polarizability 96
    Rayleigh ratio 96
    solution state 96–7
Schrödinger, Erwin 140
Schrödinger wave equation 6, 325
SDS–PAGE 47–8
secondary structure prediction of
        protein 293
    Chou–Fasman 293
sedimentation 26–35
sedimentation equilibrium 30–1, 34
    lnc vs. r^2 plot 31
    self-association 31
sedimentation velocity 28–30
    concentration dependence 28
    frictional coefficient and shape 28
    hydrodynamic modeling 28
    Lamm equation 29
    ligand binding and association 30
    Perrin coefficient 28
    polydispersity 28–9
    Svedberg coefficient (s) 28
    Svedberg equation 28
selection rules 142–3, 148, 156
selective labeling of methyl groups in
        NMR 233
selenium edge 206
semi-permeable membrane 19–20
sensitivity of absorption
        experiments 143
sequence analysis 292
    FASTA, BLAST, CLUSTAL-W 292
serine protease catalytic
        mechanism 132
SH2 domain, dynamics 300
SH2 domain, measurement of ΔH, ΔS
        and K_D by ITC 85–6
SH3 domain, NOE and rmsd data 235
signal to noise ratio 229, 239, 265,
        267, 273
simple harmonic motion 148, 321–2
    of a weight on a spring 148, 321
simulated annealing 301–2

single molecule FRET 284
single molecule observation 279–88
single particle EM 272–3
single particle imaging with X-ray
        laser 119
small-angle neutron scattering
        (SANS) 101–2
    contrast matching 101–2
    H_2O/D_2O ratio 101
small-angle X-ray scattering
        (SAXS) 99–101
    scattering cross-section 99
    scattering length 99
Snell, Willibrord 107
solid state NMR 226–8
    chemical shift anisotropy 226
    dipolar effects 226
    magic angle 228
    magic angle spinning 228
    powder spectrum 226–7
    quadrupolar effects 228
    solid state spectra of lipid
        membranes 227–8
solvation by water, simulation
        methods 300–1
    explicit solvent 300
    implicit solvent 300
    periodic boundary conditions
        300–1
solvent flattening 128
    anisotropy in EPR spectra 247–8
    anisotropy of g 247
    anisotropy of hyperfine splitting,
        A 247–8
        metmyoglobin anisotropy 247
        spin label anisotropy 248
    anisotropy of ZFS 248
spectroscopy 139–208
spin 140, 210
spin forbidden 197
spin labels 246, 249, 256
spin–spin coupling in NMR 216–7
    coupling constants (J) 216–7
    multiplet structure 216
spontaneous emission 144–5
standard states 312
statistical mechanics 303
    ergodic hypothesis 303
statistics 334
    linear regression 13, 334
    standard deviation 334
stimulated emission 144
stimulated emission depletion (STED)
        microscopy 265–6
Stokes lines in Raman spectra 156
streptavadin/biotin complex 114

structural cell biology 3
structural molecular biology 3
structural restraints from NMR
        spectra 233–4
    ^1H–^1H NOEs 233
    chemical shifts 233
    coupling constants 233
    residual dipolar coupling 233–4
    solvent exchange 234
structure comparison 296
    distance-matrix alignment
        (DALI) 296
sucrose gradient 24
super-resolution microscopy 265–6
surface enhanced Raman scattering
        (SERS) 157
surface plasmon resonance (SPR)
        111–16
    analyte 112
    association and dissociation 113
    direct and indirect coupling 114
    kinetic and thermodynamic
        measurements 115
    sensitivity 114
    sensorgram 112–3
    surface chemistry of SPR chips
        113–4
Svedberg, Theodor 26
symmetry relationships 120
synchrotron 123

T

T_1 relaxation in a lipid bilayer 220
talin subdomain binding HSQC
        titration 237
Tanaka, Koichi 55
tapping mode in AFM 275–6
TAR RNA 256
temperature factors 130–1
terminal velocity 314
tertiary protein structure
        prediction 294–5
    comparative modeling 294–5
    homology modeling 294–5
    threading/fold recognition 295
theoretical plate model in
        chromatography 37
    height equivalent 37
    Martin and Synge 37
thermal shift fluorescence
        assay 189
thermodynamics 312–3
    first law 312
    second law 312
thermolysin 200
Thomson, JJ 55

through-bond and through-space
    correlations in NMR 230–31
time of flight analyzer 58
time resolved crystallography 132
time scales and molecular motion 1,
    2, 248–9
Tiselius, Arne 44
tomography
    electron 269–70
    magnetic resonance 223
    X-ray 270
torque 211, 314
total internal reflectance fluorescence
    (TIRF) 111, 264
tracking of particles in fluorescence
    microscopy 283
    centroid fitting 283
trajectory of a particle 315
    in an electric field 315
    in a gravitational field 315
    in a magnetic field 315–6
transcription factor 180
transcriptomics 292
transient free induction decay 213
transition dipole moments 142–4, 148,
    153, 155, 163, 169–70, 177, 180,
    191, 330–1
transition probabilities 141–4, 179, 331
transmission electron microscopy
    (TEM) 266–9
transmission microscope 260
transverse relaxation optimized
    spectra (TROSY) 232–3
trigonometry 332–3
triose phosphate isomerase 242
triplet state 197
Tsien, Roger 202
Tsvet, Mikhail 36
turbidity 97
Turing, Alan 290
twinning of crystals 123, 134

two-dimensional gels 49–50
two-photon microscopy 264–5

U
unit cell 119–20, 125
units 332
Unwin, Nigel 273
urease from *H. pylori* 67–8
UV/visible spectra 162–75
    charge transfer spectra 167
    chlorophyll 166
    conjugated systems 166
    extinction coefficients 165
    nucleic acid bases 165
    peptides 165
    porphyrin 166
UV/visible monitoring of rapid
    reactions 171–3
UV/visible spectra, environmental
    effects 167–8
    solvent perturbation 167

V
validation of a molecular
    structure 131–2, 234
van Deemter equation 37–8
van der Waals forces 7–11, 297
van der Waals, Johan 9
van Laue, Max 122
van Leeuwenhoek, Antonie 258
vector representation of a wave 318
vectors 332–3
velocity 314
vibrational frequency from spring
    constant 148
viscoelasticity 22
viscometer 21–2
viscosity 21–2
    effect of macromolecules 22
    Stokes equation 21
visualization of gels 46

blotting 46
    Coomassie 46
    silver staining 46
visualization of molecules by
    computer 305
vitreous ice 269, 273
void volume 37
voltage 317
voltage clamp experiments 76–7
von Fraunhofer, Joseph 140
von Neumann, John 290

W
water as solvent 9, 300
Watson, James 125
wave mechanics 325
wave particle duality 325
waves and their summation 318
weight average molecular weight 97
Wilkins, Marc 64
Wollaston, William 140
World Wide Web 290
Wüthrich, Kurt 228

X
XAS edge spectra 203–5
X-ray absorption spectrum 204
X-ray fluorescence (XRF) 203–5
X-ray scattering 99–101
X-ray sources
    laboratory 123
    synchrotron 123
X-ray spectroscopy 203–8

Y
Y-base tRNA 185

Z
Zavoisky, Yevgeny 243
Zeiss, Carl 258
Zernicke, Frits 258, 261